POSITIVE OPERATORS

Positive Operators

by

CHARALAMBOS D. ALIPRANTIS
Purdue University, West Lafayette, U.S.A.

and

OWEN BURKINSHAW
IUPUI, Indianapolis, U.S.A.

 Springer

A C.I.P. Catalogue record for this book is available from the Library of Congress.

ISBN-13 978-90-481-7253-5
ISBN-10 1-4020-5008-9 (e-book)
ISBN-13 978-1-4020-5008-4 (e-book)

Published by Springer,
P.O. Box 17, 3300 AA Dordrecht, The Netherlands.

www.springer.com

Printed on acid-free paper

Dedication

To my high school mathematics teachers,
Ἀγγελον Ματζαβίνον και Δημήτριον Κατσώνην

Charalambos D. Aliprantis

To: Alex, Emily, Joshua, and Andrew
Grandchildren bring new joys to your life.

Owen Burkinshaw

Dedication

Contents

Foreword

This monograph is a reprint of our original book *Positive Operators* published in 1985 as volume # 119 in the Pure and Applied Mathematics series of Academic Press. With the exception of correcting several misprints and a few mathematical errors, this edition of the book is identical to the original one.

At the end of the book we have listed a collection of monographs that were published after its 1985 publication that contain material and new developments related to the subject matter of this book.

West Lafayette and Indianapolis   CHARALAMBOS D. ALIPRANTIS
May, 2006   OWEN BURKINSHAW

Historical Foreword

'Ολβιos όστιs ιστορίηs έσχε μάθησιν .

Ευριπίδηs

Wise is the person who knows history.

Euripides

Positive operators made their debut at the beginning of the nineteenth century. They were tied to integral operators (whose study triggered the birth of functional analysis) and to matrices with nonnegative entries. However, positive operators were investigated in a systematic manner much later. Their study followed closely the development of Riesz spaces. An address of F. Riesz in 1928 *On the decomposition of linear functionals* [**166**] (into their positive and negative parts), at the International Congress of Mathematicians in Bologna, Italy, marked the beginnings of the study of Riesz spaces and positive operators. The theory of Riesz spaces was developed axiomatically in the mid-1930s by H. Freudenthal [**67**] and L. V. Kantorovich [**85, 86, 87, 88, 89, 90, 91**]. Positive operators were also introduced and studied in the mid-1930s by L. V. Kantorovich, and they made their first textbook appearance in the 1940 edition of G. Birkhoff's book *Lattice Theory* [**37**]. Undoubtedly, the systematic study of positive operators was originated in the 1930s by F. Riesz, L. V. Kantorovich and G. Birkhoff.

In the 1940s and early 1950s one finds very few papers on positive operators. In this period, the main contributions came from the Soviet school (L. V. Kantorovich, M. G. Krein, A. G. Pinsker, M. A. Rutman, B. Z. Vulikh) and the Japanese school (H. Nakano, K. Yosida, T. Ogasawara, and their students). In the 1950 the book *Functional Analysis in Partially Ordered Spaces* [**92**] by L. V. Kantorovich, B. Z. Vulikh, and A. G. Pinsker appeared in the Soviet literature. This book (that has not been translated into English) contained an excellent treatment, up to that date, of positive operators and their applications.

Since the mid-1950s, research on positive operators has gained considerable momentum. From 1955 to 1970 important contributions came from T. Andô [20]–[24], C. Goffman [70], S. Kaplan [93], S. Karlin [95], P. P. Korovkin [98], M. A. Krasnoselskii, P. P. Zabreiko, E. I. Pustylnik, and P. E. Sobolevskii [100], U. Krengel [105, 106], G. Ya. Lozanovsky [119]–[123], W. A. J. Luxemburg and A. C. Zaanen [125, 130], H. Nakano [148]–[152], I. Namioka [153], A. Peressini [162], H. H. Schaefer [171], and B. Z. Vulikh [189]. Thus, by the end of the 1960s the "ground work" for the theory of positive operators was well established.

The 1970s can be characterized as the "maturity period" for the theory of positive operators. In 1974, the first monograph devoted entirely to the subject appeared in the literature. This was H. H. Schaefer's book *Banach Lattices and Positive Operators* [174], whose influence on the development of positive operators was enormous. By this time, the number of mathematicians working in the field had increased, and research was carried out in a more systematic manner. The growth of the subject has been very fast; and, in addition, applications to other disciplines have started to flourish. Another milestone for the 1970s was the breakthrough paper of P. G. Dodds and D. H. Fremlin [54] on positive compact operators. The list of contributors to the theory of positive operators in the 1970s includes Y. A. Abramovich, C. D. Aliprantis, S. J. Bernau, A. V. Buhvalov, O. Burkinshaw, D. I. Cartwright, P. G. Dodds, M. Duhoux, P. van Eldik, J. J. Grobler, D. H. Fremlin, H. P. Lotz, W. A. J. Luxemburg, M. Meyer, P. Meyer-Nieberg, R. J. Nagel, U. Schlotterbeck, H. H. Schaefer, A. R. Schep, C. T. Tucker, A. I. Veksler, A. W. Wickstead, M. Wolff, and A. C. Zaanen.

The 1980s have started very well for the theory of positive operators. A series of important papers on the subject have been written by the authors [9, 10, 11, 12, 13, 14, 15, 16, 17]. Another excellent book on positive operators was added to the literature of positive operators. This was A. C. Zaanen's book *Riesz Spaces II* [197] dealing primarily with research done up to 1980. In addition, a pool of talented young mathematicians had joined the ranks of researchers of positive operators; many have already made important contributions. Also, more and more mathematicians of the theory of Banach spaces are studying positive operators, and this was given an extra boost to the subject. In addition, the theory of positive operators has found some impressive applications in a variety of disciplines, ranging from mathematical physics to economics.

Thus, as we are progressing into the 1980s the future of positive operators looks bright. The field is alive and grows both in theory and applications. It is our hope that you, the reader, will also contribute towards this growth.

Indianapolis   CHARALAMBOS D. ALIPRANTIS
May, 1984   OWEN BURKINSHAW

Preface

Linear operators have been studied in various contexts and settings in the past. Their study is a subject of great importance both to mathematics and to its applications. The present book deals mainly with the special class of linear operators known as positive operators. A linear operator between two ordered vector spaces that carries positive elements to positive elements is referred to as a positive operator. For instance, the linear operator $T \colon C[0,1] \to C[0,1]$, defined by

$$Tf(x) = \int_0^x f(t)\,dt\,,$$

carries positive functions of $C[0,1]$ to positive functions, and is thus an example of a positive operator. The material covered in the book has some overlap with the material in the books by H. H. Schaefer [174] and A. C. Zaanen [197]. However, our attention is mostly focused on recent developments of the subject and the overlap with the above-mentioned books is kept to a minimum. On the other hand, we have intentionally covered those topics where the ingredient of positivity allowed us to obtain beautiful and elegant results.

It is well known that many linear operators between Banach spaces arising in classical analysis are in fact positive operators. For this reason, in this book positive operators are studied in the setting of Riesz spaces and Banach lattices. In order to make the book as self-sufficient as possible, some basic results from the theory of Riesz spaces and Banach lattices are included with proofs as needed. On the other hand, we assume that the reader is familiar with the elementary concepts of real analysis and functional analysis.

The material has been spread out into five chapters. Chapter 1 deals mainly with the elementary properties of positive operators. This chapter covers extension properties of positive operators, order projections, order continuous operators, and positive linear functionals. Chapter 2 studies three basic classes of operators: the components of a positive operator, the lattice homomorphisms, and the orthomorphisms. Chapter 3 considers topological aspects of vector spaces. It covers topological vector spaces, weak topologies on Banach and Riesz spaces, and locally convex-solid Riesz spaces. The fourth chapter is devoted to Banach lattices. Particular emphasis is given to Banach lattices with order continuous norms. Also, weak compactness in Banach lattices, embeddings of Banach lattices, and Banach lattices of operators are studied in this chapter. The fifth and final chapter of the book deals with compactness properties of positive operators. This is the most important (and most elegant) chapter of the book. It makes a thorough study of compact, weakly compact, and Dunford–Pettis operators on Banach lattices.

The five chapters consist of nineteen sections. Each section ends with exercises that supplement its material. There are almost 300 exercises. These exercises extend and illustrate the material of the book in a concrete manner.

We have made every effort to be as accurate as possible in crediting the major theorems of the book to their original discoverers.

CHARALAMBOS D. ALIPRANTIS
OWEN BURKINSHAW

Acknowledgments

Parts of this book were written while the authors held visiting appointments at various institutions: California Institute of Technology, Purdue University, the University of Southern California, and the Institute for Mathematics and Its Applications—University of Minnesota. We express our gratitude for the hospitality provided to us by the staffs and faculties of these institutions.

We acknowledge with many thanks the financial support we received from the National Science Foundation under grants MCS 81–00787, MCS 82–19750, and DMS 83–19594.

Finally, we would like to thank our wives, Bernadette and Betty, for their assistance, and most of all for coping with our idiosyncrasies during the difficult stages of the writing of the book.

List of Special Symbols

Numbers that follow each entry are the page numbers on which the symbol first occurs.

w^*	Weak* topology	157
x^+	Positive part of vector x	4
x^-	Negative part of vector x	4
$|x|$	Absolute value of vector x	4
$\hat{x}$	Image of x in its double dual	158
$\dot{x}$	Equivalence class of vector x	99
$x_\alpha \uparrow x$	Increasing net to x	8
$x_\alpha \downarrow x$	Decreasing net to x	8
$x_\alpha \xrightarrow{o} x$	Order convergence	33
$x_\alpha \xrightarrow{w} x$	Weak convergence	157
$[0, x]$	Order interval of x in E	12, 179
$[\![0, x]\!]$	Order interval of x in E''	176
$x \vee y$	Supremum of x and y	2
$x \wedge y$	Infimum of x and y	2
$x \perp y$	Disjoint vectors	7
X^*	Algebraic dual of vector space X	65, 139
X'	Topological dual of X	139
X''	Second (or double) dual	154
$\langle X, X' \rangle$	Dual system	145
$(X_1 \oplus X_2 \oplus \cdots)_p$	L_p-sum of the sequence of B-spaces $\{X_n\}$	183
$(\Psi, \|\|\cdot\|\|)$	Davis–Figiel–Johnson–Pelczynski space	301

Chapter 1

The Order Structure of Positive Operators

A linear operator between two ordered vector spaces that carries positive elements to positive elements is known in the literature as a positive operator. As we have mentioned in the preface, the main theme of this book is the study of positive operators. To obtain fruitful and useful results the domains and the ranges of positive operators will be taken to be Riesz spaces (vector lattices). For this reason, in order to make the material as self-sufficient as possible, the fundamental properties of Riesz spaces are discussed as they are needed.

Throughout this book the symbol $\mathbb{R}$ will denote the set of real numbers, $\mathbb{N}$ will denote the set of natural numbers, $\mathbb{Q}$ will denote the set of rational numbers, and $\mathbb{Z}$ will denote the set of integers.

1.1. Basic Properties of Positive Operators

A real vector space E is said to be an **ordered vector space** whenever it is equipped with an order relation $\geq$ (i.e., $\geq$ is a reflexive, antisymmetric, and transitive binary relation on E) that is compatible with the algebraic structure of E in the sense that it satisfies the following two axioms:

(1) If $x \geq y$, then $x + z \geq y + z$ holds for all $z \in E$.

(2) If $x \geq y$, than $\alpha x \geq \alpha y$ holds for all $\alpha \geq 0$.

An alternative notation for $x \geq y$ is $y \leq x$. A vector x in an ordered vector space E is called **positive** whenever $x \geq 0$ holds. The set of all positive vectors of E will be denoted by E^+, i.e., $E^+ := \{x \in E: x \geq 0\}$. The set E^+ of positive vectors is called the **positive cone** of E.

Definition 1.1. *An **operator** is a linear map between two vector spaces.*

That is, a mapping $T: E \to F$ between two vector spaces is called an operator if and only if $T(\alpha x + \beta y) = \alpha T(x) + \beta T(y)$ holds for all $x, y \in E$ and all $\alpha, \beta \in \mathbb{R}$. As usual, the value $T(x)$ will also be designated by Tx.

Definition 1.2. *An operator $T: E \to F$ between two ordered vector spaces is said to be **positive** (in symbols $T \geq 0$ or $0 \leq T$) if $T(x) \geq 0$ for all $x \geq 0$.*

Clearly, an operator $T: E \to F$ between two ordered vector spaces is positive if and only if $T(E^+) \subseteq F^+$ (and also if and only if $x \leq y$ implies $Tx \leq Ty$).

A **Riesz space** (or a **vector lattice**) is an ordered vector space E with the additional property that for each pair of vectors $x, y \in E$ the supremum and the infimum of the set $\{x, y\}$ both exist in E. Following the classical notation, we shall write

$$x \vee y := \sup\{x, y\} \quad \text{and} \quad x \wedge y := \inf\{x, y\}.$$

Typical examples of Riesz spaces are provided by the function spaces. A **function space** is a vector space E of real-valued functions on a set Ω such that for each pair $f, g \in E$ the functions

$$[f \vee g](\omega) := \max\{f(\omega), g(\omega)\} \quad \text{and} \quad [f \wedge g](\omega) := \min\{f(\omega), g(\omega)\}$$

both belong to E. Clearly, every function space E with the pointwise ordering (i.e., $f \leq g$ holds in E if and only if $f(\omega) \leq g(\omega)$ for all $\omega \in \Omega$) is a Riesz space. Here are some important examples of function spaces:

(a) $\mathbb{R}^\Omega$, all real-valued functions defined on a set Ω.

(b) $C(\Omega)$, all continuous real-valued functions on a topological space Ω.

(c) $C_b(\Omega)$, all bounded real-valued continuous functions on a topological space Ω.

(d) $\ell_\infty(\Omega)$, all bounded real-valued functions on a set Ω.

(e) ℓ_p $(0 < p < \infty)$, all real sequences $(x_1, x_2, \ldots)$ with $\sum_{n=1}^\infty |x_n|^p < \infty$.

The class of L_p-spaces is another important class of Riesz spaces. If (X, Σ, μ) is a measure space and $0 < p < \infty$, then $L_p(\mu)$ is the vector space of all real-valued μ-measurable functions f on X such that $\int_X |f|^p \, d\mu < \infty$. Also, $L_\infty(\mu)$ is the vector space of all real-valued μ-measurable functions f on X such that $\operatorname{esssup} |f| < \infty$. As usual, functions differing on a set of measure zero are treated as identical, i.e., $f = g$ in $L_p(\mu)$ means that $f(x) = g(x)$ for μ-almost all $x \in X$. (In other words, each $L_p(\mu)$-space consists of equivalence classes rather than functions.) It is easy to see that under the ordering $f \leq g$ whenever $f(x) \leq g(x)$ holds for μ-almost all $x \in X$, each $L_p(\mu)$ is a Riesz space.

There are several useful identities that are true in a Riesz space some of which are included in the next few results.

Theorem 1.3. *If x, y and z are elements in a Riesz space, then:*

(1) $x \vee y = -[(-x) \wedge (-y)]$ *and* $x \wedge y = -[(-x) \vee (-y)]$.

(2) $x + y = x \wedge y + x \vee y$.

(3) $x + (y \vee z) = (x + y) \vee (x + z)$ *and* $x + (y \wedge z) = (x + y) \wedge (x + z)$.

(4) $\alpha(x \vee y) = (\alpha x) \vee (\alpha y)$ *and* $\alpha(x \wedge y) = (\alpha x) \wedge (\alpha y)$ *for all $\alpha \geq 0$.*

Proof. (1) From $x \leq x \vee y$ and $y \leq x \vee y$ we get $-(x \vee y) \leq -x$ and $-(x \vee y) \leq -y$, and so $-(x \vee y) \leq (-x) \wedge (-y)$. On the other hand, if $-x \geq z$ and $-y \geq z$, then $-z \geq x$ and $-z \geq y$, and hence $-z \geq x \vee y$. Thus, $-(x \vee y) \geq z$ holds and this shows that $-(x \vee y)$ is the infimum of the set $\{-x, -y\}$. That is, $(-x) \wedge (-y) = -(x \vee y)$. To get the identity for $x \wedge y$ replace x by $-x$ and y by $-y$ in the above proven identity.

(2) From $x \wedge y \leq y$ it follows that $y - x \wedge y \geq 0$ and so $x \leq x + y - x \wedge y$. Similarly, $y \leq x + y - x \wedge y$. Consequently, we have $x \vee y \leq x + y - x \wedge y$ or $x \wedge y + x \vee y \leq x + y$. On the other hand, from $y \leq x \vee y$ we see that $x + y - x \vee y \leq x$, and similarly $x + y - x \vee y \leq y$. Thus, $x + y - x \vee y \leq x \wedge y$ so that $x + y \leq x \wedge y + x \vee y$, and the desired identity follows.

(3) Clearly, $x + y \leq x + y \vee z$ and $x + z \leq x + y \vee z$, and therefore $(x + y) \vee (x + z) \leq x + y \vee z$. On the other hand, we have $y = -x + (x + y) \leq -x + (x + y) \vee (x + z)$, and likewise $z \leq -x + (x + y) \vee (x + z)$, and so $y \vee z \leq -x + (x + y) \vee (x + z)$. Therefore, $x + y \vee z \leq (x + y) \vee (x + z)$ also holds, and thus $x + y \vee z = (x + y) \vee (x + z)$. The other identity can be proven in a similar manner.

(4) Fix $\alpha > 0$. Clearly, $(\alpha x) \vee (\alpha y) \leq \alpha(x \vee y)$. If $\alpha x \leq z$ and $\alpha y \leq z$ are both true, then $x \leq \frac{1}{\alpha} z$ and $y \leq \frac{1}{\alpha} z$ also are true, and so $x \vee y \leq \frac{1}{\alpha} z$. This implies $\alpha(x \vee y) \leq z$, and this shows that $\alpha(x \vee y)$ is the supremum of the set $\{\alpha x, \alpha y\}$. Therefore, $(\alpha x) \vee (\alpha y) = \alpha(x \vee y)$. The other identity can be proven similarly. ∎

The reader can establish in a similar manner the following general versions of the preceding formulas in (1), (3), and (4). If A is a nonempty subset of a Riesz space for which $\sup A$ exists, then:

(a) The infimum of the set $-A := \{-a : a \in A\}$ exists and

$$\inf(-A) = -\sup A.$$

(b) For each vector x the supremum of the set $x + A := \{x + a \colon a \in A\}$ exists and

$$\sup(x + A) = x + \sup A.$$

(c) For each $\alpha \geq 0$ the supremum of the set $\alpha A := \{\alpha a \colon a \in A\}$ exists and

$$\sup(\alpha A) = \alpha \sup A.$$

We have also the following useful inequality between positive vectors.

Lemma 1.4. *If $x, x_1, x_2, \ldots, x_n$ are positive elements in a Riesz space, then*

$$x \wedge (x_1 + x_2 + \cdots + x_n) \leq x \wedge x_1 + x \wedge x_2 + \cdots + x \wedge x_n.$$

Proof. Assume that x and x_1, x_2 are all positive vectors. For simplicity, let $y = x \wedge (x_1 + x_2)$. Then $y \leq x_1 + x_2$ and so $y - x_1 \leq x_2$. Also we have $y - x_1 \leq y \leq x$. Consequently $y - x_1 \leq x \wedge x_2$. This implies $y - x \wedge x_2 \leq x_1$ and since $y - x \wedge x_2 \leq y \leq x$, we infer that $y - x \wedge x_2 \leq x \wedge x_1$ or $y \leq x \wedge x_1 + x \wedge x_2$. The proof now can be completed by induction. ∎

For any vector x in a Riesz space define

$$x^+ := x \vee 0, \quad x^- := (-x) \vee 0, \quad \text{and} \quad |x| := x \vee (-x).$$

The element x^+ is called the **positive part**, x^- is called the **negative part**, and $|x|$ is called the **absolute value** of x. The vectors x^+, x^-, and $|x|$ satisfy the following important identities.

Theorem 1.5. *If x is an arbitrary vector in a Riesz space E, then:*

(1) $x = x^+ - x^-$.

(2) $|x| = x^+ + x^-$.

(3) $x^+ \wedge x^- = 0$.

Moreover, the decomposition in (1) satisfies the following minimality and uniqueness properties.

(a) *If $x = y - z$ with $y, z \in E^+$, then $y \geq x^+$ and $z \geq x^-$.*

(b) *If $x = y - z$ with $y \wedge z = 0$, then $y = x^+$ and $z = x^-$.*

Proof. (1) From Theorem 1.3 we see that

$$x = x + 0 = x \vee 0 + x \wedge 0 = x \vee 0 - (-x) \vee 0 = x^+ - x^-.$$

(2) Using Theorem 1.3 and (1), we get

$$\begin{aligned} |x| &= x \vee (-x) = (2x) \vee 0 - x = 2(x \vee 0) - x \\ &= 2x^+ - x = 2x^+ - (x^+ - x^-) = x^+ + x^-. \end{aligned}$$

(3) Note that

$$
\begin{aligned}
x^+ \wedge x^- &= (x^+ - x^-) \wedge 0 + x^- = x \wedge 0 + x^- \\
&= -[(-x) \vee 0] + x^- = -x^- + x^- = 0 \,.
\end{aligned}
$$

(a) Assume that $x = y - z$ with $y \geq 0$ and $z \geq 0$. From $x = x^+ - x^-$, we get $x^+ = x^- + y - z \leq x^- + y$, and so from Lemma 1.4 we get

$$
x^+ = x^+ \wedge x^+ \leq x^+ \wedge (x^- + y) \leq x^+ \wedge x^- + x^+ \wedge y = x^+ \wedge y \leq y \,.
$$

Similarly, $x^- \leq z$.

(b) Let $x = y - z$ with $y \wedge z = 0$. Then, using Theorem 1.3, we see that $x^+ = (y - z) \vee 0 = y \vee z - z = (y + z - y \wedge z) - z = y$. Similarly, $x^- = z$. ∎

We also have the following useful inequality regarding positive operators.

Lemma 1.6. *If $T \colon E \to F$ is a positive operator between two Riesz spaces, then for each $x \in E$ we have*

$$
|Tx| \leq T|x| \,.
$$

Proof. If $x \in E$, then $\pm x \leq |x|$ and the positivity of T yields $\pm Tx \leq T|x|$, which is equivalent to $|Tx| \leq T|x|$. ∎

A few more useful lattice identities are included in the next result.

Theorem 1.7. *If x and y are elements in a Riesz space, then we have:*

(1) $x = (x - y)^+ + x \wedge y$.

(2) $x \vee y = \frac{1}{2}(x + y + |x - y|)$ and $x \wedge y = \frac{1}{2}(x + y - |x - y|)$.

(3) $|x - y| = x \vee y - x \wedge y$.

(4) $|x| \vee |y| = \frac{1}{2}(|x + y| + |x - y|)$.

(5) $|x| \wedge |y| = \frac{1}{2}\big||x + y| - |x - y|\big|$.

(6) $|x + y| \wedge |x - y| = \big||x| - |y|\big|$.

(7) $|x + y| \vee |x - y| = |x| + |y|$.

Proof. (1) Using Theorem 1.3 we see that

$$
\begin{aligned}
x &= x \vee y - y + x \wedge y = (x - y) \vee (y - y) + x \wedge y \\
&= (x - y) \vee 0 + x \wedge y = (x - y)^+ + x \wedge y \,.
\end{aligned}
$$

(2) For the first identity note that

$$
\begin{aligned}
x + y + |x - y| &= x + y + (x - y) \vee (y - x) \\
&= [(x + y) + (x - y)] \vee [(x + y) + (y - x)] \\
&= (2x) \vee (2y) = 2(x \vee y) \,.
\end{aligned}
$$

(3) Subtract the two identities in (2).

(4) Using (2) above, we see that

$$
\begin{aligned}
|x+y| + |x-y| &= (x+y) \vee (-x-y) + |x-y| \\
&= (x+y+|x-y|) \vee (-x-y+|x-y|) \\
&= 2([x \vee y] \vee [(-x) \vee (-y)]) \\
&= 2([x \vee (-x)] \vee [y \vee (-y)]) \\
&= 2(|x| \vee |y|) .
\end{aligned}
$$

(5) Using (2) and (4) above we get

$$
\begin{aligned}
\big||x+y| - |x-y|\big| &= 2(|x+y| \vee |x-y|) - (|x+y| + |x-y|) \\
&= 2(|x| + |y|) - 2(|x| \vee |y|) \\
&= 2(|x| \wedge |y|) .
\end{aligned}
$$

(6) Notice that

$$
\begin{aligned}
&|x+y| \wedge |x-y| \\
&= [(x+y) \vee (-x-y)] \wedge [(x-y) \vee (y-x)] \\
&= \{[(x+y) \vee (-x-y)] \wedge (x-y)\} \vee \{[(x+y) \vee (-x-y)] \wedge (y-x)\} \\
&= [(x+y) \wedge (x-y)] \vee [(-x-y) \wedge (x-y)] \vee \cdots \\
&\qquad \cdots \vee [(x+y) \wedge (y-x)] \vee [(-x-y) \wedge (y-x)] \\
&= [x+y \wedge (-y)] \vee [-y+(-x) \wedge x] \vee \cdots \\
&\qquad \cdots \vee [y+x \wedge (-x)] \vee [-x+(-y) \wedge y] \\
&= \{[x+y \wedge (-y)] \vee [-x+y \wedge (-y)]\} \vee \cdots \\
&\qquad \cdots \vee \{[-y+(-x) \wedge x] \vee [y+x \wedge (-x)]\} \\
&= [x \vee (-x) + y \wedge (-y)] \vee [(-y) \vee y + x \wedge (-x)] \\
&= [|x| - |y|] \vee [|y| - |x|] = \big||x| - |y|\big| .
\end{aligned}
$$

(7) Using (3) and (5) we get

$$
\begin{aligned}
|x+y| \vee |x-y| &= \big||x+y| - |x-y|\big| + |x+y| \wedge |x-y| \\
&= 2(|x| \wedge |y|) + \big||x| - |y|\big| \\
&= 2(|x| \wedge |y|) + (|x| \vee |y| - |x| \wedge |y|) \\
&= |x| \wedge |y| + |x| \vee |y| = |x| + |y| ,
\end{aligned}
$$

and the proof is finished. ■

It should be noted that the identities in (2) above show that an ordered vector space is a Riesz space if and only if the absolute value $|x| = x \vee (-x)$ exists for each vector x.

In a Riesz space, two elements x and y are said to be **disjoint** (in symbols $x \perp y$) whenever $|x| \wedge |y| = 0$ holds. Note that according to part (5) of Theorem 1.7 we have $x \perp y$ if and only if $|x + y| = |x - y|$. Two subsets A and B of a Riesz space are called **disjoint** (denoted $A \perp B$) if $a \perp b$ holds for all $a \in A$ and all $b \in B$.

If A is a nonempty subset of a Riesz space E, then its **disjoint complement** A^{d} is defined by

$$A^{\mathrm{d}} := \left\{ x \in E \colon x \perp y \ \text{ for all } y \in A \right\}.$$

We write A^{dd} for $(A^{\mathrm{d}})^{\mathrm{d}}$. Note that $A \cap A^{\mathrm{d}} = \{0\}$.

If A and B are subsets of a Riesz space, then we shall employ in this book the following self-explanatory notation:

$$
\begin{aligned}
|A| &:= \left\{ |a| \colon \ a \in A \right\} \\
A^+ &:= \left\{ a^+ \colon \ a \in A \right\} \\
A^- &:= \left\{ a^- \colon \ a \in A \right\} \\
A \vee B &:= \left\{ a \vee b \colon \ a \in A \text{ and } b \in B \right\} \\
A \wedge B &:= \left\{ a \wedge b \colon \ a \in A \text{ and } b \in B \right\} \\
x \vee A &:= \left\{ x \vee a \colon \ a \in A \right\} \\
x \wedge A &:= \left\{ x \wedge a \colon \ a \in A \right\}
\end{aligned}
$$

The next theorem tells us that every Riesz space satisfies the infinite distributive law.

Theorem 1.8 (The Infinite Distributive Law). *Let A be a nonempty subset of a Riesz space. If $\sup A$ exists, then for each vector x the supremum of the set $x \wedge A$ exists and*

$$\sup(x \wedge A) = x \wedge \sup A.$$

Similarly, if $\inf A$ exists, then $\inf(x \vee A)$ exists for each vector x and

$$\inf(x \vee A) = x \vee \inf A.$$

Proof. Assume that $\sup A$ exists. Let $y = \sup A$ and fix some vector x. Clearly, for each $a \in A$ we have $x \wedge a \leq x \wedge y$, i.e., $x \wedge y$ is an upper bound of the set $x \wedge A$. To see that $x \wedge y$ is the least upper bound of the set $x \wedge A$, assume that some vector z satisfies $x \wedge a \leq z$ for all $a \in A$. Since for each $a \in A$ we have $a = x \wedge a + x \vee a - x \leq z + x \vee y - x$, it follows that $y \leq z + x \vee y - x$. This implies $x \wedge y = x + y - x \vee y \leq z$, and from this we see that $\sup(x \wedge A)$ exists and that $\sup(x \wedge A) = x \wedge \sup A$ holds. The other formula can be proven in a similar manner. ∎

The next result includes most of the major inequalities that are used extensively in estimations.

Theorem 1.9. *For arbitrary elements x, y, and z in a Riesz space we have the following inequalities.*

(1) $\big||x| - |y|\big| \le |x + y| \le |x| + |y|$ *(the triangle inequality).*

(2) $|x \vee z - y \vee z| \le |x - y|$ *and* $|x \wedge z - y \wedge z| \le |x - y|$ *(Birkhoff's inequalities).*

Proof. (1) Clearly, $x + y \le |x| + |y|$ and $-x - y \le |x| + |y|$ both hold. Thus, $|x + y| = (x + y) \vee (-x - y) \le |x| + |y|$.

Now observe that the inequality $|x| = \big|(x + y) - y\big| \le |x + y| + |y|$ implies $|x| - |y| \le |x + y|$. Similarly, $|y| - |x| \le |x + y|$, and hence $\big||x| - |y|\big| \le |x + y|$ is also true.

(2) Note that

$$
\begin{aligned}
x \vee z - y \vee z &= \big[(x - z) \vee 0 + z\big] - \big[(y - z) \vee 0 + z\big] \\
&= (x - z)^+ - (y - z)^+ \\
&= \big[(x - y) + (y - z)\big]^+ - (y - z)^+ \\
&\le \big[(x - y)^+ + (y - z)^+\big] - (y - z)^+ \\
&= (x - y)^+ \le |x - y| .
\end{aligned}
$$

Similarly, $y \vee z - x \vee z \le |x - y|$, and so $|x \vee z - y \vee z| \le |x - y|$. The other inequality can be proven in a similar manner. ∎

In particular, note that in any Riesz space we have

$$
\big|x^+ - y^+\big| \le |x - y| \quad \text{and} \quad \big|x^- - y^-\big| \le |x - y| .
$$

These inequalities will be employed quite often in our discussions.

A net $\{x_\alpha\}$ in a Riesz space is said to be **decreasing** (in symbols $x_\alpha \downarrow$) whenever $\alpha \succeq \beta$ implies $x_\alpha \le x_\beta$. The notation $x_\alpha \downarrow x$ means that $x_\alpha \downarrow$ and $\inf\{x_\alpha\} = x$ both hold. The meanings of $x_\alpha \uparrow$ and $x_\alpha \uparrow x$ are analogous.

The Archimedean property states that for each real number $x > 0$ the sequence $\{nx\}$ is unbounded above in $\mathbb{R}$. This is, of course, equivalent to saying that $\frac{1}{n}x \downarrow 0$ holds in $\mathbb{R}$ for each $x > 0$. Motivated by this property, a Riesz space (and in general an ordered vector space) E is called **Archimedean** whenever $\frac{1}{n}x \downarrow 0$ holds in E for each $x \in E^+$. All classical spaces of functional analysis (notably the function spaces and L_p-spaces) are Archimedean. For this reason, the focus of our work will be on the study of positive operators between Archimedean Riesz spaces. Accordingly:

- *Unless otherwise stated, throughout this book all Riesz spaces will be assumed to be Archimedean.*

The starting point in the theory of positive operators is a fundamental extension theorem of L. V. Kantorovich [**91**]. The importance of the result lies in the fact that in order for a mapping $T\colon E^+ \to F^+$ to be the restriction of a (unique) positive operator from E to F it is necessary and sufficient to be additive on E^+. The details follow.

Theorem 1.10 (Kantorovich). *Suppose that E and F are two Riesz spaces with F Archimedean. Assume also that $T\colon E^+ \to F^+$ is an **additive mapping**, that is, $T(x + y) = T(x) + T(y)$ holds for all $x, y \in E^+$. Then T has a unique extension to a positive operator from E to F. Moreover, the extension (denoted by T again) is given by*

$$T(x) = T(x^+) - T(x^-)$$

for all $x \in E$.

Proof. Let $T\colon E^+ \to F^+$ be an additive mapping. Consider the mapping $S\colon E \to F$ defined by

$$S(x) = T(x^+) - T(x^-).$$

Clearly, $S(x) = T(x)$ for each $x \in E^+$. So, the mapping S extends T to all of E. Since $x = x^+ - x^-$ for each $x \in E$, it follows that S is the only possible linear extension of T to all of E. Therefore, in order to complete the proof, we must show that S is linear. That is, we must prove that S is additive and homogeneous.

For the additivity of S start by observing that if any vector $x \in E$ can be written as a difference of two positive vectors, say $x = x_1 - x_2$ with $x_1, x_2 \in E^+$, then $S(x) = T(x_1) - T(x_2)$ holds. To see this, fix any $x \in E$ and assume that $x = x^+ - x^- = x_1 - x_2$, where $x_1, x_2 \in E^+$. Then $x^+ + x_2 = x_1 + x^-$, and so the additivity of T on E^+ yields

$$T(x^+) + T(x_2) = T(x^+ + x_2) = T(x_1 + x^-) = T(x_1) + T(x^-)$$

or $S(x) = T(x^+) - T(x^-) = T(x_1) - T(x_2)$. From this property, we can
easily establish that S is additive. Indeed, if $x, y \in E$, then note that

$$
\begin{aligned}
S(x + y) &= S\big(x^+ + y^+ - (x^- + y^-)\big) \\
&= T(x^+ + y^+) - T(x^- + y^-) \\
&= T(x^+) + T(y^+) - T(x^-) - T(y^-) \\
&= \big[T(x^+) - T(x^-)\big] + \big[T(y^+) - T(y^-)\big] \\
&= S(x) + S(y).
\end{aligned}
$$

In particular, the additivity of S implies that $S(rx) = rS(x)$ holds for all
$x \in E$ and all rational numbers r.

It remains to show that S is homogeneous. For this, we need to prove
first that S is monotone. That is, $x \geq y$ in E implies $S(x) \geq S(y)$ in F.
Indeed, if $x \geq y$, then $x - y \in E^+$, and so by the additivity of S we get

$$S(x) = S\big((x - y) + y\big) = S(x - y) + S(y) = T(x - y) + S(y) \geq S(y).$$

Now fix $x \in E^+$ and let $\lambda \geq 0$. Pick two sequences of non-negative rational
numbers $\{r_n\}$ and $\{t_n\}$ such that $r_n \uparrow \lambda$ and $t_n \downarrow \lambda$. The inequalities
$r_n x \leq \lambda x \leq t_n x$ and the monotonicity of S imply

$$r_n S(x) = S(r_n x) \leq S(\lambda x) \leq S(t_n x) = t_n S(x)$$

for each n. Using that F is Archimedean, we easily get $\lambda S(x) = S(\lambda x)$.
Finally, if $\lambda \in \mathbb{R}$ and $x \in E$, then

$$
\begin{aligned}
S(\lambda x) &= S\big(\lambda x^+ + (-\lambda)x^-\big) = S\big(\lambda x^+\big) + S\big((-\lambda)x^-\big) \\
&= \lambda S(x^+) - \lambda S(x^-) = \lambda\big[T(x^+) - T(x^-)\big] = \lambda S(x).
\end{aligned}
$$

So, S is also homogeneous, and the proof is finished. ∎

The preceding lemma is not true if F is not Archimedean.

Example 1.11. Let $\phi \colon \mathbb{R} \to \mathbb{R}$ be an additive function that is not linear, i.e.,
not of the form $\phi(x) = cx$, and let F be the lexicographic plane. Consider
the mapping $T \colon \mathbb{R}^+ \to F^+$ defined by $T(x) = \big(x, \phi(x)\big)$ for each x in $\mathbb{R}^+$.
Note that T is additive and that if T could be extended to an operator from
$\mathbb{R}$ to F, then ϕ should be linear. ∎

Thus, a mapping $T \colon E^+ \to F^+$ extends to a (unique) positive operator
from E to F if and only if T is additive on E^+. In other words, a positive
operator is determined completely by its action on the positive cone of its
domain. In the sequel, the expression "**the mapping** $T \colon E^+ \to F^+$ **defines
a positive operator**" will simply mean that T is additive on E^+ (and hence
extendable by Theorem 1.10 to a unique positive operator).

The (real) vector space of all operators from E to F will be denoted by
$\mathcal{L}(E, F)$. It is not difficult to see that $\mathcal{L}(E, F)$ under the ordering $T \geq S$

whenever $T - S$ is a positive operator (i.e., whenever $T(x) \geq S(x)$ holds for all $x \in E^+$) is an ordered vector space.

Definition 1.12. *For an operator $T\colon E \to F$ between two Riesz spaces we shall say that its modulus $|T|$ exists (or that T possesses a **modulus**) whenever*

$$|T| := T \vee (-T)$$

exists—in the sense that $|T|$ is the supremum of the set $\{-T, T\}$ in $\mathcal{L}(E, F)$.

In order to study the elementary properties of the modulus, we need a decomposition property of Riesz spaces.

Theorem 1.13 (The Decomposition Property). *If $|x| \leq |y_1 + \cdots + y_n|$ holds in a Riesz space, then there exist $x_1, \ldots, x_n$ satisfying $x = x_1 + \cdots + x_n$ and $|x_i| \leq |y_i|$ for each $i = 1, \ldots, n$. Moreover, if x is positive, then the x_i also can be chosen to be positive.*

Proof. By using induction it is enough to establish the result when $n = 2$. So, let $|x| \leq |y_1 + y_2|$.

Put $x_1 = [x \vee (-|y_1|)] \wedge |y_1|$, and observe that $|x_1| \leq |y_1|$ (and that $0 \leq x_1 \leq x$ holds if x is positive). Now put $x_2 = x - x_1$ and observe that

$$x_2 = x - [x \vee (-|y_1|)] \wedge |y_1| = [0 \wedge (x + |y_1|)] \vee (x - |y_1|).$$

On the other hand, $|x| \leq |y_1| + |y_2|$ implies $-|y_1| - |y_2| \leq x \leq |y_1| + |y_2|$, from which it follows that

$$-|y_2| = (-|y_2|) \wedge 0 \leq (x + |y_1|) \wedge 0 \leq x_2 \leq 0 \vee (x - |y_1|) \leq |y_2|.$$

Thus, $|x_2| \leq |y_2|$ also holds, and the proof is finished. ∎

An important case for the modulus to exist is described next.

Theorem 1.14. *Let $T\colon E \to F$ be an operator between two Riesz spaces such that $\sup\{|Ty|\colon |y| \leq x\}$ exists in F for each $x \in E^+$. Then the modulus of T exists and*

$$|T|(x) = \sup\{|Ty|\colon |y| \leq x\}$$

holds for all $x \in E^+$.

Proof. Define $S\colon E^+ \to F^+$ by $S(x) = \sup\{|Ty|\colon |y| \leq x\}$ for each x in E^+. Since $|y| \leq x$ implies $|\pm y| = |y| \leq x$, it easily follows that we have $S(x) = \sup\{Ty\colon |y| \leq x\}$ for each $x \in E^+$. We claim that S is additive.

To see this, let $u, v \in E^+$. If $|y| \leq u$ and $|z| \leq v$, then $|y + z| \leq |y| + |z| \leq u + v$, and so it follows from $T(y) + T(z) = T(y + z) \leq S(u + v)$ that $S(u) + S(v) \leq S(u + v)$. On the other hand, if $|y| \leq u + v$, then by Theorem 1.13 there exist y_1 and y_2 with $|y_1| \leq u$, $|y_2| \leq v$, and $y = y_1 + y_2$.

Then $T(y) = T(y_1) + T(y_2) \leq S(u) + S(v)$ holds, from which it follows that $S(u + v) \leq S(u) + S(v)$. Therefore, $S(u + v) = S(u) + S(v)$ holds. By Theorem 1.10 the mapping S defines a positive operator from E to F.

To see that S is the supremum of $\{-T, T\}$, note first that $T \leq S$ and $-T \leq S$ hold trivially in $\mathcal{L}(E, F)$. Now assume that $\pm T \leq R$ in $\mathcal{L}(E, F)$. Clearly, R is a positive operator. Fix $x \in E^+$. If $|y| \leq x$, then note that

$$Ty = Ty^+ - Ty^- \leq Ry^+ + Ry^- = R|y| \leq Rx\,.$$

Therefore, $S(x) \leq R(x)$ holds for each $x \in E^+$, and so $S = T \vee (-T)$ holds in $\mathcal{L}(E, F)$. ■

It is easy to check, but important to observe, that if the modulus of an operator $T\colon E \to F$ exists, then

$$\big|T(x)\big| \leq |T|\big(|x|\big)$$

holds for all $x \in E$.

If x and y are two vectors in a Riesz space E with $x \leq y$, then the **order interval** $[x, y]$ is the subset of E defined by

$$[x, y] := \big\{z \in E\colon\ x \leq z \leq y\big\}\,.$$

A subset A of a Riesz space is said to be **bounded above** whenever there exists some x satisfying $y \leq x$ for all $y \in A$. Similarly, a set A of a Riesz space is **bounded below** whenever there exists some x satisfying $y \geq x$ for all $y \in A$. Finally, a subset in a Riesz space is called **order bounded** if it is bounded both above and below (or, equivalently, if it is included in an order interval).

Besides $\mathcal{L}(E, F)$, a number of other important vector subspaces of $\mathcal{L}(E, F)$ will be considered. The vector subspace $\mathcal{L}_{\mathrm{b}}(E, F)$ of all order bounded operators from E to F will be of fundamental importance.

Definition 1.15. *An operator $T\colon E \to F$ between two Riesz spaces is said to be* **order bounded** *if it maps order bounded subsets of E to order bounded subsets of F.*

The vector space of all order bounded operators from E to F will be denoted $\mathcal{L}_{\mathrm{b}}(E, F)$.

An operator $T\colon E \to F$ between two Riesz spaces is said to be **regular** if it can be written as a difference of two positive operators. Of course, this is equivalent to saying that there exists a positive operator $S\colon E \to F$ satisfying $T \leq S$.

Every positive operator is order bounded. Therefore, every regular operator is likewise order bounded. Thus, if $\mathcal{L}_{\mathrm{r}}(E, F)$ denotes the vector space of all regular operators (which is the same as the vector subspace generated by the positive operators), then the following vector subspace inclusions hold:

$$\mathcal{L}_r(E, F) \subseteq \mathcal{L}_b(E, F) \subseteq \mathcal{L}(E, F).$$

Of course, $\mathcal{L}_r(E, F)$ and $\mathcal{L}_b(E, F)$ with the ordering inherited from $\mathcal{L}(E, F)$ are both ordered vector spaces. For brevity $\mathcal{L}(E, E)$, $\mathcal{L}_b(E, E)$, and $\mathcal{L}_r(E, E)$ will be denoted by $\mathcal{L}(E)$, $\mathcal{L}_b(E)$ and $\mathcal{L}_r(E)$, respectively.

The inclusion $\mathcal{L}_r(E, F) \subseteq \mathcal{L}_b(E, F)$ can be proper, as the next example of H. P. Lotz (oral communication) shows.

Example 1.16 (Lotz). Consider the operator $T\colon C[-1, 1] \to C[-1, 1]$ defined for each $f \in C[-1, 1]$ by

$$Tf(t) = f\left(\sin \tfrac{1}{t}\right) - f\left(\sin\left(t + \tfrac{1}{t}\right)\right)$$

if $0 < |t| \le 1$ and $Tf(0) = 0$. Note that the uniform continuity of f, coupled with the inequality $\left|\sin(\tfrac{1}{t}) - \sin(t + \tfrac{1}{t})\right| \le |t|$, shows that Tf is indeed continuous at zero, and so indeed $Tf \in C[-1, 1]$ for each $f \in C[-1, 1]$.

Next, observe that $T[-\mathbf{1}, \mathbf{1}] \subseteq 2[-\mathbf{1}, \mathbf{1}]$ holds, where $\mathbf{1}$ denotes the constant function one on $[-1, 1]$. Since for every $f \in C[-1, 1]$ there exists some $\lambda > 0$ with $|f| \le \lambda \mathbf{1}$, it easily follows that T is an order bounded operator.

However, we claim that T is not a regular operator. To see this, assume by way of contradiction that some positive operator $S\colon C[-1, 1] \to C[-1, 1]$ satisfies $T \le S$. We claim that for each $0 \le f \in C[-1, 1]$ we have

$$[Sf](0) \ge f(t) \text{ for all } t \in [-1, 1]. \tag{$\star$}$$

To establish this, fix $0 < f \in C[-1, 1]$, and let $0 < c < 2\pi$. Also, for each $n \in \mathbb{N}$ let $t_n = \frac{1}{c + 2n\pi}$ and note that $t_n \to 0$. Next pick some $g_n \in C[-1, 1]$ with $0 \le g_n \le f$ such that $g_n(\sin c) = f(\sin c)$ and $g_n\left(\sin(c + t_n)\right) = 0$. Therefore,

$$[Sf](t_n) \ge [Sg_n](t_n) \ge [Tg_n](t_n) = f(\sin c)$$

for all n, and so $[Sf](0) \ge f(\sin c)$ for all $0 < c < 2\pi$, i.e., $[Sf](0) \ge f(t)$ for all $t \in [-1, 1]$.

Now for each n, let $P_n = \{a_0, a_1, \ldots, a_n\}$ be a partition of $[-1, 1]$ into n subintervals. For each $1 \le i \le n$ pick some $f_i \in C[-1, 1]$ such that $0 \le f_i \le \mathbf{1}$, f_i is zero outside the interval (a_{i-1}, a_i) and $f_i\left(\frac{a_{i-1} + a_i}{2}\right) = 1$. Taking into account that $\sum_{i=1}^{n} f_i \le \mathbf{1}$, it follows from $(\star)$ that

$$[S\mathbf{1}](0) \ge \left[S\left(\sum_{i=1}^{n} f_i\right)\right](0) = \sum_{i=1}^{n} [Sf_i](0) \ge n$$

holds for each n, which is impossible. Thus, T is not a regular operator. ∎

Not every regular operator has a modulus. The next example of S. Kaplan [**94**] clarifies the situation.

Example 1.17 (Kaplan). Let c be the Riesz space of all convergent (real) sequences, i.e., $c = \{(x_1, x_2, \ldots): \lim x_n \text{ exists in } \mathbb{R}\}$. Consider the two positive operators $S, T: c \to c$ defined by

$$S(x_1, x_2, \ldots) = (x_2, x_1, x_4, x_3, x_6, x_5, \ldots)$$

and

$$T(x_1, x_2, \ldots) = (x_1, x_1, x_3, x_3, x_5, x_5, \ldots).$$

We claim the modulus of the regular operator $R = S - T$ does not exist.

To this end, assume by way of contradiction that the modulus $|R|$ exists. Let $P_n: c \to c$ be the positive operator defined by

$$P_n(x_1, \ldots, x_{n-1}, x_n, x_{n+1}, \ldots) = (x_1, \ldots, x_{n-1}, 0, x_{n+1}, \ldots).$$

Then $\pm R \leq |R|P_{2n} \leq |R|$ holds, and so $|R|P_{2n} = |R|$ holds for each n. This means that the image under $|R|$ of every element of c has its even components zero. On the other hand, if e_n is the sequence whose n^{th} component is one and every other zero and $e = (1, 1, 1, \ldots)$, then it follows from the inequalities

$$-R(e_n) \leq |R|e_n \leq |R|e$$

that the odd components of $|R|e$ are greater than or equal to one, and hence $|R|e \notin c$. Therefore, $|R|$ does not exist, as claimed. ∎

A Riesz space is called **Dedekind complete** whenever every nonempty bounded above subset has a supremum (or, equivalently, whenever every nonempty bounded below subset has an infimum). A Riesz space E is Dedekind complete if and only if $0 \leq x_\alpha \uparrow \leq x$ implies the existence of $\sup\{x_\alpha\}$. Similarly, a Riesz space is said to be **Dedekind σ-complete** if every countable subset that is bounded above has a supremum (or, equivalently, whenever $0 \leq x_n \uparrow \leq x$ implies the existence of $\sup\{x_n\}$. The L_p-spaces are examples of Dedekind complete Riesz spaces.

When F is Dedekind complete, the ordered vector space $\mathcal{L}_b(E, F)$ has the structure of a Riesz space. This important result was established first by F. Riesz [**166**] for the special case $F = \mathbb{R}$, and later L. V. Kantorovich [**90, 91**] extended it to the general setting.

Theorem 1.18 (F. Riesz–Kantorovich). *If E and F are Riesz spaces with F Dedekind complete, then the ordered vector space $\mathcal{L}_b(E, F)$ is a Dedekind complete Riesz space. Its lattice operations satisfy*

$$|T|(x) = \sup\{|Ty|: |y| \leq x\},$$
$$[S \vee T](x) = \sup\{S(y) + T(z): y, z \in E^+ \text{ and } y + z = x\}, \text{ and}$$
$$[S \wedge T](x) = \inf\{S(y) + T(z): y, z \in E^+ \text{ and } y + z = x\}$$

for all $S, T \in \mathcal{L}_b(E, F)$ and $x \in E^+$.

In addition, $T_\alpha \downarrow 0$ in $\mathcal{L}_b(E, F)$ if and only if $T_\alpha(x) \downarrow 0$ in F for each $x \in E^+$.

Proof. Fix $T \in \mathcal{L}_b(E, F)$. Since T is order bounded,

$$\sup\{|Ty|: \ |y| \leq x\} = \sup\{Ty: \ |y| \leq x\} = \sup T[-x, x]$$

exists in F for each $x \in E^+$, and so by Theorem 1.14 the modulus of T exists, and moreover

$$|T|(x) = \sup\{Ty: \ |y| \leq x\}.$$

From Theorem 1.7 we see that $\mathcal{L}_b(E, F)$ is a Riesz space.

Now let $S, T \in \mathcal{L}_b(E, F)$ and $x \in E^+$. By observing that $y, z \in E^+$ satisfy $y + z = x$ if and only if there exists some $|u| \leq x$ with $y = \frac{1}{2}(x + u)$ and $z = \frac{1}{2}(x - u)$, it follows from Theorem 1.7 that

$$
\begin{aligned}
[S \vee T](x) &= \tfrac{1}{2}\big(Sx + Tx + |S - T|x\big) \\
&= \tfrac{1}{2}\big(Sx + Tx + \sup\{(S - T)u: \ |u| \leq x\}\big) \\
&= \tfrac{1}{2}\sup\{Sx + Su + Tx - Tu: \ |u| \leq x\} \\
&= \sup\{S(\tfrac{1}{2}(x + u)) + T(\tfrac{1}{2}(x - u)): \ |u| \leq x\} \\
&= \sup\{S(y) + T(z): \ y, z \in E^+ \text{ and } y + z = x\}.
\end{aligned}
$$

The formula for $S \wedge T$ can be proven in a similar manner.

Finally, we establish that $\mathcal{L}_b(E, F)$ is Dedekind complete. To this end, assume that $0 \leq T_\alpha \uparrow \leq T$ holds in $\mathcal{L}_b(E, F)$. For each $x \in E^+$ let $S(x) = \sup\{T_\alpha(x)\}$ and note that $T_\alpha(x) \uparrow S(x)$. From $T_\alpha(x + y) = T_\alpha(x) + T_\alpha(y)$, it follows (by taking order limits) that the mapping $S: E^+ \to F^+$ is additive, and so S defines a positive operator from E to F. Clearly, $T_\alpha \uparrow S$ holds in $\mathcal{L}_b(E, F)$, proving that $\mathcal{L}_b(E, F)$ is a Dedekind complete Riesz space. ∎

From the preceding discussion it follows that when E and F are Riesz spaces with F Dedekind complete, then each order bounded operator $T: E \to F$ satisfies

$$
\begin{aligned}
T^+(x) &= \sup\{Ty: \ 0 \leq y \leq x\}, \text{ and} \\
T^-(x) &= \sup\{-Ty: \ 0 \leq y \leq x\}
\end{aligned}
$$

for each $x \in E^+$. From $T = T^+ - T^-$, it follows that $\mathcal{L}_b(E, F)$ coincides with the vector subspace generated by the positive operators in $\mathcal{L}(E, F)$. In other words, when F is Dedekind complete we have $\mathcal{L}_r(E, F) = \mathcal{L}_b(E, F)$.

Recall that a subset D of a Riesz space is said to be **directed upward** (in symbols $D \uparrow$) whenever for each pair $x, y \in D$ there exists some $z \in D$ with $x \leq z$ and $y \leq z$. The symbol $D \uparrow x$ means that D is directed upward and $x = \sup D$ holds. The meanings of $D\downarrow$ and $D\downarrow x$ are analogous. Also, the symbol $D \leq x$ means that $y \leq x$ holds for all $y \in D$.

The existence of the supremum of an upward directed subset of $\mathcal{L}_b(E, F)$ is characterized as follows.

Theorem 1.19. *Let E and F be two Riesz spaces with F Dedekind complete, and let D be a nonempty subset of $\mathcal{L}_b(E, F)$ satisfying $D\uparrow$. Then $\sup D$ exists in $\mathcal{L}_b(E, F)$ if and only if the set $\{T(x)\colon\ T \in D\}$ is bounded above in F for each $x \in E^+$. In this case,*

$$[\sup D](x) = \sup\{T(x)\colon\ T \in D\}$$

holds for all $x \in E^+$.

Proof. the "only if" part is trivial. The "if" part needs proof. So, assume that $D\uparrow$ holds in $\mathcal{L}_b(E, F)$ and that the set $\{T(x)\colon\ T \in D\}$ is bounded above in F for each $x \in E^+$. It is easy to see that without loss of generality we can assume that $D \subseteq \mathcal{L}_b^+(E, F)$. Define $S\colon E^+ \to F^+$ by

$$S(x) = \sup\{T(x)\colon\ T \in D\}\,,$$

and we claim that S is additive. To see this, let $x, y \in E^+$. Since for each $T \in D$ we have $T(x + y) = T(x) + T(y) \leq S(x) + S(y)$, we see that $S(x + y) \leq S(x) + S(y)$ holds. On the other hand, if $T_1, T_2 \in D$, then pick $T_3 \in D$ satisfying $T_1 \leq T_3$ and $T_2 \leq T_3$, and note that

$$T_1(x) + T_2(y) \leq T_3(x) + T_3(y) = T_3(x + y) \leq S(x + y)$$

implies $S(x) + S(y) \leq S(x + y)$. Therefore, $S(x + y) = S(x) + S(y)$ holds, and so S is additive. By Theorem 1.10 the mapping S defines a positive operator from E to F, and a routine argument shows that $S = \sup D$ holds in $\mathcal{L}_b(E, F)$. ∎

Our next objective is to describe the lattice operations of $\mathcal{L}_b(E, F)$ in terms of directed sets. To do this, we need a result from the theory of Riesz spaces known as the **Riesz Decomposition Property**; it is due to F. Riesz [167].

Theorem 1.20 (The Riesz Decomposition Property). *Let $x_1, \ldots, x_n$ and $y_1, \ldots, y_m$ be positive vectors in a Riesz space. If*

$$\sum_{i=1}^{n} x_i = \sum_{j=1}^{m} y_j$$

holds, then there exists a finite subset $\{z_{ij}\colon\ i = 1, \ldots, n; j = 1, \ldots, m\}$ of positive vectors such that

$$x_i = \sum_{j=1}^{m} z_{ij}\,, \quad for\ each\ i = 1, \ldots, n,$$

and

$$y_j = \sum_{i=1}^{n} z_{ij}, \quad \text{for each } j = 1, \ldots, m.$$

Proof. We shall use induction on m. For $m = 1$ the desired conclusion follows from Theorem 1.13. Thus, assume the result to be true for some m and all $n = 1, 2, \ldots$. Let

$$\sum_{i=1}^{n} x_i = \sum_{j=1}^{m+1} y_j,$$

where the vectors x_i and the y_j are all positive. Since $\sum_{j=1}^{m} y_j \leq \sum_{i=1}^{n} x_i$ holds, it follows from Theorem 1.13 that there exist vectors $u_1, \ldots, u_n$ satisfying $0 \leq u_i \leq x_i$ for each $i = 1, \ldots, n$ and $\sum_{i=1}^{n} u_i = \sum_{j=1}^{m} y_j$. Therefore, from our induction hypothesis, there exists a set of positive vectors $\{z_{ij} : i = 1, \ldots, n; j = 1, \ldots, m\}$ such that:

$$u_i = \sum_{j=1}^{m} z_{ij} \text{ for } i = 1, \ldots, n \quad \text{and} \quad y_j = \sum_{i=1}^{n} z_{ij} \text{ for } j = 1, \ldots, m.$$

For each $i = 1, \ldots, n$ put $z_{i,m+1} = x_i - u_i \geq 0$ and note that the collection of positive vectors $\{z_{ij} : i = 1, \ldots, n; j = 1, \ldots, m+1\}$ satisfies

$$x_i = \sum_{j=1}^{m+1} z_{ij} \text{ for } i = 1, \ldots, n \quad \text{and} \quad y_j = \sum_{i=1}^{n} z_{ij} \text{ for } j = 1, \ldots, m+1.$$

Thus, the conclusion is valid for $m+1$ and all $n = 1, 2, \ldots$, and the proof is finished. ∎

We are now in a position to express the lattice operations of $\mathcal{L}_b(E, F)$ in terms of directed sets.

Theorem 1.21. *If E and F are two Riesz spaces with F Dedekind complete, then for all $S, T \in \mathcal{L}_b(E, F)$ and each $x \in E^+$ we have:*

(1) $\left\{ \sum_{i=1}^{n} S(x_i) \vee T(x_i) : x_i \in E^+ \text{ and } \sum_{i=1}^{n} x_i = x \right\} \uparrow [S \vee T](x).$

(2) $\left\{ \sum_{i=1}^{n} S(x_i) \wedge T(x_i) : x_i \in E^+ \text{ and } \sum_{i=1}^{n} x_i = x \right\} \downarrow [S \wedge T](x).$

(3) $\left\{ \sum_{i=1}^{n} |T(x_i)| : x_i \in E^+ \text{ and } \sum_{i=1}^{n} x_i = x \right\} \uparrow |T|(x).$

Proof. (1) Consider the set

$$D = \left\{ \sum_{i=1}^{n} S(x_i) \vee T(x_i) \colon \ x_i \in E^+ \text{ for each } i \text{ and } \sum_{i=1}^{n} x_i = x \right\}.$$

Since $\sum_{i=1}^{n} x_i = x$ with each $x_i \in E^+$ implies

$$\sum_{i=1}^{n} S(x_i) \vee T(x_i) \leq \sum_{i=1}^{n} \left[(S \vee T)x_i \right] \vee \left[(S \vee T)x_i \right] = [S \vee T](x),$$

we see that $D \leq [S \vee T](x)$. On the other hand, if $D \leq u$ holds, then for each $y, z \in E^+$ with $y + z = x$ we have

$$S(y) + T(z) \leq S(y) \vee T(y) + S(z) \vee T(z) \leq u,$$

and consequently

$$[S \vee T](x) = \sup \left\{ S(y) + T(z) \colon \ y, z \in E^+ \text{ and } y + z = x \right\} \leq u.$$

Thus, $\sup D = [S \vee T](x)$, and it remains to be shown that D is directed upward.

To this end, let $x = \sum_{i=1}^{n} x_i = \sum_{j=1}^{m} y_j$ with all the x_i and y_j in E^+. By Theorem 1.20 there exists a finite collection $\{z_{ij} \colon \ i = 1, \ldots, n; \ j = 1, \ldots, m\}$ of positive vectors such that

$$x_i = \sum_{j=1}^{m} z_{ij}, \ \text{ for each } \ i = 1, \ldots, n,$$

and

$$y_j = \sum_{i=1}^{n} z_{ij}, \ \text{ for each } \ j = 1, \ldots, m.$$

In particular, we have $\sum_{i=1}^{n} \sum_{j=1}^{m} z_{ij} = x$. On the other hand, using the lattice identity $x \vee y = \frac{1}{2}(x + y + |x - y|)$, we see that

$$\sum_{i=1}^{n} S(x_i) \vee T(x_i)$$

$$= \frac{1}{2} \sum_{i=1}^{n} \left[S(x_i) + T(x_i) + |S(x_i) - T(x_i)| \right]$$

$$= \frac{1}{2} \sum_{i=1}^{n} \left[\sum_{j=1}^{m} S(z_{ij}) + \sum_{j=1}^{m} T(z_{ij}) + \left| \sum_{j=1}^{m} \{ S(z_{ij}) - T(z_{ij}) \} \right| \right]$$

$$\leq \frac{1}{2} \sum_{i=1}^{n} \left[\sum_{j=1}^{m} \{ S(z_{ij}) + T(z_{ij}) + |S(z_{ij}) - T(z_{ij})| \} \right]$$

$$= \sum_{i=1}^{n} \sum_{j=1}^{m} S(z_{ij}) \vee T(z_{ij}) .$$

Similarly,

$$\sum_{j=1}^{m} S(y_j) \vee T(y_j) \leq \sum_{i=1}^{n} \sum_{j=1}^{m} S(z_{ij}) \vee T(z_{ij})$$

holds, and so D is directed upward.

(2) Use (1) in conjunction with the identity $T \wedge S = -\left[(-S) \vee (-T) \right]$.

(3) Use (1) and the identity $|T| = T \vee (-T)$. ∎

The next result presents an interesting local approximation property of positive operators.

Theorem 1.22. *Let $T \colon E \to F$ be a positive operator between two Riesz spaces with F Dedekind σ-complete. Then for each $x \in E^{+}$ there exists a positive operator $S \colon E \to F$ such that:*

(1) $0 \leq S \leq T$.

(2) $S(x) = T(x)$.

(3) $S(y) = 0$ *for all $y \perp x$.*

Proof. Let $x \in E^{+}$ be fixed and define $S \colon E^{+} \to F^{+}$ by

$$S(y) = \sup \{ T(y \wedge nx) \colon n = 1, 2, \dots \} .$$

(The supremum exists since F is Dedekind σ-complete and the sequence $\{ T(y \wedge nx) \}$ is bounded above in F by Ty.) We claim that S is additive.

To see this, let $y, z \in E^{+}$. From $(y + z) \wedge nx \leq y \wedge nx + z \wedge nx$ we get

$$T\big((y + z) \wedge nx \big) \leq T(y \wedge nx) + T(z \wedge nx) \leq S(y) + S(z) ,$$

and so $S(y + z) \leq S(y) + S(z)$. On the other hand, for each m and n we have $y \wedge nx + z \wedge mx \leq (y + z) \wedge (n + m)x$, and thus

$$T(y \wedge nx) + T(z \wedge mx) \leq T(y + z) \wedge (n + m)x \leq S(y + z)$$

holds for all n and m. This implies $S(y) + S(z) \leq S(y + z)$, and hence $S(y + z) = S(y) + S(z)$, so that S is additive.

By Theorem 1.10 the mapping S extends uniquely to all of E as a positive operator. Now it is a routine matter to verify that the operator S satisfies the desired properties. ∎

As an application of the preceding result let us derive some formulas that are in some sense the "dual" formulas to those stated after Theorem 1.18.

Theorem 1.23. *If* $T: E \to F$ *is a positive operator between two Riesz spaces with* F *Dedekind* σ-*complete, then for each* $x \in E$ *we have:*

$$T(x^+) = \max\{S(x): S \in \mathcal{L}(E, F) \text{ and } 0 \leq S \leq T\}.$$
$$T(x^-) = \max\{-S(x): S \in \mathcal{L}(E, F) \text{ and } 0 \leq S \leq T\}.$$
$$T(|x|) = \max\{S(x): S \in \mathcal{L}(E, F) \text{ and } -T \leq S \leq T\}.$$

Proof. (1) Let $x \in E$ be fixed. By Theorem 1.22 there exists a positive operator $R: E \to F$ such that $0 \leq R \leq T$, $R(x^+) = T(x^+)$, and $R(x^-) = 0$. Therefore, $T(x^+) = R(x)$. On the other hand, if $S \in \mathcal{L}(E, F)$ satisfies $0 \leq S \leq T$, then we have $S(x) \leq S(x^+) \leq T(x^+)$, and the conclusion follows.

(2) Apply (1) to the identity $x^- = (-x)^+$.

(3) If the operator $S: E \to F$ satisfies $-T \leq S \leq T$, then

$$S(x) = S(x^+) - S(x^-) \leq T(x^+) + T(x^-) = T(|x|)$$

holds. On the other hand, according to Theorem 1.22, there exist two positive operators $R_1, R_2: E \to F$ bounded by T such that:

(a) $R_1(x^+) = T(x^+)$ and $R_1(x^-) = 0$.
(b) $R_2(x^-) = T(x^-)$ and $R_2(x^+) = 0$.

Then the operator $S = R_1 - R_2$ satisfies $-T \leq S \leq T$ and $T(|x|) = S(x)$, and the desired conclusion follows. ∎

Now let $\{E_i: i \in I\}$ be a family of Riesz spaces. Then it is not difficult to check that the **Cartesian product** ΠE_i, under the ordering $\{x_i\} \geq \{y_i\}$ whenever $x_i \geq y_i$ holds in E_i for each $i \in I$, is a Riesz space. Clearly, if $x = \{x_i\}$ and $y = \{y_i\}$ are vectors of ΠE_i, then

$$x \vee y = \{x_i \vee y_i\} \quad \text{and} \quad x \wedge y = \{x_i \wedge y_i\}.$$

The **direct sum** $\Sigma \oplus E_i$ (or more formally $\Sigma_{i \in I} \oplus E_i$) is the vector subspace of ΠE_i consisting of all vectors $x = \{x_i\}$ for which $x_i = 0$ holds for all but a finite number of indices i. With the pointwise algebraic and lattice operations $\Sigma \oplus E_i$ is a Riesz subspace of ΠE_i (and hence a Riesz space in its own right). Note that if, in addition, each E_i is Dedekind complete, then ΠE_i and $\Sigma \oplus E_i$ are likewise both Dedekind complete Riesz spaces.

It is not difficult to see that every operator $T: \Sigma \oplus E_i \to \Sigma \oplus F_j$ between two direct sums of families of Riesz spaces can be represented by a matrix $T = [T_{ji}]$, where $T_{ji}: E_i \to F_j$ are operators defined appropriately. Sometimes it pays to know that the algebraic and lattice operations represented by matrices are the pointwise ones. The next result (whose easy proof is left for the reader) clarifies the situation.

Theorem 1.24. *Let $\{E_i: i \in I\}$ and $\{F_j: j \in J\}$ be two families of Riesz spaces with each F_j Dedekind complete. If $S = [S_{ji}]$ and $T = [T_{ji}]$ are order bounded operators from $\Sigma \oplus E_i$ to $\Sigma \oplus F_j$, then*

(1) $S + T = [S_{ji} + T_{ji}]$ *and* $\lambda S = [\lambda S_{ji}]$, *and*

(2) $S \vee T = [S_{ji} \vee T_{ji}]$ *and* $S \wedge T = [S_{ji} \wedge T_{ji}]$

hold in $\mathcal{L}_b(\Sigma \oplus E_i, \Sigma \oplus F_j)$.

Exercises

1. Let E be an Archimedean Riesz space and let $A \subseteq \mathbb{R}$ be nonempty and bounded above. Show that for each $x \in E^+$ the supremum of the set $Ax := \{\alpha x: \alpha \in A\}$ exists and $\sup(Ax) = (\sup A)x$.

2. Show that in a Riesz space $x \perp y$ implies
   (a) $\alpha x \perp \beta y$ for all $\alpha, \beta \in \mathbb{R}$, and
   (b) $|x + y| = |x| + |y|$.
   Use the conclusion in (b) to establish that if in a Riesz space the nonzero vectors $x_1, \ldots, x_n$ are pairwise disjoint, then $x_1, \ldots, x_n$ are linearly independent. [*Hint:* If $|x| \wedge |y| = 0$, then

   $$|x + y| \geq ||x| - |y|| = |x| \vee |y| - |x| \wedge |y|$$
   $$= |x| \vee |y| + |x| \wedge |y| = |x| + |y| \geq |x + y| .]$$

3. In this exercise we ask you to complete the missing details in Example 1.11. Let G be the **lexicographic plane**. (That is, we consider $G = \mathbb{R}^2$ as a Riesz space under the **lexicographic ordering** $(x_1, x_2) \geq (y_1, y_2)$ whenever either $x_1 > y_1$ or else $x_1 = y_1$ and $x_2 \geq y_2$.) Also, let $\phi: \mathbb{R} \to \mathbb{R}$ be an additive function that is not linear (i.e., not of the form $\phi(x) = cx$).
   Show that the mapping $T: \mathbb{R}^+ \to G^+$ defined by

   $$T(x) = (x, \phi(x))$$

is additive but that it cannot be extended to a positive operator from $\mathbb{R}$ to G. Why does this not contradict Theorem 1.10?

4. Let E and F be two Riesz spaces with F Dedekind complete, and let $\mathcal{A}$ be a nonempty subset of $\mathcal{L}_b(E, F)$. Show that $\sup \mathcal{A}$ exists in $\mathcal{L}_b(E, F)$ if and only if for each $x \in E^+$ the set $\{(\bigvee_{i=1}^n T_i)x : T_1, \ldots, T_n \in \mathcal{A}\}$ is bounded above in F.

5. Consider the positive operators $S, T : L_1[0, 1] \to L_1[0, 1]$ defined by

$$S(f) = f \quad \text{and} \quad T(f) = \left[\int_0^1 f(x)\, dx\right] \cdot \mathbf{1}\,,$$

where $\mathbf{1}$ is the constant function one. Show that $S \wedge T = 0$.

6. Let E and F be two Riesz spaces with F Dedekind complete. Then for each $T \in \mathcal{L}_b(E, F)$ and each $x \in E^+$ show that:

$$\begin{aligned} T^+(x) &= \sup\{(Ty)^+ :\ 0 \le y \le x\}\,. \\ T^-(x) &= \sup\{(Ty)^- :\ 0 \le y \le x\}\,. \end{aligned}$$

7. Let $T : E \to F$ be a positive operator between two Riesz spaces with F Dedekind complete. If $x, y \in E$, then show that:
   (a) $T(x \vee y) = \max\{Rx + Sy :\ R, S \in \mathcal{L}_b^+(E, F) \text{ and } R + S = T\}$.
   (b) $T(x \wedge y) = \min\{Rx + Sy :\ R, S \in \mathcal{L}_b^+(E, F) \text{ and } R + S = T\}$.

8. If $0 < p < 1$, then show that the only positive operator from $L_p[0, 1]$ to $C[0, 1]$ is the zero operator.

9. Consider the continuous function $g : [0, 1] \to [0, 1]$ defined by $g(x) = x$ if $0 \le x \le \frac{1}{2}$ and $g(x) = \frac{1}{2}$ if $\frac{1}{2} < x \le 1$. Now define the operator $T : C[0, 1] \to C[0, 1]$ by $[Tf](x) = f(g(x)) - f(\frac{1}{2})$.
   Show that T is a regular operator whose modulus does not exist.

10. Let $T : C[0, 1] \to C[0, 1]$ be the regular operator defined by

$$[Tf](x) = f(\sin x) - f(\cos x)\,.$$

Show that T^+ and T^- both exist and that

$$[T^+ f](x) = f(\sin x) \quad \text{and} \quad [T^- f](x) = f(\cos x)\,.$$

11. For each $n \ge 2$ fix a continuous function $e_n : [0, 1] \to [0, 1]$ such that:
    (a) $0 \le e_n \le 1$.
    (b) $e_n = 0$ outside $\left[\frac{1}{2} + \frac{1}{n+1}, \frac{1}{2} + \frac{1}{n}\right]$.
    (c) $e_n(x) = 1$ for some $x \in \left[\frac{1}{2} + \frac{1}{n+1}, \frac{1}{2} + \frac{1}{n}\right]$.
    Now define the operator $T : C[0, 1] \to C[0, 1]$ by

$$Tf = \sum_{n=2}^\infty \left[\int_0^1 f(x) \sin(n\pi x)\, dx\right] e_n\,.$$

Show that T is indeed an operator from $C[0, 1]$ to $C[0, 1]$, that T is a regular operator, and that its modulus does not exist.

12. Prove Theorem 1.24.

1.2. Extensions of Positive Operators

In this section we shall gather some basic extension theorems for operators, and, in particular, for positive operators.

A function $p\colon G \to F$, where G is a (real) vector space and F is an ordered vector space, is called **sublinear** whenever

(a) $p(x+y) \leq p(x) + p(y)$ for all $x, y \in G$, and

(b) $p(\lambda x) = \lambda p(x)$ for all $x \in G$ and all $\lambda \geq 0$.

The next result is the most general version of the classical Hahn–Banach extension theorem. This theorem plays a fundamental role in modern analysis and without any doubt it will be of great importance to us here. It is due to H. Hahn [**74**] and S. Banach [**30**].

Theorem 1.25 (Hahn–Banach). *Let G be a (real) vector space, F a Dedekind complete Riesz space, and let $p\colon G \to F$ be a sublinear function. If H is a vector subspace of G and $S\colon H \to F$ is an operator satisfying $S(x) \leq p(x)$ for all $x \in H$, then there exists some operator $T\colon G \to F$ such that:*

(1) *$T = S$ on H, i.e., T is a linear extension of S to all of G.*

(2) *$T(x) \leq p(x)$ holds for all $x \in G$.*

Proof. The critical step is to show that S has a linear extension satisfying (2) on an arbitrary vector subspace generated by H and one extra vector. If this is done, then an application of Zorn's lemma guarantees the existence of an extension of S to all of G with the desired properties.

To this end, let $x \notin H$, and let $V = \{y + \lambda x\colon y \in H \text{ and } \lambda \in \mathbb{R}\}$. If $T\colon V \to F$ is a linear extension of S, then

$$T(y + \lambda x) = S(y) + \lambda T(x)$$

must hold true for all $y \in H$ and all $\lambda \in \mathbb{R}$. Put $z = T(x)$. To complete the proof, we must establish the existence of some $z \in F$ such that

$$S(y) + \lambda z \leq p(y + \lambda x) \qquad (\star)$$

holds for all $y \in H$ and $\lambda \in \mathbb{R}$. For $\lambda > 0$, $(\star)$ is equivalent to

$$S(y) + z \leq p(y + x)$$

for all $y \in H$, while for $\lambda < 0$ the inequality $(\star)$ is equivalent to

$$S(y) - z \leq p(y - x)$$

for all $y \in H$. The last two inequalities certainly will be satisfied by a choice of z for which

$$S(y) - p(y - x) \leq z \leq p(u + x) - S(u) \qquad (\star\star)$$

holds for all $y, u \in H$.

To see that there exists some $z \in F$ satisfying $(\star\star)$, start by observing that for each $y, u \in H$ we have

$$\begin{aligned} S(y) + S(u) &= S(y + u) \leq p(y + u) = p\big(y - x + (u + x)\big) \\ &\leq p(y - x) + p(u + x), \end{aligned}$$

and so

$$S(y) - p(y - x) \leq p(u + x) - S(u)$$

holds for all $y, u \in H$. This inequality in conjunction with the Dedekind completeness of F guarantees that both suprema

$$s = \sup\big\{S(y) - p(y - x) \colon y \in H\big\} \quad \text{and} \quad t = \inf\big\{p(u + x) - S(u) \colon u \in H\big\}$$

exist in F, and satisfy $s \leq t$. Now any $z \in F$ satisfying $s \leq z \leq t$ (for instance $z = s$) satisfies $(\star\star)$, and hence $(\star)$. This complete the proof of the theorem. ∎

Recall that a vector subspace G of a Riesz space E is said to be a **Riesz subspace** (or a **vector sublattice**) whenever G is closed under the lattice operations of E, i.e., whenever for each pair $x, y \in G$ the vector $x \vee y$ (taken in E) belongs to G.

As a first application of the Hahn–Banach extension theorem we present the following useful extension property of positive operators.

Theorem 1.26. *Let $T \colon E \to F$ be a positive operator between two Riesz spaces with F Dedekind complete. Assume also that G is a Riesz subspace of E and that $S \colon G \to F$ is an operator satisfying $0 \leq Sx \leq Tx$ for all $x \in G^{+}$. Then S can be extended to a positive operator from E to F such that $0 \leq S \leq T$ holds in $\mathcal{L}(E, F)$.*

Proof. Define $p \colon E \to F$ by $p(x) = T(x^{+})$, and note that p is sublinear and satisfies $S(x) \leq p(x)$ for all $x \in G$. By Theorem 1.25 there exists a linear extension of S to all of E (which we denote by S again) satisfying $S(x) \leq p(x)$ for all $x \in E$. Now if $x \in E^{+}$, then

$$-S(x) = S(-x) \leq p(-x) = T\big((-x)^{+}\big) = T(0) = 0,$$

and so $0 \leq S(x) \leq p(x) = T(x)$ holds, as desired. ∎

The rest of the section is devoted to extension properties of positive operators. The first result of this kind informs us that a positive operator whose domain is a Riesz subspace extends to a positive operator if and only if it is dominated by a monotone sublinear mapping. As usual, a mapping $f \colon E \to F$ between two ordered vector spaces is called **monotone** whenever $x \leq y$ in E implies $f(x) \leq f(y)$ in F.

Theorem 1.27. *Let E and F be Riesz spaces with F Dedekind complete. If G is a Riesz subspace of E and $T\colon G \to F$ is a positive operator, then the following statements are equivalent.*

(1) *T extends to a positive operator from E to F.*

(2) *T extends to an order bounded operator from E to F.*

(3) *There exists a monotone sublinear mapping $p\colon E \to F$ satisfying $T(x) \leq p(x)$ for all $x \in G$.*

Proof. (1) $\Longrightarrow$ (2) Obvious.

(2) $\Longrightarrow$ (3) Let $S \in \mathcal{L}_b(E, F)$ satisfy $S(x) = T(x)$ for all $x \in G$. Then the mapping $p\colon E \to F$ defined by $p(x) = |S|(x^+)$ is monotone, sublinear and satisfies

$$T(x) \leq T(x^+) = S(x^+) \leq |S|(x^+) = p(x)$$

for all $x \in G$.

(3) $\Longrightarrow$ (1) Let $p\colon E \to F$ be a monotone sublinear mapping satisfying $T(x) \leq p(x)$ for all $x \in G$. Then the formula $q(x) = p(x^+)$ defines a sublinear mapping from E to F such that

$$T(x) \leq T(x^+) \leq p(x^+) = q(x)$$

holds for all $x \in G$. Thus, by the Hahn–Banach Extension Theorem 1.25 there exists an extension $R \in \mathcal{L}(E, F)$ of T satisfying $R(x) \leq q(x)$ for all $x \in E$. In particular, if $x \in E^+$, then the relation

$$-R(x) = R(-x) \leq q(-x) = p\big((-x)^+\big) = p(0) = 0$$

implies $R(x) \geq 0$. That is, R is a positive linear extension of T to all of E, and the proof is finished. ∎

A subset A of a Riesz space is called **solid** whenever $|x| \leq |y|$ and $y \in A$ imply $x \in A$. A solid vector subspace of a Riesz space is referred to as an **ideal**. From the lattice identity $x \vee y = \frac{1}{2}(x + y + |x - y|)$, it follows immediately that every ideal is a Riesz subspace.

The next result deals with restrictions of positive operators to ideals.

Theorem 1.28. *If $T\colon E \to F$ is a positive operator between two Riesz spaces with F Dedekind complete, then for every ideal A of E the formula*

$$T_A(x) = \sup\{T(y)\colon y \in A \text{ and } 0 \leq y \leq x\}, \quad x \in E^+,$$

defines a positive operator from E to F. Moreover, we have:

(a) *$0 \leq T_A \leq T$.*

(b) *$T_A = T$ on A and $T_A = 0$ on A^d.*

(c) *If B is another ideal with $A \subseteq B$, then $T_A \leq T_B$ holds.*

Proof. Note first that

$$T_A(x) = \sup\{T(x \wedge y): \ y \in A^+\}$$

holds for all $x \in E^+$. According to Theorem 1.10 it suffices to show that T_A is additive on E^+.

To this end, let $x, y \in E^+$. If $z \in A^+$, then the inequality

$$(x + y) \wedge z \leq x \wedge z + y \wedge z$$

implies that $T((x + y) \wedge z) \leq T(x \wedge z) + T(y \wedge z) \leq T_A(x) + T_A(y)$, and hence

$$T_A(x + y) \leq T_A(x) + T_A(y).$$

On the other hand, the inequality $x \wedge u + y \wedge v \leq (x + y) \wedge (u + v)$ implies

$$T_A(x) + T_A(y) \leq T_A(x + y).$$

Therefore, $T_A(x + y) = T_A(x) + T_A(y)$ holds, so that T_A is additive on E^+.

Properties (1)–(3) are now easy consequences of the formula defining the operator T_A. ■

As mentioned before, if G is a vector subspace of an ordered vector space and F is another ordered vector space, then it is standard to call an operator $T: G \to F$ **positive** whenever $0 \leq x \in G$ implies $0 \leq T(x) \in F$.

Now consider a positive operator $T: G \to F$, where G is a vector subspace of an ordered vector space E and F is a Dedekind complete Riesz space. We shall denote by $\mathcal{E}(T)$ the collection of all positive extensions of T to all of E. That is,

$$\mathcal{E}(T) := \{S \in \mathcal{L}(E, F): \ S \geq 0 \text{ and } S = T \text{ on } G\}.$$

The set $\mathcal{E}(T)$ is always a convex subset of $\mathcal{L}(E, F)$, i.e. $\lambda S + (1 - \lambda)R \in \mathcal{E}(T)$ holds for all $S, R \in \mathcal{E}(T)$ and all $0 \leq \lambda \leq 1$. The set $\mathcal{E}(T)$ might happen to be empty. The next example presents such a case.

Example 1.29. Let $E = L_p[0, 1]$ with $0 < p < 1$ and let $G = L_1[0, 1]$. Clearly, $G \subseteq E$ and G is an ideal of E. (Here $f \geq g$ means that $f(x) \geq g(x)$ holds for almost all x with respect to the Lebesgue measure.)

Now consider the operator $T: G \to \mathbb{R}$ defined by

$$T(f) = \int_0^1 f(x)\, dx.$$

We claim that T does not have a positive linear extension to all of E. To see this, assume by way of contradiction that T is extendable to a positive operator from E to $\mathbb{R}$. In particular, this implies that if $f \in E$ is defined by $f(x) = \frac{1}{x}$, then the set of real numbers

$$D = \{T(g): \ g \in G \text{ and } 0 \leq g \leq f\}$$

is bounded. on the other hand, if $g_n = f\chi_{(\frac{1}{n},1)}$, then $T(g_n) = \ln n \in D$ holds for each n. Therefore, D must be unbounded, a contradiction. Consequently, in this case we have $\mathcal{E}(T) = \emptyset$. ∎

A positive operator $T: G \to F$ (where G is a vector subspace of an ordered vector space E) is said to have a **smallest extension** whenever there exists some $S \in \mathcal{E}(T)$ satisfying $S \leq R$ for all $R \in \mathcal{E}(T)$, in which case S is called the **smallest extension** of T. In other words, T has a smallest extension if and only if $\min \mathcal{E}(T)$ exists in $\mathcal{L}(E, F)$.

It turns out that an extendable positive operator whose domain is an ideal always has a smallest extension.

Theorem 1.30. *Let E and F be two Riesz spaces with F Dedekind complete, let A be an ideal of E, and let $T: A \to F$ be a positive operator. If $\mathcal{E}(T) \neq \emptyset$, then T has a smallest extension. Moreover, if in this case $S = \min \mathcal{E}(T)$, then*

$$S(x) = \sup\{Ty: \ y \in A \ and \ 0 \leq y \leq x\}$$

holds for all $x \in E^+$.

Proof. Since T has (at least) one positive extension, the formula

$$T_A(x) = \sup\{T(y): \ y \in A \ \text{and} \ 0 \leq y \leq x\}, \quad x \in E^+,$$

defines a positive operator from E to F satisfying $T_A = T$ on A, and so $T_A \in \mathcal{E}(T)$. (See the proof of Theorem 1.28.)

Now if $S \in \mathcal{E}(T)$, then $S = T$ holds on A, and hence $T_A = S_A \leq S$. Therefore, $T_A = \min \mathcal{E}(T)$ holds, as desired. ∎

For a positive operator $T: E \to F$ with F Dedekind complete, Theorem 1.30 implies that for each ideal A of E the positive operator T_A is the smallest extension of the restriction of T to A.

Among the important points of a convex set are its extreme points. Recall that a vector e of a convex set C is said to be an **extreme point** of C whenever the expression $e = \lambda x + (1 - \lambda)y$ with $x, y \in C$ and $0 < \lambda < 1$ implies $x = y = e$.

The extreme points of the convex set $\mathcal{E}(T)$ have been characterized by Z. Lipecki, D. Plachky, and W. Thomsen [**116**] as follows.

Theorem 1.31 (Lipecki–Plachky–Thomsen). *Let E and F be two Riesz spaces with F a Dedekind complete. If G is a vector subspace of E and $T: G \to F$ is a positive operator, then for an operator $S \in \mathcal{E}(T)$ the following statements are equivalent:*

(1) *S is an extreme point of $\mathcal{E}(T)$.*

(2) *For each $x \in E$ we have $\inf\{S(|x - y|): \ y \in G\} = 0$.*

Proof. $(1) \implies (2)$ Let S be an extreme point of $\mathcal{E}(T)$. Define the mapping $p: E \to F$ for each $x \in E$ by

$$p(x) = \inf\big\{ S(|x - y|): \ y \in G \big\}.$$

Clearly, p is a sublinear mapping that satisfies $0 \le p(x) = p(-x) \le S|x|$ for all $x \in E$, and $p(y) = 0$ for each $y \in G$.

Next, we claim that $p(x) = 0$ holds for all $x \in E$. To see this, assume by way of contradiction that $p(x) > 0$ holds for some $x \in E$. Define the operator $R: \{\lambda x: \ \lambda \in \mathbb{R}\} \to F$ by $R(\lambda x) = \lambda p(x)$, and note that $R(\lambda x) \le p(\lambda x)$ holds. So, by the Hahn–Banach Extension Theorem 1.25, the operator R has a linear extension to all of E (which we shall denote by R again) such that $R(z) \le p(z)$ holds for all $z \in E$; clearly, $R \ne 0$. It is easy to see that $|R(z)| \le p(z)$ for all $z \in E$, and so $R(y) = 0$ for all $y \in G$. Since for each $z \ge 0$ we have $R(z) \le p(z) \le S(|z|) = S(z)$ and

$$-R(z) = R(-z) \le p(-z) \le S(|-z|) = S(z),$$

it easily follows that $S - R \ge 0$ and $S + R \ge 0$ both hold. Thus, $S - R$ and $S + R$ both belong to $\mathcal{E}(T)$. Now the identity

$$S = \tfrac{1}{2}(S - R) + \tfrac{1}{2}(S + R),$$

in conjunction with $S - R \ne S$ and $S + R \ne S$, shows that S is not an extreme point of $\mathcal{E}(T)$, a contradiction. Thus, $p(x) = 0$ holds for each $x \in E$, as desired.

$(2) \implies (1)$ Let S satisfy (2) and assume that $S = \lambda Q + (1 - \lambda)R$ with $Q, R \in \mathcal{E}(T)$ and $0 < \lambda < 1$. Then for each $x, y \in E$ we have

$$|Q(x) - Q(y)| \le Q|x - y| = \big(\tfrac{1}{\lambda} S - \tfrac{1-\lambda}{\lambda} R\big)|x - y| \le \tfrac{1}{\lambda} S|x - y|.$$

In particular, if $x \in E$ and $y \in G$, then from $S(y) = Q(y) = T(y)$ it follows that

$$|S(x) - Q(x)| \le |S(x) - S(y)| + |Q(y) - Q(x)| \le \big(1 + \tfrac{1}{\lambda}\big) S|x - y|.$$

Taking into account our hypothesis, the last inequality yields $S(x) = Q(x)$ for each $x \in E$, and this shows that S is an extreme point of $\mathcal{E}(T)$. ∎

Let us say that a vector subspace G of an ordered vector space E is **majorizing** E whenever for each $x \in E$ there exists some $y \in G$ with $x \le y$ (or, equivalently, if for each $x \in E$ there exists some $y \in G$ with $y \le x$).

It is important to know that every positive operator whose domain is a majorizing vector subspace and whose values are in a Dedekind complete Riesz space always has a positive extension. This is a classical result due to L. V. Kantorovich [**90**].

Theorem 1.32 (Kantorovich). *Let E and F be two ordered vector spaces with F a Dedekind complete Riesz space. If G is a majorizing vector subspace of E and $T: G \to F$ is a positive operator, then T has a positive linear extension to all of E.*

Proof. Fix $x \in E$ and let $y \in G$ satisfy $x \le y$. Since G is majorizing there exists a vector $u \in G$ with $u \le x$. Hence, $u \le y$ and the positivity of T implies $T(u) \le T(y)$ for all $y \in G$ with $x \le y$. In particular, it follows that $\inf\{T(y): y \in G \text{ and } x \le y\}$ exists in F for each $x \in E$. Thus, a mapping $p: E \to F$ can be defined via the formula

$$p(x) = \inf\{T(y): y \in G \text{ and } x \le y\}.$$

Clearly, $T(x) = p(x)$ holds for each $x \in G$ and an easy argument shows that p is also sublinear.

Now, by the Hahn–Banach Extension Theorem 1.25, the operator T has a linear extension S to all of E satisfying $S(z) \le p(z)$ for each $z \in E$. If $z \in E^+$, then $-z \le 0$, and so from

$$-S(z) = S(-z) \le p(-z) \le T(0) = 0,$$

we see that $S(z) \ge 0$. This shows that S is a positive linear extension of T to all of E. ∎

It is a remarkable fact that in case the domain of a positive operator T is a majorizing vector subspace, then the convex set $\mathcal{E}(T)$ is not merely nonempty but it also has extreme points. This result is due to Z. Lipecki [115].

Theorem 1.33 (Lipecki). *Let E and F be two Riesz spaces with F Dedekind complete. If G is a majorizing vector subspace of E and $T: G \to F$ is a positive operator, then the nonempty convex set $\mathcal{E}(T)$ has an extreme point.*

Proof. According to Theorem 1.31 we must establish the existence of some $S \in \mathcal{E}(T)$ satisfying

$$\inf\{S(|x - y|): y \in G\} = 0$$

for all $x \in E$.

Start by considering pairs (H, S) where H is a vector subspace majorizing E and $S: H \to F$ is a positive operator. For every such pair (H, S) define $p_{H,S}: E \to F$ by

$$p_{H,S}(x) = \inf\{S(y): y \in H \text{ and } x \le y\}.$$

It should be clear that $p_{H,S}$ is a sublinear mapping satisfying $p_{H,S}(y) = S(y)$ for every $y \in H$. In addition, if (H_1, S_1) and (H_2, S_2) satisfy $H_1 \subseteq H_2$ and $S_2 = S_1$ on H_1, then $p_{H_2,S_2}(x) \le p_{H_1,S_1}(x)$ holds for all $x \in E$.

Now let $\mathcal{C}$ be the collection of all pairs (H, S) such that:

(1) H is a vector subspace of E with $G \subseteq H$ (and so H majorizes E).

(2) $S \colon H \to F$ is a positive operator with $S = T$ on G.

(3) $\inf\{p_{H,S}(|x - y|) \colon y \in G\} = 0$ holds in F for all $x \in H$.

In view of $(G, T) \in \mathcal{C}$, the set $\mathcal{C}$ is nonempty. Moreover, if we define a binary relation $\geq$ on $\mathcal{C}$ by letting $(H_2, S_2) \geq (H_1, S_1)$ whenever $H_2 \supseteq H_1$ and $S_2 = S_1$ on H_1, then $\geq$ is an order relation on $\mathcal{C}$. By a routine argument we can verify that every chain of $\mathcal{C}$ has an upper bound in $\mathcal{C}$. Therefore, by Zorn's lemma the collection $\mathcal{C}$ has a maximal element, say (M, R). The rest of the proof is devoted to proving that $M = E$. (If this is done, then $R = p_{M,R}$ must be the case, which by Theorem 1.31 shows that R must be an extreme point of $\mathcal{E}(T)$.)

To this end, assume by way of contradiction that there exists some vector x that does not belong to M. Consider $H = \{u + \lambda x \colon u \in M \text{ and } \lambda \in \mathbb{R}\}$, and then define $S \colon H \to F$ by $S(u + \lambda x) = R(u) + \lambda p_{M,R}(x)$. Clearly, M is a proper subspace of H, $S = R$ holds on M, and $S \colon H \to F$ is a positive operator. (For the positivity of S let $u + \lambda x \geq 0$ with $u \in M$. For $\lambda > 0$ the inequality $x \leq -\frac{u}{\lambda}$ implies $p_{M,R}(x) \geq -R(\frac{u}{\lambda})$, and consequently $S(u + \lambda x) = R(u) + \lambda p_{M,R}(x) \geq 0$. The case $\lambda < 0$ is similar, while the case $\lambda = 0$ is trivial.) Finally, we verify that (H, S) satisfies (3). First, observe that by the sublinearity of $p_{H,S}$ the set

$$V = \{y \in E | \ \inf\{p_{H,S}(|y - z|) \colon z \in M\} = 0\}$$

is a vector subspace of E satisfying $M \subseteq V$. Also, from

$$
\begin{aligned}
0 \ &\leq \ \inf\{p_{H,S}(|x - z|) \colon z \in M\} \\
&\leq \ \inf\{p_{H,S}(z - x) \colon z \in M \text{ and } x \leq z\} \\
&= \ \inf\{R(z) - p_{M,R}(x) \colon z \in M \text{ and } x \leq z\} \\
&= \ \inf\{R(z) \colon z \in M \text{ and } x \leq z\} - p_{M,R}(x) = 0,
\end{aligned}
$$

we see that $x \in V$, and hence $H \subseteq V$. Now for arbitrary $u \in H$, $z \in M$, and $v \in G$ we have

$$
\begin{aligned}
p_{H,S}(|u - v|) \ &\leq \ p_{H,S}(|u - z|) + p_{H,S}(|v - z|) \\
&\leq \ p_{H,S}(|u - z|) + p_{M,R}(|v - z|),
\end{aligned}
$$

and so from $(M, R) \in \mathcal{C}$ and $u \in H \subseteq V$, it follows that

$$\inf\{p_{H,S}(|u - v|) \colon v \in G\} = 0$$

holds for all $u \in H$.

Thus, $(H, S) \in \mathcal{C}$. However, $(H, S) \geq (M, R)$ and $(H, S) \neq (M, R)$ contradict the maximality of (M, R). Therefore, $M = E$ must be true, as required. ∎

Exercises

1. Let E and F be two Riesz spaces with F Dedekind complete, and let A be an ideal of E. For each $T \in \mathcal{L}_b(E, F)$ let $\mathcal{R}(T)$ denote the restriction of T to A. Show that the positive operator $\mathcal{R} \colon \mathcal{L}_b(E, F) \to \mathcal{L}_b(A, F)$ satisfies

$$\mathcal{R}(S \vee T) = \mathcal{R}(S) \vee \mathcal{R}(T) \quad \text{and} \quad \mathcal{R}(S \wedge T) = \mathcal{R}(S) \wedge \mathcal{R}(T)$$

   for all $S, T \in \mathcal{L}_b(E, F)$. [1]

2. For two arbitrary solid sets A and B of a Riesz space show that:
   (a) $A + B$ is a solid set.
   (b) If $0 \le c \in A + B$ holds, then there exist $0 \le a \in A$ and $0 \le b \in B$ with $c = a + b$.

3. Let $T \colon E \to F$ be a positive operator between two Riesz spaces with F Dedekind complete. If two ideals A and B of E satisfy $A \perp B$, then show that:
   (a) $T_A \wedge T_B = 0$.
   (b) The ideal $A + B$ satisfies $T_{A+B} = T_A + T_B = T_A \vee T_B$.

4. As usual, ℓ_∞ denotes the Riesz space of all bounded real sequence, and c the Riesz subspace of ℓ_∞ consisting of all convergent sequences. If $\phi \colon c \to \mathbb{R}$ is the positive operator defined by

$$\phi(x_1, x_2, \ldots) = \lim_{n \to \infty} x_n \,,$$

   then show that ϕ has a positive linear extension to all of ℓ_∞.

1.3. Order Projections

In this section we shall study a special class of positive operators known as order (or band) projections. Before starting our discussion, let us review a few properties of order dense Riesz subspaces. Recall that a Riesz subspace G of a Riesz space E is said to be **order dense** in E whenever for each $0 < x \in E$ (i.e., $0 \le x$ and $x \ne 0$) there exists some $y \in G$ with $0 < y \le x$.

The following characterization of order dense Riesz subspaces in Archimedean Riesz spaces will be used freely in this book.

Theorem 1.34. *A Riesz subspace G of an Archimedean Riesz space E is order dense in E if and only if for each $x \in E^+$ we have*

$$\{y \in G \colon 0 \le y \le x\} \uparrow x \,.$$

Proof. If $\sup\{y \in G \colon 0 \le y \le x\} = x$ holds in E for each $x \in E^+$, then G is clearly order dense in E. For the converse, assume that G is order dense

---

[1] An operator between spaces of operators is referred to as a **transformer**. So, the operator $\mathcal{R}$ is an example of a transformer.

in E, and let $x \in E^+$. Assume by way of contradiction that some $z \in E$ satisfies $z < x$ and $y \leq z$ for each $y \in G$ with $0 \leq y \leq x$. Then, by the order denseness of G in E, there exists some $u \in G$ with $0 < u \leq x - z$. From $0 \leq u \leq x$ we see that $u \leq z$, and therefore $0 < 2u = u + u \leq (x - z) + z = x$. By induction, $0 < nu \leq x$ holds for each n, contradicting the Archimedean property of E. Thus, $\{y \in G \colon 0 \leq y \leq x\} \uparrow x$ holds in E, and the proof is finished. ∎

Consider an order dense Riesz subspace G of a Riesz space E. It is useful to know that the embedding of G into E preserves arbitrary suprema and infima. The result (whose straightforward proof is left for the reader) is stated next.

Theorem 1.35. *Let G be either an ideal or an order dense Riesz subspace of a Riesz space E, and let $D \subseteq G^+$ satisfy $D \downarrow$. Then $D \downarrow 0$ holds in G if and only if $D \downarrow 0$ holds in E.*

Recall that a subset A of a Riesz space is called *solid* whenever $|x| \leq |y|$ and $y \in A$ imply $x \in A$. A solid vector subspace is called an *ideal*. From Theorem 1.13 it readily follows that if A and B are solid subsets of a Riesz space, then their **algebraic sum**

$$A + B := \{a + b \colon a \in A \text{ and } b \in B\}$$

is likewise a solid set. In particular, the algebraic sum of two ideals also is an ideal.

The next theorem describes the basic properties of order dense ideals. Keep in mind that the disjoint complement of an arbitrary nonempty set of a Riesz space is always an ideal.

Theorem 1.36. *For an ideal A of a Riesz space E we have the following.*

(1) *The ideal A is order dense in E if and only if $A^d = \{0\}$.*

(2) *The ideal $A \oplus A^d$ is order dense in E.*

(3) *The ideal A is order dense in A^{dd}.*

Proof. (1) Let A be order dense in E and let $x \in A^d$. If $x \neq 0$ holds, then there exists some $y \in A$ with $0 < y \leq |x|$. This implies $y \in A \cap A^d = \{0\}$, a contradiction. Thus, $A^d = \{0\}$ holds.

For the converse, assume that $A^d = \{0\}$ holds and let $0 < x \in E$. If $y \wedge x = 0$ holds for all $y \in A^+$, then $x \in A^d = \{0\}$ also must be the case. Thus, $y \wedge x > 0$ must be true for some $y \in A^+$. But then $y \wedge x \in A$ and $0 < y \wedge x \leq x$ show that A is order dense in E.

(2) If $x \perp A \oplus A^{\mathrm{d}}$, then $x \perp A$ and $x \perp A^{\mathrm{d}}$ both hold. Therefore, $x \in A^{\mathrm{d}} \cap A^{\mathrm{dd}} = \{0\}$. This shows that $(A \oplus A^{\mathrm{d}})^{\mathrm{d}} = \{0\}$. By part (1) the ideal $A \oplus A^{\mathrm{d}}$ is order dense in E.

(3) This follows immediately from part (1). ∎

A net $\{x_\alpha\}$ of a Riesz space is said to be **order convergent** to a vector x (in symbols $x_\alpha \xrightarrow{o} x$) whenever there exists another net $\{y_\alpha\}$ with the same index set satisfying $y_\alpha \downarrow 0$ and $|x_\alpha - x| \leq y_\alpha$ for all indices α (abbreviated as $|x_\alpha - x| \leq y_\alpha \downarrow 0$). A subset A of a Riesz space is said to be **order closed** whenever $\{x_\alpha\} \subseteq A$ and $x_\alpha \xrightarrow{o} x$ imply $x \in A$.

Lemma 1.37. *A solid subset A of a Riesz space is order closed if and only if $\{x_\alpha\} \subseteq A$ and $0 \leq x_\alpha \uparrow x$ imply $x \in A$.*

Proof. Assume that a solid set A of a Riesz space has the stated property and let a net $\{x_\alpha\} \subseteq A$ satisfy $x_\alpha \xrightarrow{o} x$. Pick a net $\{y_\alpha\}$ with the same index net satisfying $y_\alpha \downarrow 0$ and $|x_\alpha - x| \leq y_\alpha$ for each α. Now note that we have $(|x| - y_\alpha)^+ \leq |x_\alpha|$ for each α and $0 \leq (|x| - y_\alpha)^+ \uparrow |x|$, and from this it follows that $x \in A$. That is, A is order closed. ∎

An order closed ideal is referred to as a **band**. Thus, according to Lemma 1.37 an ideal A is a band if and only if $\{x_\alpha\} \subseteq A$ and $0 \leq x_\alpha \uparrow x$ imply $x \in A$ (or, equivalently, if and only if $D \subseteq A^+$ and $D \uparrow x$ imply $x \in A$). In the early developments of Riesz spaces a band was called a normal subspace (G. Birkhoff [**36**], S. Bochner and R. S. Phillips [**39**]), while F. Riesz was calling a band a *famille complète*.

Let A be a nonempty subset of a Riesz space E. Then the **ideal generated** by A is the smallest (with respect to inclusion) ideal that includes A. A moment's thought reveals that this ideal is

$$E_A = \left\{ x \in E \colon \exists\, x_1, \ldots, x_n \in A \text{ and } \lambda \in \mathbb{R}^+ \text{ with } |x| \leq \lambda \sum_{i=1}^{n} |x_i| \right\}.$$

The ideal generated by a vector $x \in E$ will be denoted by E_x. By the preceding discussion we have

$$E_x = \left\{ y \in E \colon \exists\, \lambda > 0 \text{ with } |y| \leq \lambda |x| \right\}.$$

Every ideal of the form E_x is referred to as a **principal ideal**.

Similarly, the **band generated** by a set A is the smallest band that includes the set A. Such a band always exists (since it is the intersection of the family of all bands that include A, and E is one of them.) Clearly, the band generated by A coincides with the band generated by the ideal generated by A. The band generated by a vector x is called the **principal band** generated by x and is denoted by B_x.

The band generated by an ideal is described as follows.

Theorem 1.38. *If A is an ideal of a Riesz space E, then the band generated by A is precisely the vector subspace:*

$$\{x \in E \colon \exists \{x_\alpha\} \subseteq A^+ \text{ with } 0 \leq x_\alpha \uparrow |x|\}.$$

In particular, every ideal is order dense in the band it generates.

Moreover, the principal band B_x generated by a vector x is given by

$$B_x = \{y \in E \colon |y| \wedge n|x| \uparrow |y|\}.$$

Proof. Let $B = \{x \in E \colon \exists \{x_\alpha\} \subseteq A^+ \text{ with } 0 \leq x_\alpha \uparrow |x|\}$. Clearly, every band containing A must include B. Thus, in order to establish our result it is enough to show that B is a band.

To this end, let $x, y \in B$. Pick two nets $\{x_\alpha\} \subseteq A^+$ and $\{y_\beta\} \subseteq A^+$ with $0 \leq x_\alpha \uparrow |x|$ and $0 \leq y_\beta \uparrow |y|$. From

$$|x + y| \wedge (x_\alpha + y_\beta) \uparrow_{(\alpha,\beta)} |x + y| \wedge (|x| + |y|) = |x + y|$$

and

$$|\lambda| x_\alpha \uparrow |\lambda x|,$$

we see that B is a vector subspace. Also, if $|z| \leq |x|$ holds, then from $\{|z| \wedge x_\alpha\} \subseteq A$ and $0 \leq |z| \wedge x_\alpha \uparrow |z| \wedge |x| = |z|$, it follows that $z \in B$. Hence, B is an ideal. Finally, to see that B is a band, let $\{x_\alpha\} \subseteq B$ satisfy $0 \leq x_\alpha \uparrow x$. Put $D = \{y \in A \colon \exists \alpha \text{ with } 0 \leq y \leq x_\alpha\}$. Then $D \subseteq A^+$ and $D \uparrow x$ hold. Therefore, $x \in B$ and so B is a band.

To establish the identity for B_x, let $y \in B_x$. By the above, there exists a net $\{x_\alpha\} \subseteq E_x$ with $0 \leq x_\alpha \uparrow |y|$. Now given an index α there exists some n with $x_\alpha \leq n|x|$, and so $x_\alpha \leq |y| \wedge n|x| \leq |y|$ holds. This easily implies $|y| \wedge n|x| \uparrow |y|$, and our conclusion follows. ∎

From Theorem 1.8 it follows that A^d is always a band. It is important to know that the band generated by a set A is precisely A^{dd}.

Theorem 1.39. *The band generated by a nonempty subset A of an Archimedean Riesz space is precisely A^{dd} (and hence if A is a band, then $A = A^{dd}$ holds).*

Proof. We mentioned before that the band generated by A is the same as the band generated by the ideal generated by A. Therefore, we can assume that A is an ideal. By part (3) of Theorem 1.36 we know that A is order dense in A^{dd}, and hence (by Theorem 1.34) for each $x \in A^{dd}$ there exists a net $\{x_\alpha\} \subseteq A$ with $0 \leq x_\alpha \uparrow |x|$. This easily implies that A^{dd} is the smallest band including A. ∎

A useful condition under which an ideal is necessarily a band is presented next.

Theorem 1.40. *Let A and B be two ideals in a Riesz space E such that $E = A \oplus B$. Then A and B are both bands satisfying $A = B^d$ and $B = A^d$ (and hence $A = A^{dd}$ and $B = B^{dd}$ both hold).*

Proof. Note first that for each $a \in A$ and $b \in B$ we have

$$|a| \wedge |b| \in A \cap B = \{0\},$$

and so $A \perp B$. In particular, $A \subseteq B^d$.

On the other hand, if $x \in B^d$, then write $x = a+b$ with $a \in A$ and $b \in B$, and note that $b = x - a \in B \cap B^d = \{0\}$ implies $x = a \in A$. Thus, $B^d \subseteq A$, and so $A = B^d$ holds. This shows that A is a band. By the symmetry of the situation $B = A^d$ also holds. ∎

A band B in a Riesz space E that satisfies $E = B \oplus B^d$ is referred to as a **projection band**. The next result characterizes the ideals that are projection bands.

Theorem 1.41. *For an ideal B in a Riesz space E the following statements are equivalent.*

(1) *B is a projection band, i.e., $E = B \oplus B^d$ holds.*

(2) *For each $x \in E^+$ the supremum of the set $B^+ \cap [0, x]$ exists in E and belongs to B.*

(3) *There exists an ideal A of E such that $E = B \oplus A$ holds.*

Proof. (1) $\Longrightarrow$ (2) Let $x \in E^+$. Choose the (unique) vectors $0 \le y \in B$ and $0 \le z \in B^d$ such that $x = y + z$. If $u \in B^+$ satisfies $u \le x = y + z$, then it follows from $0 \le (u - y)^+ \le z \in B^d$ and $(u - y)^+ \in B$ that $(u - y)^+ = 0$. Thus, $u \le y$, and so y is an upper bound of the set $B^+ \cap [0, x]$. Since $y \in B \cap [0, x]$, we see that $y = \sup\{u \in B^+ : u \le x\} = \sup B \cap [0, x]$ in E.

(2) $\Longrightarrow$ (3) Fix some $x \in E^+$, and let $u = \sup B \cap [0, x]$. Clearly, u belongs to B. Put $y = x - u \ge 0$. If $0 \le w \in B$, then $0 \le y \wedge w \in B$, and moreover from $0 \le u + y \wedge w \in B$ and

$$u + y \wedge w = (u + y) \wedge (u + w) = x \wedge (u + w) \le x,$$

it follows that $u + y \wedge w \le u$. Hence, $y \wedge w = 0$ holds, and so $y \in B^d$. From $x = u + y$ we see that $E = B \oplus B^d$, and therefore (3) holds with $A = B^d$.

(3) $\Longrightarrow$ (1) This follows from Theorem 1.40. ∎

Not every band is a projection band, and a Riesz space in which every band is a projection band is referred to as a Riesz space with the **projection**

property. From the preceding theorem it should be clear that in a Dedekind complete Riesz space every band is a projection band. This was proven by F. Riesz [**166**] is one of his early fundamental papers on Riesz spaces. Because it guarantees an abundance of order projections, we state it next as a separate theorem.

Theorem 1.42 (F. Riesz). *If B is a band in a Dedekind complete Riesz space E, then $E = B \oplus B^{\mathrm{d}}$ holds.*

As usual, an operator $P\colon V \to V$ on a vector space is called a **projection** if $P^2 = P$. If a projection P is defined on a Riesz space and P is also a positive operator, then P will be referred to as a **positive projection**.

Now let B be a projection band in a Riesz space E. Thus, $E = B \oplus B^{\mathrm{d}}$ holds, and so every vector $x \in E$ has a unique decomposition $x = x_1 + x_2$, where $x_1 \in B$ and $x_2 \in B^{\mathrm{d}}$. Then it is easy to see that a projection $P_B\colon E \to E$ is defined via the formula

$$P_B(x) := x_1 .$$

Clearly, P_B is a positive projection. Any projection of the form P_B is called an **order projection** (or a **band projection**). Thus, the order projections are associated with the projection bands in a one-to-one fashion.

Theorem 1.43. *If B is a projection band of a Riesz space E, then*

$$P_B(x) = \sup\{y \in B\colon\ 0 \le y \le x\} = \sup B \cap [0, x]$$

holds for all $x \in E^+$.

Proof. Let $x \in E^+$. Then (by Theorem 1.41) $u = \sup\{y \in B\colon\ 0 \le y \le x\}$ exists and belongs to B. We claim that $u = P_B(x)$.

To see this, write $x = x_1 + x_2$ with $0 \le x_1 \in B$ and $0 \le x_2 \in B^{\mathrm{d}}$, and note that $0 \le x_1 \le x$ implies $0 \le x_1 \le u$. Thus, $0 \le u - x_1 \le x - x_1 = x_2$, and hence $u - x_1 \in B^{\mathrm{d}}$, Since $u - x_1 \in B$ and $B \cap B^{\mathrm{d}} = \{0\}$, we see that $u = x_1$, as claimed. ∎

Among projections the order projections are characterized as follows.

Theorem 1.44. *For an operator $T\colon E \to E$ on a Riesz space the following statements are equivalent.*

   (1) *T is an order projection.*

   (2) *T is a projection satisfying $0 \le T \le I$ (where, of course, I is the identity operator on E).*

   (3) *T and $I - T$ have disjoint ranges, i.e., $Tx \perp y - Ty$ holds for all $x, y \in E$.*

Proof. $(1) \Longrightarrow (2)$ Obvious.

$(2) \Longrightarrow (3)$ Let $x, y \in E^+$. Put $z = Tx \wedge (I - T)y$. From the inequality $0 \le z \le (I - T)y$ it follows that $0 \le Tz \le T(I - T)y = (T - T^2)y = 0$, and so $Tz = 0$. Similarly, $(I - T)z = 0$, and hence $z = (I - T)z + Tz = 0$ holds. This shows that T and $I - T$ have disjoint ranges.

$(3) \Longrightarrow (1)$ Let A and B be the ideals generated by the ranges of T and $I - T$, respectively. By our hypothesis it follows that $A \perp B$, and from $x = Tx + (I - T)x$ we see that $E = A \oplus B$. But then, by Theorem 1.40 both A and B are projection bands of E. Now the identity

$$P_A x - Tx = P_A x - P_A Tx = P_A(x - Tx) = 0$$

shows that $T = P_A$ holds. Thus, T is an order projection, and the proof is finished. ■

A positive projection need not be an order projection. For instance, consider the operator $T \colon L_1[0, 1] \to L_1[0, 1]$ defined by

$$T(f) = \left[\int_0^1 f(x)\, dx \right] \cdot \mathbf{1},$$

where $\mathbf{1}$ denotes the constant function one. Clearly, $0 \le T = T^2$ holds, and its is not difficult to see that T is not an order projection.

The basic properties of order projections are summarized in the next theorem.

Theorem 1.45. *If A and B are projection bands in a Riesz space E, then A^{d}, $A \cap B$, and $A + B$ are likewise projection bands. Moreover, we have:*

(1) $P_{A^{\mathrm{d}}} = I - P_A$.

(2) $P_{A \cap B} = P_A P_B = P_B P_A$.

(3) $P_{A+B} = P_A + P_B - P_A P_B$.

Proof. (1) From $E = A \oplus A^{\mathrm{d}}$ it follows that $A^{\mathrm{dd}} = A$ holds (see Theorem 1.40), and so A^{d} is a projection band. The identity $P_{A^{\mathrm{d}}} = I - P_A$ should be obvious.

(2) To see that $A \cap B$ is a projection band note first that the identity $B \cap [0, x] = [0, P_B x]$ implies $A \cap B \cap [0, x] = A \cap [0, P_B x]$ for each $x \in E^+$. Consequently,

$$P_A P_B x = \sup A \cap [0, P_B x] = \sup A \cap B \cap [0, x]$$

holds for each $x \in E^+$, which (by Theorem 1.41) shows that $A \cap B$ is a projection band and that $P_A P_B = P_{A \cap B}$ holds. Similarly, $P_B P_A = P_{A \cap B}$.

(3) Assume at the beginning that the two projection bands A and B satisfy $A \perp B$. Let $x \in E^{+}$. If $0 \le a + b \in A + B$ satisfies $a + b \le x$, then clearly $a \in A \cap [0, x]$ and $b \in B \cap [0, x]$, and so $a + b \le P_A x + P_B x \in A + B$ holds. This shows that

$$\sup(A + B) \cap [0, x] = P_A x + P_B x \in A + B,$$

and hence by Theorem 1.41 the ideal $A + B$ is a projection band. Also, $P_{A+B} = P_A + P_B$ holds.

For the general case observe that $A + B = (A \cap B^d) \oplus B$. Now using the preceding case, we get

$$\begin{aligned} P_{A+B} &= P_{(A \cap B^d) \oplus B} = P_{A \cap B^d} + P_B = P_A P_{B^d} + P_B \\ &= P_A(I - P_B) + P_B = P_A - P_A P_B + P_B \\ &= P_A + P_B - P_{A \cap B}, \end{aligned}$$

and the proof is finished. ∎

An immediate consequence of statement (2) of the preceding theorem is that two arbitrary order projections mutually commute.

A useful comparison property of order projections is described next.

Theorem 1.46. *If A and B are projection bands in a Riesz space, then the following statements are equivalent.*

(1) $A \subseteq B$.

(2) $P_A P_B = P_B P_A = P_A$.

(3) $P_A \le P_B$.

Proof. (1) $\Longrightarrow$ (2) Let $A \subseteq B$. Then from Theorem 1.45 it follows that

$$P_A P_B = P_B P_A = P_{A \cap B} = P_A.$$

(2) $\Longrightarrow$ (3) For each $0 \le x$ we have $P_A x = P_B P_A x \le P_B x$, and so $P_A \le P_B$ holds.

(3) $\Longrightarrow$ (1) If $0 \le x \in A$, then it follows from $0 \le x = P_A x \le P_B x \in B$ that $x \in B$. Therefore, $A \subseteq B$ holds, as required. ∎

A vector x in a Riesz space E is said to be a **projection vector** whenever the principal band B_x generated by x (i.e., $B_x = \{y \in E: |y| \wedge n|x| \uparrow |y|\}$) is a projection band. If every vector in a Riesz space is a projection vector, then the Riesz space is said to have the **principal projection property**. For a projection vector x we shall write P_x for the order projection onto the band B_x.

Theorem 1.47. *A vector x in a Riesz space is a projection vector if and only if* $\sup\{y \wedge n|x|: \ n \in \mathbb{N}\}$ *exists for each $y \geq 0$. In this case*

$$P_x(y) = \sup\{y \wedge n|x|: \ n \in \mathbb{N}\}$$

holds for all $y \geq 0$.

Proof. Let $y \geq 0$. We claim that the sets $B_x \cap [0, y]$ and $\{y \wedge n|x|: \ n \in \mathbb{N}\}$ have the same upper bounds. To see this, note first that

$$\{y \wedge n|x|: \ n \in \mathbb{N}\} \subseteq B_x \cap [0, y]$$

holds. Now let $y \wedge n|x| \leq u$ for all n. If $z \in B_x \cap [0, y]$, then by Theorem 1.38 we have $z \wedge n|x| \uparrow z$. In view of $z \wedge n|x| \leq y \wedge n|x| \leq u$, we see that $z \leq u$, and so the two sets have the same upper bounds. Now to finish the proof invoke Theorems 1.41 and 1.43. ∎

From the preceding theorem it follows immediately that in a Dedekind σ-complete Riesz space every principal band is a projection band. If $x, y \geq 0$ are projection vectors in a Riesz space, then note that the formulas of Theorem 1.45 take the form

$$P_{x \wedge y} = P_x P_y = P_y P_x \quad \text{and} \quad P_{x+y} = P_x + P_y - P_{x \wedge y}.$$

A vector $e > 0$ in a Riesz space E is said to be a **weak order unit** whenever the band generated by e satisfies $B_e = E$ (or, equivalently, whenever for each $x \in E^+$ we have $x \wedge ne \uparrow x$). Clearly, every vector $0 < x \in E$ is a weak order unit in the band it generates. Also, note that a vector $e > 0$ in an Archimedean Riesz space is a weak order unit if and only if $x \perp e$ implies $x = 0$.

Projection vectors satisfy the following useful properties.

Theorem 1.48. *In a Riesz space E the following statements hold:*

(1) *If u, v, and w are projection vectors satisfying $0 \leq w \leq v \leq u$, then for each $x \in E$ we have $(P_u - P_v)x \perp (P_v - P_w)x$.*

(2) *If $0 \leq u_\alpha \uparrow u$ holds in E with u and all the u_α projection vectors, then $P_{u_\alpha}(x) \uparrow P_u(x)$ holds for each $x \in E^+$.*

Proof. (1) By Theorem 1.46 we have $P_w \leq P_v \leq P_u$ and so if $x \in E$, then

$$
\begin{aligned}
0 &\leq \big|(P_u - P_v)x\big| \wedge \big|(P_v - P_w)x\big| \\
&\leq (P_u - P_v)|x| \wedge (P_v - P_w)|x| \\
&\leq \big[P_u|x| - P_v(P_u|x|)\big] \wedge P_v(P_u|x|) = 0.
\end{aligned}
$$

(2) Let $x \in E^+$. Clearly, $P_{u_\alpha}(x) \uparrow \leq P_u(x)$. Thus, $P_u(x)$ is an upper bound for the net $\{P_{u_\alpha}(x)\}$, and we claim it is the least upper bound.

To see this, assume $P_{u_\alpha}(x) \le y$ for all α. Hence, $x \wedge n u_\alpha \le y$ holds for all α and n. Consequently, $u_\alpha \uparrow u$ implies $x \wedge nu \le y$ for all n, and therefore $P_u(x) = \sup\{x \wedge nu: \ n \in \mathbb{N}\} \le y$. Hence, $P_u(x)$ is the least upper bound of $\{P_{u_\alpha}(x)\}$, and thus $P_{u_\alpha}(x) \uparrow P_u(x)$.   ■

Let e be a positive vector of a Riesz space E. A vector $x \in E^+$ is said to be a **component** of e whenever $x \wedge (e - x) = 0$. The collection of all components of e will be denoted by $\mathcal{C}_e$, i.e.,

$$\mathcal{C}_e := \{x \in E^+: \ x \wedge (e - x) = 0\}.$$

Clearly, $x \in \mathcal{C}_e$ implies $e - x \in \mathcal{C}_e$. Also, $P_B e \in \mathcal{C}_e$ for each projection band B.

Under the partial ordering induced by E, the set of components $\mathcal{C}_e$ is a Boolean algebra,[2] consisting precisely of the extreme points of the order interval $[0, e]$. The details follow.

Theorem 1.49. *For a positive vector e in a Riesz space E we have:*

(1) *If $x, y \in \mathcal{C}_e$ and $x \le y$ holds, then $y - x \in \mathcal{C}_e$.*

(2) *If $x_1, x_2, y_1, y_2 \in \mathcal{C}_e$ satisfy the inequalities $x_1 \le x_2 \le y_1 \le y_2$, then $x_2 - x_1 \perp y_2 - y_1$.*

(3) *If $x, y \in \mathcal{C}_e$, then $x \vee y$ and $x \wedge y$ both belong to $\mathcal{C}_e$ (and so $\mathcal{C}_e$ is a Boolean algebra with smallest element 0 and largest element e).*

(4) *If E is Dedekind complete, then for every non-empty subset C of $\mathcal{C}_e$ the elements $\sup C$ and $\inf C$ both belong to $\mathcal{C}_e$ (and so in this case $\mathcal{C}_e$ is a Dedekind complete Boolean algebra).*

(5) *The set of components $\mathcal{C}_e$ of e is precisely the set of all extreme points of the convex set $[0, e]$.[3]*

Proof. (1) It follows immediately from the inequalities

$$\begin{aligned} 0 \ &\le \ (y - x) \wedge \big[e - (y - x)\big] = (y - x) \wedge \big[(e - y) + x\big] \\ &\le \ (y - x) \wedge (e - y) + (y - x) \wedge x \\ &\le \ y \wedge (e - y) + (e - x) \wedge x = 0 + 0 = 0 . \end{aligned}$$

(2) Note that $0 \le (x_2 - x_1) \wedge (y_2 - y_1) \le y_1 \wedge (e - y_1) = 0$.

---

[2] Recall that a **Boolean algebra** $\mathcal{B}$ is a distributive lattice with smallest and largest elements that is complemented. That is, $\mathcal{B}$ is a partially ordered set that is a distributive lattice with a smallest element 0 and a largest element e such that for each $a \in \mathcal{B}$ there exists a (necessarily unique) element $a' \in \mathcal{B}$ (called the *complement* of a) satisfying $a \wedge a' = 0$ and $a \vee a' = e$. A Boolean algebra $\mathcal{B}$ is **Dedekind complete** if every nonempty subset of $\mathcal{B}$ has a supremum.

[3] Recall that a vector u in a convex set C is said to be an **extreme point** of C if it follows from $u = \lambda v + (1 - \lambda)w$ with $v, w \in C$ and $0 < \lambda < 1$ that $v = w = u$.

(3) Let $x, y \in C_e$. Then, using the distributive laws, we see that

$$
\begin{aligned}
(x \vee y) \wedge (e - x \vee y) &= (x \vee y) \wedge \left[(e - x) \wedge (e - y)\right] \\
&= \left[x \wedge (e - x) \wedge (e - y)\right] \vee \left[y \wedge (e - x) \wedge (e - y)\right] \\
&= 0 \vee 0 = 0 \,,
\end{aligned}
$$

and

$$
\begin{aligned}
(x \wedge y) \wedge (e - x \wedge y) &= (x \wedge y) \wedge \left[(e - x) \vee (e - y)\right] \\
&= \left[x \wedge y \wedge (e - x)\right] \vee \left[x \wedge y \wedge (e - y)\right] \\
&= 0 \vee 0 = 0 \,.
\end{aligned}
$$

(4) Now assume that E is Dedekind complete and let C be a nonempty set of components of e. Put $u = \sup C$ and $v = \inf C$. Then, using the infinite distributive laws, we get

$$
\begin{aligned}
0 \le u \wedge (e - u) &= [\sup C] \wedge (e - u) = \sup\{c \wedge (e - u) \colon \ c \in C\} \\
&\le \sup\{c \wedge (e - c) \colon \ c \in C\} = 0 \,.
\end{aligned}
$$

Similarly, we have

$$
\begin{aligned}
0 \le v \wedge (e - v) &= = v \wedge \left(e - \inf C\right) = v \wedge \sup\{e - c \colon \ c \in C\} \\
&= \sup\{v \wedge (e - c) \colon \ c \in C\} \\
&\le \sup\{c \wedge (e - c) \colon \ c \in C\} = 0 \,.
\end{aligned}
$$

(5) Assume first that an element $x \in [0, e]$ is an extreme point of $[0, e]$. Let $y = x \wedge (e - x) \ge 0$. We must show that $y = 0$. Clearly, $0 \le x - y \le e$ and $0 \le x + y \le e$, and from the convex combination $x = \frac{1}{2}(x - y) + \frac{1}{2}(x + y)$ we get $x - y = x$. So $y = 0$, as desired.

For the converse, assume that $v \in C_e$ and let $v = \lambda x + (1 - \lambda)y$, where $x, y \in [0, e]$ and $0 < \lambda < 1$. From $v \wedge (e - v) = 0$, it follows that $x \wedge (e - v) = 0$, and so from part (1) of Lemma 1.4 we get

$$
x = x \wedge e = x \wedge \left[(v + (e - v)\right] \le x \wedge v + x \wedge (e - v) = x \wedge v \le v \,.
$$

Similarly, $y \le v$. Now if either $x < v$ or $y < v$ were true, then

$$
v = \lambda x + (1 - \lambda)y < \lambda v + (1 - \lambda)v = v
$$

also would be true, which is impossible. Hence $x = y = v$ holds, and so v is an extreme point of $[0, e]$. This completes the proof of the theorem. ∎

When E has the principal projection property, Y. A. Abramovich [1] has described the lattice operations of $\mathcal{L}_b(E, F)$ in terms of components as follows.

Theorem 1.50 (Abramovich). *If a Riesz space E has the principal projection property and F is a Dedekind complete Riesz space, then for each pair $S, T \in \mathcal{L}_b(E, F)$ and each $x \in E^+$ we have:*

$$[S \vee T](x) = \sup\{S(y) + T(z): \ y \wedge z = 0 \ and \ y + z = x\}.$$
$$[S \wedge T](x) = \inf\{S(y) + T(z): \ y \wedge z = 0 \ and \ y + z = x\}.$$

Proof. Notice that the first formula follows from the second by using the identity $S \vee T = -[(-S) \wedge (-T)]$. Also, if the second formula is true for the special case $S \wedge T = 0$, then it is true in general. This claim follows easily from the identity $(S - S \wedge T) \wedge (T - S \wedge T) = 0$. To complete the proof, assume that $S \wedge T = 0$ in $\mathcal{L}_b(E, F)$. Fix $x \in L^+$ and put

$$u = \inf\{S(y) + T(x - y): \ y \wedge (x - y) = 0\}.$$

We must show that $u = 0$.

To this end, fix any $0 \le y \in E^+$ satisfy $0 \le y \le x$. Let P denote the order projection of E onto the band generated by $(2y - x)^+$ and put $z = P(x)$. From $x \le 2y + (x - 2y)^+$ and $(x - 2y)^+ \wedge (2y - x)^+ = 0$, it follows that $P(x) \le 2P(y) + P((x - 2y)^+) = 2P(y) \le 2y$. Therefore,

$$z \le 2y. \tag{$\star$}$$

Also, from $(2y - x)^+ \le (2x - x)^+ = x$ we see that

$$2y - x \le (2y - x)^+ = P((2y - x)^+) \le P(x) = z,$$

and consequently

$$x - z \le 2(x - y). \tag{$\star\star$}$$

Now combining $(\star)$ and $(\star\star)$, we get

$$0 \le u \le S(z) + T(x - z) \le 2[S(y) + T(x - y)], \tag{$\star\star\star$}$$

for all elements $y \in E^+$ with $0 \le y \le x$. Taking into consideration that (according to Theorem 1.18) we have $\inf\{S(y) + T(x - y): \ 0 \le y \le x\} = 0$, it follows from $(\star\star\star)$ that $u = 0$, and the proof is finished. ■

It should be noted that Theorem 1.50 is false without assuming that E has the principal projection property. For instance, let $E = C[0, 1]$, $F = \mathbb{R}$, and let $S, T: E \to F$ be defined by $S(f) = f(0)$ and $T(f) = f(1)$. Then $S \wedge T = 0$ holds, while

$$\inf\{S(f) + T(g): \ f \wedge g = 0 \ and \ f + g = 1\}$$
$$= \inf\{S(f) + T(1 - f): \ f = 0 \ or \ f = 1\} = 1.$$

When E has the principal projection property, the lattice operations of $\mathcal{L}_b(E, F)$ also can be expressed in terms of directed sets involving components as follows.

Theorem 1.51. *Assume that E has the principal projection property and that F is Dedekind complete. Then for all $S, T \in \mathcal{L}_b(E, F)$ and $x \in E^+$ we have:*

(1) $\left\{ \sum_{i=1}^{n} S(x_i) \vee T(x_i) \colon x_i \wedge x_j = 0 \text{ for } i \neq j \text{ and } \sum_{i=1}^{n} x_i = x \right\} \uparrow [S \vee T](x).$

(2) $\left\{ \sum_{i=1}^{n} S(x_i) \wedge T(x_i) \colon x_i \wedge x_j = 0 \text{ for } i \neq j \text{ and } \sum_{i=1}^{n} x_i = x \right\} \downarrow [S \wedge T](x).$

(3) $\left\{ \sum_{i=1}^{n} |T(x_i)| \colon x_i \wedge x_j = 0 \text{ for } i \neq j \text{ and } \sum_{i=1}^{n} x_i = x \right\} \uparrow |T|(x).$

Proof. Since (2) and (3) follow from (1) by using the usual lattice identities $S \wedge T = -[(-S) \vee (-T)]$ and $|T| = T \vee (-T)$, we prove only the first formula. Put

$$ D = \left\{ \sum_{i=1}^{n} S(x_i) \vee T(x_i) \colon x_i \wedge x_j = 0 \text{ for } i \neq j \text{ and } \sum_{i=1}^{n} x_i = x \right\}, $$

where $x \in E^+$ is fixed, and note that $\sup D \leq [S \vee T](x)$ holds in F. On the other hand, if $y, z \in E^+$ satisfy $y \wedge z = 0$ and $y + z = x$, then the relation

$$ S(y) + T(z) \leq S(y) \vee T(y) + S(z) \vee T(z) \in D, $$

in conjunction with Theorem 1.50, shows that $\sup D = [S \vee T](x)$ holds. Therefore, what remains to be shown is that D is directed upward.

To this end, let $\{x_1, \ldots, x_n\}$ and $\{y_1, \ldots, y_m\}$ be two subsets of E^+ each of which is pairwise disjoint such that $\sum_{i=1}^{n} x_i = \sum_{j=1}^{m} y_j = x$. Then note that the finite set $\{x_i \wedge y_j \colon i = 1, \ldots, n; j = 1, \ldots, m\}$ is pairwise disjoint and

$$ \sum_{i=1}^{n} \sum_{j=1}^{m} x_i \wedge y_j = \sum_{i=1}^{n} x_i \wedge \left[\sum_{j=1}^{m} y_j \right] = \sum_{i=1}^{n} x_i \wedge x = \sum_{i=1}^{n} x_i = x. $$

In addition, we have

$$ \sum_{i=1}^{n} S(x_i) \vee T(x_i) = \sum_{i=1}^{n} S\left(x_i \wedge \sum_{j=1}^{n} y_j \right) \vee T\left(x_i \wedge \sum_{j=1}^{m} y_j \right) $$

$$ = \sum_{i=1}^{n} \left[\sum_{j=1}^{n} S(x_i \wedge y_j) \right] \vee \left[\sum_{j=1}^{m} T(x_i \wedge y_j) \right] $$

$$ \leq \sum_{i=1}^{n} \sum_{j=1}^{m} S(x_i \wedge y_j) \vee T(x_i \wedge y_j), $$

and, similarly,

$$\sum_{j=1}^{m} S(y_j) \vee T(y_j) \le \sum_{i=1}^{n} \sum_{j=1}^{m} S(x_i \wedge y_j) \vee T(x_i \wedge y_j).$$

Therefore, $D \uparrow [S \vee T](x)$ holds. ∎

The final result of this section deals with retracts of Riesz spaces. Let us say that a Riesz subspace G of a Riesz space E is a **retract** (or that E is **retractable** on G) whenever there exists a positive projection $P \colon E \to E$ whose range is G.

Theorem 1.52. *For a Riesz subspace G of a Riesz space E we have the following:*

(1) *If G is a retract of E and E is Dedekind complete, then G is a Dedekind complete Riesz space in its own right.*

(2) *If G is Dedekind complete in its own right and G majorizes E, then G is a retract of E.*

Proof. (1) Let $P \colon E \to E$ be a positive projection whose range is the Riesz subspace G, and let $0 \le x_\alpha \uparrow \le x$ in G. Then there exists some $y \in E$ with $0 \le x_\alpha \uparrow y \le x$ in E, and so $0 \le x_\alpha = Px_\alpha \le Py$ holds in G for each α. On the other hand, if for some $z \in G$ we have $0 \le x_\alpha \le z$ for all α, then $y \le z$, and hence $Py \le Pz = z$. In other words, $0 \le x_\alpha \uparrow Py$ holds in G, which proves that G is a Dedekind complete Riesz space.

(2) Apply Theorem 1.32 to the identity operator $I \colon G \to G$. ∎

Exercises

1. For two nets $\{x_\alpha\}$ and $\{y_\beta\}$ in a Riesz space satisfying $x_\alpha \xrightarrow{o} x$ and $y_\beta \xrightarrow{o} y$ establish the following properties.
   (a) If $x_\alpha \xrightarrow{o} u$, then $u = x$ (and so the order limits whenever they exist are uniquely determined).
   (b) $\lambda x_\alpha + \mu y_\beta \xrightarrow{o} \lambda x + \mu y$ for all $\lambda, \mu \in \mathbb{R}$.
   (c) $|x_\alpha| \xrightarrow{o} |x|$.
   (d) $x_\alpha \vee y_\beta \xrightarrow{o} x \vee y$ and $x_\alpha \wedge y_\beta \xrightarrow{o} x \wedge y$.
   (e) $(x_\alpha - y_\beta)^+ \xrightarrow{o} (x - y)^+$.
   (f) If $x_\alpha \le z$ holds for all $\alpha \succeq \alpha_0$, then $x \le z$.

2. Show that the intersection of two order dense ideals is also an order dense ideal.

3. Let $0 \le y \le x \le e$ hold in a Riesz space. If y is a component of x and x is a component of e, then show that y is a component of e.

4. If **1** denotes the constant function one on $[0, 1]$, then compute C_1 in:

   (a) $C[0, 1]$;   (b) $L_1[0, 1]$;   (c) $\ell_\infty[0, 1]$.

5. Show that in an Archimedean Riesz space a vector $e > 0$ is a weak order unit if and only if $x \perp e$ implies $x = 0$.

6. If E has the principal projection property, then show that $P_{x^+}(x) = x^+$ holds for all $x \in E$.

7. Let E be a Riesz space satisfying the principal projection property, let $0 \leq y \leq x$, and let $\epsilon \in \mathbb{R}$. If P denotes the order projection onto the band generated by $(y - \epsilon x)^+$, then show that $\epsilon P(x) \leq y$ holds.

8. If A and B are two projection bands in a Riesz space E, then show that:
   (a) $P_{A \cap B}(x) = P_A(x) \wedge P_B(x)$ holds for all $x \in E^+$.
   (b) $P_{A+B}(x) = P_A(x) \vee P_B(x)$ holds for all $x \in E^+$.
   (c) $P_{A+B} = P_A + P_B$ holds if and only if $A \perp B$.

9. If P and Q are order projections on a Riesz space E, then show that

$$P(x) \wedge Q(y) = PQ(x \wedge y)$$

   for all $x, y \in E^+$.

10. For an order projection P on a Riesz space E establish the following:
    (a) $|Px| = P(|x|)$ holds for all $x \in E$.
    (b) If D is a nonempty subset of E for which $\sup D$ exists in E, then $\sup P(D)$ exists in E and $\sup P(D) = P(\sup D)$.

11. Let E and F be two Riesz spaces with F Dedekind complete. Show that:
    (a) If P is an order projection on E and Q is an order projection on F, then the operator (transformer) $T \mapsto QTP$ is an order projection on $\mathcal{L}_b(E, F)$.
    (b) If P_1, P_2 are order projections on E and Q_1, Q_2 are order projections on F, then

$$(Q_1 T P_1) \wedge (Q_2 S P_2) = Q_1 Q_2 (T \wedge S) P_1 P_2$$

   holds in $\mathcal{L}_b(E, F)$ for all $S, T \in \mathcal{L}_b^+(E, F)$.

12. Let E and F be two Riesz spaces with F Dedekind complete. Show that:
    (a) If P is an order projection on E, then $|TP| = |T|P$ holds for all $T \in \mathcal{L}_b(E, F)$.
    (b) If Q is an order projection on F, then $|QT| = Q|T|$ holds for all $T \in \mathcal{L}_b(E, F)$.

1.4. Order Continuous Operators

In this section the basic properties of order continuous operators will be studied. Our discussion starts with their definition introduced by T. Ogasawara around 1940; see the work of M. Nakamura [**146**]. Recall that a net $\{x_\alpha\}$ in a Riesz space is order convergent to some vector x, denoted

$x_\alpha \xrightarrow{\ o\ } x$, whenever there exists another net $\{y_\alpha\}$ with the same index set satisfying $|x_\alpha - x| \le y_\alpha \downarrow 0$.

Definition 1.53. *An operator* $T\colon E \to F$ *between two Riesz spaces is said to be:*

(a) **Order continuous,** *if* $x_\alpha \xrightarrow{\ o\ } 0$ *in* E *implies* $Tx_\alpha \xrightarrow{\ o\ } 0$ *in* F.

(b) **σ-order continuous,** *if* $x_n \xrightarrow{\ o\ } 0$ *in* E *implies* $Tx_n \xrightarrow{\ o\ } 0$ *in* F.

It is useful to note that a positive operator $T\colon E \to F$ between two Riesz spaces is order continuous if and only if $x_\alpha \downarrow 0$ in E implies $Tx_\alpha \downarrow 0$ in F (and also if and only if $0 \le x_\alpha \uparrow x$ in E implies $Tx_\alpha \uparrow Tx$ in F.) In the terminology of directed sets a positive operator $T\colon E \to F$ is, of course, order continuous if and only if $D \downarrow 0$ in E implies $T(D) \downarrow 0$ in F. Similar observations hold true for positive σ-order continuous operators.

Lemma 1.54. *Every order continuous operator is order bounded.*

Proof. Let $T\colon E \to F$ be an order continuous operator and let $x \in E^+$. If we consider the order interval $[0, x]$ as a net $\{x_\alpha\}$, where $x_\alpha = \alpha$ for each $\alpha \in [0, x]$, then $x_\alpha \downarrow 0$. So, by the order continuity of T, there exists a net $\{y_\alpha\}$ of F with the same index $[0, x]$ such that $|Tx_\alpha| \le y_\alpha \downarrow 0$. Consequently, if $\alpha \in [0, x]$, then we have $|T\alpha| = |Tx_\alpha| \le y_\alpha \le y_x$, and this shows that $T[0, x]$ is an order bounded subset of F. ∎

A σ-order continuous operator need not be order continuous, as the next example shows.

Example 1.55. Let E be the vector space of all Lebesgue integrable real-valued functions defined on $[0, 1]$. Note that two functions that differ at one point are considered to be different. Under the pointwise ordering (i.e., $f \ge g$ means $f(x) \ge g(x)$ for all $x \in [0, 1]$), E is a Riesz space—in fact, it is a function space. Also, note that $f_\alpha \uparrow f$ holds in E if and only if $f_\alpha(x) \uparrow f(x)$ holds in $\mathbb{R}$ for all $x \in [0, 1]$.

Now define the operator $T\colon E \to \mathbb{R}$ by

$$T(f) = \int_0^1 f(x)\,dx\,.$$

Clearly, T is a positive operator, and from the Lebesgue dominated convergence theorem it easily follows that T is σ-order continuous. However, T is not order continuous.

To see this, note first that if $\mathcal{F}$ denotes the collection of all finite subsets of $[0, 1]$, then the net $\{\chi_\alpha\colon \alpha \in \mathcal{F}\} \subseteq E$ (where χ_α is the characteristic function of α) satisfies $\chi_\alpha \uparrow 1$ (= the constant function one). On the other hand, observe that $T(\chi_\alpha) = 0 \not\to T(1) = 1$. ∎

The order continuous operators have a number of nice characterizations.

Theorem 1.56. *For an order bounded operator $T\colon E \to F$ between two Riesz spaces with F Dedekind complete, the following statements are equivalent.*

(1) *T is order continuous.*

(2) *If $x_\alpha \downarrow 0$ holds in E, then $Tx_\alpha \xrightarrow{\ o\ } 0$ holds in F.*

(3) *If $x_\alpha \downarrow 0$ holds in E, then $\inf\{|Tx_\alpha|\} = 0$ in F.*

(4) *T^+ and T^- are both order continuous.*

(5) *$|T|$ is order continuous.*

Proof. $(1) \implies (2)$ and $(2) \implies (3)$ are obvious.

$(3) \implies (4)$ It is enough to show that T^+ is order continuous. To this end, let $x_\alpha \downarrow 0$ in E. Let $T^+ x_\alpha \downarrow z \geq 0$ in F. We have to show that $z = 0$. Fix some index β and put $x = x_\beta$.

Now for each $0 \leq y \leq x$ and each $\alpha \succeq \beta$ we have

$$0 \leq y - y \wedge x_\alpha = y \wedge x - y \wedge x_\alpha \leq x - x_\alpha,$$

and consequently

$$T(y) - T(y \wedge x_\alpha) = T(y - y \wedge x_\alpha) \leq T^+(x - x_\alpha) = T^+x - T^+x_\alpha,$$

from which it follows that

$$0 \leq z \leq T^+x_\alpha \leq T^+x + |T(y \wedge x_\alpha)| - Ty \qquad (\star)$$

holds for all $\alpha \succeq \beta$ and all $0 \leq y \leq x$. Now since for each fixed vector $0 \leq y \leq x$ we have $y \wedge x_\alpha \downarrow_{\alpha \succeq \beta} 0$, it then follows from our hypothesis that $\inf_{\alpha \succeq \beta}\{|T(y \wedge x_\alpha)|\} = 0$, and hence from $(\star)$ we see that $0 \leq z \leq T^+x - Ty$ holds for all $0 \leq y \leq x$. In view of $T^+x = \sup\{Ty\colon 0 \leq y \leq x\}$, the latter inequality yields $z = 0$, as desired.

$(4) \implies (5)$ The implication follows from the identity $|T| = T^+ + T^-$.

$(5) \implies (1)$ The implication follows easily from the lattice inequality $|Tx| \leq |T|(|x|)$. ∎

The reader can formulate by himself the analogue of Theorem 1.56 for σ-order continuous operators.

The collection of all order continuous operators of $\mathcal{L}_b(E, F)$ will be denoted by $\mathcal{L}_n(E, F)$; the subscript n is justified by the fact that the order continuous operators are also known as normal operators. That is,

$$\mathcal{L}_n(E, F) := \{T \in \mathcal{L}_b(E, F)\colon T \text{ is order continuous}\}.$$

Similarly, $\mathcal{L}_c(E, F)$ will denote the collection of all order bounded operators from E to F that are σ-order continuous. That is,

$$\mathcal{L}_c(E, F) := \{T \in \mathcal{L}_b(E, F) \colon \ T \text{ is } \sigma\text{-order continuous}\}.$$

Clearly, $\mathcal{L}_n(E, F)$ and $\mathcal{L}_c(E, F)$ are both vector subspaces of $\mathcal{L}_b(E, F)$, and moreover $\mathcal{L}_n(E, F) \subseteq \mathcal{L}_c(E, F)$ holds. When F is Dedekind complete T. Ogasawara [156] has shown that both $\mathcal{L}_n(E, F)$ and $\mathcal{L}_c(E, F)$ are bands of $\mathcal{L}_b(E, F)$. The details follow.

Theorem 1.57 (Ogasawara). *If E and F are Riesz spaces with F Dedekind complete, then $\mathcal{L}_n(E, F)$ and $\mathcal{L}_c(E, F)$ are both bands of $\mathcal{L}_b(E, F)$.*

Proof. We shall establish that $\mathcal{L}_n(E, F)$ is a band of $\mathcal{L}_b(E, F)$. That $\mathcal{L}_c(E, F)$ is a band can be proven in a similar manner.

Note first that if $|S| \leq |T|$ holds in $\mathcal{L}_b(E, F)$ with $T \in \mathcal{L}_n(E, F)$, then from Theorem 1.56 it follows that $S \in \mathcal{L}_n(E, F)$. That is, $\mathcal{L}_n(E, F)$ is an ideal of $\mathcal{L}_b(E, F)$.

To see that the ideal $\mathcal{L}_n(E, F)$ is a band, let $0 \leq T_\lambda \uparrow T$ in $\mathcal{L}_b(E, F)$ with $\{T_\lambda\} \subseteq \mathcal{L}_n(E, F)$, and let $0 \leq x_\alpha \uparrow x$ in E. Then for each fixed index λ we have

$$0 \leq T(x - x_\alpha) \leq (T - T_\lambda)(x) + T_\lambda(x - x_\alpha),$$

and $x - x_\alpha \downarrow 0$, in conjunction with $T_\lambda \in \mathcal{L}_n(E, F)$, implies

$$0 \leq \inf_\alpha\{T(x - x_\alpha)\} \leq (T - T_\lambda)(x)$$

for all λ. From $T - T_\lambda \downarrow 0$ we see that $\inf_\alpha\{T(x - x_\alpha)\} = 0$, and hence $T(x_\alpha) \uparrow T(x)$. Thus, $T \in \mathcal{L}_n(E, F)$, and the proof is finished. ∎

Now consider two Riesz spaces E and F with F Dedekind complete. The band of all operators in $\mathcal{L}_b(E, F)$ that are disjoint from $\mathcal{L}_c(E, F)$ will be denoted by $\mathcal{L}_s(E, F)$, i.e., $\mathcal{L}_s(E, F) := \mathcal{L}_c^d(E, F)$, and its nonzero members will be referred to as **singular operators**. Since $\mathcal{L}_b(E, F)$ is a Dedekind complete Riesz space (see Theorem 1.18), it follows from Theorem 1.42 that $\mathcal{L}_c(E, F)$ is a projection band, and so

$$\mathcal{L}_b(E, F) = \mathcal{L}_c(E, F) \oplus \mathcal{L}_s(E, F)$$

holds. In particular, each operator $T \in \mathcal{L}_b(E, F)$ has a unique decomposition $T = T_c + T_s$, where $T_c \in \mathcal{L}_c(E, F)$ and $T_s \in \mathcal{L}_s(E, F)$. The operator T_c is called the σ-**order continuous component** of T, and T_s is called the **singular component** of T. Similarly,

$$\mathcal{L}_b(E, F) = \mathcal{L}_n(E, F) \oplus \mathcal{L}_\sigma(E, F),$$

where $\mathcal{L}_\sigma(E, F) := \mathcal{L}_n^d(E, F)$. Thus, every operator $T \in \mathcal{L}_b(E, F)$ also has a unique decomposition $T = T_n + T_\sigma$, where $T_n \in \mathcal{L}_n(E, F)$ and $T_\sigma \in \mathcal{L}_\sigma(E, F)$. The operator T_n is called the **order continuous component** of T.

The next examples shows that $\mathcal{L}_c(E, F) = \{0\}$ is possible.

Example 1.58. For each $1 < p < \infty$ we have

$$\mathcal{L}_c(C[0, 1], L_p[0, 1]) = \{0\}.$$

That is, the zero operator is the only σ-order continuous positive operator from $C[0, 1]$ to $L_p[0, 1]$.

To establish this, we need to show first that the only positive σ-order continuous operator from $C[0, 1]$ to $\mathbb{R}$ is the zero operator. To this end, let $\phi \colon C[0, 1] \to \mathbb{R}$ be a positive σ-order continuous operator.

Let $\{r_1, r_2, \ldots\}$ be an enumeration of all rational numbers of $[0, 1]$. For each pair $m, n \in \mathbb{N}$ choose some $x_{m,n} \in C[0, 1]$ such that:

(a) $0 \leq x_{m,n}(t) \leq 1$ for all $t \in [0, 1]$.

(b) $x_{m,n}(r_n) = 1$.

(c) $x_{m,n}(t) = 0$ for all $t \in [0, 1]$ with $|t - r_n| > \frac{1}{2^{n+m}}$.

Put $y_{m,n} = \bigvee_{i=1}^n x_{m,i}$, and note that for each fixed m we have $y_{m,n} \uparrow_n$ in $C[0, 1]$. In view of $y_{m,n}(r_n) = 1$, it follows that $y_{m,n} \uparrow_n 1$ ($=$ the constant function one). Since ϕ is a positive σ-order continuous operator, we see that $\phi(y_{m,n}) \uparrow_n \phi(1)$ holds in $\mathbb{R}$ for each fixed m.

Put $\epsilon > 0$. For each m choose some $n_m \in \mathbb{N}$ with $\phi(1) - \phi(y_{m,n_m}) < \frac{1}{2^m}\epsilon$, and then put $z_n = \bigwedge_{m=1}^n y_{m,n_m}$. Clearly, $z_n \downarrow$ holds in $C[0, 1]$, and since each set $\{t \in [0, 1]\colon y_{m,n}(t) > 0\}$ has Lebesgue measure less that $\frac{1}{2^m}$, it follows that $z_n \downarrow 0$. Now the inequalities

$$\begin{aligned} 0 \; \leq \; \phi(1) - \phi(z_n) &= \phi(1 - z_n) = \phi\left(\bigvee_{m=1}^n (1 - y_{m,n_m})\right) \\ &\leq \phi\left(\sum_{m=1}^n (1 - y_{m,n_m})\right) = \sum_{m=1}^n \phi(1 - y_{m,n_m}) < \epsilon, \end{aligned}$$

in conjunction with $\phi(z_n) \downarrow 0$, imply $0 \leq \phi(1) \leq \epsilon$ for all $\epsilon > 0$. Therefore, $\phi(1) = 0$, and from this we see that $\phi = 0$.

Now let $T\colon C[0, 1] \to L_p[0, 1]$ be a positive σ-order continuous operator. Then for each fixed $0 \leq g \in L_q[0, 1]$, where $\frac{1}{p} + \frac{1}{q} = 1$, the positive operator $\psi\colon C[0, 1] \to \mathbb{R}$ defined by

$$\psi(f) = \int_0^1 g(t) [Tf(t)] \, dt$$

is σ-order continuous. Hence, by the previous case $\int_0^1 g(t)[Tf(t)]\,dt = 0$ for all $g \in L_q[0,1]$ and all $f \in C[0,1]$. The latter easily implies $T = 0$, as claimed. ∎

If x and y are vectors in a Riesz space and ϵ is any real number, then from the identity $x - y = (1 - \epsilon)x + (\epsilon x - y)$ we see that

$$x - y \le (1 - \epsilon)x + (\epsilon x - y)^+ .$$

This simple inequality is useful in many contexts and was introduced by T. Andô [125, Note XIV]. In the sequel it will be referred to as **Andô's inequality**.

The σ-order continuous and order continuous components of a positive operator are described by formulas as follows.

Theorem 1.59. *Let E and F be two Riesz spaces with F Dedekind complete. If $T\colon E \to F$ is a positive operator, then*

(1) $T_c(x) = \inf\{\sup T(x_n)\colon\ 0 \le x_n \uparrow x\}$, *and*

(2) $T_n(x) = \inf\{\sup T(x_\alpha)\colon\ 0 \le x_\alpha \uparrow x\}$

hold for each $x \in E^+$.[4]

Proof. We prove the formula for T_n and leave the identical arguments for T_c to the reader.

For each positive operator $S\colon E \to F$ define $S^\star\colon E^+ \to F^+$ by

$$S^\star(x) = \inf\{\sup S(x_\alpha)\colon\ 0 \le x_\alpha \uparrow x\}, \quad x \in E^+ .$$

Clearly, $0 \le S^\star(x) \le S(x)$ holds for all $x \in E^+$, and $S^\star(x) = S(x)$ whenever $S \in \mathcal{L}_n(E,F)$. Moreover, it is not difficult to see that $S^\star$ is additive on E^+, and hence (by Theorem 1.10), $S^\star$ extends to a positive operator from E to F. On the other hand, it is easy to see that $S \mapsto S^\star$, from $\mathcal{L}_b^+(E,F)$ to $\mathcal{L}_b^+(E,F)$, is likewise additive, i.e., $(S_1 + S_2)^\star = S_1^\star + S_2^\star$ holds, and hence $S \mapsto S^\star$ defines a positive operator from $\mathcal{L}_b(E,F)$ to $\mathcal{L}_b(E,F)$. From the inequality $0 \le S^\star \le S$ we also see that $S \mapsto S^\star$ is order continuous, i.e., $S_\alpha \downarrow 0$ in $\mathcal{L}_b(E,F)$ implies $S_\alpha^\star \downarrow 0$.

---

[4] These formulas have an interesting history. When $F = \mathbb{R}$, the formula for T_c is due to W. A. J. Luxemburg and A. C. Zaanen [130, Note VI, Theorem 20.4, p. 663], and for the same case, the formula for T_n is due to W. A. J. Luxemburg [125]. When $\mathcal{L}_n(F, \mathbb{R})$ separates the points of F, the formulas were established by C. D. Aliprantis [6]. In 1975 A. R. Schep announced the validity of the formulas in the general setting and later published his proof in [176]. An elementary proof for the T_c formula also was obtained by P. van Eldik in [59]. The proof presented here is due to the authors [12].

Now let $T\colon E \to F$ be a fixed positive operator. It is enough to show that $T^\star$ is order continuous. If this is done, then the inequality $T^\star \le T$ implies $T^\star = (T^\star)_n \le T_n$, and since $T_n \le T^\star$ is trivially true, we see that $T_n = T^\star$. To this end, let $0 \le y_\lambda \uparrow y$ in E. We must show that $T^\star(y - y_\lambda) \downarrow 0$.

Fix $0 < \epsilon < 1$, and let T_λ denote the operator defined in Theorem 1.28 that agrees with T on the ideal generated by $(\epsilon y - y_\lambda)^+$ and vanishes on $(\epsilon y - y_\lambda)^-$. Clearly, $T \ge T_\lambda {\downarrow} \ge 0$, and $T_\lambda(y_\lambda - \epsilon y)^+ = 0$ holds for all λ. Let $T_\lambda \downarrow R$ in $\mathcal{L}_b(E, F)$. From $0 \le (y_\lambda - \epsilon y)^+ \uparrow (1 - \epsilon)y$ and $R(y_\lambda - \epsilon y)^+ = 0$ for each λ, we see that $R^\star(y) = 0$. From Andô's inequality

$$0 \le y - y_\lambda \le (1 - \epsilon) + (\epsilon - y_\lambda)^+,$$

it follows that

$$0 \le T^\star(y - y_\lambda) \le (1 - \epsilon)T^\star(y) + T^\star(\epsilon y - y_\lambda)^+. \tag{$\dagger$}$$

Now since $0 \le x \le (\epsilon y - y_\lambda)^+$ implies $T(x) = T_\lambda(x)$, we see that

$$
\begin{aligned}
T^\star(\epsilon y - y_\lambda)^+ &= \inf\big\{\sup T(x_\alpha)\colon\ 0 \le x_\alpha \uparrow (\epsilon y - y_\lambda)^+\big\} \\
&= \inf\big\{\sup T_\lambda(x_\alpha)\colon\ 0 \le x_\alpha \uparrow (\epsilon y - y_\lambda)^+\big\} \\
&= T_\lambda^\star(\epsilon y - y_\lambda)^+ \le T_\lambda^\star(y),
\end{aligned}
$$

and so, substituting into ($\dagger$), we obtain

$$0 \le T^\star(y - y_\lambda) \le (1 - \epsilon)T^\star(y) + T_\lambda^\star(y). \tag{$\dagger\dagger$}$$

From $T_\lambda \downarrow R$ and the order continuity of $S \mapsto S^\star$, it follows that $T_\lambda^\star \downarrow R^\star$. In particular, $T_\lambda^\star(y) \downarrow R^\star(y) = 0$, and so from ($\dagger\dagger$) we see that

$$0 \le \inf_\lambda\big\{T^\star(y - y_\lambda)\big\} \le (1 - \epsilon)T^\star(y)$$

holds for all $0 < \epsilon < 1$. Hence, $T^\star(y - y_\lambda) \downarrow 0$, as desired. ∎

Consider an order bounded operator $T\colon E \to F$ between two Riesz spaces with F Dedekind complete. Then the **null ideal** N_T of T is defined by

$$N_T := \big\{x \in E\colon\ |T|(|x|) = 0\big\}.$$

Note that N_T is indeed an ideal of E. The disjoint complement of N_T is referred to as the **carrier** of T and is denoted by C_T. That is,

$$C_T := N_T^d = \big\{x \in E\colon\ x \perp N_T\big\}.$$

Clearly, $|T|$ is strictly positive on C_T, i.e., $0 < x \in C_T$ implies $0 < |T|(x)$.

When an order bounded operator is, in addition, order continuous, then it is easy to see that its null ideal is a band. However, the converse is false.

Example 1.60. Consider an infinite set X, and let $X_\infty = X \cup \{\infty\}$ be the one-point compactification of X considered equipped with the discrete topology. Thus, a function $f \colon X \to \mathbb{R}$ belongs to $C(X_\infty)$ if and only if there exists some constant c (depending upon f) such that for each $\epsilon > 0$ we have $|f(x) - c| < \epsilon$ for all but a finite number of x, in which case $c = f(\infty)$.

Now fix a countable subset $\{x_1, x_2, \ldots\}$ of X, and then define the operator $T \colon C(X_\infty) \to \mathbb{R}$ by

$$T(f) = f(\infty) + \sum_{n=1}^{\infty} 2^{-n} f(x_n).$$

Clearly, T is a positive operator, and

$$N_T = \{f \in C(X_\infty) \colon\; f(x_n) = 0 \;\text{ for }\; n = 1, 2, \ldots\}.$$

Since $f_\alpha \uparrow f$ holds in $C(X_\infty)$ if and only if $f_\alpha(x) \uparrow f(x)$ holds in $\mathbb{R}$ for all $x \in X$ (why?), it follows that N_T is a band of $C(X_\infty)$. On the other hand, we claim that T is not order continuous.

To see this, consider the net $\{\chi_\alpha\} \subseteq C(X_\infty)$, where α runs over the collection of all finite subsets of X. Then $0 \le \chi_\alpha \uparrow \mathbf{1}$ holds in $C(X_\infty)$, while $T(\chi_\alpha) \not\to T(\mathbf{1})$. Also, it is interesting to observe that if X is countable, then T is necessarily σ-order continuous! ∎

In terms of null ideals the order and σ-order continuous operators are characterized as follows. (Recall that an ideal A of a Riesz space is said to be a σ-**ideal** whenever $\{x_n\} \subseteq A$ and $0 \le x_n \uparrow x$ imply $x \in A$.)

Theorem 1.61. *For an order bounded operator $T \colon E \to F$ between two Riesz spaces with F Dedekind complete we have the following.*

(1) *T is order continuous if and only if the null ideal N_S is a band for every operator S in the ideal $\mathcal{A}_T$ generated by T is $\mathcal{L}_b(E, F)$.*

(2) *T is σ-order continuous if and only if the null ideal N_S is a σ-ideal for each $S \in \mathcal{A}_T$.*

Proof. We shall only prove (1) since the proof of (2) is similar. The "only if" part follows immediately from Theorem 1.56. For the "if" part (in view of Theorem 1.56) we can assume that $T \ge 0$. Let $0 \le x_\alpha \uparrow x$ in E, and let $0 \le Tx_\alpha \uparrow y \le Tx$ in F. We must show that $y = Tx$ holds.

To this end, let $0 < \epsilon < 1$. For each α, let T_α be the operator given by Theorem 1.28 that agrees with T on the ideal generated by $(\epsilon x - x_\alpha)^+$ and vanishes on $(\epsilon x - x_\alpha)^-$. Clearly, $T \ge T_\alpha \downarrow \ge 0$, and $T_\alpha(\epsilon x - x_\alpha)^- = 0$ for each α. Let $T_\alpha \downarrow S \ge 0$ in $\mathcal{L}_b(E, F)$, and note that $S \in \mathcal{A}_T$. Also, $S(\epsilon x - x_\alpha)^- = 0$ holds for each α, and so $\{(\epsilon x - x_\alpha)^-\} \subseteq N_S$. On the other hand, $0 \leq (\epsilon x - x_\alpha)^- \uparrow (1 - \epsilon)x$ holds in E, and hence, since by our hypothesis N_S is a band, $x \in N_S$. Therefore, $Sx = 0$. Now the relation

$$0 \leq T(\epsilon x - x_\alpha)^+ = T_\alpha(\epsilon x - x_\alpha)^+ \leq T_\alpha(x),$$

in conjunction with Andô's inequality $0 \leq x - x_\alpha \leq (1 - \epsilon)x + (\epsilon x - x_\alpha)^+$, yields

$$0 \leq Tx - y \leq T(x - x_\alpha) \leq (1 - \epsilon)Tx + T(\epsilon x - x_\alpha)^+ \leq (1 - \epsilon)Tx + T_\alpha(x).$$

Taking into consideration that $T_\alpha(x) \downarrow S(x) = 0$, the preceding inequality yields $0 \leq Tx - y \leq (1 - \epsilon)Tx$ for all $0 < \epsilon < 1$. Hence, $y = Tx$ holds, as required. ∎

To illustrate the previous theorem, consider the operator $T \colon C(X_\infty) \to R$ of Example 1.60 defined by

$$T(f) = f(\infty) + \sum_{n=1}^{\infty} 2^{-n} f(x_n).$$

As we have seen before, $N_T = \{f \in C(X_\infty) \colon f(x_n) = 0 \text{ for } n = 1, 2, \dots\}$, and this shows that N_T is a band of $C(X_\infty)$. On the other hand, if $S \colon C(X_\infty) \to \mathbb{R}$ is defined by

$$S(f) = f(\infty),$$

then S is a positive operator satisfying $0 \leq S \leq T$. Clearly, the null ideal of S is given by $N_S = \{f \in C(X_\infty) \colon f(\infty) = 0\}$. Now note that the net $\{\chi_\alpha\}$ of all characteristic functions of the finite subsets of X satisfies $\{\chi_\alpha\} \subseteq N_S$ and $\chi_\alpha \uparrow 1$. Since $1 \notin N_S$, we see that N_S is not a band of $C(X_\infty)$, in accordance with part (1) of Theorem 1.61.

Consider two Riesz spaces E and F with F Dedekind complete. An operator $T \in \mathcal{L}_b(E, F)$ is said to have **zero carrier** whenever $C_T = \{0\}$ (or, equivalently, whenever N_T is order dense in E). It is easy to check that the zero operator is the only order continuous operator with zero carrier. On the other hand, If $T \in \mathcal{L}_b(E, F)$ has a zero carrier, then $T \perp \mathcal{L}_n(E, F)$, that is, $T \in \mathcal{L}_\sigma(E, F)$. (To see this, write $T = T_n + T_\sigma$, and note that $|T| = |T_n| + |T_\sigma|$; see Exercise 2 of Section 1.1. Therefore, $N_T \subseteq N_{T_n}$ holds, and so by the order denseness of N_T we see that $N_{T_n} = E$. That is, $T_n = 0$ and so $T = T_\sigma \in \mathcal{L}_\sigma(E, F)$.) From $|T + S| \leq |T| + |S|$, it follows that $N_T \cap N_S \subseteq N_{T+S}$, and using the fact that the intersection of two order dense ideals is an order dense ideal (why?), we see that the operators of $\mathcal{L}_b(E, F)$ with zero carriers form an ideal. The next theorem tells us that this ideal is always order dense in $\mathcal{L}_\sigma(E, F)$.

Theorem 1.62. *Let E and F be two Riesz spaces with F Dedekind complete. Then the ideal*

$$\{T \in \mathcal{L}_b(E, F) \colon C_T = \{0\}\}$$

is order dense in $\mathcal{L}_\sigma(E, F)$.

Proof. We have mentioned before that the set $\{T \in \mathcal{L}_b(E, F): C_T = \{0\}\}$ is an ideal that is included in $\mathcal{L}_\sigma(E, F)$. For the order denseness assume that $0 < T \in \mathcal{L}_\sigma(E, F)$.

Since T is not order continuous, there exists (by Theorem 1.61) an operator $0 < S \leq T$ such that N_S is not a band. Denote by B the band generated by N_S. Let R be the operator determined by Theorem 1.28 such that $R = S$ on B and $R = 0$ on B^d. Clearly, $N_S \subseteq N_R$ and $0 < R \leq S$. On the other hand, since $R = 0$ holds on $B^d = N_S^d = C_S$, we see that $N_S \oplus C_S \subseteq N_R$, and this (in view of Theorem 1.36) shows that N_R is order dense in E. Thus, R has zero carrier. Now to complete the proof note that $0 < R \leq T$ holds. ∎

The preceding theorem shows that $\mathcal{L}_\sigma(E, F) = \{0\}$ holds (or, equivalently, $\mathcal{L}_b(E, F) = \mathcal{L}_n(E, F)$) if and only if every nonzero operator from E to F has a nonzero carrier. Thus, in view of Theorem 1.61 we see that the following theorem of the authors [12] holds.

Theorem 1.63 (Aliprantis–Burkinshaw). *For a pair of Riesz spaces E and F with F Dedekind complete, the following statements are equivalent.*

(1) *Every order bounded operator from E to F is order continuous, i.e., $\mathcal{L}_b(E, F) = \mathcal{L}_n(E, F)$.*

(2) *Every nonzero order bounded operator from E to F has a nonzero carrier.*

(3) *The null ideal of every order bounded operator from E to F is a band.*

The next result tells us when a positive operator is order continuous on a given ideal.

Theorem 1.64. *Let $T: E \to F$ be a positive operator between two Riesz spaces with F Dedekind complete, and let A be an ideal of E. Then the operator T is order (resp. σ-order) continuous on A if and only if T_A is an order (resp. σ-order) continuous operator.*

Proof. We establish the result for the "order continuous" case; the "σ-order continuous" case can be proven in a similar fashion. Recall that for each $x \in E^+$ the operator T_A is given (according to Theorem 1.28) by

$$T_A(x) = \sup\{T(y): \ y \in A \text{ and } 0 \leq y \leq x\}.$$

Since $T_A = T$ holds on A, it should be obvious that if T_A is an order continuous operator, then T must be order continuous on A. For the converse, assume that T is order continuous on A, and let $0 \leq x_\alpha \uparrow x$ in E. Let

$T_A(x_\alpha) \uparrow z \leq T_A(x)$. Now fix $y \in A \cap [0, x]$. Then $0 \leq y \wedge x_\alpha \uparrow y$ holds in A, and so $T(y \wedge x_\alpha) \uparrow T(y)$ holds in F. From

$$T(y \wedge x_\alpha) = T_A(y \wedge x_\alpha) \leq z \leq T_A(x),$$

it follows that $T(y) \leq z \leq T_A(x)$ holds for all $y \in A \cap [0, x]$. Hence,

$$T_A(x) = \sup T(A \cap [0, x]) \leq z \leq T_A(x),$$

and so $z = T_A(x)$, proving that T_A is an order continuous operator. ∎

The final result of this section is an extension theorem for positive order continuous operators and is due to A. I. Veksler [188].

Theorem 1.65 (Veksler). *Let G be an order dense majorizing Riesz subspace of a Riesz space E, and let F be Dedekind complete. If $T: G \to F$ is a positive order continuous operator, then the formula*

$$T(x) = \sup\{T(y): \ y \in G \ and \ 0 \leq y \leq x\}, \quad x \in E^+,$$

defines a unique order continuous linear extension of T to all of E.

Proof. Since G majorizes E, it is easy to see that

$$S(x) = \sup\{T(y): \ y \in G \ \text{and} \ 0 \leq y \leq x\}$$

exists in F for each $x \in E^+$. Also, note that if $\{x_\alpha\} \subseteq G$ satisfies $0 \leq x_\alpha \uparrow x$, then $T(x_\alpha) \uparrow S(x)$ holds. Indeed, if $0 \leq y \in G$ satisfies $0 \leq y \leq x$, then $0 \leq x_\alpha \wedge y \uparrow y$ holds in G, and so by the order continuity of $T: G \to F$ we see that

$$T(y) = \sup\{T(x_\alpha \wedge y)\} \leq \sup\{T(x_\alpha)\} \leq S(x).$$

This easily implies that $T(x_\alpha) \uparrow S(x)$.

Now let $x, y \in E^+$. Pick two nets $\{x_\alpha\}$ and $\{y_\beta\}$ of G^+ with $0 \leq x_\alpha \uparrow x$ and $0 \leq y_\beta \uparrow y$ (see Theorem 1.34). Then $0 \leq x_\alpha + y_\beta \uparrow x + y$ holds, and so by the above discussion

$$T(x_\alpha) + T(y_\beta) = T(x_\alpha + y_\beta) \uparrow S(x + y).$$

From $T(x_\alpha) \uparrow S(x)$ and $T(y_\beta) \uparrow S(y)$, we get $S(x + y) = S(x) + S(y)$. That is, $S: E^+ \to F^+$ is additive, and thus by Theorem 1.10 it extends uniquely to a positive operator from E to F. Clearly, S is an extension of T.

Finally, it remains to be shown that S is order continuous. To this end, let $0 \leq x_\alpha \uparrow x$ in E. Put

$$D = \{y \in G^+: \ \text{there exists some } \alpha \text{ with } y \leq x_\alpha\},$$

and note that $\sup T(D) \leq \sup\{S(x_\alpha)\} \leq S(x)$ holds in F. Since G is order dense in E, it is easy to see that $D \uparrow x$ holds. Thus, by the above discussion $\sup T(D) = S(x)$, and so $S(x_\alpha) \uparrow S(x)$, proving that S is order continuous. The proof of the theorem is now complete. ∎

Exercises

1. A Riesz space is said to have the **countable sup property**, if whenever an arbitrary subset D has a supremum, then there exists an at most countable subset C of D with $\sup C = \sup D$.
   (a) Show that if F is an Archimedean Riesz space with the countable sup property and $T\colon E \to F$ is a **strictly positive operator** (i.e., $x > 0$ implies $Tx > 0$), then E likewise has the countable sup property.
   (b) Let $T\colon E \to F$ be a positive operator between two Riesz spaces with E having the countable sup property. Then show that T is order continuous if and only if T is σ-order continuous.

2. Let E be Dedekind σ-complete, and let F be **super Dedekind complete** (i.e., let F be Dedekind complete with the countable sup property), and let $T\colon E \to F$ be a positive σ-order continuous operator. Show that:
   (a) C_T is a super Dedekind complete Riesz space and that T restricted to C_T is strictly positive and order continuous.
   (b) C_T is a projection band.
   (c) T is order continuous if and only if N_T is a band.

3. Let E and F be two Riesz spaces with F Dedekind complete. Consider the band $\mathcal{L}_{c\sigma}(E,F) := \mathcal{L}_c(E,F) \cap \mathcal{L}_\sigma(E,F)$, and note that

   $$\mathcal{L}_b(E,F) = \mathcal{L}_n(E,F) \oplus \mathcal{L}_{c\sigma}(E,F) \oplus \mathcal{L}_s(E,F).$$

   Thus, every operator $T \in \mathcal{L}_b(E,F)$ has a unique decomposition of the form $T = T_n + T_{c\sigma} + T_s$, where $T_n \in \mathcal{L}_n(E,F)$, $T_{c\sigma} \in \mathcal{L}_{c\sigma}(E,F)$, and $T_s \in \mathcal{L}_s(E,F)$. Clearly, $T_c = T_n + T_{c\sigma}$ and $T_\sigma = T_{c\sigma} + T_s$ hold.
   If F is super Dedekind complete and $T \in \mathcal{L}_c(E,F)$, then prove the following statements.
   (a) $T \in \mathcal{L}_{c\sigma}(E,F)$ if and only if $C_T = \{0\}$ (or, equivalently, if and only if N_T is order dense in E).
   (b) $N_T \oplus C_T \subseteq N_{T_{c\sigma}}$.
   (c) The largest ideal of E on which T is order continuous is the order dense ideal $N_{T_{c\sigma}}$.

4. Let $T\colon E \to F$ be a positive operator between two Riesz spaces with F Dedekind complete. Then show that:
   (a) In the formula

   $$T_n(x) = \inf\{\sup T(x_\alpha)\colon \ 0 \le x_\alpha \uparrow x\},$$

   the greatest lower bound is attained for each $x \in E^+$ if and only if N_{T_σ} is order dense in E.
   (b) In the formula

   $$T_c(x) = \inf\{\sup T(x_n)\colon \ 0 \le x_n \uparrow x\},$$

   the greatest lower bound is attained for each $x \in E^+$ if and only if N_{T_s} is super order dense in E. (Recall that an ideal A in a Riesz space E is said to be **super order dense** whenever for each $x \in E^+$ there exists a sequence $\{x_n\} \subseteq A$ such that $0 \le x_n \uparrow x$.)

5. Let E and F be two Riesz spaces with F Dedekind complete. Show that for each $T \in \mathcal{L}_b(E, F)$ the ideal N_{T_σ} (resp. N_{T_s}) is the largest ideal of E on which T is order (resp. σ-order) continuous.

6. Consider the operator T of Example 1.60. Show that T is σ-order continuous if and only if X is an uncountable set.

7. For a pair of Riesz spaces E and F with F Dedekind complete show that the following statements are equivalent.
   (a) Every order bounded operator from E to F is σ-order continuous.
   (b) The null ideal of every order bounded operator from E to F is a σ-ideal.

8. Let $T\colon E \to E$ be an order continuous positive operator on a Riesz space, and let $\{T_\alpha\}$ be a net of positive order continuous operators from E to E satisfying $T_\alpha(x) \uparrow T(x)$ in E for each $x \in E^+$. Show that:
   (a) If $0 \le x_\lambda \uparrow x$ in E, then $T_\alpha(x_\lambda) \uparrow_{\alpha,\lambda} T(x)$ holds in E.
   (b) If $x \in E^+$, then $T_\alpha^k(x) \uparrow T^k(x)$ holds in E for each k.
   Also, establish the sequential analogues of the above statements.

9. Let $T\colon E \to F$ be a positive operator between two Riesz spaces with F Dedekind complete. Then show that the components T_σ and T_s of T for each $x \in E^+$ are given by the formulas

$$T_\sigma(x) = \sup\{\inf T(x_\alpha)\colon \ x \ge x_\alpha \downarrow 0\}$$

and

$$T_s(x) = \sup\{\inf T(x_n)\colon \ x \ge x_n \downarrow 0\}.$$

10. Show that an order bounded operator $T\colon E \to F$ between two Riesz spaces with F Dedekind complete is order continuous if and only if $T \perp S$ holds for each operator $S \in \mathcal{L}_b(E, F)$ with $C_S = \{0\}$.

11. As usual, if $\{x_\alpha\}$ is an order bounded net in a Dedekind complete Riesz space, then we define

$$\limsup x_\alpha := \bigwedge_\alpha \bigvee_{\beta \succeq \alpha} x_\beta \quad \text{and} \quad \liminf x_\alpha := \bigvee_\alpha \bigwedge_{\beta \succeq \alpha} x_\beta .$$

   (a) Show that in a Dedekind complete Riesz space an order bounded net $\{x_\alpha\}$ satisfies $x_\alpha \xrightarrow{o} x$ if and only if $x = \limsup x_\alpha = \liminf x_\alpha$.
   (b) If $T\colon E \to F$ is a positive operator between two Riesz spaces with F Dedekind complete, then show that

$$T_c(x) = \inf\{\liminf T(x_n)\colon \ 0 \le x_n \le x \text{ and } x_n \xrightarrow{o} x\}$$

and

$$T_n(x) = \inf\{\liminf T(x_\alpha)\colon \ 0 \le x_\alpha \le x \text{ and } x_\alpha \xrightarrow{o} x\}$$

   hold for each $x \in E^+$.

12. For two Riesz spaces E and F with F Dedekind complete establish the following:
   (a) If A is an ideal of E, then its **annihilator**

$$A^\circ := \{T \in \mathcal{L}_b(E, F)\colon \ T = 0 \text{ on } A\}$$

is a band of $\mathcal{L}_b(E, F)$.

(b) If $\mathcal{A}$ is an ideal of $\mathcal{L}_b(E, F)$, then its **inverse annihilator**

$$^\circ\mathcal{A} := \{x \in E \colon\ T(x) = 0 \text{ for each } T \in \mathcal{A}\}$$

is an ideal of E.

(c) Every order bounded operator from E to F is order continuous (i.e., $\mathcal{L}_b(E, F) = \mathcal{L}_n(E, F)$ holds) if and only if for every order dense ideal A of E we have $A^\circ = \{0\}$.

13. Consider two Riesz spaces E and F with F Dedekind complete. As usual, we say that $\mathcal{L}_b(E, F)$ **separates the points** of E whenever for each $x \neq 0$ in E there exists some $T \in \mathcal{L}_b(E, F)$ with $T(x) \neq 0$.

Show that if $\mathcal{L}_b(E, F)$ separates the points of E and $(^\circ\mathcal{B})^\circ = \mathcal{B}$ holds for each band $\mathcal{B}$ of $\mathcal{L}_b(E, F)$ (for notation see the preceding exercise), then every order bounded operator from E to F is order continuous.

1.5. Positive Linear Functionals

Let E be a Riesz space. A linear functional $f \colon E \to \mathbb{R}$ is said to be **positive** whenever $f(x) \geq 0$ holds for each $x \in E^+$. Also, a linear functional f is called **order bounded** if f maps order bounded subsets of E to bounded subsets of $\mathbb{R}$. The vector space $E^\sim$ of all order bounded linear functionals on E is called the **order dual** of E, i.e., $E^\sim = \mathcal{L}_b(E, \mathbb{R})$. Since $\mathbb{R}$ is a Dedekind complete Riesz space, it follows at once from Theorem 1.18 that $E^\sim$ is precisely the vector space generated by the positive linear functionals. Moreover, $E^\sim$ is a Dedekind complete Riesz space. Recall that $f \geq g$ in $E^\sim$ means $f(x) \geq g(x)$ for all $x \in E^+$. Also, note that if $f, g \in E^\sim$ and $x \in E^+$, then according to Theorem 1.18 we have:

(1) $f^+(x) = \sup\{f(y) \colon\ 0 \leq y \leq x\}$.

(2) $f^-(x) = \sup\{-f(y) \colon\ 0 \leq y \leq x\}$.

(3) $|f|(x) = \sup\{|f(y)| \colon\ |y| \leq x\}$.

(4) $[f \vee g](x) = \sup\{f(y) + g(z) \colon\ y, z \in E^+ \text{ and } y + z = x\}$.

(5) $[f \wedge g](x) = \inf\{f(y) + g(z) \colon\ y, z \in E^+ \text{ and } y + z = x\}$.

Observe that from formula (5) the following important characterization of disjointness in $E^\sim$ holds: For $f, g \in E^\sim$ we have $f \perp g$ if and only if for each $\epsilon > 0$ and each $x \in E^+$ there exist $y, z \in E^+$ with $y + z = x$ and $|f|(y) < \epsilon$ and $|g|(z) < \epsilon$.

The order dual $E^\sim$ may happen to be trivial. For instance, if $0 < p < 1$, then it has been shown by M. M. Day that the Riesz space $E = L_p[0, 1]$ satisfies $E^\sim = \{0\}$; see our book [**7**, Theorem 5.24, p. 128]. In this book, Riesz spaces with trivial order dual will be of little interest. As a matter of fact, we are interested in Riesz spaces whose order duals separate the points of the spaces. Recall that the expression $E^\sim$ **separates the points** of E means that for each $x \neq 0$ there exists some $f \in E^\sim$ with $f(x) \neq 0$. Since E is a Riesz space, it is easy to see that $E^\sim$ separates the points of E if and only if for each $0 < x \in E$ there exists some $0 < f \in E^\sim$ with $f(x) \neq 0$.

Theorem 1.66. *If $E^\sim$ separates the points of the Riesz space E, then a vector $x \in E$ satisfies $x \geq 0$ if and only if $f(x) \geq 0$ holds for all $0 \leq f \in E^\sim$.*

Proof. Clearly, if $x \geq 0$ holds, then $f(x) \geq 0$ likewise holds for every $0 \leq f \in E^\sim$.

For the converse, assume that some vector $x \in E$ satisfies $f(x) \geq 0$ for all $0 \leq f \in E^\sim$. If $0 \leq f \in E^\sim$ is fixed, then by Theorem 1.23 there exists some $0 \leq g \leq f$ with $f(x^-) = -g(x)$. Since by our hypothesis $g(x) \geq 0$ holds, it follows that $0 \leq f(x^-) = -g(x) \leq 0$, and so $f(x^-) = 0$ holds for all $0 \leq f \in E^\sim$. Since $E^\sim$ separates the points of E, we see that $x^- = 0$. Consequently, $x = x^+ - x^- = x^+ \geq 0$ holds, and the proof is finished. ∎

Besides the order dual of a Riesz space, we shall need to consider the bands of order continuous and σ-order continuous linear functionals.

Let E be a Riesz space. The vector space $\mathcal{L}_n(E, \mathbb{R})$ of all order continuous linear functionals on E will be denoted by $E_n^\sim$. Similarly, the vector space $\mathcal{L}_c(E, \mathbb{R})$ of all σ-order continuous linear functionals on E will be denoted by $E_c^\sim$. That is,

$$E_n^\sim := \mathcal{L}_n(E, \mathbb{R}) \quad \text{and} \quad E_c^\sim := \mathcal{L}_c(E, \mathbb{R}).$$

Note that a positive linear functional f on E is order continuous if and only if $x_\alpha \downarrow 0$ in E implies $f(x_\alpha) \downarrow 0$ in $\mathbb{R}$. Likewise, f is σ-order continuous if and only if for every sequence $\{x_n\}$ with $x_n \downarrow 0$ we have $f(x_n) \downarrow 0$ in $\mathbb{R}$. Clearly, we have

$$E_n^\sim \subseteq E_c^\sim \subseteq E^\sim.$$

By Theorem 1.57 both $E_c^\sim$ and $E_n^\sim$ are bands of $E^\sim$. The band $E_n^\sim$ will be referred to as the **order continuous dual** of E, and the band $E_c^\sim$ as the **σ-order continuous dual** of E.

Here are two examples of Riesz spaces and their duals. (For a justification of their duals see Section 4.1.)

(1) Let $1 \leq p < \infty$ and $\frac{1}{p} + \frac{1}{q} = 1$.
   (a) If $E = \ell_p$, then $E^\sim = E_c^\sim = E_n^\sim = \ell_q$; and
   (b) if $E = L_p[0, 1]$, then $E^\sim = E_c^\sim = E_n^\sim = L_q[0, 1]$.
(2) Consider $E = C[0, 1]$. Then $E_c^\sim = E_n^\sim = \{0\}$, and $E^\sim$ is the Riesz space of all regular Borel measures on $[0, 1]$.

Recall that the null ideal of an arbitrary linear functional $f \in E^\sim$ is the ideal $N_f := \{x \in E: |f|(|x|) = 0\}$, and its carrier is the band $C_f := N_f^d$.

H. Nakano [**150**, Theorem 20.1, p. 74] has shown that two linear function-als in $E_n^\sim$ are disjoint if and only if their carriers are disjoint sets. This remarkable result is stated next.

Theorem 1.67 (Nakano). *If E is Archimedean, then for a pair $f, g \in E_n^\sim$ the following statements are equivalent.*

(1) $f \perp g$.

(2) $C_f \subseteq N_g$.

(3) $C_g \subseteq N_f$.

(4) $C_f \perp C_g$.

Proof. Without loss of generality we can assume that $0 \leq f, g \in E_n^\sim$.

(1) $\Longrightarrow$ (2) Let $0 \leq x \in C_f = N_f^d$, and let $\epsilon > 0$. In view of $f \wedge g = 0$, there exists a sequence $\{x_n\} \subseteq E^+$ satisfying

$$0 \leq x_n \leq x \quad \text{and} \quad f(x_n) + g(x - x_n) < 2^{-n}\epsilon \text{ for all } n.$$

Put $y_n = \bigwedge_{i=1}^n x_i$, and note that $y_n \downarrow 0$ in E. Indeed, if $0 \leq y \leq y_n$ holds for all n, then $0 \leq f(y) \leq f(y_n) < 2^{-n}\epsilon$ also holds for all n, and consequently $f(y) = 0$. Thus, $y \in C_f \cap N_f = \{0\}$, and so $y = 0$.

Now since $0 \leq g \in E_n^\sim$, we see that $g(x - y_n) \uparrow g(x)$. On the other hand, from

$$0 \leq g(x - y_n) = g\left(\bigvee_{i=1}^n (x - x_i)\right) \leq \sum_{i=1}^n g(x - x_i) < \epsilon,$$

it follows that $0 \leq g(x) \leq \epsilon$ holds for all $\epsilon > 0$. Thus, $g(x) = 0$, so that $C_f \subseteq N_g$ holds.

(2) $\Longrightarrow$ (3) Since N_f is a band, it follows from $C_f = N_f^d \subseteq N_g$ and Theorem 1.39 that

$$C_g = N_g^d \subseteq N_f^{dd} = N_f.$$

(3) $\Longrightarrow$ (4) Since $C_g \subseteq N_f$ is true by our hypothesis and $N_f \perp C_f$, we see that $C_g \perp C_f$ holds.

(4) $\Longrightarrow$ (1) From $C_f \perp C_g$ it follows that $C_g \subseteq C_f^d = N_f^{dd} = N_f$. Now if $0 \leq x = y + z \in N_g \oplus C_g$, then

$$0 \leq [f \wedge g](x) = [f \wedge g](y) + [f \wedge g](z) \leq g(y) + f(z) = 0,$$

and thus $f \wedge g = 0$ holds on the order dense ideal $N_g \oplus C_g$ (see Theorem 1.36). Since $f \wedge g \in E_n^\sim$, it follows that $[f \wedge g](x) = 0$ holds for all $x \in E$, and the proof is finished. ∎

It should be noted that the above proof of the implication (4) $\Longrightarrow$ (1) shows that the following general result is true.

- *If two positive order continuous operators S and T satisfy $C_S \perp C_T$, then $S \perp T$.*

However, as the next example shows, the converse is not true.

Example 1.68. Let $A = \left[0, \frac{1}{2}\right]$ and $B = \left[\frac{1}{2}, 1\right]$ and consider the two positive operators $S, T \colon L_1[0,1] \to L_1[0,1]$ defined by

$$S(f) = \left[\int_0^1 f(x)\, dx\right]\chi_A \quad \text{and} \quad T(f) = \left[\int_0^1 f(x)\, dx\right]\chi_B \,.$$

The Lebesgue dominated convergence theorem shows that S and T are both order continuous operators. On the other hand, note that if $0 \le f \in L_1[0,1]$, then we have

$$0 \le [S \wedge T](f) \le S(f) \wedge T(f) = \left[\int_0^1 f(x)\, dx\right] \cdot \chi_A \wedge \chi_B = 0\,,$$

and so $S \wedge T = 0$ holds in $\mathcal{L}_b(L_1[0,1])$.

Finally, note that $N_S = N_T = \{0\}$, and so $C_S = C_T = L_1[0,1]$, proving that C_S and C_T are not disjoint sets. ∎

If E is a Riesz space, then its order dual $E^{\sim}$ is again a Riesz space. Thus, we can consider the Riesz space of all order bounded linear functionals on $E^{\sim}$. The **second order dual** $E^{\sim\sim}$ of E is the order dual of $E^{\sim}$, that is, $E^{\sim\sim} := (E^{\sim})^{\sim}$. For each $x \in E$ an order bounded linear functional $\widehat{x}$ can be defined on $E^{\sim}$ via the formula

$$\widehat{x}(f) := f(x)\,, \quad f \in E^{\sim}\,.$$

Clearly, $x \ge 0$ implies $\widehat{x} \ge 0$. Also, since $f_\alpha \downarrow 0$ in $E^{\sim}$ holds if and only if $\widehat{x}(f_\alpha) = f_\alpha(x) \downarrow 0$ for all $x \in E^+$, it easily follows that each $x \in E$ defines an order continuous linear functional on $E^{\sim}$. Thus, a positive operator $x \mapsto \widehat{x}$ can be defined from E to $E^{\sim\sim}$. This operator is called the **canonical embedding** of E into $E^{\sim\sim}$. The canonical embedding always preserves finite suprema and infima, and when $E^{\sim}$ separates the points of E, it is also one-to-one. The details follow.

Theorem 1.69. *Let E be a Riesz space. Then the canonical embedding $x \mapsto \widehat{x}$ is a lattice preserving operator (from E to $E^{\sim\sim}$).*

In particular, if $E^{\sim}$ separates the points of E, then $x \mapsto \widehat{x}$ is also one-to-one (and hence, in this case E, identified with its canonical image in $E^{\sim\sim}$, can be considered as a Riesz subspace of $E^{\sim\sim}$).

Proof. Only the preservation of the lattice operations needs verification. To this end, let $x \in E$ and $0 \le f \in E^{\sim}$. Applying Theorems 1.18 and 1.23

consecutively, we see that

$$(\widehat{x})^{+}(f) = \sup\{\widehat{x}(g): \ g \in E^{\sim} \text{ and } 0 \le g \le f\}$$
$$= \sup\{g(x): \ g \in E^{\sim} \text{ and } 0 \le g \le f\}$$
$$= f(x^{+}) = \widehat{(x^{+})}(f).$$

That is, $(\widehat{x})^{+} = \widehat{(x^{+})}$ holds. Now by using the lattice identity

$$x \vee y = (x - y)^{+} + y = -\left[(-x) \wedge (-y)\right],$$

we see that the canonical embedding $x \mapsto \widehat{x}$ preserves finite suprema and infima. ■

It should be noted that the canonical embedding of E into $E^{\sim\sim}$ does not necessarily preserve infinite suprema and infima; see Exercise 10 at the end of this section. In the sequel the vectors of a Riesz space E will play a double role. Besides being the vectors of E, they also will be considered (by identifying x with $\widehat{x}$) as order bounded linear functionals on $E^{\sim}$.

Now let E be a Riesz space, and let A be an ideal of $E^{\sim}$. Then it is easy to see that for each $x \in E$, the restriction of $\widehat{x}$ to A defines an order continuous linear functional (and hence order bounded) on A. Therefore, there exists a natural embedding $x \mapsto \widehat{x}$ of E into $A_{n}^{\sim}$ defined by

$$\widehat{x}(f) := f(x), \quad f \in A.$$

As in Theorem 1.69 we can see that the natural embedding $x \mapsto \widehat{x}$, from E into $A_{n}^{\sim}$, is lattice preserving and is one-to-one if and only if the ideal A separates the points of E.

When A consists of order continuous linear functionals, H. Nakano [150, Theorem 22.6, p. 83] has shown (among other things) that $x \mapsto \widehat{x}$ preserves arbitrary suprema and infima. The details are included in the next theorem.

Theorem 1.70 (Nakano). *Let E be an Archimedean Riesz space, and let A be an ideal of $E_{n}^{\sim}$. Then the embedding $x \mapsto \widehat{x}$ is an order continuous lattice preserving operator from E to $A_{n}^{\sim}$ whose range is an order dense Riesz subspace of $A_{n}^{\sim}$.*

Proof. To see that $x \mapsto \widehat{x}$ is order continuous, note that if $x_{\alpha} \downarrow 0$ holds in E, then $\widehat{x}_{\alpha}(f) = f(x_{\alpha}) \downarrow 0$ holds for each $0 \le f \in A$, and so $\widehat{x}_{\alpha} \downarrow 0$ holds in $A_{n}^{\sim}$. That is, $x \mapsto \widehat{x}$ is an order continuous operator.

Now let us establish that the range of $x \mapsto \widehat{x}$ is an order dense Riesz subspace of $A_{n}^{\sim}$. To this end, let $0 < \phi \in A_{n}^{\sim}$. Pick some $0 < f \in C_{\phi}$, and then choose $0 < x \in C_{f}$. Clearly, $f(x) > 0$. If $\widehat{x} \wedge \phi = 0$ holds, then by Theorem 1.67 we have $\widehat{x}(C_{\phi}) = \{0\}$, and so $\widehat{x}(f) = f(x) = 0$, a contradiction. Thus, $\widehat{x} \wedge \phi > 0$ holds, and hence, by replacing ϕ with $\widehat{x} \wedge \phi$, we can assume that $0 < \phi \le \widehat{x}$ holds in $A_{n}^{\sim}$ for some $x \in E$. Next fix some

$0 < \epsilon < 1$ with $\psi = (\phi - \epsilon x)^+ > 0$. Choose some $0 < g \in C_\psi$, and then select some $0 < y \in C_g$. We claim that the vector $z = y \wedge \epsilon x \in E$ satisfies $0 < \widehat{z} \leq \phi$ in $A_n^\sim$.

To see that $\widehat{z} > 0$ holds, note that if $\widehat{z} = \widehat{y} \wedge \epsilon \widehat{x} = 0$, then $\widehat{y} \wedge \widehat{x} = 0$, and so in view of $0 \leq \psi \leq \widehat{x}$, we see that $\widehat{y} \wedge \psi = 0$. By Theorem 1.67 we have $\widehat{y}(C_\psi) = \{0\}$, and hence $\widehat{y}(f) = f(y) = 0$, a contradiction. Thus, $\widehat{z} > 0$.

Finally, let us establish that $\widehat{z} \leq \phi$ holds. To this end, assume by way of contradiction that $\omega = (\widehat{z} - \phi)^+ > 0$. Choose $0 < h \in C_\omega$, and note that, in view of $0 < \omega \leq (\epsilon \widehat{x} - \phi)^+ = (\phi - \epsilon \widehat{x})^-$, we have $\omega \perp \psi$ and so by Theorem 1.67 we get $C_\omega \perp C_\psi$. In particular, $h \perp g$ holds, and by applying Theorem 1.67 once more, we get $h(C_g) = \{0\}$. Therefore,

$$0 < \omega(h) = (\widehat{z} - \phi)^+(h) \leq \widehat{z}(h) \leq \widehat{y}(h) = h(y) = 0$$

holds, which is impossible. Hence, $\widehat{z} \leq \phi$, and the proof is complete. ∎

As an application of Theorem 1.70, we shall characterize the perfect Riesz spaces. A Riesz space E is said to be **perfect** whenever the natural embedding $x \mapsto \widehat{x}$ from E to $(E_n^\sim)_n^\sim$ is one-to-one and onto. Clearly, every perfect Riesz space must be Dedekind complete. H. Nakano [**150**, Section 24] has characterized the perfect Riesz spaces as follows.

Theorem 1.71 (Nakano). *A Riesz space E is a perfect Riesz space if and only if the following two conditions hold:*

(1) *$E_n^\sim$ separates the points of E.*

(2) *Whenever a net $\{x_\alpha\} \subseteq E$ satisfies $0 \leq x_\alpha \uparrow$ and $\sup\{f(x_\alpha)\} < \infty$ for each $0 \leq f \in E_n^\sim$, then there exists some $x \in E$ satisfying $0 \leq x_\alpha \uparrow x$ in E.*

Proof. Assume that E is a perfect Riesz space, i.e., assume that $x \mapsto \widehat{x}$ from E to $(E_n^\sim)_n^\sim$ is one-to-one and onto. Then, clearly, $E_n^\sim$ separates the points of E. On the other hand, if a net $\{x_\alpha\} \subseteq E^+$ satisfies $0 \leq x_\alpha \uparrow$ and $\phi(f) = \sup\{f(x_\alpha)\} < \infty$ for each $0 \leq f \in E_n^\sim$, then it easily follows that the mapping $\phi \colon (E_n^\sim)^+ \to \mathbb{R}^+$ is additive, and hence ϕ defines a positive linear functional on $E_n^\sim$. In view of $\widehat{x}_\alpha \uparrow \phi$ in $(E_n^\sim)^\sim$, it follows (from Theorem 1.57) that $\phi \in (E_n^\sim)_n^\sim$. Pick some $x \in E$ with $\phi = \widehat{x}$, and note that $0 \leq x_\alpha \uparrow x$ holds in E.

For the converse assume that E satisfies the two conditions. Then, by Theorem 1.70, the operator $x \mapsto \widehat{x}$ from E to $(E_n^\sim)_n^\sim$ is order continuous, one-to-one, and lattice preserving whose range is order dense in $(E_n^\sim)_n^\sim$. Now let $0 \leq \phi \in (E_n^\sim)_n^\sim$. Pick a net $\{x_\alpha\} \subseteq E^+$ with $0 \leq \widehat{x}_\alpha \uparrow \phi$ in $(E_n^\sim)_n^\sim$. Then $\{x_\alpha\}$ satisfies condition (2), and so there exists some $x \in E$ with $0 \leq x_\alpha \uparrow x$

in E. It follows that $0 \leq \widehat{x}_\alpha \uparrow \widehat{x}$ holds in $(E_n^\sim)_n^\sim$, and thus $\phi = \widehat{x}$, proving that $x \mapsto \widehat{x}$ is also onto. $\blacksquare$

The (order bounded) finite rank operators will be of great importance. If $f \in E^\sim$ and $u \in F$, then the symbol $f \otimes u$ will denote the order bounded operator of $\mathcal{L}_b(E, F)$ defined by

$$[f \otimes u](x) := f(x)u$$

for each $x \in E$. Every operator of the form $f \otimes u$ is referred to as a **rank one operator**. Note that if $f \in E_n^\sim$ (resp. $f \in E_c^\sim$), then $f \otimes u$ is an order (resp. σ-order) continuous operator. Every operator $T \colon E \to F$ of the form $T = \sum_{i=1}^n f_i \otimes u_i$, where $f_i \in E^\sim$ and $u_i \in F$ $(i = 1, \ldots, n)$, is called a **finite rank operator**. In general, if G is a vector subspace of $E^\sim$, then we define

$$G \otimes F := \Big\{ T \in \mathcal{L}(E, F) \colon \exists n, \ f_i \in G, \ u_i \in F \ (1 \leq i \leq n) \text{ with } T = \sum_{i=1}^n f_i \otimes u_i \Big\}.$$

Clearly, $G \otimes F$ is a vector subspace of $\mathcal{L}_b(E, F)$.

The next theorem describes some basic lattice properties of the rank one operators.

Theorem 1.72. *For a pair of Riesz spaces E and F we have the following:*

(1) *If $0 \leq f \in E^\sim$ and $u, v \in F$, then $(f \otimes u) \vee (f \otimes v)$ and $(f \otimes u) \wedge (f \otimes v)$ both exist in $\mathcal{L}(E, F)$ and*

$$(f \otimes u) \vee (f \otimes v) = f \otimes (u \vee v)$$

and

$$(f \otimes u) \wedge (f \otimes v) = f \otimes (u \wedge v).$$

(2) *If $0 \leq u \in F$ and $f, g \in E^\sim$, then $(f \otimes u) \vee (g \otimes u)$ and $(f \otimes u) \wedge (g \otimes u)$ both exist in $\mathcal{L}(E, F)$ and*

$$(f \otimes u) \vee (g \otimes u) = (f \vee g) \otimes u$$

and

$$(f \otimes u) \wedge (g \otimes u) = (f \wedge g) \otimes u.$$

(3) *If $f \in E^\sim$ and $u \in F$, then the modulus of $f \otimes u$ exists in $\mathcal{L}(E, F)$ and*

$$\big| f \otimes u \big| = |f| \otimes |u|.$$

Proof. (1) Let $0 \leq f \in E^\sim$, and let $u, v \in F$. Clearly, $f \otimes u \leq f \otimes (u \vee v)$ and $f \otimes v \leq f \otimes (u \vee v)$ both hold. On the other hand, if some $T \in \mathcal{L}(E, F)$ satisfies $f \otimes u \leq T$ and $f \otimes v \leq T$, then for each $x \in E^+$ we have

$$\begin{aligned}[f \otimes (u \vee v)](x) &= f(x)(u \vee v) = [f(x)u] \vee [f(x)v] \\ &\leq T(x) \vee T(x) = T(x).\end{aligned}$$

That is, $f \otimes (u \vee v) \leq T$ holds in $\mathcal{L}_b(E, F)$, and so $f \otimes (u \vee v)$ is the least upper bound of $f \otimes u$ and $f \otimes v$ in $\mathcal{L}(E, F)$, as required. The other case can be proven in a similar manner.

(2) Fix $u \in F^+$ and $f, g \in E^\sim$. Clearly, $f \otimes u \leq (f \vee g) \otimes u$ and $g \otimes u \leq (f \vee g) \otimes u$. Now let $T \in \mathcal{L}(E, F)$ satisfy $f \otimes u \leq T$ and $g \otimes u \leq T$. Observe that if $y, z \in E^+$ satisfy $y + z = x$, then

$$[f \otimes u](y) + [g \otimes u](z) \leq T(y) + T(z) = T(x)$$

holds. Thus, for each $x \in E^+$ we have

$$
\begin{aligned}
[(f \vee g) \otimes u](x) &= [(f \vee g)(x)] \cdot u \\
&= [\sup\{f(y) + g(z) : y, z \in E^+ \text{ and } y + z = x\}] \cdot u \\
&= \sup\{f(y)u + g(z)u : y, z \in E^+ \text{ and } y + z = x\} \\
&= \sup\{[f \otimes u](y) + [g \otimes u](z) : y, z \in E^+ \text{ and } y + z = x\} \\
&\leq T(x).
\end{aligned}
$$

Therefore, $(f \vee g) \otimes u$ is the least upper bound of $f \otimes u$ and $g \otimes u$ in $\mathcal{L}(E, F)$. The other formula can be proven in a similar fashion.

(3) For each $x \in E^+$ we have

$$
\begin{aligned}
\pm[f \otimes u](x) &= \pm[f(x) \cdot u] \leq |f(x)u| = |f(x)| \cdot |u| \\
&\leq |f|(x) \cdot |u| = [|f| \otimes |u|](x),
\end{aligned}
$$

and so $\pm[f \otimes u] \leq |f| \otimes |u|$. Now assume that some $T \in \mathcal{L}(E, F)$ satisfies

$$f \otimes u \leq T \quad \text{and} \quad -[f \otimes u] \leq T.$$

Let $x \in E^+$. If $f(x) < 0$, then $[f \otimes |u|](x) \leq T(x)$ holds trivially. On the other, if $f(x) \geq 0$, then we have

$$[f \otimes |u|](x) = f(x)|u| = [f(x)u] \vee [-f(x)u] \leq T(x).$$

Therefore, $f \otimes |u| \leq T$ holds. By the symmetry of the situation we have $(-f) \otimes |u| \leq T$. Thus, by part (2) we see that

$$|f| \otimes |u| = [f \otimes |u|] \vee [(-f) \otimes |u|] \leq T.$$

Consequently, $|f| \otimes |u|$ is the least upper bound of $f \otimes u$ and $-f \otimes u$. That is, $|f \otimes u| = |f| \otimes |u|$ holds in $\mathcal{L}(E, F)$. ∎

Recall that the **algebraic dual** V^* of a vector space V is the vector space consisting of all linear functionals on V. For an operator $T : V \to W$ between two vector spaces its **algebraic adjoint** (or **transpose**) $T^* : W^* \to V^*$ is the operator defined by

$$[T^* f](v) = f(Tv)$$

for all $f \in W^*$ and $v \in V$. In standard duality notation this identity is written as

$$\langle T^*f, v \rangle = \langle f, Tv \rangle.$$

Clearly, if $S \colon V \to W$ is another operator and $\alpha \in \mathbb{R}$, then

$$(S + T)^* = S^* + T^* \quad \text{and} \quad (\alpha T)^* = \alpha T^*.$$

When $T \colon E \to F$ is an order bounded operator between two Riesz spaces, then T^* carries $F^\sim$ into $E^\sim$. Indeed, if A is an order bounded subset of E and $f \in F^\sim$, then it follows from $[T^*f](A) = f(T(A))$ that $[T^*f](A)$ is a bounded subset of $\mathbb{R}$, and so $T^*f \in E^\sim$. The restriction of T^* to $F^\sim$ is called the (**order**) **adjoint** of T and will be denoted by T'. That is, $T' \colon F^\sim \to E^\sim$ satisfies

$$\langle T'f, x \rangle = \langle f, Tx \rangle$$

for all $f \in F^\sim$ and $x \in E$. Note that if T is a positive operator, then its adjoint T' is likewise a positive operator.

The adjoint of an order bounded operator between two Riesz spaces is always order bounded and order continuous. The details follow.

Theorem 1.73. *If $T \colon E \to F$ is an order bounded operator between two Riesz spaces, then its (order) adjoint $T' \colon F^\sim \to E^\sim$ is order bounded and order continuous.*

Proof. Assume that $T \colon E \to F$ is an order bounded operator. We shall first establish that $T' \colon F^\sim \to E^\sim$ is order bounded.

To this end, let $0 \le f \in F^\sim$. Consider the set

$$D = \left\{ \sum_{i=1}^{n} |T'f_i| \colon \ f_i \ge 0 \text{ for each } i \text{ and } \sum_{i=1}^{n} f_i = f \right\}.$$

We claim that $D \uparrow$ holds in $E^\sim$. To see this, let $f_1, \ldots, f_n \in F_+^\sim$ and $g_1, \ldots, g_m \in F_+^\sim$ satisfy $\sum_{i=1}^{n} f_i = \sum_{j=1}^{m} g_j = f$. By Theorem 1.20 there exist linear functionals $h_{ij} \in F_+^\sim$ $(i = 1, \ldots, n; \ j = 1, \ldots, m)$ such that

$$f_i = \sum_{j=1}^{m} h_{ij} \text{ for } i = 1, \ldots, n \quad \text{and} \quad g_j = \sum_{i=1}^{n} h_{ij} \text{ for } j = 1, \ldots, m.$$

Clearly, $\sum_{i=1}^{n} \sum_{j=1}^{m} h_{ij} = f$. On the other hand, we have

$$\sum_{i=1}^{n} |T'f_i| = \sum_{i=1}^{n} \left| \sum_{j=1}^{m} T'h_{ij} \right| \le \sum_{i=1}^{n} \sum_{j=1}^{m} |T'h_{ij}|,$$

and similarly

$$\sum_{j=1}^{m} |T'g_j| \le \sum_{i=1}^{n} \sum_{j=1}^{m} |T'h_{ij}|.$$

The above show that $D \uparrow$ holds in $E^\sim$.

Now let $x \in E^+$. Since T is order bounded, there exists some $u \in F^+$ satisfying $|Ty| \le u$ for all $|y| \le x$. Consequently, if $f_1, \ldots, f_n \in F^\sim_+$ satisfy $\sum_{i=1}^n f_i = f$, then we have

$$
\begin{aligned}
\Big\langle \sum_{i=1}^n |T'f_i|, x \Big\rangle &= \sum_{i=1}^n \sup\{\langle T'f_i, y\rangle : |y| \le x\} \\
&= \sum_{i=1}^n \sup\{\langle f_i, Ty\rangle : |y| \le x\} \\
&\le \sum_{i=1}^n \langle f_i, u\rangle = f(u), \qquad\qquad (\star)
\end{aligned}
$$

which shows that the set $\{\phi(x) : \phi \in D\}$ is bounded above in $\mathbb{R}$ for each $x \in E^+$. By Theorem 1.19 the supremum $h = \sup D$ exists in $E^\sim$. Now if $0 \le g \le f$, then $|T'g| \le |T'g| + |T'(f - g)| \le h$ holds in $E^\sim$, which shows that $T'[0, f] \subseteq [-h, h]$. Therefore, $T' \colon F^\sim \to E^\sim$ is order bounded.

Finally, we show that T' is order continuous. To this end, let $f_\alpha \downarrow 0$ in $F^\sim$, and let $x \in E^+$ be fixed. Pick some $u \in F^+$ with $|Ty| \le u$ for all $|y| \le x$. From $(\star)$ and part (3) of Theorem 1.21 we see that $\big[|T'|f\big](x) \le f(u)$ holds for all $0 \le f \in F^\sim$. In particular, we have $\big[|T'|f_\alpha\big](x) \le f_\alpha(u) \downarrow 0$, and so $\big[|T'|f_\alpha\big](x) \downarrow 0$ holds for each $x \in E^+$, i.e., $|T'|f_\alpha \downarrow 0$ holds in $E^\sim$. Therefore, $|T'|$ is order continuous, and so T' is likewise order continuous. The proof of the theorem is now complete. ∎

It is interesting to know that the converse of the preceding theorem is false. That is, there are operators $T \colon E \to F$ between Riesz spaces that are not order bounded, while their algebraic adjoints carry $F^\sim$ into $E^\sim$ and are order bounded and order continuous. For instance, the operator $T \colon L_1[0, 1] \to c_0$ defined by

$$
T(f) = \left(\int_0^1 f(x) \sin x \, dx, \int_0^1 f(x) \sin 2x \, dx, \ldots \right),
$$

is not order bounded, while

$$
T' \colon c_0^\sim = \ell_1 \to L_1^\sim[0, 1] = L_\infty[0, 1]
$$

(where $\langle T'(x_1, x_2, \ldots), f\rangle = \sum_{n=1}^\infty x_n \int_0^1 f(x) \sin nx \, dx$) is order bounded and order continuous. For details see Exercise 10 of Section 5.1.

Consider an order bounded operator $T \colon E \to F$ between two Riesz spaces. By Theorem 1.73 we know that $T' \colon F^\sim \to E^\sim$ is likewise order bounded, and so (since $E^\sim$ is Dedekind complete) the modulus of T' exists. On the other hand, if the modulus of T also exists, then it follows from $\pm T \le |T|$ that $\pm T' \le |T|'$. That is, whenever the modulus of T exists, then

$|T'| \leq |T|'$ holds. The strict inequality $|T'| < |T|'$ may very well happen, as the next example shows.

Example 1.74. Consider the operator $T: \ell_1 \to \ell_\infty$ defined by

$$T(x_1, x_2, \ldots) = (x_1 - x_2, x_2 - x_3, x_3 - x_4, \ldots).$$

Clearly, T is a regular operator, and an easy argument shows that

$$|T|(x_1, x_2, \ldots) = \sup\{T(y_1, y_2, \ldots): |(y_1, y_2, \ldots)| \leq (x_1, x_2, \ldots)\}$$
$$= (x_1 + x_2, x_2 + x_3, x_3 + x_4, \ldots)$$

holds for all $0 \leq (x_1, x_2, \ldots) \in \ell_1$.

Next consider the Riesz subspace c of ℓ_∞ consisting of all convergent sequences. Clearly, c majorizes ℓ_∞, and moreover the formula

$$\phi(x_1, x_2, \ldots) = \lim_{n \to \infty} x_n, \quad (x_1, x_2, \ldots) \in c,$$

defines a positive linear functional on c. By Theorem 1.32 the positive linear functional ϕ has a positive linear extension to all of ℓ_∞, which we denote by ϕ again. Put $e = (1, 1, \ldots)$, and note that

$$\langle |T|'\phi, e \rangle = \langle \phi, |T|e \rangle = \phi(2, 2, \ldots) = 2.$$

Now let $\psi \in \ell_\infty^\sim$ satisfy $|\psi| \leq \phi$. Note that if $(x_1, x_2, \ldots) \in \ell_\infty$ satisfies $\lim_{n \to \infty} x_n = 0$, then the relation

$$|\psi(x)| \leq |\psi|(|x|) \leq \phi(|x|) = \lim_{n \to \infty} |x_n| = 0,$$

implies $\psi(x) = 0$. Therefore, $[T'\psi](x) = \psi(Tx) = 0$ holds for all $x \in \ell_1$. In other words, $T'\psi = 0$ holds for all $|\psi| \leq \phi$, and so by Theorem 1.14 we see that

$$|T'|\phi = \sup\{|T'\psi|: |\psi| \leq \phi\} = 0.$$

Thus, $0 = \langle |T'|\phi, e \rangle \neq \langle |T|'\phi, e \rangle = 2$, and consequently the operator T satisfies $|T'| < |T|'$. ■

To continue our discussion we need a simple lemma.

Lemma 1.75. *If $T: E \to F$ is an order bounded operator between two Riesz spaces, then for each $0 \leq f \in F^\sim$ and each $x \in E^+$ we have*

$$\langle f, |Tx| \rangle \leq \langle |T'|f, x \rangle.$$

Proof. Fix $0 \leq f \in F^\sim$ and $x \in E^+$. Then by Theorem 1.23 there exists some $g \in F^\sim$ with $|g| \leq f$ and $\langle f, |Tx| \rangle = \langle g, Tx \rangle$. Thus,

$$\langle f, |Tx| \rangle = \langle g, Tx \rangle \leq \langle T'g, x \rangle \leq \langle |T'||g|, x \rangle \leq \langle |T'|f, x \rangle,$$

as desired. ■

Although $|T'|$ and $|T|'$ need not be equal, they do agree on the order continuous linear functionals. This important result is due to U. Krengel [**105**] and J. Synnatzschke [**181**] and is stated next.

Theorem 1.76 (Krengel–Synnatzschke). *If $T\colon E \to F$ is an order bounded operator between two Riesz spaces with F Dedekind complete, then*

$$|T'|f = |T|'f$$

holds for all $f \in F_n^\sim$.

Proof. Let $0 \le f \in F_n^\sim$ be fixed. We already know that $|T'|f \le |T|'f$ holds. On the other hand, if $0 \le x \in E$, then from Theorem 1.21 and Lemma 1.75 we see that

$$
\begin{aligned}
\langle |T|'f, x \rangle &= \langle f, |T|x \rangle \\
&= \Big\langle f, \sup\Big\{ \sum_{i=1}^n |Tx_i|\colon\ x_i \in E^+ \text{ and } \sum_{i=1}^n x_i = x \Big\} \Big\rangle \\
&= \sup\Big\{ \sum_{i=1}^n \langle f, |Tx_i| \rangle\colon\ x_i \in E^+ \text{ and } \sum_{i=1}^n x_i = x \Big\} \\
&\le \Big\{ \sum_{i=1}^n \langle |T'|f, x_i \rangle\colon\ x_i \in E^+ \text{ and } \sum_{i=1}^n x_i = x \Big\} \\
&= \langle |T'|f, x \rangle,
\end{aligned}
$$

and so $|T|'f \le |T'|f$. Therefore, $|T'|f = |T|'f$ holds for all $f \in F_n^\sim$. ∎

When is every order bounded linear functional on a Riesz space σ-order continuous?

As we shall see, this question is closely related to the following question regarding a σ-order continuity property of the map $T \mapsto T^2$, from $\mathcal{L}_b(E)$ to $\mathcal{L}_b(E)$. When does $0 \le T_n \uparrow T$ in $\mathcal{L}_b(E)$ imply $T_n^2 \uparrow T^2$?

In general, $0 \le T_n \uparrow T$ does not imply $T_n^2 \uparrow T^2$, even if T and all the T_n are rank one operators.

Example 1.77. Let $E = \ell_\infty$, the Dedekind complete Riesz space of all bounded real-valued sequences, and consider the Riesz subspace c of E consisting of all convergent sequences. Clearly, c majorizes E and the formula $f(x) = \lim x_n$ defines a positive linear functional on c. By Theorem 1.32 the positive linear functional has a positive linear extension to all of E (which we denote by f again.)

Now let $u_n = (1, 1, \ldots, 1_n, 0, 0, \ldots)$ and $e = (1, 1, \ldots)$. Put $T_n = f \otimes u_n$, $T = f \otimes e$, and note that $0 \le T_n \uparrow T$ holds in $\mathcal{L}_b(E)$. On the other hand, it is not difficult to see that $T_n^2 = 0$ for each n and $T^2 = T$. So, $T_n^2 \not\uparrow T^2$. ∎

In contrast to the preceding example, observe that $T_n \downarrow 0$ in $\mathcal{L}_b(E)$ implies $T_n^2 \downarrow 0$. (To see this, note that $0 \leq T_n^2(x) \leq T_n(T_1 x)$ for all $x \in E^+$.)

Example 1.77 can be used to establish the existence of a Dedekind complete Riesz space E with the property that for each k, there exists a sequence $\{T_n\}$ of positive operators on E such that $0 \leq T_n^i \uparrow_n T^i$ holds for each $i = 1, \ldots, k$ and $T_n^{k+1} \nearrow T^{k+1}$. The next example is taken from [18].

Example 1.78. Let f, u_n, and e be as they were defined in Example 1.77, and let $E = (\ell_\infty)^{\mathbb{N}}$ ($=$ the Dedekind complete Riesz space of all ℓ_∞-valued sequences).

Now let k be fixed, and define the positive operators

$$T_n(x_1, x_2, \ldots) = \big(f(x_k)u_n, x_1, \ldots, x_{k-1}, 0, 0, \ldots\big),$$

and

$$T(x_1, x_2, \ldots) = \big(f(x_k)e, x_1, \ldots, x_{k-1}, 0, 0, \ldots\big).$$

Then it is a routine matter to verify that

$$0 \leq T_n^i \uparrow T^i \text{ for each } i = 1, \ldots, k \quad \text{and} \quad T_n^{k+1} \nearrow T^{k+1}$$

hold in $\mathcal{L}_b(E)$. ∎

The next result of C. D. Aliprantis, O. Burkinshaw and P. Kranz [18] characterizes the Riesz spaces on which every positive linear functional is σ-order continuous.

Theorem 1.79 (Aliprantis–Burkinshaw–Kranz). *For a Riesz space E whose order dual separates the points of E the following statements are equivalent:*

(a) $E_c^\sim = E^\sim$, *i.e., every positive linear functional on E is σ-order continuous.*

(b) *Whenever $T: E \to E$ is a positive operator and a sequence $\{T_n\}$ of positive operators from E to E satisfies $T_n(x) \uparrow T(x)$ in E for each $x \in E^+$, then $T_n^2(x) \uparrow T^2(x)$ likewise holds in E for each $x \in E^+$.*

Proof. (1) $\Longrightarrow$ (2) Let $0 \leq T_n(x) \uparrow T(x)$ for each $x \in E^+$, and let $y \in E^+$ be fixed. Clearly, $0 \leq T_n^2(y) \uparrow \leq T^2(y)$ holds in E. To see that $T^2(y)$ is the least upper bound of the sequence $\{T_n^2(y)\}$, let $T_n^2(y) \leq z$ hold in E for all n. Then for each $0 \leq f \in E^\sim$ we have $f(T_n^2(y)) \leq f(z)$ for all n.

On the other hand, it follows that for each $0 \leq f \in E^\sim$ the sequence $\{f \circ T_n\} \subseteq E^\sim = E_c^\sim$ satisfies $0 \leq f \circ T_n \uparrow f \circ T$ in $E^\sim$. Thus,

$$f(T_n^2(y)) = [f \circ T_n](T_n y) \uparrow [f \circ T](Ty) = f(T^2(y)),$$

and so $f(T^2(y)) \leq f(z)$ holds for all $0 \leq f \in E^\sim$. Since $E^\sim$ separates the points of E, it follows from Theorem 1.66 that $T^2(y) \leq z$. Therefore, $T_n^2(y) \uparrow T^2(y)$ holds in E for each $y \in E^+$.

(2) $\Longrightarrow$ (1) Fix $0 \leq f \in E^\sim$, and let $0 \leq x_n \uparrow x$ in E. Then we have $0 \leq [f \otimes x_n](y) \uparrow [f \otimes x](y)$ for all $y \in E^+$, and so by our hypothesis

$$[f \otimes x_n]^2(y) = f(x_n)[f(y)x_n] \uparrow [f \otimes x]^2(y) = f(x)[f(y)x]$$

also holds for all $y \in E^+$. Now an easy argument shows that $f(x_n) \uparrow f(x)$, and hence f is σ-order continuous. Therefore, $E_c^\sim = E^\sim$ holds. ∎

Since $E^\sim$ is Dedekind complete, every band of $E^\sim$ is a projection band (see Theorem 1.42). The rest of the section is devoted to deriving formulas for the order projections of $E^\sim$.

Theorem 1.80. *Let E be a Riesz space and let $\phi \in E^\sim$. If P_ϕ denotes the order projection of $E^\sim$ onto the band generated by ϕ, then for each $x \in E^+$ and each $0 \leq f \in E^\sim$ we have*

$$[P_\phi f](x) = \sup_{\epsilon > 0} \inf\{f(y): \ 0 \leq y \leq x \ and \ |\phi|(x-y) < \epsilon\}.$$

Proof. We can assume that $0 \leq \phi \in E^\sim$. Fix $x \in E^+$ and $0 \leq f \in E^\sim$, and put

$$r = \sup_{\epsilon > 0} \inf\{f(y): \ 0 \leq y \leq x \ and \ |\phi|(x-y) < \epsilon\}.$$

Fix $\epsilon > 0$. Since $f \wedge n\phi \uparrow P_\phi f$ (Theorem 1.47), there exists some k with $(P_\phi f - f \wedge k\phi)(x) < \epsilon$. Now let $0 < \delta < \epsilon$, and let $0 \leq y \leq x$ satisfy $\phi(x-y) < \delta$. Then we have

$$\begin{aligned}[P_\phi f](x) &= (P_\phi f - f \wedge k\phi)(x) + (f \wedge k\phi)(x) < \epsilon + (f \wedge k\phi)(x)\\ &\leq \epsilon + k\phi(x-y) + f(y) < \epsilon + k\delta + f(y),\end{aligned}$$

and consequently

$$\begin{aligned}[P_\phi f](x) &\leq \epsilon + k\delta + \inf\{f(y): \ 0 \leq y \leq x \ and \ \phi(x-y) < \delta\}\\ &\leq \epsilon + k\delta + r\end{aligned}$$

holds for all $0 < \delta < \epsilon$. Thus, $[P_\phi f](x) \leq \epsilon + r$ holds for all $\epsilon > 0$, and therefore $[P_\phi f](x) \leq r$.

For the reverse inequality, let $\epsilon > 0$. Since $(f - P_\phi f) \wedge \phi = 0$, for each $0 < \delta < \epsilon$ there exists some $0 \leq z \leq x$ with $(f - P_\phi f)(z) + \phi(x-z) < \delta$. This implies $f(z) < \delta - \phi(x-z) + [P_\phi f](z) < \delta + [P_\phi f](x)$. In particular, we have

$$\inf\{f(y): \ 0 \leq y \leq x \ and \ \phi(x-y) < \epsilon\} \leq f(z) < \delta + [P_\phi f](x)$$

for all $\epsilon > 0$. This implies that $r \leq [P_\phi f](x)$, and hence $[P_\phi f](x) = r$. ∎

The next theorem presents a formula for P_ϕ in terms of increasing sequences and is due to W. A. J. Luxemburg [**125**, Note XV].

Theorem 1.81 (Luxemburg). *Let E be a Riesz space and let $\phi \in E^\sim$. Then for each $x \in E^+$ and $0 \le f \in E^\sim$ we have*

$$[P_\phi f](x) = \inf\big\{\sup f(x_n): \ 0 \le x_n \uparrow \le x \ \text{and} \ |\phi|(x - x_n) \downarrow 0\big\}.$$

Proof. We can assume that $0 \le \phi \in E^\sim$. Fix $x \in E^+$ and $0 \le f \in E^\sim$ and put

$$r = \inf\big\{\sup f(x_n): \ 0 \le x_n \uparrow \le x \ \text{and} \ |\phi|(x - x_n) \downarrow 0\big\}.$$

Let $0 \le x_n \uparrow \le x$ satisfy $\phi(x - x_n) \downarrow 0$. Then for each n and k we have

$$
\begin{aligned}
[P_\phi f](x) - f(x_n) &\le [P_\phi f](x - x_n) \\
&\le (P_\phi f - f \wedge k\phi)(x) + (f \wedge k\phi)(x - x_n) \\
&\le (P_\phi f - f \wedge k\phi)(x) + k\phi(x - x_n),
\end{aligned}
$$

and so, taking limits with respect to n, we get

$$[P_\phi f](x) - \sup f(x_n) \le (P_\phi f - f \wedge k\phi)(x)$$

for all k. Since $f \wedge k\phi \uparrow P_\phi f$, it follows that $[P_\phi f](x) \le \sup f(x_n)$, and from this we see that $[P_\phi f](x) \le r$.

Now let $\epsilon > 0$. Since $(f - P_\phi f) \wedge \phi = 0$ holds, for each n there exists some $0 \le y_n \le x$ with $(f - P_\phi f)(y_n) + \phi(x - y_n) < \epsilon 2^{-n}$. Put $x_n = \bigvee_{i=1}^n y_i$, and note that $0 \le x_n \uparrow \le x$. From $0 \le \phi(x - x_n) \le \phi(x - y_n) \to 0$, we see that $\phi(x - x_n) \downarrow 0$. Also, note that $0 \le (f - P_\phi f)(x_n) \le \sum_{i=1}^n (f - P_\phi f)(y_i) < \epsilon$ holds. Therefore,

$$r \le \sup f(x_n) \le \sup(f - P_\phi f)(x_n) + \sup[P_\phi f](x_n) \le \epsilon + [P_\phi f](x)$$

holds for all $\epsilon > 0$, and so $r \le [P_\phi f](x)$. Consequently, $[P_\phi f](x) = r$ holds, and the proof is finished. ∎

A formula, due to the authors [**16**], describing the order projection onto an arbitrary band of $E^\sim$ is presented next.

Theorem 1.82 (Aliprantis–Burkinshaw). *Let E be a Riesz space and let B be a band of $E^\sim$. If P_B denotes the order projection of $E^\sim$ onto B, then for each $x \in E^+$ and $0 \le f \in E^\sim$ we have*

$$[P_B f](x) = \sup_{\substack{\epsilon > 0 \\ \phi \in B^+}} \inf\big\{f(y): \ 0 \le y \le x \ \text{and} \ \phi(x - y) < \epsilon\big\}.$$

Proof. Fix $x \in E^+$ and $0 \le f \in E^\sim$, and put

$$r = \sup_{\substack{\epsilon > 0 \\ \phi \in B^+}} \inf\big\{f(y): \ 0 \le y \le x \ \text{and} \ \phi(x - y) < \epsilon\big\}.$$

Note that for each $\phi \in B^+$ we have $P_\phi \leq P_B$. Thus from Theorem 1.80 it easily follows that $r \leq [P_B f](x)$. Now let $\psi = P_B f$. Then $\psi \in B^+$, and so by Theorem 1.80 we have

$$
\begin{aligned}
[P_B f](x) &= [P_\psi \psi](x) \leq [P_\psi f](x) \\
&= \sup_{\epsilon > 0} \inf \{ f(y) : \ 0 \leq y \leq x \text{ and } \psi(x - y) < \epsilon \} \leq r .
\end{aligned}
$$

Thus, $[P_B f](x) = r$ holds, as desired.   ∎

In view of Theorem 1.81, it might be expected that the following formula also holds:

$$
[P_B f](x) = \inf \{ \sup f(x_n) : \ 0 \leq x_n \uparrow \leq x \text{ and } \phi(x - x_n) \downarrow 0 \ \forall \phi \in B^+ \} .
$$

Unfortunately, such formula is not true. For an example, let E be the Riesz space of all Lebesgue integrable (real-valued) functions on $[0, 1]$ with the pointwise ordering. (Note that two functions differing at one point are considered to be different.) Since $x_\alpha \downarrow 0$ in E implies $x_\alpha(t) \downarrow 0$ for each $t \in [0, 1]$, it follows that the point evaluations $x \mapsto x(t)$ are all order continuous positive linear functionals on E. This implies that $E_n^\sim$ separates the points of E. Now consider the positive linear functional $f \colon E \to \mathbb{R}$ defined by

$$
f(x) = \int_0^1 x(t) \, dt .
$$

According to Example 1.55, the linear functional f is σ-order continuous but not order continuous. If $B = E_n^\sim$, then

$$
\begin{aligned}
&\inf \{ \sup f(x_n) : \ 0 \leq x_n \uparrow \leq 1 \text{ and } \phi(1 - x_n) \downarrow 0 \text{ for all } \phi \in B^+ \} \\
&\quad = \inf \{ \sup f(x_n) : \ 0 \leq x_n \uparrow 1 \} = f(1) = 1 .
\end{aligned}
$$

On the other hand, it is not difficult to see that $[P_B f](1) < 1$ must hold.

Finally, we close this section by presenting necessary and sufficient conditions for a linear functional to belong to a principal band of $E^\sim$.

Theorem 1.83. *Let E be a Riesz space and let $f \in E^\sim$. Then for an order bounded linear functional $g \in E^\sim$ the following statements are equivalent.*

(1) *g belongs to the principal band generated by f in $E^\sim$.*

(2) *For each $x \in E^+$ and $\epsilon > 0$ there exists some $\delta > 0$ such that whenever $|y| \leq x$ satisfies $|f|(|y|) < \delta$, then $|g|(|y|) < \epsilon$ holds.*

(3) *If an order bounded sequence $\{x_n\}$ of E satisfies $\lim |f|(|x_n|) = 0$, then $\lim g(x_n) = 0$.*

(4) *If $0 \leq x_n \uparrow \leq x$ and $\lim |f|(x - x_n) = 0$, then $\lim g(x - x_n) = 0$.*

Proof. (1) $\Longrightarrow$ (2) Let $x \in E^+$ and $\epsilon > 0$. Since $|g| \wedge k|f| \uparrow |g|$ holds in $E^\sim$ (Theorem 1.47), there exists some k with $(|g| - |g| \wedge k|f|)(x) < \epsilon$. If $|y| \le x$ satisfies $|f|(|y|) \le \frac{\epsilon}{k}$, then we have

$$
\begin{aligned}
|g|(|y|) &= (|g| - |g| \wedge k|f|)(|y|) + (|g| \wedge k|f|)(|y|) \\
&\le (|g| - |g| \wedge k|f|)(x) + k|f|(|y|) < \epsilon + \epsilon = 2\epsilon \, .
\end{aligned}
$$

(2) $\Longrightarrow$ (3) and (3) $\Longrightarrow$ (4) are obvious.

(4) $\Longrightarrow$ (1) Write $g = \phi + \psi$, with $\phi \in B_f$ and $\psi \perp f$. Fix $x \in E^+$ and $\epsilon > 0$. Now let $0 \le y \le x$. Since $\psi \perp f$ holds, for each n there exists some $0 \le y_n \le y$ with $|\psi|(y_n) + |f|(y - y_n) < 2^{-n}\epsilon$. Then $x_n = \bigvee_{i=1}^{n} y_i$ satisfies $0 \le x_n \uparrow \le y$ and $|\psi|(x_n) \le \sum_{i=1}^{n} |\psi|(y_i) < \epsilon$. On the other hand, the inequalities $|f|(y - x_n) \le |f|(y - y_n) \le 2^{-n}\epsilon$ imply $|f|(y - x_n) \downarrow 0$. Hence, by our hypothesis $\lim g(y - x_n) = 0$. In particular, note that

$$
\begin{aligned}
g(y) &= \lim_{n\to\infty} g(x_n) = \lim_{n\to\infty} \big[\phi(x_n) + \psi(x_n)\big] \\
&\le \limsup_{n\to\infty} \big[|\phi|(x) + |\psi|(x_n)\big] \le |\phi|(x) + \epsilon \, .
\end{aligned}
$$

Since $\epsilon > 0$ is arbitrary, we see that $g(y) \le |\phi|(x)$ holds for all $0 \le y \le x$. Therefore,

$$
g^+(x) = \sup\{g(y)\colon \ 0 \le y \le x\} \le |\phi|(x)
$$

holds for all $x \in E^+$. Hence, $g^+ \in B_f$. Similarly, $g^- \in B_f$, and therefore $g = g^+ - g^- \in B_f$, and the proof is finished. ∎

Exercises

1. Show that if $f\colon E \to \mathbb{R}$ is a σ-order continuous linear functional on an Archimedean Riesz space, then f is order bounded.

2. Consider an Archimedean Riesz space E. If $f \in E_n^\sim$ and $g \in E^\sim$, then show that the following statements are equivalent.
   (a) $f \perp g$.
   (b) $C_g \subseteq N_f$.
   (c) $C_g \perp C_f$.

3. Establish the following properties of perfect Riesz spaces.
   (a) Every band of a perfect Riesz space is a perfect Riesz space in its own right.
   (b) If F is a perfect Riesz space, then $\mathcal{L}_b(E, F)$ is likewise a perfect Riesz space for each Riesz space E. (In particular, the order dual of every Riesz space is a perfect Riesz space.)
   (c) If E is a perfect Riesz space, then $E^{\sim\sim}$ is retractable on E.

4. Let E and F be two Riesz spaces such that $E_n^\sim = E^\sim$ and $F^\sim$ separates the points of F. Then show that every positive operator from E to F

is order continuous. (Also, state and prove the corresponding result for σ-order continuous operators.)

5. Let E be a Riesz space. If $0 \leq f \in E^\sim$, then show that

$$f(x \vee y) = \sup\{g(x) + h(y): \ g \wedge h = 0 \ \text{and} \ g + h = f\},$$

and

$$f(x \wedge y) = \inf\{g(x) + h(y): \ g \wedge h = 0 \ \text{and} \ g + h = f\}$$

hold for all $x, y \in E$.

6. Let E and F be two Riesz spaces with F Dedekind complete, and let $f, g \in E^\sim$ and $x, y \in F$. Show that if either $f \perp g$ in $E^\sim$ or $x \perp y$ in F, then $f \otimes x \perp g \otimes y$ holds in $\mathcal{L}_b(E, F)$.

7. Let E and F be two Riesz spaces with F Dedekind complete. If $x, y \in F$ and $f \in E^\sim$, then prove the following identities in $\mathcal{L}_b(E, F)$:
    (a) $(f \otimes x) \vee (f \otimes y) = f^+ \otimes (x \vee y) - f^- \otimes (x \wedge y)$.
    (b) $(f \otimes x)^+ = f^+ \otimes x^+ + f^- \otimes x^-$.
    (c) $(f \otimes x)^- = f^+ \otimes x^- + f^- \otimes x^+$.

8. Let $T: E \to F$ be an order continuous operator between two Riesz spaces such that $E^\sim$ separates the points of E and let $0 < y \in E$ and $x \in F^+$. If $Ty \wedge x = 0$ holds in F, then show that there exists some $0 < f \in E^\sim$ with $T \wedge (f \otimes x) = 0$.

9. Show that the singular linear functionals on $L_\infty[0, 1]$ separate the points of $L_\infty[0, 1]$. [*Hint:* Assume that $f > 0$. Fix some $\epsilon > 0$ such that $A = \{x \in [0, 1]: \ f(x) \geq \epsilon\}$ has positive measure. Write $A = \bigcup_{n=1}^\infty A_n$ with $\{A_n\}$ pairwise disjoint and each A_n having positive measure, and let $f_n = f \chi_{A_n}$. Now consider the Riesz subspace

$$G = \Big\{ \sum_{n=1}^\infty \alpha_n f_n: \ (\alpha_1, \alpha_2, \ldots) \in c \Big\},$$

and define the positive linear functional $\phi: G \to \mathbb{R}$ by

$$\phi\Big(\sum_{n=1}^\infty \alpha_n f_n \Big) = \lim_{n \to \infty} \alpha_n .$$

Since $\phi(g) \leq \frac{1}{\epsilon} \text{ess sup} \, |g|$ holds for all $g \in G$, it follows from Theorem 1.27 that ϕ has a positive linear extension to all of $L_\infty[0, 1]$. Now note that $\phi_s(f) > 0$.]

10. If $E = C[0, 1]$, then show that:
    (a) $E_n^\sim = E_c^\sim = \{0\}$.
    (b) The canonical embedding of E into $E^{\sim\sim}$ does not preserve infinite suprema and infima. [*Hint:* Consider the linear functional $\phi: C[0, 1] \to \mathbb{R}$ defined by $\phi(f) = f(1)$, and let $f_n(x) = x^n$. Then $f_n \downarrow 0$ holds in E, while $\hat{f}_n(\phi) = 1$ for all n.]

11. Let $E = L_0[0, 1]$. That is, let E be the Riesz space of all Lebesgue measurable functions on $[0, 1]$ with the ordering $f \geq g$ if $f(x) \geq g(x)$ holds for almost all x. Show that $E^\sim = \{0\}$.

12. For each $a \in [0,1]$ define the positive linear functional $\delta_a \colon C[0,1] \to \mathbb{R}$ by
$\delta_a(f) = f(a)$ for each $f \in C[0,1]$.

    If $T \colon C[0,1] \to C[0,1]$ is the regular operator defined by

$$[Tf](x) = f(x) - f(x^2),$$

then establish the following properties:

  (a) The modulus of T exists and is given by the formula

$$[|T|f](x) = f(x) + f(x^2).$$

  (b) $|T'|\delta_a = |T|'\delta_a$ holds for each $0 < a < 1$.
  (c) $|T'|\delta_0 < |T|'\delta_0$ and $|T'|\delta_1 < |T|'\delta_1$.

13. Let E be a Dedekind complete Riesz space such that $E^\sim$ separates its points. Show that $E_c^\sim = E^\sim$ holds if and only if $T_n \downarrow T \geq 0$ in $\mathcal{L}_b(E)$ implies $T_n^2 \downarrow T^2$.

14. State and prove the analogue of Theorem 1.79 for order continuous linear functionals.

15. Let E be a Riesz space, and let $\phi \in E^\sim$. Then show that for each $x \in E^+$ and $0 \leq f \in E^\sim$ we have

$$[P_\phi f](x) = \inf\left\{\sup f(x_n) \colon 0 \leq x_n \leq x \text{ and } \lim_{n \to \infty} |\phi|(x - x_n) = 0\right\}.$$

16. Let E be a Riesz space and let $\phi \in E^\sim$. If Q denotes the order projection of $E^\sim$ onto the band $\{\phi\}^{\mathrm{d}}$, then show that for each $x \in E^+$ and each $0 \leq f \in E^\sim$ we have

$$\begin{aligned}
[Qf](x) &= \inf_{\epsilon > 0} \sup \left\{f(y) \colon 0 \leq y \leq x \text{ and } |\phi|(y) < \epsilon\right\} \\
&= \sup \left\{\inf f(x_n) \colon 0 \leq x_n \downarrow\leq x \text{ and } |\phi|(x_n) \downarrow 0\right\}.
\end{aligned}$$

17. Let E be a Riesz space, and let f be in the band generated by g in $E^\sim$. Show that $C_f \subseteq C_g$.

18. Let E be a Riesz space, and let $f \in E_n^\sim$. If $g \in E^\sim$, then show that f is in the band generated by g if and only if $|g|(|x|) = 0$ implies $f(x) = 0$.

19. Let E be a Riesz space, and let B be the band generated in $E^\sim$ by a sequence $\{\phi_1, \phi_2, \dots\} \subseteq E^\sim$. Show that for each $x \in E^+$ and each $0 \leq f \in E^\sim$ we have

$$[P_B f](x) = \inf\left\{\sup f(x_n) \colon 0 \leq x_n \uparrow\leq x \text{ and } |\phi_k|(x - x_n) \downarrow 0 \text{ for all } k\right\}.$$

20. Let E be a Riesz space and let $g \in E_c^\sim$. Then show that

$$[P_g f](x) = f_c(x)$$

holds for all $x \in C_g$ and all $f \in E^\sim$.

21. (Nakano [**150**]) Let E be a Dedekind σ-complete Riesz space, and let $f \colon E \to \mathbb{R}$ be a linear functional. If some sequence $\{f_n\}$ of $E^\sim$ converges pointwise to f (i.e., $f_n(x) \to f(x)$ holds for all $x \in E$), then show that $f \in E^\sim$. Can the Dedekind σ-completeness be replaced by Archimedeaness?

22. A sequence $\{u_n\}$ in a Riesz space is said to be **disjoint** whenever $n \neq m$ implies $u_n \perp u_m$.

Let $\{x_n\}$ be a disjoint sequence in a Riesz space E. If $\{f_n\}$ is a sequence of $E^{\sim}$, then show that there exists a disjoint sequence $\{g_n\}$ of $E^{\sim}$ (which is positive if the sequence $\{f_n\}$ is positive) with $|g_n| \leq |f_n|$ and $g_n(x_n) = f_n(x_n)$ for all n. [*Hint:* Consider each x_n as a vector of $(E^{\sim})_n^{\sim}$, and let $N_n = \{h \in E^{\sim}: |h|(|x_n|) = 0\}$ and $C_n = N_n^{\mathrm{d}}$. Since N_n is a band, we have $E^{\sim} = N_n \oplus C_n$. The disjointness condition $x_n \perp x_m$ implies (by Theorem 1.67) $C_n \perp C_m$ $(n \neq m)$. Now let g_n be the projection of f_n onto the band C_n.]

Chapter 2

Components,
Homomorphisms,
and Orthomorphisms

The basic properties of positive operators were investigated in the previous chapter. In this chapter we shall study the lattice behavior of three specific classes of positive operators.

The first section of this chapter deals with the components of a positive operator. If $T\colon E \to F$ is a positive operator between Riesz spaces with F Dedekind complete, then a positive operator $S\colon E \to F$ is said to be a *component* of T whenever $S \wedge (T - S) = 0$ holds in $\mathcal{L}_b(E, F)$. The operators of the form QTP, where Q is an order projection on F and P is an order projection on E, are the simplest components of T, and they are called *elementary components*. We shall see that the elementary components are the "building blocks" for all components, and to a larger extend the building blocks for all positive operators in the ideal generated by T.

The second class of positive operators under investigation is the class of lattice homomorphisms. An operator $T\colon E \to F$ between two Riesz spaces is said to be a *lattice homomorphism* whenever it preserves the lattice operations, that is, whenever $T(x \vee y) = T(x) \vee T(y)$ holds for all $x, y \in E$. The remarkable properties of lattice homomorphisms are unraveled in the second section of this chapter.

An operator $T\colon E \to F$ between two Riesz spaces is said to *preserve disjointness* whenever $x \perp y$ in E implies $Tx \perp Ty$ in F. Among the operators preserving disjointness are the band preserving operators on a

Riesz space, i.e., the operators that leave all bands invariant. The order
bounded band preserving operators are called *orthomorphisms*, and they
are the subject of discussion in the third section of this chapter.

2.1. The Components of a Positive Operator

Let $T\colon E \to F$ be a positive operator between two Riesz spaces with F
Dedekind complete. The purpose of this section is to describe the components of T. Recall that a positive operator $S\colon E \to F$ is a **component** of T
whenever $S \wedge (T - S) = 0$. As usual, $\mathcal{C}_T$ will denote the Dedekind complete
Boolean algebra of all components of T; see Theorem 1.49. That is,

$$\mathcal{C}_T = \big\{ S \in \mathcal{L}_{\mathrm{b}}(E, F)\colon \ S \wedge (T - S) = 0 \big\}.$$

When a positive operator is defined by a matrix, its components have
a simple description. The details are included in the next theorem whose
proof follows immediately from Theorem 1.24.

Theorem 2.1. *Let $\{E_i\colon \ i \in I\}$ and $\{F_j\colon \ j \in J\}$ be two families of Riesz
spaces with each F_j Dedekind complete, and let $T = [T_{ji}]$ and $S = [S_{ji}]$ be
two positive operators from $\sum \oplus E_i$ to $\sum \oplus F_j$. Then S is a component of
T if and only if each S_{ji} is a component of T_{ji} for all $i \in I$ and all $j \in J$.*

Now fix an order projection P on E and an order projection Q on the
Dedekind complete Riesz space F.

Then the operator $\Pi\colon \mathcal{L}_{\mathrm{b}}(E, F) \to \mathcal{L}_{\mathrm{b}}(E, F)$ defined by $\Pi(T) = QTP$
satisfies $0 \leq \Pi(T) \leq T$ for each $T \in \mathcal{L}_{\mathrm{b}}^+(E, F)$ and $\Pi^2 = \Pi$. Thus, by Theorem 1.44 the operator Π is an order projection on $\mathcal{L}_{\mathrm{b}}(E, F)$. In particular,
if T is a positive operator, then for each order projection Q on F and each
order projection P on E the operator QTP is a component of T. Any component of the form QTP will be referred to as an **elementary component**
of T. Any component of the form $\bigvee_{i=1}^n Q_i T P_i$ (i.e., any finite supremum
of elementary components) will be called a **simple component** of T. We
shall denote the collection of all simple components of T by $\mathcal{S}_T$.

It is not difficult to see that if Π_1 and Π_2 are two order projections on
a Riesz space, then

$$\Pi_1 u \wedge \Pi_2 v = \Pi_1 \Pi_2 (u \wedge v)$$

holds for all $u, v \geq 0$. Thus, if Q_1, Q_2 are order projections on F and P_1,
P_2 are order projections on E, then the above identity applied to the order
projections $\Pi_1(T) = Q_1 T P_1$ and $\Pi_2(T) = Q_2 T P_2$ on $\mathcal{L}_{\mathrm{b}}(E, F)$ shows that

$$(Q_1 S P_1) \wedge (Q_2 T P_2) = Q_1 Q_2 (S \wedge T) P_1 P_2$$

holds for all $S, T \in \mathcal{L}_b^+(E, F)$. In particular, it follows that if either $Q_1 Q_2 = 0$ or $P_1 P_2 = 0$, then $Q_1 T P_1 \perp Q_2 S P_2$ holds for all $S, T \in \mathcal{L}_b^+(E, F)$. This observation shows that

$$T - QTP = (I - Q)T + QT(I - P) = [(I - Q)T] \vee [QT(I - P)].$$

Now let $T : E \rightarrow F$ be a positive operator between two Riesz spaces with F Dedekind complete. If $\bigvee_{i=1}^{n} Q_i T P_i$ and $\bigvee_{j=1}^{m} R_j T S_J$ are two simple components of T, then by the preceding discussion we see that

$$\left(\bigvee_{i=1}^{n} Q_i T P_i \right) \vee \left(\bigvee_{j=1}^{m} R_j T S_j \right) \in \mathcal{S}_T,$$

$$\left(\bigvee_{i=1}^{n} Q_i T P_i \right) \wedge \left(\bigvee_{j=1}^{m} R_j T S_j \right) = \bigvee_{i=1}^{n} \bigvee_{j=1}^{m} Q_i R_j T P_i S_j \in \mathcal{S}_T,$$

and

$$T - \bigvee_{i=1}^{n} Q_i T P_i = \bigwedge_{i=1}^{n} (T - Q_i T P_i) = \bigwedge_{i=1}^{n} [(I - Q_i)T \vee Q_i T(I - P_i)] \in \mathcal{S}_T.$$

Therefore, $\mathcal{S}_T$ is a Boolean subalgebra of the (Dedekind complete) Boolean of all components $\mathcal{C}_T$ of T.

Our next objective is to show how to obtain (in certain cases) every component of T from the simple components. This will show, of course, that the elementary components are the "building blocks" for the components of an operator. We start with a lemma.

Lemma 2.2. *Let E have the principal projection property, let F be Dedekind complete, and let $S, T : E \rightarrow F$ be two positive operators satisfying $0 \leq S \leq T$. If $x_1, \ldots, x_n \in E^+$ are pairwise disjoint vectors and $x = x_1 + \cdots + x_n$, then there exist order projections $P_1, \ldots, P_n$ on E and order projections $Q_1, \ldots, Q_n$ on F such that:*

(1) $P_i P_j = 0$ *for* $i \neq j$.
(2) $P_1 x + \cdots + P_n x = x$.
(3) $\left| S - \sum_{i=1}^{n} Q_i T P_i \right| x \leq \sum_{i=1}^{n} S x_i \wedge (T - S) x_i$.

Proof. Let $x_1, \ldots, x_n \in E^+$ be pairwise disjoint such that $x_1 + \cdots + x_n = x$. Denote by P_i ($1 \leq i \leq n$) the order projection of E onto the band generated by x_i. From $x_i \wedge x_j = 0$ ($i \neq j$), it follows that $P_i P_j = 0$ ($i \neq j$). In addition, $P_i x = x_i$ holds for each i, and so $P_1 x + \cdots + P_n x = x_1 + \cdots + x_n = x$.

Now let Q_i ($1 \leq i \leq n$) denote the order projection of F onto the band generated by $(2Sx_i - Tx_i)^+$. Clearly, $Q_i [(2S - T)x_i] = (2Sx_i - Tx_i)^+$. Now notice that

$$\left|S - \sum_{i=1}^{n} Q_i T P_i\right| x \leq \left[S - \sum_{i=1}^{n} Q_i S P_i\right] x + \left[\sum_{i=1}^{n} Q_i (T - S) P_i\right] x$$

$$= Sx - \sum_{i=1}^{n} Q_i \left[(2S - T)P_i x\right]$$

$$= Sx - \sum_{i=1}^{n} (2Sx_i - Tx_i)^+$$

$$= Sx - \sum_{i=1}^{n} \left[Sx_i - (T - S)x_i\right]^+$$

$$= Sx - \sum_{i=1}^{n} \left[Sx_i - Sx_i \wedge (T - S)x_i\right]$$

$$= \sum_{i=1}^{n} Sx_i \wedge (T - S)x_i \,,$$

and the proof is finished. ∎

The next result shows that every component of a positive operator can be approximated by elementary components. This will be the basic tool for this section.

Theorem 2.3. *Let* $T \colon E \to F$ *be a positive operator between two Riesz spaces where* E *has the principal projection property and* F *is Dedekind complete. If* S *is a component of* T, *then given* $x \in E^+$, $0 \leq f \in F_n^\sim$, *and* $\epsilon > 0$, *there exist pairwise disjoint order projections* $P_1, \dots, P_n$ *on* E *and order projections* $Q_1, \dots, Q_n$ *on* F *such that the simple component* $\sum_{i=1}^{n} Q_i T P_i$ $(= \bigvee_{i=1}^{n} Q_i T P_i)$ *of* T *satisfies*

$$\left\langle f, \left|S - \sum_{i=1}^{n} Q_i T P_i\right| x\right\rangle < \epsilon.$$

Proof. Let $x \in E^+$, $0 \leq f \in F_n^\sim$, let $\epsilon > 0$ be fixed, and let S be a component of T. By Theorem 1.51 we have

$$\left\{\sum_{i=1}^{n} Sx_i \wedge (T - S)x_i \colon x_i \geq 0, \; x_i \perp x_j, \; \sum_{i=1}^{n} x_i = x\right\} \downarrow \left[S \wedge (T - S)\right]x = 0,$$

and so, by the order continuity of f, there exist pairwise disjoint vectors $x_1, \dots, x_n \in E^+$ with $x_1 + \cdots + x_n = x$ and $\left\langle f, \sum_{i=1}^{n} Sx_i \wedge (T - S)x_i\right\rangle < \epsilon$.

Now by Lemma 2.2 there exist pairwise disjoint order projections $P_1, \ldots, P_n$ on E and order projections $Q_1, \ldots, Q_n$ on F such that

$$\left| S - \sum_{i=1}^{n} Q_i T P_i \right| x \leq \sum_{i=1}^{n} S x_i \wedge (T - S) x_i .$$

Therefore, $\langle f, |S - \sum_{i=1}^{n} Q_i T P_i| x \rangle < \epsilon$. On the other hand, since the order projections $P_1, \ldots, P_n$ are pairwise disjoint operators, the elementary components $Q_1 T P_1, \ldots, Q_n T P_n$ are also pairwise disjoint, and consequently we have $\sum_{i=1}^{n} Q_i T P_i = \bigvee_{i=1}^{n} Q_i T P_i \in \mathcal{S}_T$. ∎

Let E be a Riesz space. Then for a subset A of E we shall employ the following notation:

$$A^{\uparrow} := \{ x \in E \colon \exists \text{ a sequence } \{x_n\} \subseteq A \text{ with } x_n \uparrow x \}, \text{ and}$$

$$A^{\uparrow} := \{ x \in E \colon \exists \text{ a net } \{x_\alpha\} \subseteq A \text{ with } x_\alpha \uparrow x \}.$$

The meanings of $A^{\downarrow}$ and $A^{\downarrow}$ are analogous. As usual, also we shall write $A^{\uparrow\downarrow} = (A^{\uparrow})^{\downarrow}$, $A^{\uparrow\uparrow\downarrow} := ((A^{\uparrow})^{\uparrow})^{\downarrow}$, etc. Clearly, $A^{\uparrow\uparrow} = A^{\uparrow}$ and $A^{\downarrow\downarrow} = A^{\downarrow}$.

Now consider a positive operator $T \colon E \to F$ where F is Dedekind complete. Since $\mathcal{S}_T$ is a Boolean algebra, it is closed under finite suprema and infima. In particular, all "ups and downs" of $\mathcal{S}_T$ (for instance $(\mathcal{S}_T)^{\uparrow}$ and $(\mathcal{S}_T)^{\uparrow\downarrow}$) are likewise closed under finite suprema and infima (and hence they are also directed upward and downward). As we shall see, a finite process of "ups and downs" of simple components suffices to generate all components of a positive operator.

But first, we shall approximate pointwise an arbitrary component of a positive operator T by a component of $(\mathcal{S}_T)^{\uparrow\downarrow}$ as follows.

Lemma 2.4. *Let* $T \colon E \to F$ *be a positive operator between two Riesz spaces where* E *has the principal projection property and* F *is Dedekind complete. If* S *is a component of* T, *then given* $x \in E^+$ *and* $0 \leq f \in F_n^{\sim}$, *there exists a component* $C \in (\mathcal{S}_T)^{\uparrow\downarrow}$ *satisfying* $\langle f, |S - C|x \rangle = 0$.

Proof. Fix $x \in E^+$ and $0 \leq f \in F_n^{\sim}$. Then for each $n \in \mathbb{N}$ there exists (according to Theorem 2.3) some $B_n \in \mathcal{S}_T$ with $\langle f, |S - B_n|x \rangle < 2^{-n}$. For each n put $C_n = \bigvee_{j=n}^{\infty} B_j$, and note that in view of $\bigvee_{j=n}^{n+i} B_j \uparrow_i C_n$ we have $\{C_n\} \subseteq (\mathcal{S}_T)^{\uparrow}$. Now from the inequality

$$\left| S - \bigvee_{j=n}^{n+i} B_j \right| = \left| \bigvee_{j=n}^{n+i} (B_j - S) \right| \leq \sum_{j=n}^{n+i} |S - B_j|,$$

we see that

$$\left\langle f, \left|S - \bigvee_{j=n}^{n+i} B_j\right|x\right\rangle \leq \sum_{j=n}^{\infty}\langle f, |S - B_j|x\rangle < 2^{1-n}.$$

The latter combined with $\left|S - \bigvee_{j=n}^{n+i} B_j\right| \xrightarrow{o} |S - C_n|$ and the fact that $f \in F_n^{\sim}$ shows that for each n we have

$$\langle f, |S - C_n|x\rangle \leq 2^{1-n}. \tag{$\star$}$$

Finally, let $C_n \downarrow C$. Then clearly $C \in (\mathcal{S}_T)^{\uparrow\downarrow}$ (of course, $C = \limsup C_n$). On the other hand, in view of $|S - C_n| \xrightarrow{o} |S - C|$ and $f \in F_n^{\sim}$, it follows from $(\star)$ that $\langle f, |S - C|x\rangle = 0$, and the proof is finished. ∎

A better "pointwise" approximation by components of $(\mathcal{S}_T)^{\uparrow\downarrow}$ is achieved in the next lemma.

Lemma 2.5. *Let $T\colon E \to F$ be a positive operator between two Riesz spaces, where E has the principal projection property and F is Dedekind complete such that $F_n^{\sim}$ separates the points of F. If S is a component of T, then given $x \in E^+$ and $0 \leq f \in F_n^{\sim}$ there exists a component $C \in (\mathcal{S}_T)^{\uparrow\downarrow}$ with*

$$0 \leq C \leq S \qquad and \qquad \langle f, Cx\rangle = \langle f, Sx\rangle.$$

Proof. Fix $x \in E^+$ and $0 \leq f \in F_n^{\sim}$, and let

$$D = \{C \in (\mathcal{S}_T)^{\uparrow\downarrow}\colon \langle f, |S - C|x\rangle\}.$$

By Lemma 2.4 the set D is nonempty, and an easy argument shows that D is directed downward. Let $D \downarrow C_0$. Clearly, $C_0 \in (\mathcal{S}_T)^{\uparrow\downarrow\downarrow} = (\mathcal{S}_T)^{\uparrow\downarrow}$, and hence $\langle f, |S - C_0|x\rangle = 0$, i.e., $C_0 \in D$.

We claim that $0 \leq C_0 \leq S$. To this end, let $0 \leq g \in F_n^{\sim}$ and $y \in E^+$. By Lemma 2.4, there is some $R \in (\mathcal{S}_T)^{\uparrow\downarrow}$ with $\langle f + g, |S - R|(x + y)\rangle = 0$. This implies $R \in D$, and so $C_0 \leq R$. On the other hand, we have

$$\langle g, (C_0 - S)^+ y\rangle \leq \langle f + g, |R - S|(x + y)\rangle = 0.$$

Since $0 \leq g \in F_n^{\sim}$ and $y \in E^+$ are arbitrary, we see that $(C_0 - S)^+ = 0$. Hence, $0 \leq C_0 \leq S$, and clearly $\langle f, C_0 x\rangle = \langle f, Sx\rangle$ holds. ∎

We now come to the main result of this section that characterizes the components of a positive operator. It is due to B. de Pagter [159] and has its origins in the work of H. Leinfelder [111]. However, the proof as well as the approach in this section are due to the authors [14].

Theorem 2.6 (de Pagter). *If $T\colon E \to F$ is a positive operator between two Riesz spaces where E has the principal projection property and F is Dedekind complete such that $F_n^{\sim}$ separates the points of F, then*

$$\mathcal{C}_T = (\mathcal{S}_T)^{\uparrow\downarrow\uparrow} = (\mathcal{S}_T)^{\downarrow\uparrow\downarrow}.$$

Proof. Let $S \in \mathcal{C}_T$ be fixed, and let

$$\mathcal{B} = \left\{ C \in (\mathcal{S}_T)^{\uparrow\downarrow} \colon\ 0 \le C \le S \right\}.$$

Clearly, $\mathcal{B} \uparrow$ holds, and by Lemma 2.5 we know that $\mathcal{B} \ne \emptyset$. Let $\mathcal{B} \uparrow C_0$, and note that $C_0 \in (\mathcal{S}_T)^{\uparrow\downarrow\uparrow}$. Also, observe that $0 \le C_0 \le S$ holds. On the other hand, if $0 \le f \in F_n^\sim$ and $x \in E^+$ are arbitrary, then by Lemma 2.5 there exists some $C \in \mathcal{B}$ with $\langle f, Cx \rangle = \langle f, Sx \rangle$. Taking into account that $0 \le C \le C_0 \le S$ must be true, we see that $\langle f, C_0 x \rangle = \langle f, Sx \rangle$ also must be the case, and from this we see that $C_0 = S$. Thus, $S = C_0 \in (\mathcal{S}_T)^{\uparrow\downarrow\uparrow}$ holds, and so $\mathcal{C}_T = (\mathcal{S}_T)^{\uparrow\downarrow\uparrow}$. For the other equality note that a component S of T satisfies $S \in (\mathcal{S}_T)^{\uparrow\downarrow\uparrow}$ if and only if $T - S \in (\mathcal{S}_T)^{\downarrow\uparrow\downarrow}$. ∎

We saw previously that a pair of order projections P on E and Q on F gives rise to the component QTP for each positive operator T. Thus, an abundance of order projections on E and F guarantees that each positive operator has an abundance of components. However, as we shall see next, even without nontrivial order projections on E a positive operator always has a plethora of components.

By Theorem 1.28, if F is Dedekind complete, then for each ideal A of E and each positive operator $T \colon E \to F$ a new positive operator T_A can be defined via the formula

$$T_A(x) := \sup\{T(y) \colon\ y \in A \text{ and } 0 \le y \le x\}, \quad x \in E^+.$$

As we shall see next, T_A is actually a component of T.

Now let the ideal A be fixed. Then a mapping $T \mapsto T_A$ is defined from $\mathcal{L}_b^+(E, F)$ to $\mathcal{L}_b^+(E, F)$. Since for each $x \in E^+$ the set $\{y \in A \colon\ 0 \le y \le x\}$ is directed upward, it is not difficult to see that

$$(S + T)_A(x) = S_A(x) + T_A(x)$$

holds for all $S, T \in \mathcal{L}_b^+(E, F)$ and all $x \in E^+$. This shows, of course, that the mapping $T \mapsto T_A$ is additive on $\mathcal{L}_b^+(E, F)$, and thus, by Theorem 1.10, it is extendable to a positive operator from $\mathcal{L}_b(E, F)$ to $\mathcal{L}_b(E, F)$.

Remarkably, the positive operator $T \mapsto T_A$ is an order projection on $\mathcal{L}_b(E, F)$. The details follow.

Theorem 2.7. *If F is Dedekind complete and A is an ideal of E, then the operator $T \mapsto T_A$, from $\mathcal{L}_b(E, F)$ to $\mathcal{L}_b(E, F)$, is an order projection.*

In particular, if $T \colon E \to F$ is a positive operator, then T_A is a component of T for each ideal A of E.

Proof. The inequality $0 \le T_A \le T$ for each $T \in \mathcal{L}_b^+(E, F)$ shows that the operator $T \mapsto T_A$ is bounded by the identity operator of $\mathcal{L}_b(E, F)$. Also, in view of $T_A = T$ on A, it follows that $(T_A)_A = T_A$ holds. Thus, statement (2) of Theorem 1.44 is true and shows that the operator $T \mapsto T_A$ is an order projection. ∎

So far, we have seen that the elementary components of a positive operator are the building blocks for all components. In turn, we shall show next that the components of a positive operator are the building blocks for all positive operators dominated by the operator. To do this, we need to recall a few basic facts from the theory of Riesz spaces.

Consider a Riesz space E and a vector $0 < x \in E$. An x-**step function** is any vector $s \in E$ for which there exist pairwise disjoint components $x_1, \ldots, x_n$ of x with $x_1 + \cdots + x_n = x$ and real numbers $\alpha_1, \ldots, \alpha_n$ satisfying

$$s = \sum_{i=1}^{n} \alpha_i x_i .$$

(This last expression is referred to as representation of s as an x-step function.) Clearly, every component of x is an x-step function. Also, it is easy to see that every x-step function belongs to the ideal generated by x. Now if u and v are x-step functions with representations

$$u = \sum_{i=1}^{n} \alpha_i x_i \quad \text{and} \quad v = \sum_{j=1}^{m} \beta_j y_j ,$$

then

$$u = \sum_{i=1}^{n} \sum_{j=1}^{m} \alpha_i (x_i \wedge y_j) \quad \text{and} \quad v = \sum_{i=1}^{n} \sum_{j=1}^{m} \beta_j (x_i \wedge y_j)$$

are also representations of u and v as x-step functions. In particular, it follows that:

$$u + v = \sum_{i=1}^{n} \sum_{j=1}^{m} (\alpha_i + \beta_j)(x_i \wedge y_j) .$$

$$\lambda u = \sum_{i=1}^{n} \lambda \alpha_i x_i .$$

$$|u| = \sum_{i=1}^{n} |\alpha_i| x_i .$$

Therefore, the collection of all x-step functions is a Riesz subspace of E.

The term "step function" is justified by the following example from the theory of integration. Let $E = L_1[0, 1]$ and consider the constant function **1**. It is easy to see that the components of **1** are precisely the characteristic functions of the Lebesgue measurable subsets of $[0, 1]$. Consequently, a **1**-step function as defined above is exactly a step function in the sense of the theory of integration.

When E has the principal projection property, every vector in the ideal generated by a vector $x > 0$ can be "approximated uniformly" by x-step functions. This can be seen from the following fundamental spectral theorem of H. Freudenthal [**67**].

Theorem 2.8 (Freudenthal's Spectral Theorem). *Let E be a Riesz space with the principal projection property and let $0 < x \in E$. Then for every $y \in E_x$ there exists a sequence $\{u_n\}$ of x-step functions satisfying*

$$0 \le y - u_n \le \tfrac{1}{n}x \ \text{ for each } \ n \ \text{ and } \ u_n \uparrow y \,.$$

Proof. Without loss of generality we can assume that $0 < y \le x$ holds. For each $0 \le \alpha \le 1$ denote by P_α the order projection of E onto the band generated by $(y - \alpha x)^+$. The family $\{P_\alpha \colon 0 \le \alpha \le 1\}$ of order projections has the following properties:

(a) $P_0 = P_y$ and $P_1 = 0$.

(b) If $0 \le \alpha \le \beta \le 1$, then $0 \le P_\beta \le P_\alpha \le I$.

(c) If $0 \le \alpha \le \beta \le 1$ holds, then

$$\alpha(P_\alpha - P_\beta)x \le (P_\alpha - P_\beta)y \le \beta(P_\alpha - P_\beta)x \,.$$

Indeed, from

$$
\begin{aligned}
(P_\alpha - P_\beta)y - \alpha(P_\alpha - P_\beta)x &= (I - P_\beta)P_\alpha(y - \alpha x) \\
&= (I - P_\beta)(y - \alpha x)^+ \ge 0\,,
\end{aligned}
$$

it follows that $\alpha(P_\alpha - P_\beta)x \le (P_\alpha - P_\beta)y$. Also, from

$$
\begin{aligned}
(P_\alpha - P_\beta)y - \beta(P_\alpha - P_\beta)x &= P_\alpha(y - \beta x) - P_\beta(y - \beta x) \\
&= P_\alpha(y - \beta x) - (y - \beta x)^+ \\
&\le P_\alpha(y - \beta x)^+ - (y - \beta x)^+ \le 0\,,
\end{aligned}
$$

we see that $(P_\alpha - P_\beta)y \le \beta(P_\alpha - P_\beta)x$.

(d) If $0 \le \alpha \le \beta \le \alpha_1 \le \beta_1 \le 1$ holds, then $(P_\alpha - P_\beta)x$ and $(P_{\alpha_1} - P_{\beta_1})x$ are two disjoint components of x.

To see this, note that

$$0 \le (P_\alpha - P_\beta)x \wedge (P_{\alpha_1} - P_{\beta_1})x \le (x - P_\beta x) \wedge P_\beta x = 0\,,$$

and

$$
\begin{aligned}
0 &\le (P_\alpha - P_\beta)x \wedge \left[x - (P_\alpha - P_\beta)x\right] \\
&\le P_\alpha x \wedge (x - P_\alpha x) + (x - P_\beta x) \wedge P_\beta x = 0\,.
\end{aligned}
$$

Now let $\epsilon > 0$. Since the collection of all x-step functions is a Riesz subspace, it is enough to establish the existence of some x-step function u with $0 \le y - u \le \epsilon x$. To this end, let $0 = \alpha_0 < \alpha_1 < \cdots < \alpha_n = 1$ be a partition of $[0,1]$ with mesh less than ϵ. Put $x_i = (P_{\alpha_{i-1}} - P_{\alpha_i})x$ for each $i = 1, \ldots, n$, and note that $x_1 + \cdots + x_n \leq x$ and that $x_i \wedge x_j = 0$ holds for $i \neq j$. On the other hand, using (c) above, it is easy to see that the x-step function $u = \sum_{i=1}^n \alpha_{i-1} x_i$ satisfies

$$
u = \sum_{i=1}^n \alpha_{i-1} x_i \ \leq \ \sum_{i=1}^n (P_{\alpha_{i-1}} - P_{\alpha_i})y = y \leq \sum_{i=1}^n \alpha_i x_i
$$

$$
\leq \ \sum_{i=1}^n (\alpha_{i-1} + \epsilon) x_i \leq u + \epsilon x,
$$

and so $0 \leq y - u \leq \epsilon x$ holds, as desired. ∎

For a number of interesting applications of Freudenthal's spectral theorem the reader is referred to Chapter 6 of the book by W. A. J. Luxemburg and A. C. Zaanen [**132**].

For positive operators, the following special case of Theorem 2.8 will be very useful. It indicates that a property of a positive operator is inherited by operators it dominates whenever its components inherit the same property.

Theorem 2.9. *Let $S, T: E \to F$ be positive operators between two Riesz spaces with F Dedekind complete. If $0 \leq S \leq T$ holds, then for each $\epsilon > 0$ there exist positive real numbers $\alpha_1, \ldots, \alpha_n$ and components $C_1, \ldots, C_n$ of T satisfying*

$$
0 \leq S - \sum_{i=1}^n \alpha_i C_i \leq \epsilon T.
$$

In particular, there exists a sequence $\{S_n\}$ of T-step functions satisfying

$$
0 \leq S - S_n \leq \tfrac{1}{n} T \quad \text{for each} \quad n \quad \text{and} \quad 0 \leq S_n \uparrow S.
$$

The preceding theorem combined with Theorem 2.3 shows that the following approximation theorem also is true.

Theorem 2.10. *Let E be a Riesz space with the principal projection property, let F be Dedekind complete, and let $S, T: E \to F$ be two positive operators such that $0 \leq S \leq T$. Then, given $x \in E^+$, $0 \leq f \in F_n^\sim$, and $\epsilon > 0$, there exist order projections $P_1, \ldots, P_n$ on E, order projections $Q_1, \ldots, Q_n$ on F and positive real numbers $\alpha_1, \cdots, \alpha_n$ such that*

$$
\left\langle f, \left| S - \sum_{i=1}^n \alpha_i Q_i T P_i \right| x \right\rangle < \epsilon.
$$

In the sequel, we shall use the following notation: If $\mathcal{Q} = (Q_1, \ldots, Q_n)$ is an n-tuple of order projections on F, $\mathcal{P} = (P_1, \ldots, P_n)$ is an n-tuple of order projections on E, and $T\colon E \to F$ is a positive operator, then

$$T_{\mathcal{Q},\mathcal{P}} := \bigvee_{i=1}^{n} Q_i T P_i\,.$$

Our next theorem characterizes the operators that lie in the band generated by another operator and is due to the authors [14].

Theorem 2.11 (Aliprantis–Burkinshaw). *Let E be a Riesz space with the principal projection property, and let F be Dedekind complete such that $F_{\mathrm{n}}^{\sim}$ separates the points of F. Then for two positive operators $S, T\colon E \to F$ the following statements are equivalent:*

(1) *T is in the principal band generated by S in $\mathcal{L}_{\mathrm{b}}(E, F)$.*

(2) *Given $x \in E^{+}$, $0 \le f \in F_{\mathrm{n}}^{\sim}$, and $\epsilon > 0$, there exists some $\delta > 0$ such that $\langle f, S_{\mathcal{Q},\mathcal{P}} x\rangle < \delta$ implies $\langle f, T_{\mathcal{Q},\mathcal{P}} x\rangle < \epsilon$.*

(3) *Given $x \in E^{+}$, $0 \le f \in F_{\mathrm{n}}^{\sim}$, and $\epsilon > 0$, there exists some $\delta > 0$ such that $\langle f, (S - S_{\mathcal{Q},\mathcal{P}})x\rangle < \delta$ implies $\langle f, (T - T_{\mathcal{Q},\mathcal{P}})x\rangle < \epsilon$.*

Proof. In view of the identity

$$S - \bigvee_{i=1}^{n} Q_i S P_i = \bigwedge_{i=1}^{n} \left[(I - Q_i)S \vee Q_i(I - P_i) \right],$$

it is easy to see that (2) and (3) are equivalent statements. We shall establish next the equivalence of (1) and (2).

(1) $\Longrightarrow$ (2) Assume that T is in the band generated by S. Then by Theorem 1.47 we have $T \wedge nS \uparrow T$. Let $x \in E^{+}$, $0 \le f \in F_{\mathrm{n}}^{\sim}$, and $\epsilon > 0$ be fixed.

From $[T \wedge nS](x) \uparrow T(x)$ we see that there exists some k satisfying $\langle f, (T - T \wedge kS)x\rangle < \epsilon$. Now observe that whenever $\mathcal{P} = (P_1, \ldots, P_n)$ and $\mathcal{Q} = (Q_1, \ldots, Q_n)$ are n-tuples of order projections on E and F, respectively, then we have

$$
\begin{aligned}
T_{\mathcal{Q},\mathcal{P}} &= \bigvee_{i=1}^{n} Q_i T P_i = \bigvee_{i=1}^{n} Q_i \big[(T - T \wedge kS) + T \wedge kS \big] P_i \\
&\le \bigvee_{i=1}^{n} Q_i (T - T \wedge kS) P_i + k \bigvee_{i=1}^{n} Q_i S P_i \le T - T \wedge kS + k S_{\mathcal{Q},\mathcal{P}}\,,
\end{aligned}
$$

and so

$$\langle f, T_{\mathcal{Q},\mathcal{P}} x\rangle \le \langle f, (T - T \wedge kS)x\rangle + k\langle f, S_{\mathcal{Q},\mathcal{P}} x\rangle < \epsilon + k\langle f, S_{\mathcal{Q},\mathcal{P}} x\rangle\,.$$

In particular, if $\langle f, S_{\mathcal{Q},\mathcal{P}} x\rangle < \delta = \frac{\epsilon}{k}$, then $\langle f, T_{\mathcal{Q},\mathcal{P}} x\rangle < 2\epsilon$, and the conclusion follows.

(2) $\Longrightarrow$ (1) Write $T = T_1 + T_2$, where T_1 is in the band generated by S and $T_2 \perp S$. It is enough to show that $T_2 = 0$.

To this end, let $x \in E^+$, $0 \le f \in F_n^\sim$, and $\epsilon > 0$ be fixed. Choose $0 < \delta < \epsilon$ so that statement (2) holds. Also, by Theorem 1.51 there exist pairwise disjoint elements $x_1, \ldots, x_n$ of E^+ with $x_1 + \cdots + x_n = x$ and $\langle f, \sum_{i=1}^n Sx_i \wedge T_2x_i \rangle < \delta$. From $0 \le T_2 \le S + T_2$ and Lemma 2.2, it follows that there exist pairwise disjoint order projections $P_1, \ldots, P_n$ on E and order projections $Q_1, \ldots, Q_n$ on F such that

$$\left| T_2 - \sum_{i=1}^n Q_i(T_2 + S)P_i \right| x \le \sum_{i=1}^n T_2x_i \wedge Sx_i \,.$$

Next, recall that in a Riesz space $u \perp v$ implies $|u| + |v| = |u - v|$; see Theorem 1.7. Taking into account $T_2 \perp S$, the last observation yields

$$\left| T_2 - \sum_{i=1}^n Q_iT_2P_i \right| + \left| \sum_{i=1}^n Q_iSP_i \right| = \left| T_2 - \sum_{i=1}^n Q_i(T_2 + S)P_i \right| .$$

By the above, we have $\langle f, \sum_{i=1}^n Q_iSP_ix \rangle < \delta$, and so $\langle f, \sum_{i=1}^n Q_iTP_ix \rangle < \epsilon$. Therefore,

$$\langle f, T_2x \rangle \le \left\langle f, \left(T_2 - \sum_{i=1}^n Q_iT_2P_i \right)x \right\rangle + \left\langle f, \sum_{i=1}^n Q_iTP_ix \right\rangle < \delta + \epsilon < 2\epsilon \,.$$

Since $0 \le f \in F_n^\sim$ and $\epsilon > 0$ are arbitrary, it follows that $T_2 = 0$, and the proof is finished. ∎

The final result of this section presents a formula for the order projection onto the band generated by a positive operator and is due to the authors [14].

Theorem 2.12 (Aliprantis–Burkinshaw). *Let $S \colon E \to F$ be a positive operator between two Riesz spaces, where E has the principal projection property and F is Dedekind complete such that $F_n^\sim$ separates the points of F. If $\mathbf{P}$ denotes the order projection of $\mathcal{L}_b(E, F)$ onto the band generated by S, then for each $x \in E^+$, each $0 \le f \in F_n^\sim$, and each positive operator $T \colon E \to F$, we have*

$$\langle f, \mathbf{P}Tx \rangle = \sup_{\epsilon > 0} \inf \{ \langle f, T_{Q,P}x \rangle \colon \ \langle f, (S - S_{Q,P})x \rangle < \epsilon \} \,.$$

Proof. Let $T \colon E \to F$ be a positive operator, and let $0 \le f \in F_n^\sim$ and $x \in E^+$. Put

$$r = \sup_{\epsilon > 0} \inf \{ \langle f, T_{Q,P}x \rangle \colon \ \langle f, (S - S_{Q,P})x \rangle < \epsilon \} \,,$$

and for simplicity write $R = \mathbf{P}T$.

Fix $\epsilon > 0$. Since $(T - R) \perp S$ and $0 \leq S \leq S + (T - R)$ for each $0 < \delta < \epsilon$, there exists (by Lemma 2.2) order projections $P_1, \ldots, P_n$ on E and other projections $Q_1, \ldots, Q_n$ on F such that:

(a) $P_i P_j = 0$ for $i \neq j$.

(b) $P_1 x + \cdots + P_n x = x$.

(c) $\langle f, |S - \sum_{i=1}^n Q_i [S + (T - R)] P_i | x \rangle$
$= \langle f, (S - \sum_{i=1}^n Q_i S P_i) x \rangle + \langle f, \sum_{i=1}^n Q_i (T - R) P_i x \rangle < \delta.$

Consequently, we have

$$\inf \{ \langle f, T_{Q,P} x \rangle : \ \langle f, (S - S_{Q,P}) x \rangle < \epsilon \}$$
$$\leq \sum_{i=1}^n \langle f, Q_i T P_i x \rangle$$
$$\leq \sum_{i=1}^n \langle f, Q_i (T - R) P_i x \rangle + \sum_{i=1}^n \langle f, R P_i x \rangle$$
$$< \delta + \langle f, R x \rangle$$

holds for all $0 < \delta < \epsilon$. This implies

$$\inf \{ \langle f, T_{Q,P} x \rangle : \ \langle f, (S - S_{Q,P}) x \rangle < \epsilon \} \leq \langle f, R x \rangle$$

for each $\epsilon > 0$, and so $r \leq \langle f, R x \rangle$.

On the other hand, if $\epsilon > 0$ is given, then by Theorem 2.11 there exists some $\delta > 0$ such that $\langle f, (S - S_{Q,P}) x \rangle < \delta$ implies $\langle f, (R - R_{Q,P}) x \rangle < \epsilon$. Thus, if $\langle f, (S - S_{Q,P}) x \rangle < \delta$, then

$$\langle f, R x \rangle \ = \ \langle f, (R - R_{Q,P}) x \rangle + \langle f, R_{Q,P} x \rangle$$
$$\leq \ \langle f, (R - R_{Q,P}) x \rangle + \langle f, T_{Q,P} x \rangle$$
$$< \ \epsilon + \langle f, T_{Q,P} x \rangle.$$

Therefore,

$$\langle f, R x \rangle \leq \epsilon + \inf \{ \langle f, T_{Q,P} x \rangle : \ \langle f, (S - S_{Q,P}) x \rangle < \delta \} \leq \epsilon + r.$$

Since $\epsilon > 0$ is arbitrary, we see that $\langle f, R x \rangle \leq r$ also holds, and therefore $r = \langle f, \mathbf{P} T x \rangle$, as desired. ∎

Exercises

1. If $T \colon \ell_2 \to \ell_2$ is the **shift operator**, defined by

$$T(x_1, x_2, \ldots) = (0, x_1, x_2, \ldots),$$

then describe its components.

2. If E is a Dedekind complete Riesz space, then show that the order projections of E are precisely the components of the identity operator. [*Hint*: $S^2 \wedge (I - S)^2 = S^2 \wedge (I - 2S + S^2) = S \wedge (I - S) + S^2 - S$.]

3. Let A be an ideal in a Dedekind complete Riesz space. If B is the band generated by A, then show that $I_A = P_B$.

4. Let E and F be two Riesz spaces with F Dedekind complete. If A is an ideal of E, then show that the projection band $\mathcal{B} = (A^\circ)^{\mathrm{d}}$ (for notation see Exercise 12 of Section 1.4) satisfies $P_\mathcal{B}(T) = T_A$ for each $T \in \mathcal{L}_b(E, F)$.

5. Let $T: E \to F$ be a positive operator between two Riesz spaces with F Dedekind complete. If S is a simple component of T, then show that there exists pairwise disjoint elementary components $Q_1 T P_1, \ldots, Q_n T P_n$ such that

$$S = Q_1 T P_1 + \cdots + Q_n T P_n .$$

6. Show that $S \in (\mathcal{S}_T)^{\downarrow\downarrow}$ if and only if $T - S \in (\mathcal{S}_T)^{\downarrow\uparrow}$.

7. (de Pagter [**159**]) Recall that a linear functional f on a Riesz space is said to be **strictly positive** whenever $x > 0$ implies $f(x) > 0$.

Let E be a Riesz space with the principal projection property, and let F be Dedekind complete. Assume further that E has a weak order unit and that F admits a strictly positive order continuous linear functional. Then show that for each order continuous positive operator $T: E \to F$ we have

$$\mathcal{C}_T = (\mathcal{S}_T)^{\downarrow\downarrow} = (\mathcal{S}_T)^{\downarrow\uparrow} .$$

8. Let $S: E \to F$ be a positive operator between two Riesz spaces, where E has the principal projection property and F is Dedekind complete with $F_n^\sim$ separating the points of F. Let $\mathbf{Q}$ denote the order projection of $\mathcal{L}_b(E, F)$ onto the band disjoint from S (i.e., onto $\mathcal{B}_S^{\mathrm{d}}$). Then show that for each $x \in E^+$, each $0 \le f \in F_n^\sim$, and each positive operator $T: E \to F$ we have

$$\langle f, \mathbf{Q}Tx \rangle = \inf_{\epsilon > 0} \sup\{\langle f, T_{\mathcal{Q},\mathcal{P}}x \rangle : \ \langle f, S_{\mathcal{Q},\mathcal{P}}x \rangle < \epsilon\} .$$

9. Consider the Riesz space $E = L_p[0, 1]$ with $0 < p < 1$, and recall that $E^\sim = \{0\}$. If $T: L_p[0, 1] \to L_p[0, 1]$ is a positive operator, then show that the Boolean algebra of components $\mathcal{C}_T$ satisfies

$$\mathcal{C}_T = (\mathcal{S}_T)^{\downarrow\downarrow} = (\mathcal{S}_T)^{\downarrow\uparrow} .$$

[*Hint*: Replace the linear functionals in the proofs of Theorem 2.3 and Lemma 2.4 by

$$\rho(u) = \int_0^1 \frac{|u(t)|}{1 + |u(t)|} \, dt .]$$

2.2. Lattice Homomorphisms

In this section a special class of positive operators will be studied. These are the operators that preserve the lattice operations, and they are known as *lattice* (or *Riesz*) *homomorphisms*.

Definition 2.13. *An operator* $T\colon E \to F$ *between two Riesz spaces is said to be a **lattice** (or **Riesz**) **homomorphism** whenever* $T(x \vee y) = T(x) \vee T(y)$ *holds for all* $x, y \in E$.

Observe that every lattice homomorphism $T\colon E \to F$ is necessarily a positive operator. Indeed, if $x \in E^+$, then

$$T(x) = T(x \vee 0) = T(x) \vee T(0) = [T(x)]^+ \geq 0$$

holds in F. Also, it is important to note that the range of a lattice homomorphism is a Riesz subspace.

The elementary characterizations of lattice homomorphisms are presented next.

Theorem 2.14. *For an operator* $T\colon E \to F$ *between two Riesz spaces the following statements are equivalent.*

(1) T *is a lattice homomorphism.*

(2) $T(x^+) = (Tx)^+$ *for all* $x \in E$.

(3) $T(x \wedge y) = T(x) \wedge T(y)$ *for all* $x, y \in E$.

(4) *If* $x \wedge y = 0$ *in* E, *then* $T(x) \wedge T(y) = 0$ *holds in* F.

(5) $T(|x|) = |T(x)|$ *for all* $x \in E$.

Proof. (1) $\Longrightarrow$ (2) Obvious.

(2) $\Longrightarrow$ (3) Note that

$$\begin{aligned} T(x \wedge y) &= T\big(x - (x - y)^+\big) = T(x) - T(x - y)^+ \\ &= T(x) - (Tx - Ty)^+ = T(x) \wedge T(y)\,. \end{aligned}$$

(3) $\Longrightarrow$ (4) If $x \wedge y = 0$, then $T(x) \wedge T(y) = T(x \wedge y) = T(0) = 0$.

(4) $\Longrightarrow$ (5) In view of $x^+ \wedge x^- = 0$, we have

$$\begin{aligned} |T(x)| &= |T(x^+) - T(x^-)| = T(x^+) \vee T(x^-) - T(x^+) \wedge T(x^-) \\ &= T(x^+) \vee T(x^-) = T(x^+) + T(x^-) \\ &= T(x^+ + x^-) = T\big(|x|\big)\,. \end{aligned}$$

(5) $\Longrightarrow$ (1) Observe that

$$T(x \vee y) \; = \; T\big(\tfrac{1}{2}[x + y + |x - y|]\big) = \tfrac{1}{2}\big[T(x) + T(y) + T(|x - y|)\big]$$
$$= \; \tfrac{1}{2}\big[T(x) + T(y) + |T(x) - T(y)|\big] = T(x) \vee T(y),$$

and the proof is finished. ∎

Two immediate consequences of the preceding result are the following.

(a) Every order projection is an order continuous lattice homomorphism.

(b) The kernel of every lattice homomorphism is an ideal.[1]

Also, combining Theorem 1.10 and Theorem 2.14, we see that a mapping $T\colon E^+ \to F^+$ extends to a (unique) lattice homomorphism from E to F if and only if T is additive on E^+ and $x \wedge y = 0$ implies $T(x) \wedge T(y) = 0$.

A lattice homomorphism which is in addition one-to-one is referred to as a **lattice** (or **Riesz**) **isomorphism**. Two Riesz spaces E and F are called **Riesz** (or **lattice**) **isomorphic** whenever there exists a lattice isomorphism from E onto F. If two Riesz spaces E and F are lattice isomorphic, then from the Riesz space point of view E and F can be considered as identical.

Among the positive operators that are onto, the lattice isomorphisms are characterized as follows.

Theorem 2.15. *Assume that an operator $T\colon E \to F$ between two Riesz spaces is one-to-one and onto. Then T is a lattice isomorphism if and only if T and T^{-1} are both positive operators.*

Proof. If T is a lattice isomorphism, then clearly T and T^{-1} are both positive operators. For the converse assume that T and T^{-1} are both positive operators and that $x, y \in E$. From $x \leq x \vee y$ and $y \leq x \vee y$, it follows $T(x) \leq T(x \vee y)$ and $T(y) \leq T(x \vee y)$ holds, and so

$$T(x) \vee T(y) \leq T(x \vee y). \qquad (\star)$$

Similarly, we see that $T^{-1}(u) \vee T^{-1}(v) \leq T^{-1}(u \vee v)$ for all $u, v \in F$. For $u = T(x)$ and $v = T(y)$ the last inequality yields $x \vee y \leq T^{-1}(T(x) \vee T(y))$, and by applying T it follows that $T(x \vee y) \leq T(x) \vee T(y)$. This combined with $(\star)$ implies $T(x \vee y) = T(x) \vee T(y)$ so that the operator T is a lattice homomorphism. ∎

Our next goal is to investigate the relationship between lattice homomorphisms and interval preserving operators. An operator $T\colon E \to F$ between two Riesz spaces is said to be **interval preserving** whenever T is a positive

---

[1] Recall that if $T\colon V \to W$ is an operator between vector spaces, then its **kernel** is the vector subspace of V defined by $\mathrm{Ker}\,(T) := \{x \in V \colon T(x) = 0\} = T^{-1}(\{0\})$.

operator and $T[0, x] = [0, Tx]$ holds for each $x \in E^+$. The concept of an interval preserving operator first appeared (with a different terminology) in the works of D. Maharam [134, 135].

Note that the range of an interval preserving operator is an ideal. The converse of the latter is, of course, false. For instance, the positive operator $T \colon \mathbb{R}^2 \to \mathbb{R}^2$, defined by $T(x, y) = (x + y, y)$, is onto, and hence its range is an ideal. However, $0 \le (0, 1) \le (2, 1) = T(1, 1)$ holds, and there is no vector $0 \le (x, y) \le (1, 1)$ satisfying $T(x, y) = (0, 1)$.

The concepts of interval preserving operators and lattice homomorphisms are independent. It is easy to see that every positive linear functional is interval preserving. Thus, for instance, $\phi(f) = \int_0^1 f(t)\, dt$, from $C[0, 1]$ to $\mathbb{R}$, is interval preserving but fails to be a lattice homomorphism. On the other hand, the operator $T \colon C[0, 1] \to L_1[0, 1]$ defined by

$$[Tf](t) = tf(t), \quad t \in [0, 1],$$

is a lattice homomorphism, which is not interval preserving. To see the latter, note first that $0 \le t\left|\sin\left(\frac{1}{t}\right)\right| \le t = [T\mathbf{1}](t)$ holds for all $t \in [0, 1]$, where $\mathbf{1}$ denotes the constant function one. If T were interval preserving, then there should exist some $f \in C[0, 1]$ with $0 \le f \le 1$ and $tf(t) = t\left|\sin\left(\frac{1}{t}\right)\right|$ for all $t \in [0, 1]$. However, this implies that $\lim_{t \to 0+}\left|\sin\left(\frac{1}{t}\right)\right|$ exists, which is absurd.

The first duality property between lattice homomorphisms and interval preserving operators is included in the next theorem and is due to W. Arendt [26].

Theorem 2.16 (Arendt). *For a Dedekind complete Riesz space F and a positive operator $R \colon E \to G$ between two Riesz spaces we have the following.*

(1) *If R is a lattice homomorphism, then the operator $T \mapsto TR$ from $\mathcal{L}_b(G, F)$ to $\mathcal{L}_b(E, F)$ is interval preserving.*

(2) *If R is interval preserving, then the operator $T \mapsto TR$ from $\mathcal{L}_b(G, F)$ to $\mathcal{L}_b(E, F)$ is a lattice homomorphism.*

Proof. (1) Clearly, $T \mapsto TR$ from $\mathcal{L}_b(G, F)$ to $\mathcal{L}_b(E, F)$ is a positive operator. Fix $0 \le T \in \mathcal{L}_b(G, F)$, and let $0 \le S \in \mathcal{L}_b(E, F)$ satisfy $0 \le S \le TR$. We have to show the existence of some $0 \le T_1 \le T$ with $T_1 R = S$.

To see this, define the operator T_1 on the range of R by $T_1(Rx) = Sx$, and note that T_1 is well defined. Indeed, if $Rx = Ry$ holds, then it follows from

$$|Sx - Sy| \le S|x - y| \le TR|x - y| = T|Rx - Ry| = 0$$

that $Sx = Sy$. On the other hand, note that for each $x \in E$ we have

$$T_1(Rx) = Sx \le Sx^+ \le TRx^+ = T(Rx)^+$$

and that $p(y) = T(y^+)$ $(y \in G)$ is a sublinear mapping. Consequently, by the Hahn–Banach Theorem 1.25, T_1 has a linear extension to all of G (which we denote by T_1 again) satisfying $T_1(y) \leq T(y^+)$ for all $y \in G$. Clearly, $T_1 \leq T$ holds, and moreover, if $y \in G^+$, then

$$-T_1(y) = T_1(-y) \leq T\big((-y)^+\big) = T(0) = 0$$

so that $T_1(y) \geq 0$. That is, T_1 is also a positive operator. Therefore, $0 \leq T_1 \leq T$ holds, and by our construction $T_1 R = S$, as desired.

(2) Let $x \in E^+$, and let $S, T \colon G \to F$ be two order bounded operators. Since $[0, Rx] = R[0, x]$ holds, it is easy to see that $y, z \in [0, Rx]$ satisfy $y + z = Rx$ if and only if there exist $u, v \in [0, x]$ with $u + v = x$, $y = Ru$, and $z = Rv$. Thus,

$$
\begin{aligned}
[SR \vee TR](x) &= \sup\{SR(u) + TR(v) \colon u, v \in E^+ \text{ and } u + v = x\} \\
&= \sup\{S(y) + T(z) \colon y, z \in G^+ \text{ and } y + z = Rx\} \\
&= (S \vee T)R(x),
\end{aligned}
$$

and so $SR \vee TR = (S \vee T)R$ holds in $\mathcal{L}_b(E, F)$. This shows that the operator $T \mapsto TR$ is a lattice homomorphism, and the proof is finished. ∎

In the previous theorem we considered the lattice behavior of the operator $T \mapsto TR$ involving composition on the right. The next theorem deals with the lattice behavior of the operator $T \mapsto RT$ involving composition on the left and is due to C. D. Aliprantis, O. Burkinshaw, and P. Kranz [18].

Theorem 2.17. *For a positive operator $R \colon G \to F$ between two Dedekind complete Riesz spaces and an arbitrary Riesz space E we have the following.*

(1) *If R is an order continuous lattice homomorphism, then the operator $T \mapsto RT$, from $\mathcal{L}_b(E, G)$ to $\mathcal{L}_b(E, F)$, is a lattice homomorphism.*

(2) *If the operator $T \mapsto RT$, from $\mathcal{L}_b(E, G)$ to $\mathcal{L}_b(E, F)$, is a lattice homomorphism and $F^\sim \neq \{0\}$, then R is a lattice homomorphism.*

Proof. (1) Let $S, T \in \mathcal{L}_b(E, G)$. Then, using the fact that R is an order continuous lattice homomorphism, Theorem 1.21 yields

$$
\begin{aligned}
R(S \vee T)(x) &= R\Big(\sup\Big\{\sum_{i=1}^{n} S(x_i) \vee T(x_i) \colon x_i \in E^+ \text{ and } \sum_{i=1}^{n} x_i = x\Big\}\Big) \\
&= \sup\Big\{R\Big(\sum_{i=1}^{n} S(x_i) \vee T(x_i)\Big) \colon x_i \in E^+ \text{ and } \sum_{i=1}^{n} x_i = x\Big\} \\
&= \sup\Big\{\sum_{i=1}^{n} RS(x_i) \vee RT(x_i) \colon x_i \in E^+ \text{ and } \sum_{i=1}^{n} x_i = x\Big\} \\
&= [RS \vee RT](x)
\end{aligned}
$$

for each $x \in E^+$. Thus, $R(S \vee T) = RS \vee RT$ holds, and so $T \mapsto RT$ is a lattice homomorphism.

(2) To see that R is a lattice homomorphism, fix some $0 < f \in F^\sim$, and let $x \wedge y = 0$ in G. Then from Theorem 1.72 we see that

$$
\begin{aligned}
f \otimes (Rx \wedge Ry) &= (f \otimes Rx) \wedge (f \otimes Ry) = R[f \otimes x] \wedge R[f \otimes y] \\
&= R[(f \otimes x) \wedge (f \otimes y)] = R[f \otimes (x \wedge y)] = 0.
\end{aligned}
$$

This easily implies $Rx \wedge Ry = 0$, and so R is a lattice homomorphism. ∎

Let $T \colon E \to F$ be an order bounded operator between two Riesz spaces. By Theorem 1.73 we know that the (order) adjoint operator $T' \colon F^\sim \to E^\sim$ is order bounded and order continuous. Thus, the adjoint operator defines an operator (transformer) $T \mapsto T'$ from $\mathcal{L}_b(E, F)$ to $\mathcal{L}_b(F^\sim, E^\sim)$. Clearly, $T \mapsto T'$ is a positive operator. Also by restricting each T' to $F_n^\sim$, we can consider $T \mapsto T'$ as a positive operator from $\mathcal{L}_b(E, F)$ to $\mathcal{L}_b(F_n^\sim, E^\sim)$. In this case, $T \mapsto T'$ is an order continuous lattice homomorphism. The details follow.

Theorem 2.18. *Let E and F be two Riesz spaces with F Dedekind complete. Then the positive operator $T \mapsto T'$, from $\mathcal{L}_b(E, F)$ to $\mathcal{L}_b(F_n^\sim, E^\sim)$, is an order continuous lattice homomorphism.*

Proof. For the order continuity of $T \mapsto T'$, let $T_\alpha \downarrow 0$ in $\mathcal{L}_b(E, F)$. Then for each $0 \leq f \in F_n^\sim$ and $x \in E^+$ we have

$$
\langle T_\alpha' f, x \rangle = \langle f, T_\alpha x \rangle = f(T_\alpha x) \downarrow 0.
$$

Therefore $T_\alpha' f \downarrow 0$ holds in $E^\sim$ for each $0 \leq f \in F_n^\sim$. This shows that $T_\alpha' \downarrow 0$ holds in $\mathcal{L}_b(F_n^\sim, E^\sim)$, and thus the positive operator $T \mapsto T'$, from $\mathcal{L}_b(E, F)$ to $\mathcal{L}_b(F_n^\sim, E^\sim)$, is order continuous.

Now from Theorem 1.76 we see that $|T'| = |T|'$ holds in $\mathcal{L}_b(F_n^\sim, E^\sim)$ for each $T \in \mathcal{L}_b(E, F)$. This shows that $T \mapsto T'$, from $\mathcal{L}_b(E, F)$ to $\mathcal{L}_b(F_n^\sim, E^\sim)$, is a lattice homomorphism, and the proof is finished. ∎

The adjoint of an interval preserving operator is always a lattice homomorphism.

Theorem 2.19. *If* $T: E \to F$ *is an interval preserving operator between two Riesz spaces, then* $T': F^\sim \to E^\sim$ *is a lattice homomorphism.*

Proof. Note first that $T'f = f \circ T$ holds for all $f \in F^\sim$. Now by part (2) of Theorem 2.16 the operator $f \mapsto f \circ T = T'f$ is a lattice homomorphism. Therefore, if $f \wedge g = 0$ holds in $F^\sim$, then

$$T'f \wedge T'g = (f \circ T) \wedge (g \circ T) = (f \wedge g) \circ T = 0$$

holds in $E^\sim$. Therefore, T' is a lattice homomorphism. ∎

The next theorem presents another duality relationship between lattice homomorphisms and interval preserving operators. This result appeared first in the paper of J. Kim [**97**], whereas the theorem is attributed to T. Andô by H. P. Lotz [**118**].

Theorem 2.20 (Kim–Andô). *If* $F^\sim$ *separates the points of* F, *then a positive operator* $T: E \to F$ *is a lattice homomorphism if and only if* $T': F^\sim \to E^\sim$ *is interval preserving.*

Proof. Note that $T': F^\sim \to E^\sim$ satisfies $T'f = f \circ T$ for all $f \in F^\sim$. Thus, if T is a lattice homomorphism, then by part (1) of Theorem 2.16 we see that the positive operator $f \mapsto f \circ T = T'f$ is interval preserving.

For the converse assume that T' is interval preserving, and let $x \in E$. If $0 \leq f \in F^\sim$, then from Theorem 1.23 we see that

$$
\begin{aligned}
f\big((Tx)^+\big) &= \max\{g(Tx): \ g \in F^\sim \text{ and } 0 \leq g \leq f\} \\
&= \max\{[T'g](x): \ g \in F^\sim \text{ and } 0 \leq g \leq f\} \\
&= \max\{h(x): \ h \in E^\sim \text{ and } 0 \leq h \leq T'f\} \\
&= [T'f](x^+) = f\big(T(x^+)\big).
\end{aligned}
$$

Since $F^\sim$ separates the points of F, it follows that $(Tx)^+ = T(x^+)$ holds for all $x \in E$, and so T is a lattice homomorphism. ∎

Now we turn our discussion to σ-order continuous lattice homomorphisms. The first result characterizes the onto lattice homomorphisms that are σ-order continuous. Recall that an ideal A is called a σ-**ideal** whenever $\{x_n\} \subseteq A$ and $0 \leq x_n \uparrow x$ imply $x \in A$.

Theorem 2.21. *For an onto lattice homomorphism $T\colon E \to F$ we have the following.*

(1) *T is σ-order continuous if and only if its kernel $\mathrm{Ker}\,(T)$ is a σ-ideal of E.*

(2) *T is order continuous if and only if $\mathrm{Ker}\,(T)$ is a band of E.*

Proof. We shall establish the validity of (1) and leave the proof of (2) for the reader. Start by observing that the kernel of a lattice homomorphism is always an ideal, which, in case the lattice homomorphism is σ-order continuous, is also a σ-ideal.

Now assume that $\mathrm{Ker}\,(T) = \{x \in E\colon\ T(x) = 0\}$ is a σ-ideal, and that $x_n \downarrow 0$ holds in E. Let $0 \le y \le T(x_n)$ hold in F for each n. Since T is an onto lattice homomorphism, there exists some $z \in E^+$ with $T(z) = y$. Next, note that $0 \le (z - x_n)^+ \uparrow z$ holds in E and that

$$T(z - x_n)^+ = (Tz - Tx_n)^+ = (y - Tx_n)^+ = 0$$

implies that $(z - x_n)^+ \in \mathrm{Ker}\,(T)$ for each n. Since $\mathrm{Ker}\,(T)$ is a σ-ideal, z must belong to $\mathrm{Ker}\,(T)$, and so $y = T(z) = 0$. Thus, $T(x_n) \downarrow 0$ holds in F, and the proof of (1) is finished. ∎

The lattice homomorphisms are closely related to ideals. For every ideal A of a Riesz space E the canonical projection of E onto the Riesz space E/A is a lattice homomorphism.

To see this, let us recall first why E/A is a Riesz space. The **equivalence class** determined by x in E/A will be denoted by $\dot{x} = x + A$. Clearly, $x \mapsto \dot{x}$, from E to E/A, is a linear operator called the **canonical projection** of E onto E/A. In E/A we introduce a relation $\ge$ by letting $\dot{x} \ge \dot{y}$ whenever there exist $x_1 \in \dot{x}$ (i.e., $x_1 - x \in A$) and $y_1 \in \dot{y}$ with $x_1 \ge y_1$. The relation $\ge$ satisfies the following properties:

(1) $\dot{x} \ge \dot{x}$ for all $x \in E$.

(2) $\dot{x} \ge \dot{y}$ and $\dot{y} \ge \dot{x}$ imply $\dot{x} = \dot{y}$.

To see this, pick $x_1, x_2 \in \dot{x}$ and $y_1, y_2 \in \dot{y}$ with $x_1 \ge y_1$ and $y_2 \ge x_2$. Since A is an ideal, it follows from

$$0 \le x_1 - y_1 \le (x_1 - y_1) + (y_2 - x_2) = (x_1 - x_2) + (y_2 - y_1) \in A + A = A$$

that $x_1 - y_1 \in A$. Therefore, $\dot{x} - \dot{y} = \dot{x}_1 - \dot{y}_1 = (x_1 - y_1)\dot{} = 0$, and so $\dot{x} = \dot{y}$.

(3) $\dot{x} \ge \dot{y}$ and $\dot{y} \ge \dot{z}$ imply $\dot{x} \ge \dot{z}$.

Indeed, if $x_1 \in \dot{x}$, $y_1, y_2 \in \dot{y}$, and $z_1 \in \dot{z}$ satisfy $x_1 \ge y_1$ and $y_2 \ge z_1$, then we have

$$z_1 \le y_2 \le y_2 + (x_1 - y_1) = x_1 + (y_2 - y_1)\,.$$

Since $x_1 + (y_2 - y_1) \in \dot{x}$, we see that $\dot{x} \geq \dot{z}$ must hold.

(4) $\dot{x} \geq \dot{y}$ implies $\dot{x} + \dot{z} \geq \dot{y} + \dot{z}$ for all $z \in E$.

(5) $\dot{x} \geq \dot{y}$ implies $\alpha\dot{x} \geq \alpha\dot{y}$ for all $\alpha \geq 0$.

Consequently, E/A under the relation $\geq$ is an ordered vector space. In fact, the quotient vector space E/A is itself a Riesz space.

Theorem 2.22. *If A is an ideal of a Riesz space E, then the quotient vector space E/A is a Riesz space and the canonical projection of E onto E/A is a lattice homomorphism.*

Proof. Let $x \in E$. From $x \leq x^+$ and $0 \leq x^+$ we see that $\dot{x} \leq (x^+)^{\cdot}$ and $0 \leq (x^+)^{\cdot}$ hold in E/A. On the other hand, assume $\dot{x} \leq \dot{y}$ and $0 \leq \dot{y}$ in E/A. Pick $x_1 \in \dot{x}$, $y_1 \in \dot{y}$, and $y_2 \in \dot{y}$ with $x_1 \leq y_1$ and $0 \leq y_2$. Then the relations

$$\begin{aligned} x &= x_1 + (x - x_1) \leq y_1 \vee y_2 + (x - x_1)^+ \\ &= y_2 + (y_1 - y_2)^+ + (x - x_1)^+ \in y_2 + A \end{aligned}$$

and

$$y_2 + (y_1 - y_2)^+ + (x - x_1)^+ \geq 0$$

imply $x^+ \leq y_2 + (y_1 - y_2)^+ + (x - x_1)^+$, which in turn guarantees that $(x^+)^{\cdot} \leq \dot{y}_2 = \dot{y}$. Therefore, $(\dot{x})^+ = (x^+)^{\cdot}$ holds in E/A. This establishes that E/A is a Riesz space and that the canonical projection $x \mapsto \dot{x}$ is a lattice homomorphism. ∎

According to Theorem 2.21, the canonical projection of E onto E/A is σ-order continuous if and only if A is a σ-ideal. Similarly, the canonical projection is order continuous if and only if A is a band.

When is E/A Archimedean? Before answering this question, we need the concept of relative uniform convergence in Riesz spaces as introduced by E. H. Moore [143]. A sequence $\{x_n\}$ in a Riesz space is said to be **relatively uniformly convergent** to x whenever there exist some $u > 0$ and a sequence $\{\epsilon_n\}$ of real numbers with $\epsilon_n \downarrow 0$ such that $|x_n - x| \leq \epsilon_n u$ holds for all n. A subset A of a Riesz space is said to be **uniformly closed** whenever for each sequence $\{x_n\} \subseteq A$ that is relatively uniformly convergent to some x we have $x \in A$.

We are now in the position to answer when E/A is Archimedean. The result is due to W. A. J. Luxemburg and L. C. Moore, Jr. [127].

Theorem 2.23 (Luxemburg–Moore). *Let A be an ideal of a (not necessarily Archimedean) Riesz space E. Then the quotient Riesz space E/A is Archimedean if and only if A is a uniformly closed ideal of E.*

Proof. Assume first that E/A is Archimedean. Let $\{x_n\} \subseteq A$ converge relatively uniformly to x. Pick $u > 0$ and $\epsilon_n \downarrow 0$ such that $|x_n - x| \le \epsilon_n u$ holds for each n. Since the canonical projection $x \mapsto \dot{x}$ of E onto E/A is a lattice homomorphism, it follows that

$$|\dot{x}| = |x|\dot{} \le |x - x_n|\dot{} + |x_n|\dot{} = |x - x_n|\dot{} \le \epsilon_n \dot{u}$$

holds for all n. The Archimedeaness of E/A implies $\dot{x} = 0$, and so $x \in A$. Therefore, A is a uniformly closed ideal.

For the converse assume that A is uniformly closed and that $0 \le n\dot{x} \le \dot{y}$ holds in E/A for each n. We can assume that $0 \le x \le y$ in E. Now let $x_n = \left(x - \frac{1}{n}y\right)^+$, and note that $\dot{x}_n = \left(\dot{x} - \frac{1}{n}\dot{y}\right)^+ = \frac{1}{n}(n\dot{x} - \dot{y})^+ = 0$ implies $\{x_n\} \subseteq A$. On the other hand, the inequality

$$|x_n - x| = \left|\left(x - \tfrac{1}{n}y\right)^+ - x^+\right| \le \left|\left(x - \tfrac{1}{n}y\right) - x\right| = \tfrac{1}{n}y$$

shows that $\{x_n\}$ converges relatively uniformly to x. Since A is uniformly closed, it follows that $x \in A$, and hence $\dot{x} = 0$. That is, E/A is an Archimedean Riesz space, and the proof is finished. ∎

Since every σ-ideal in an Archimedean Riesz space E is uniformly closed, the preceding theorem shows that for every σ-ideal A the Riesz space E/A is Archimedean.

For the first time in our discussion we shall need the concept of the Dedekind completion of an Archimedean Riesz space. So, let us pause to discuss briefly this notion. A Dedekind complete Riesz space L is said to be a **Dedekind completion** of the Riesz space E whenever E is Riesz isomorphic to a majorizing order dense Riesz subspace of L (which we identify with E). Clearly, only Archimedean Riesz spaces can have Dedekind completions. Also, any two Dedekind completions of a Riesz space are necessarily Riesz isomorphic (why?). That is, the Dedekind completion of a Riesz space (if it exists) is uniquely determined up to lattice isomorphism.

It is a classical result that every Archimedean Riesz space E has a Dedekind completion, which we shall denote by E^δ. This result is stated next, and we refer the reader for a proof to [**132**, p. 191], [**189**, p. 109], or [**162**, p. 151].

Theorem 2.24. *Every Archimedean Riesz space has a (unique) Dedekind completion.*

In other words, if E is an Archimedean Riesz space, then there exists a Dedekind complete Riesz space E^δ (uniquely determined up to lattice isomorphism) such that:

(a) E is Riesz isomorphic to a Riesz subspace of E^δ (which we identify with E).

(b) For each $x \in E^\delta$ we have

$$x = \sup\{y \in E\colon\ y \le x\} = \inf\{z \in E\colon\ x \le z\}.$$

It is interesting to know that there are Riesz spaces E with the property that an arbitrary lattice homomorphism with domain E is automatically σ-order continuous (or order continuous). These Riesz spaces were investigated first by C. T. Tucker [184, 185, 186], and later by D. H. Fremlin [65, 66].

Definition 2.25 (Fremlin). *A Riesz space E is said to have the σ-order continuity property whenever every positive operator from E to an arbitrary Archimedean Riesz space is automatically σ-order continuous.*

Similarly, E has the order continuity property whenever every positive operator from E to an arbitrary Archimedean Riesz space is automatically order continuous.

The next result of C. T. Tucker [186] and D. H. Fremlin [65] characterizes the σ-order continuity property. For its proof we shall need the following simple property.

- *If $T\colon E \to F$ is an order bounded operator between two Riesz spaces with F Dedekind complete, then the null ideal of T is uniformly closed.*

Indeed, if $\{x_n\} \subseteq N_T$ satisfies $|x - x_n| \le \epsilon_n u$ with $\epsilon_n \downarrow 0$, the inequalities

$$0 \le |T|(|x|) = |T|(|x| - |x_n|) \le |T|(|x - x_n|) \le \epsilon_n |T| u$$

easily imply that $|T|(|x|) = 0$.

Theorem 2.26 (Tucker–Fremlin). *For a Riesz space E the following statements are equivalent.*

(1) *E has the σ-order continuity property.*

(2) *Every lattice homomorphism from E to any Archimedean Riesz space is σ-order continuous.*

(3) *Every uniformly closed ideal of E is a σ-ideal.*

Proof. (1) $\Longrightarrow$ (2) Every lattice homomorphism is a positive operator, and the desired conclusion follows.

(2) $\Longrightarrow$ (3) Let A be a uniformly closed ideal of E. Then by Theorem 2.23 the Riesz space E/A is Archimedean. Since the canonical projection of E onto E/A is a lattice homomorphism, by our hypothesis it must be σ-order continuous. Therefore, by Theorem 2.21 its kernel, i.e., the ideal A, must be a σ-ideal.

(3) $\Longrightarrow$ (1) Let F be an Archimedean Riesz space, and let $T\colon E \to F$ be a positive operator. Since F is an order dense Riesz subspace of its Dedekind

completion F^δ, we see that $T\colon E \to F$ is σ-order continuous if and only if $T\colon E \to F^\delta$ is σ-order continuous. This means that we can assume without loss of generality that F is Dedekind complete.

Now let $S\colon E \to F$ be an order bounded operator satisfying $|S| \le T$. By the discussion preceding this theorem, N_S is uniformly closed, and hence by our assumption N_S is a σ-ideal. Thus, the null ideal of every operator in $\mathcal{A}_T$ (the ideal generated by T in $\mathcal{L}_b(E, F)$) is a σ-ideal, and consequently by Theorem 1.61 the operator T is σ-order continuous. $\blacksquare$

Recall that a Riesz space E is said to have the **countable sup property** whenever $\sup A$ exists in E, then there exists an at most countable subset B of A with $\sup B = \sup A$. A Riesz space has, of course, the countable sup property if and only if $0 \le D \uparrow x$ implies the existence of an at most countable subset C of D with $\sup C = x$.

The following result characterizes the Archimedean Riesz spaces with the countable sup property.

Theorem 2.27. *An Archimedean Riesz space has the countable sup property if and only if every σ-ideal is a band.*

Proof. If a Riesz space has the countable sup property, then clearly every σ-ideal is a band. For the converse assume that in an Archimedean Riesz space E every σ-ideal is a band. Let $0 \le D \uparrow x$ hold in E. We must show that there exists an at most countable subset C of D with $\sup C = x$. The proof will be based upon the following property:

- *For each $0 < \epsilon < 1$ there exists an at most countable subset C of D with*
$$\inf\{(\epsilon x - c)^+\colon c \in C\} = 0. \tag{$\star$}$$

If this established, then for $\epsilon_n = 1 - \frac{1}{n}$ $(n = 2, 3, \ldots)$ choose an at most countable subset C_n of D with $\inf\{(\epsilon_n x - c)^+\colon c \in C_n\} = 0$, and consider the at most countable subset $C = \bigcup_{n=2}^\infty C_n$ of D. Now let $0 \le y \le x - c$ hold for all $c \in C$. Then from Andô's inequality

$$0 \le y \le x - c \le (1 - \epsilon_n)x + (\epsilon_n x - c)^+ = \tfrac{1}{n}x + (\epsilon_n x - c)^+,$$

we see that

$$0 \le y \le \tfrac{1}{n}x + \inf\{(\epsilon_n x - c)^+\colon c \in C_n\} = \tfrac{1}{n}x$$

for all n. Therefore, $y = 0$, and so $\inf\{x - c\colon c \in C\} = 0$. In other words, $\sup C = x$, proving that E has the countable sup property.

To establish $(\star)$ fix $0 < \epsilon < 1$, and let

$$A = \{y \in E\colon \exists \text{ an at most countable } C \subseteq D \text{ with}$$
$$\inf\{|y| \wedge (\epsilon x - c)^+\colon c \in C\} = 0\}.$$

We claim that A is a band. According to our hypothesis, to establish this claim it suffices to show that A is a σ-ideal.

Clearly, A is a solid set and $\alpha y \in A$ for all $y \in A$ and $\alpha \in \mathbb{R}$. On the other hand, if $y, z \in A$, then pick two at most countable subsets U and V of D with

$$\inf\{|y| \wedge (\epsilon x - u)^+ : u \in U\} = \inf\{|z| \wedge (\epsilon x - v)^+ : v \in V\} = 0.$$

Since D is directed upward, for each $u \in U$ and $v \in V$ there exists some $w_{u,v} \in D$ with $u \vee v \leq w_{u,v}$. From the inequality

$$|y + z| \wedge (\epsilon x - w_{u,v})^+ \leq |y| \wedge (\epsilon x - u)^+ + |z| \wedge (\epsilon x - v)^+,$$

it follows that the at most countable set $W = \{w_{u,v} : u \in U \text{ and } v \in V\}$ of D satisfies $\inf\{|y + z| \wedge (\epsilon x - w)^+ : w \in W\} = 0$. Thus, $y + z \in A$, and so A is an ideal of E.

To see that A is a σ-ideal, let $\{y_n\} \subseteq A$ satisfy $0 \leq y_n \uparrow y$ in E. For each n pick an at most countable subset C_n of D with

$$\inf\{y_n \wedge (\epsilon x - c)^+ : c \in C_n\} = 0,$$

and consider the at most countable subset $C = \bigcup_{n=1}^{\infty} C_n$ of D. Clearly, if $0 \leq u \leq y \wedge (\epsilon x - c)^+$ for all $c \in C$, then $0 \leq y_n \wedge u \leq y_n \wedge (\epsilon x - c)^+$ also holds for all $c \in C_n$, and so $y_n \wedge u = 0$ for all n. In view of $y_n \wedge u \uparrow y \wedge u = u$, we see that $u = 0$. Therefore, $\inf\{y \wedge (\epsilon x - c)^+ : c \in C\} = 0$, and so $y \in A$. That is, A is a σ-ideal, and hence A is a band.

Now note that $(u - \epsilon x)^+ \in A$ holds for all $u \in D$, and in view of

$$\{(u - \epsilon x)^+ : u \in D\} \uparrow (1 - \epsilon)x,$$

we see that $x \in A$. Therefore, there exists an at most countable subset C of D satisfying $\inf\{x \wedge (\epsilon x - c)^+ : c \in C\} = \inf\{(\epsilon x - c)^+ : c \in C\} = 0$, and the proof is finished. ∎

We are now in the position to characterize the Riesz spaces with the order continuity property.

Theorem 2.28 (Tucker–Fremlin). *For an Archimedean Riesz space E the following statements are equivalent.*

(1) *E has the order continuity property.*

(2) *Every lattice homomorphism from E into any Archimedean Riesz space is order continuous.*

(3) *Every uniformly closed ideal of E is a band.*

(4) *E has the σ-order continuity property and the countable sup property.*

Proof. (1) $\Longrightarrow$ (2) Obvious.

(2) $\Longrightarrow$ (3) Let A be a uniformly closed ideal of E. By Theorem 2.23 the Riesz space E/A is Archimedean. Now the canonical projection of E onto E/A is a lattice homomorphism, and hence by our hypothesis it must be also order continuous. By Theorem 2.21 the ideal A is a band.

(3) $\Longrightarrow$ (4) By Theorem 2.26 the Riesz space E has the σ-order continuity property. Also, since every σ-ideal is uniformly closed, it follows that every σ-ideal of E is a band. Thus, by Theorem 2.27, the Riesz space E must have the countable sup property too.

(4) $\Longrightarrow$ (1) Obvious. ∎

Now our discussion turns to extension properties of lattice homomorphisms. The first result of this type states that a lattice homomorphism whose domain is a majorizing Riesz subspace has always a lattice homomorphic extension to the whole space. This interesting result was proven by Z. Lipecki [115] and W. A. J. Luxemburg and A. R. Schep [129].

Theorem 2.29 (Lipecki–Luxemburg–Schep). *Let E and F be two Riesz spaces with F Dedekind complete. If G is a majorizing Riesz subspace of E and $T: G \to F$ is a lattice homomorphism, then T extends to all of E as a lattice homomorphism.*

Proof. Let $T: G \to F$ be a lattice homomorphism. By Theorem 1.33 the convex set $\mathcal{E}(T)$ has extreme points; let S be such a point. Then, by Theorem 1.31, we have $\inf\{S|x - y|: \ y \in G\} = 0$ for each $x \in E$.

Now fix $x \in E$. Then for each $y \in G$, the fact that $T: G \to F$ is a lattice homomorphism implies $S|y| = T|y| = |Ty| = |Sy|$, and so

$$
\begin{aligned}
S|x| &\leq S|x - y| + S|y| = S|x - y| + |Sy| \\
&\leq S|x - y| + |Sy - Sx| + |Sx| \leq 2S|x - y| + |Sx|
\end{aligned}
$$

holds for all $y \in G$. From this, taking the infimum, it follows that

$$
S|x| \leq |Sx| \leq S|x|
$$

holds for all $x \in E$. So, S is a lattice homomorphic extension of T. ∎

The above proof is due to Z. Lipecki [115]. Later the converse of this theorem will be established. That is, it will be shown (Theorem 2.51) that the extreme points of $\mathcal{E}(T)$ are precisely the lattice homomorphic extensions of T. As an illustration of this property we present the following example.

Example 2.30. Let $E = C[0, 1]$, $F = \mathbb{R}$, and let $G = \{\lambda \mathbf{1}: \ \lambda \in \mathbb{R}\}$, where $\mathbf{1}$ denotes the constant function one. Clearly, G is a Riesz subspace majorizing E. Define $T\colon G \to \mathbb{R}$ by $T(\lambda\mathbf{1}) = \lambda$, and note that T is a lattice homomorphism.

Using the Riesz representation theorem it is easy to see that $\mathcal{E}(T)$ can be identified with the convex set of all probability measures on $[0, 1]$; see, for instance [**8**, Section 38]. The extreme points of $\mathcal{E}(T)$ are precisely the Dirac measures $\{\delta_t\colon t \in [0, 1]\}$, and these are precisely the lattice homomorphisms that extend T to all of E. In other words, $S \in \mathcal{E}(T)$ is a lattice homomorphic extension of T if and only if there exists some $t \in [0, 1]$ such that

$$S(f) = f(t) = \int f\, d\delta_t$$

holds for all $f \in C[0, 1]$. ∎

The next result from the theory of Riesz spaces deals with order dense Riesz subspaces, and is very useful in many contexts.

Theorem 2.31. *Let G be an order dense Riesz subspace of an Archimedean Riesz space E. If G is Dedekind complete in its own right, then G is an ideal of E.*

Proof. Assume that $0 \le x \le y \in G$ and $x \in E$. Since E is Archimedean and G is order dense in E, there exists a net $\{x_\alpha\} \subseteq G$ with $0 \le x_\alpha \uparrow x$ in E; see Theorem 1.34. Also, by the Dedekind completeness of G, we have $0 \le x_\alpha \uparrow z$ in G for some $z \in G$. Now, by Theorem 1.35, also we have $x_\alpha \uparrow z$ in E. Thus, $x = z \in G$ holds, and this shows that G is an ideal of E. ∎

Recall that a Riesz space is called **laterally complete** whenever every subset of pairwise disjoint positive vectors has a supremum. For the next important extension theorem that deals with lattice homomorphisms having values in laterally complete we shall need the following simple property.

An order continuous lattice homomorphism preserves arbitrary suprema and infima.

To see this, let $T\colon E \to F$ be an order continuous lattice homomorphism, and let $x = \sup D$ in E. If C denotes the collection of all finite suprema of D, then (since T is a lattice homomorphism) the collection of all finite suprema of $T(D)$ is precisely $T(C)$. In view of $C \uparrow x$, the order continuity of T implies $T(C) \uparrow T(x)$, and from this it follows that $T(x) = \sup T(D)$ holds. It should be noted that the converse is also true. That is, a lattice homomorphism is order continuous if and only if it preserves arbitrary suprema and infima.

Theorem 2.32. *Let $T\colon G \to F$ be an order continuous lattice homomorphism from a Dedekind complete Riesz space G to an Archimedean laterally*

complete Riesz space F. If G is an order dense Riesz subspace of an Archimedean Riesz space E, then the formula

$$T(x) = \sup\{T(y)\colon \ y \in G \ \text{and} \ 0 \le y \le x\}, \quad x \in E^+,$$

defines an extension of T from E to F, which is an order continuous lattice
homomorphism.

In addition, if T is an order continuous lattice isomorphism (into), then
so is its extension.

Proof. The difficult part of the proof is to establish that for each $x \in E^+$
the supremum

$$\sup\{T(y)\colon \ y \in G \ \text{and} \ 0 \le y \le x\}$$

exists in F. If this is done, then the rest of the proof can be completed by
repeating the proof of Theorem 1.65. Since G does not majorize E and F
need not be Dedekind complete, the existence of the above supremum is by
no means obvious.

To prove the existence of the supremum, fix $0 \le x \in E$ and consider the
nonempty set $D = \{y \in G\colon 0 \le y \le x\}$. The existence of $\sup T(D)$ will be
proven by steps, which are also of some independent interest in their own
right. Keep in mind that (according to Theorem 2.31) G is an ideal of E.

Step 1. *For each $0 < e \in G$ there exists some $w \in G$ with $0 < w \le e$ and a
positive integer n satisfying $P_w(y) \le nw$ for all $y \in D$, where P_w
is the order projection of G onto the band generated by w in G.*

Fix $0 < e \in G$. Since E is Archimedean, there exists some integer n
with $(ne - x)^+ > 0$. Pick some $v \in G$ with $0 < v \le (ne - x)^+$, and note that
$v \perp (y - ne)^+$ holds for all $y \in D$. From $0 < v \le ne$ we see that $v \wedge e > 0$,
and so the vector $w = P_v(e) \in G$ satisfies $0 < w \le e$.

Now for each $y \in D$ the relation

$$\left[P_v(y) - nw\right]^+ = P_v(y - ne)^+ \in B_v \cap B_v^d = \{0\}$$

implies that $\left[P_v(y) - nw\right]^+ = 0$, and hence $P_v(y) \le nw$ holds for all $y \in D$.
Finally, from $P_w(y) = \sup\{y \wedge kw\colon \ k = 1, 2, \ldots\} \in B_v$, it follows that

$$P_w(y) = P_v\big(P_w(y)\big) = P_w\big(P_v(y)\big) \le nP_w(w) = nw$$

holds for all $y \in D$.

Step 2. *There exists a maximal disjoint collection $\{e_i\colon \ i \in I\}$ of nonzero
positive vectors of G such that $P_{e_i}(y) \le e_i$ holds for all $i \in I$ and
all $y \in D$.*

Zorn's lemma, in conjunction with Step 1 and the order denseness of G
in E guarantees the existence of a maximal disjoint collection of nonzero

positive vectors $\{w_i \colon\; i \in I\}$, such that for each i there exists a positive integer n_i (depending upon i) satisfying $P_{w_i}(y) \le n_i w_i$ for all $y \in D$. If $e_i = n_i w_i$, then $\{e_i \colon\; i \in I\}$ satisfies the desired properties.

Step 3. *There exists a pairwise disjoint set $\{y_i \colon\; i \in I\} \subseteq D$ satisfying $x = \sup\{y_i\}$ in E.*

Let $\{e_i \colon\; i \in I\}$ be as in Step 2. Since G is Dedekind complete and $P_{e_i}(y) \le e_i$ holds for all $y \in D$, it follows that $y_i = \sup\{P_{e_i}(y) \colon\; y \in D\}$ exists in G and satisfies $y_i \le e_i$ for each $i \in I$. Now let $y \in D$. If for some $0 \le w \in G$ we have $P_{e_i}(y) \le y - w$ for all i, then $0 \le w \le y - P_{e_i}(y) \in B_{e_i}^{\mathrm{d}}$ holds for each i, and so $w \wedge e_i = 0$ for all i. Since $\{e_i \colon\; i \in I\}$ is a maximal disjoint set of nonzero positive vectors, it follows that $w = 0$, and therefore $y = \sup\{P_{e_i}(y) \colon\; i \in I\}$ holds in G (and hence in E) for each $y \in D$.

Now note that for each $y \in D$ we have $P_{e_i}(y) \le y \le x$, and so $y_i \le x$ holds for each i. On the other hand, if $y_i \le w$ holds in E for each i, then $P_{e_i}(y) \le w$ also holds for all $y \in D$ and all i, and so by the above discussion

$$y = \sup\{P_{e_i}(y) \colon\; i \in I\} \le w$$

in E for all $y \in D$. So, $x \le w$, and therefore $x = \sup\{y_i\}$ holds in E.

Step 4. *If $\{y_i \colon\; i \in I\}$ is a pairwise disjoint subset of D with $x = \sup\{y_i\}$, then $\sup T(D) = \sup\{T(y_i)\}$ holds in F.*

Since $T \colon G \to F$ is a lattice homomorphism, $\{T(y_i) \colon\; i \in I\}$ is a pairwise disjoint subset of F^+. By the lateral completeness of the Riesz space F, the vector $s = \sup\{T(y_i) \colon\; i \in I\}$ exists in F, and we claim that we have $s = \sup T(D)$. To see this, let $y \in D$. Since T is an order continuous lattice homomorphism, it follows from

$$y = y \wedge x = y \wedge \sup\{y_i \colon\; i \in I\} = \sup\{y \wedge y_i \colon\; i \in I\}$$

that

$$
\begin{aligned}
T(y) &= \sup\{T(y \wedge y_i) \colon\; i \in I\} = \sup\{T(y) \wedge T(y_i) \colon\; i \in I\} \\
&= T(y) \wedge \sup\{T(y_i) \colon\; i \in I\} = T(y) \wedge s \le s \,.
\end{aligned}
$$

Thus, $s = \sup T(D)$ holds, and the proof is finished. ∎

We shall close this section with two theorems dealing with lattice homomorphisms on $C(X)$-spaces.

Theorem 2.33. *Let X be a Hausdorff compact topological space. Then for a nonzero linear functional $\phi \colon C(X) \to \mathbb{R}$ the following statements are equivalent.*

(1) *ϕ is a lattice homomorphism.*

(2) *There exists a unique constant $c > 0$ and a unique point $x \in X$ such that $\phi = c\delta_x$, i.e., $\phi(f) = cf(x)$ holds for all $f \in C(X)$.*

Proof. (1) $\Longrightarrow$ (2) Assume that ϕ is a lattice homomorphism. Let μ be the unique Borel measure on X satisfying $\phi(f) = \int_X f\, d\mu$ for all $f \in C(X)$; see [8, Section 38].

If x and t are two distinct points in the support of μ, then pick two disjoint open sets V and W with $x \in V$ and $t \in W$. Next choose some $f \in C(X)$ with $f(x) = 1$ and $f = 0$ on V^c, and some $g \in C(X)$ with $g(t) = 1$ and $g = 0$ on W^c. Clearly, $f \wedge g = 0$ holds. On the other hand, it is easy to check that $\phi(f) \wedge \phi(g) > 0$ also holds, which is impossible. Therefore, the support of μ consists precisely of one point, say x. Now note that if $c = \mu(\{x\}) > 0$, then

$$\phi(f) = \int_X f\, d\mu = f(x)\mu(\{x\}) = cf(x)$$

holds for all $f \in C(X)$.

(2) $\Longrightarrow$ (1) Obvious. ∎

The positive operators between spaces of continuous functions that are lattice homomorphisms are characterized as follows.

Theorem 2.34. *Let X and Y be two Hausdorff compact topological spaces. Then for a nonzero positive operator $T\colon C(X) \to C(Y)$ the following statements are equivalent.*

(1) *T is a lattice homomorphism.*

(2) *There exists a unique positive function $g \in C(Y)$ and a function $\xi\colon Y \to X$ (which is continuous on $\{y \in Y\colon g(y) > 0\}$) satisfying*

$$[Tf](y) = g(y)f\big(\xi(y)\big)$$

for all $y \in Y$ and all $f \in C(X)$.

Proof. (1) $\Longrightarrow$ (2) Assume that T is a lattice homomorphism. From

$$[Tf](y) = \delta_y(Tf) = (\delta_y \circ T)f\,,$$

we see that the linear functional $\delta_y \circ T\colon C(X) \to \mathbb{R}$ is a lattice homomorphism for each $y \in Y$. Thus, by Theorem 2.33, for each $y \in Y$ there exists a unique constant $g(y) \geq 0$ and some (not necessarily unique) $\xi(y) \in X$ satisfying

$$[Tf](y) = (\delta_y \circ T)f = g(y)f\big(\xi(y)\big)\,.$$

Clearly, $g = T\mathbf{1} \in C(Y)$.

Now let a net $\{y_\alpha\} \subseteq Y$ satisfy $y_\alpha \to y$ in Y, $g(y) > 0$, and $g(y_\alpha) > 0$ for all α. Then $g(y_\alpha) \to g(y)$. On the other hand, from

$$g(y_\alpha)f\big(\xi(y_\alpha)\big) = [Tf](y_\alpha) \longrightarrow [Tf](y) = g(y)f\big(\xi(y)\big),$$

it follows that $f\big(\xi(y_\alpha)\big) \to f\big(\xi(y)\big)$ holds for all $f \in C(X)$. This implies that $\xi(y_\alpha) \to \xi(y)$, and so ξ is continuous on $\{y \in Y : g(y) > 0\}$.

$(2) \implies (1)$ Obvious. ■

Exercises

1. (Jameson [**78**]) If $T \colon E \to F$ is an onto operator between two Riesz spaces, then show that the following statements are equivalent:
   (a) T is a lattice homomorphism.
   (b) The kernel of T (i.e., $\{x \in E : Tx = 0\}$) is an ideal and $T(E^+) = F^+$.

2. A sequence $\{x_n\}$ in a Riesz space is said to be **uniformly Cauchy** whenever there exists some $u > 0$ such that for each $\epsilon > 0$ we have $|x_n - x_m| \leq \epsilon u$ for all n and m sufficiently large. A Riesz space is called **uniformly complete** whenever every uniformly Cauchy sequence is relatively uniformly convergent. Prove the following statements due to W. A. J. Luxemburg and L. C. Moore, Jr. [**127**].
   (a) If $T \colon E \to F$ is a σ-order continuous lattice homomorphism from a Dedekind σ-complete Riesz space E onto a Riesz space F, then F is Dedekind σ-complete.
   (b) If $T \colon E \to F$ is a lattice homomorphism from a uniformly complete Riesz space E onto a Riesz space F, then F is uniformly complete.

3. For a positive operator $R \colon E \to F$ between two Riesz spaces establish the following:
   (a) If for each Dedekind complete Riesz space G the operator $T \mapsto TR$, from $\mathcal{L}_b(F, G)$ to $\mathcal{L}_b(E, G)$, is a lattice homomorphism, then the operator $R' \colon F^\sim \to E^\sim$ is a lattice homomorphism.
   (b) Assume that E is Dedekind complete, $E_n^\sim$ separates the points of E, and R is order continuous. Then for each Dedekind complete Riesz space G the operator $T \mapsto TR$, from $\mathcal{L}_b(F, G)$ to $\mathcal{L}_b(E, G)$, is a lattice homomorphism if and only if R is interval preserving.

4. This exercise deals with the lattice properties of the mapping $T \mapsto T^2$ from $\mathcal{L}_b(E)$ to $\mathcal{L}_b(E)$. Assume that E is Dedekind complete.
   (a) Give an example of two operators $S, T \in \mathcal{L}_b(E)$ with $S \wedge T = 0$ and $S^2 \wedge T^2 \neq 0$.
   (b) Give an example of two operators $S, T \in \mathcal{L}_b(E)$ with $S^2 \wedge T^2 = 0$ and $S \wedge T \neq 0$.
   (c) Show that if $S^2 \wedge T^2 = 0$, then $(S \wedge T)^2 = 0$.
   (d) Assume that $S, T \in \mathcal{L}_b(E)$ satisfy $S \wedge T = 0$. Show that if either $T(E) \perp S(E)$ or $S + T$ is interval preserving, then we have $S^n \wedge T^n = 0$ for $n = 1, 2, \ldots$.

(e) Present an example of two interval preserving operators such that $S^n \wedge T^n = 0$ for all n and with $S + T$ not interval preserving.

5. A vector $e > 0$ in a Riesz space E is called:
   (a) A **discrete vector**, whenever the ideal generated by e coincides with the vector subspace generated by e, i.e., $E_e = \{\lambda e : \lambda \in \mathbb{R}\}$.
   (b) An **atom**, if $x \wedge y = 0$ and $x, y \in [0, e]$ imply either $x = 0$ or $y = 0$.

   Establish the following.
   (i) In an Archimedean Riesz space a positive vector is an atom if and only if it is a discrete vector.
   (ii) If E is a Riesz space and $0 < \phi \in E^\sim$, then ϕ is a lattice homomorphism (from E to $\mathbb{R}$) if and only if ϕ is a discrete vector of $E^\sim$.

6. (Phelps [**163**]; Ellis [**60**]) Consider two compact Hausdorff topological spaces X and Y, and let $\mathcal{C}$ be the convex set of all positive operators $T : C(X) \to C(Y)$ satisfying $T(1_X) = 1_Y$. Show that for a positive operator $T \in \mathcal{C}$ the following statements are equivalent.
   (a) T is an extreme point of $\mathcal{C}$.
   (b) T is an algebra homomorphism, i.e., $T(fg) = T(f)T(g)$ holds for all $f, g \in C(X)$.
   (c) T is a lattice homomorphism.
   (d) There exists a unique continuous function $\xi : Y \to X$ such that $Tf = f \circ \xi$ holds for all $f \in C(X)$.

7. Consider an $(m \times n)$-matrix $A = [a_{ij}]$ with nonnegative elements. Show that A (as an operator from $\mathbb{R}^n$ to $\mathbb{R}^m$) is a lattice homomorphism if and only if there exists a diagonal matrix C with nonnegative elements and a $(0, 1)$-stochastic matrix[2] B satisfying $A = BC$.

8. (Kutty–Quinn [**109**]) Show that for an Archimedean Riesz space E the following two statements are equivalent.
   (a) E has the projection property.
   (b) If $x \in E^+$ and $y \in E^\delta$ satisfy $x \wedge (x - y) = 0$, then $y \in E$.

9. For a Riesz space E show that:
   (a) E is Dedekind σ-complete if and only if E has the principal projection property and is uniformly complete. (See Exercise 2 above for the definition of a uniformly complete Riesz space.)
   (b) E is Dedekind complete if and only if E has the projection property and is uniformly complete. [*Hint*: Use Theorem 2.8.]

2.3. Orthomorphisms

In this section a special class of operators called orthomorphisms will be studied. The remarkable properties of these operators have been investigated in various ways by many authors. H. Nakano [**148**] was the first to introduce these operators, under the name *dilatators*, by means of their commutativity with order projections. The positive orthomorphisms in the

---

[2] A $(0, 1)$-**stochastic matrix** is a $(0, 1)$-matrix having exactly one 1 in each row.

works of G. Birkhoff are referred to as *essentially positive operators* [**37**, p. 396]. P. F. Conrad and J. E. Diem [**50**] called them *polar preserving endomorphisms*, while A. Bigard and K. Keimel [**34**] introduced their present name *orthomorphisms*. Other names used for orthomorphisms in the literature are *stabilisateurs* (M. Meyer [**137, 138, 139**]) and *multiplication operators* (R. C. Buck [**44**] and A. W. Wickstead [**191**]).

Definition 2.35. *An operator $T\colon E \to E$ on a Riesz space is said to be* ***band preserving*** *whenever T leaves all bands of E invariant, i.e., whenever $T(B) \subseteq B$ holds for each band B of E.*

The next theorem characterizes the band preserving operators. Keep in mind that all Riesz spaces are assumed to be Archimedean.

Theorem 2.36. *For an operator $T\colon E \to E$ on an Archimedean Riesz space the following statements are equivalent.*

(1) *T is band preserving.*

(2) *$x \perp y$ implies $Tx \perp y$.*

(3) *For each $x \in E$ we have $Tx \in B_x$.*

Proof. (1) $\Longrightarrow$ (2) Let $x \perp y$. Then $y \perp B_x$ holds. From our hypothesis we have $T(B_x) \subseteq B_x$, and so $y \perp T(B_x)$. In particular, we have $Tx \perp y$.

(2) $\Longrightarrow$ (3) Let $x \in E$. Then for each $y \in B_x^{\mathrm{d}}$ we have $x \perp y$, and so $Tx \perp y$ holds for all $y \in B_x^{\mathrm{d}}$. Therefore, $Tx \in B_x^{\mathrm{dd}} = B_x$.

(3) $\Longrightarrow$ (1) Let B be a band of E. If $x \in B$, then $B_x \subseteq B$ holds, and so $Tx \in B_x \subseteq B$. That is, $T(B) \subseteq B$ so that T is band preserving. $\blacksquare$

In case E has the principal projection property, the band preserving operators are precisely the ones that commute with the order projections. This property was used by H. Nakano [**148, 151**] to define the concept of a *dilatator*.

Theorem 2.37. *If E has the principal projection property, then an operator $T\colon E \to E$ is band preserving if and only if T commutes with every order projection of E.*

Proof. Assume first that T commutes with every order projection. Then for each $x \in E$ we have

$$T(x) = TP_x(x) = P_x T(x) \in B_x\,,$$

and so by Theorem 2.36 the operator T is band preserving.

For the converse suppose that T is band preserving, and let P_B be an order projection. If $x \in E$, then write $x = y + z \in B \oplus B^{\mathrm{d}}$, and note that

$T(y) \in B$ and $T(z) \in B^d$. Thus,

$$P_B T(x) = P_B Ty + P_B Tz = Ty = T P_B(x)$$

holds for all $x \in E$, and so $P_B T = T P_B$, as desired. ∎

A band preserving operator need not be order bounded as the next example of M. Meyer [**139**] shows.

Example 2.38 (Meyer). Let E denote the vector space of all real-valued functions f defined on the interval $[0,1)$ for which there exists a partition $0 = x_0 < x_1 < \cdots < x_n = 1$ (depending upon f) such that f is linear (i.e., of the form $f(x) = mx + b$) on each interval $[x_{i-1}, x_i)$ for each $1 \leq i \leq n$. It is easy to check that (under the pointwise algebraic and lattice operations) E is an Archimedean Riesz space.

Let $f \in E$. Pick a partition $0 = x_0 < x_1 < \cdots < x_n = 1$ for which f is linear on each $[x_{i-1}, x_i)$; let $f(x) = m_i x + b_i$ for each $x \in [x_{i-1}, x_i)$, $1 \leq i \leq n$. Next define $[Tf](x) = m_i$ for each $x \in [x_{i-1}, x_i)$, $1 \leq i \leq n$. It should be clear that $f \mapsto Tf$ defines an operator from E to E.

It is a routine matter to verify that T is not order bounded. However, it is not difficult to see that $f \perp g$ implies $Tf \perp g$, which shows that T is a band preserving operator. ∎

Recall that an operator $T \colon E \to F$ between two Riesz spaces is said to **preserve disjointness** whenever $x \perp y$ in E implies $Tx \perp Ty$ in F. Clearly, every band preserving operator preserves disjointness. The following basic result of M. Meyer [**137**] describes an important property of disjointness preserving operators; its proof below is due to S. J. Bernau [**31**].

Lemma 2.39 (Meyer). *Let E and F be two Riesz spaces with F Archimedean. If an order bounded operator $T \colon E \to F$ preserves disjointness, then*

$$(Tx)^+ \wedge (Ty)^- = 0$$

holds for all $x, y \in E^+$.

Proof. Fix $x, y \in E^+$, and let $0 < \epsilon < 1$. Since T is order bounded, there exists some $w \in F$ with $|Tz| \leq w$ for each $0 \leq z \leq x$. Let

$$\Lambda = \left\{ \lambda \geq 0 \colon \; [(Tx)^+ \wedge (Ty)^- - \epsilon w]^+ \perp T((\lambda x - y)^+) \right\}.$$

In view of $0 \leq y \leq (y - \epsilon x)^+ + \epsilon x$, it follows from Theorem 1.13 that there exist $0 \leq y_1 \leq (y - \epsilon x)^+$ and $0 \leq y_2 \leq \epsilon x$ with $y = y_1 + y_2$. Clearly, $|Ty_2| \leq \epsilon w$, and so

$$[(Tx)^+ \wedge (Ty)^- - \epsilon w]^+ \leq \left(|Ty| - \epsilon w\right)^+$$
$$\leq \left(|Ty_1| + |Ty_2| - \epsilon w\right)^+ \leq |Ty_1|.$$

Since $y_1 \perp (\delta x - y)^+$ holds in E for each $0 \le \delta \le \epsilon$, our hypothesis implies $Ty_1 \perp T(\delta x - y)^+$ in F for all $0 \le \delta \le \epsilon$, and so from the last inequality we get $[0, \epsilon] \subseteq \Lambda$.

Now suppose $\lambda \in \Lambda$ satisfies $\lambda \ge \epsilon$. Put $\alpha = \max\{1, \frac{1}{\lambda}\}$, and note that $\alpha \ge 1$, $\alpha\lambda \ge 1$, and $\alpha\epsilon \le 1$. Recalling that in any Riesz space we have $2(u \wedge v) \le u + v$, we get $u^+ \wedge v^+ = (u \wedge v) \vee 0 \le 2(u \wedge v)^+ \le (u + v)^+$. Therefore, using the last inequality, we see that

$$
\begin{aligned}
(Tx)^+ \wedge (Ty)^- &\le (\alpha\lambda Tx)^+ \wedge (\alpha Ty)^- = \alpha\big[(T\lambda x)^+ \wedge (Ty)^-\big] \\
&= \alpha\big[(T\lambda x)^+ \wedge (-Ty)^+\big] \le \alpha\big[T(\lambda x - y)\big]^+ \qquad (\star) \\
&\le \alpha\big[\big|T(\lambda x - y)^+\big| + \big|T(y - \lambda x)^+\big|\big].
\end{aligned}
$$

Next, fix any δ with $0 \le \delta \le \epsilon^2$, and then write $(y - \lambda x)^+ = y_1 + y_2$, where $0 \le y_1 \le [y - (\lambda + \delta)x]^+$ and $0 \le y_2 \le \delta x$. Clearly, $\alpha|Ty_2| \le \alpha\epsilon^2 w \le \epsilon w$ holds, and so from $(\star)$ we get

$$
\begin{aligned}
\big[(Tx)^+ \wedge (Ty)^- - \epsilon w\big]^+ &\le \big[\alpha\big|T(\lambda x - y)^+\big| + \alpha|Ty_1| + \alpha|Ty_2| - \epsilon w\big]^+ \\
&\le \alpha\big|T(\lambda x - y)^+\big| + \alpha|Ty_1|. \qquad (\star\star)
\end{aligned}
$$

Since $\lambda \in \Lambda$, we have $\big[(Tx)^+ \wedge (Ty)^- - \epsilon w\big]^+ \perp T(\lambda x - y)^+$, and hence it follows from $(\star\star)$ that $\big[(Tx)^+ \wedge (Ty)^- - \epsilon w\big]^+ \le \alpha|Ty_1|$. On the other hand, $y_1 \perp [(\lambda + \delta)x - y]^+$ implies $Ty_1 \perp T\big((\lambda + \delta)x - y\big)^+$, and consequently $\big[(Tx)^+ \wedge (Ty)^- - \epsilon w\big]^+ \perp T\big((\lambda + \delta)x + y\big)^+$. Therefore, $\lambda + \delta \in \Lambda$ holds for each δ with $0 \le \delta \le \epsilon^2$, and hence $[\lambda, \lambda + \epsilon^2] \subseteq \Lambda$. This implies $[0, \epsilon + n\epsilon^2] \subseteq \Lambda$ holds for all $n = 1, 2, \ldots$, and consequently $\Lambda = [0, \infty)$.

Finally, fix some $v \in F$ with $|Tz| \le v$ for all $0 \le z \le y$. Since for each $\delta > 0$ we have $x = (x - \delta y)^+ + x \wedge \delta y$, we get

$$
\big[(Tx)^+ \wedge (Ty)^- - \epsilon w\big]^+ \le (Tx)^+ \le \big|T(x - \delta y)^+\big| + \delta v,
$$

and in view of $\big[(Tx)^+ \wedge (Ty)^- - \epsilon w\big]^+ \perp \delta T\big(\frac{1}{\delta}x - y\big)^+ = T(x - \delta y)^+$, we see that $\big[(Tx)^+ \wedge (Ty)^- - \epsilon w\big]^+ \le \delta v$ holds for all $\delta > 0$. Since F is Archimedean, the latter implies $\big[(Tx)^+ \wedge (Ty)^- - \epsilon w\big]^+ = 0$. Since $0 < \epsilon < 1$ is arbitrary, the Archimedeaness of F once more yields $(Tx)^+ \wedge (Ty)^- = 0$, and the proof is finished. ∎

Every order bounded operator that preserves disjointness has a modulus.

Theorem 2.40. *If an order bounded operator $T: E \to F$ between two Riesz spaces with F Archimedean preserves disjointness, then its modulus exists and*

$$
|T|(|x|) = |T(|x|)| = |T(x)|
$$

holds for all $x \in E$.

Proof. Let $T\colon E \to F$ be an order bounded and disjointness preserving operator between two Riesz spaces with F Archimedean. We claim that $0 \le y \le x$ implies $|Ty| \le |Tx|$. Indeed, if $0 \le y \le x$, then

$$
\begin{aligned}
T(x) &= T(x-y) + T(y) \\
&= \left[(T(x-y))^+ + (Ty)^+ \right] - \left[(T(x-y))^- + (Ty)^- \right],
\end{aligned}
$$

and each positive term is (by Lemma 2.39) disjoint from each negative term. Consequently, from Theorem 1.5 we have

$$
\begin{aligned}
|Tx| &= \left[T(x-y) \right]^+ + (Ty)^+ + \left[T(x-y) \right]^- + (Ty)^- \\
&= \left| T(x-y) \right| + |Ty| \ge |Ty|.
\end{aligned}
$$

Now let $|y| \le x$ in E. Using the fact that $Ty^+ \perp Ty^-$, the above discussion yield

$$
|Ty| = |Ty^+ - Ty^-| = |Ty^+ + Ty^-| = \left| T(|y|) \right| \le |Tx|,
$$

and so $|Tx| = \sup\{ |Ty| \colon |y| \le x \}$ holds for each $x \in E^+$. By Theorem 1.14 the modulus of T exists, and

$$
|T|(x) = \left| T(x) \right|
$$

holds for all $x \in E^+$.

Finally, if $x \in E$, then $T(x^+) \perp T(x^-)$, and so

$$
\begin{aligned}
\left| T(|x|) \right| &= |T|(|x|) = |T|(x^+) + |T|(x^-) = |T(x^+)| + |T(x^-)| \\
&= \left| T(x^+) - T(x^-) \right| = \left| T(x) \right|,
\end{aligned}
$$

and the proof is finished. ∎

An order bounded band preserving operator is known as an *orthomorphism*.

Definition 2.41. *An **orthomorphism** is a band preserving operator that is also order bounded.*

Thus, by Theorem 2.36 an order bounded operator $T\colon E \to F$ is an orthomorphism if and only if $x \perp y$ implies $Tx \perp y$. An orthomorphism which is at the same time a positive operator is referred to as a *positive orthomorphism*. Note that every positive orthomorphism is necessarily a lattice homomorphism. Also, it should be noted that every orthomorphism preserves disjointness. For positive orthomorphisms, Definition 2.41 is due to A. Bigard and K. Keimel [34], while M. Meyer [137] was the first to define an orthomorphism as above.

Now let $T\colon E \to F$ be an operator between two Riesz spaces. Then the set

$$
\{ x \in E \colon T[0, |x|] \text{ is order bounded} \}
$$

is clearly nonempty and solid. Also, an easy application of Theorem 1.13 shows that this set is a vector subspace of E, and hence it is an ideal. Obviously, this is the largest ideal (with respect to inclusion) on which the operator T is order bounded, and it will be referred to as the **ideal of order boundedness** of T.

The next result, due to Y. A. Abramovich [**3**] and B. de Pagter [**160**], characterizes the band preserving operators that are orthomorphisms.

Theorem 2.42 (Abramovich–de Pagter). *The ideal of order boundedness of a band preserving operator on an Archimedean Riesz space is always a band.*

In particular, a band preserving operator is an orthomorphism if and only if the operator is order bounded on an order dense ideal.

Proof. Let $T: E \to E$ be a band preserving operator on an Archimedean Riesz space, let A be its ideal of order boundedness, and let B be the band generated by A. We have to show that T is order bounded on B, and for this it is enough to establish that $0 \le y \le x \in B$ implies $|Ty| \le |Tx|$. To this end, note that $T(B) \subseteq B$ holds and the operator $T: A \to B$ is disjointness preserving. Therefore, if $0 \le u \le v \in A$, then (by Theorem 2.40) we have $|Tu| \le |Tv|$.

Now let $0 \le y \le x \in B$, and suppose by way of contradiction that $(|Ty| - |Tx|)^+ > 0$. From $T(B) \subseteq B$ we see that $(|Ty| - |Tx|)^+ \in B$, and so there exists some $z \in A$ with $0 < z \le (|Ty| - |Tx|)^+$. Put $u_n = nz \wedge y$, $v_n = nz \wedge x$, and note that $0 \le u_n \le v_n \in A$ implies $|Tu_n| \le |Tv_n|$ for all n. Now from

$$y - u_n = (y - nz)^+ \perp (nz - y)^+ \ge (nz - x)^+$$

and

$$x - v_n = (x - nz)^+ \perp (nz - x)^+,$$

we see that $T(y - u_n) \perp (nz - x)^+$ and $T(x - v_n) \perp (nz - x)^+$. Therefore, the relation

$$
\begin{aligned}
(|Ty| - |Tx|)^+ &= \left| (|Ty| - |Tx|)^+ - (|Tu_n| - |Tv_n|)^+ \right| \\
&\le \left\| |Ty| - |Tx| - (|Tu_n| - |Tv_n|) \right| \\
&\le \left| T(y - u_n) \right| + \left| T(x - v_n) \right| \perp (nz - x)^+,
\end{aligned}
$$

implies $(nz - x)^+ \perp (|Ty| - |Tx|)^+$. The latter, combined with the inequality $(nz - x)^+ \le n(|Ty| - |Tx|)^+$, shows that $(nz - x)^+ = 0$ for all n. That is, $0 < nz \le x$ holds for all $n \in \mathbb{N}$, contradicting the Archimedean property of E, and the proof is finished. ∎

The disjointness preserving operators do not form a vector space. For instance, consider the two operators $S, T: C[0,1] \to C[0,1]$ defined by

$$S(f) = f(0) \cdot 1 \quad \text{and} \quad T(f) = f(1) \cdot 1 \,.$$

Then S and T are both disjointness preserving (in fact, both are lattice homomorphisms), and an easy argument shows that $S+T$ does not preserve disjointness.

In contrast with the disjointness preserving operators, it is easy to check that the collection of all orthomorphisms on a Riesz space E forms a vector space. This vector space will be denoted by $\text{Orth}(E)$, i.e.,

$$\text{Orth}(E) := \{T \in \mathcal{L}_b(E): \ x \perp y \text{ implies } Tx \perp y\} \,.$$

Clearly, $\text{Orth}(E)$ is a vector subspace of $\mathcal{L}_b(E)$, and $\text{Orth}(E)$ with the ordering inherited from $\mathcal{L}_b(E)$ is an ordered vector space in its own right.

The ordered vector space $\text{Orth}(E)$ has the following remarkable property: *It is a Riesz space under the pointwise algebraic and lattice operations.* This important result is due to A. Bigard and K. Keimel [**34**], and P. F. Conrad and J. E. Diem [**50**].

Theorem 2.43 (Bigard–Keimel–Conrad–Diem). *If E is an Archimedean Riesz space, then $\text{Orth}(E)$ is an Archimedean Riesz space under the pointwise algebraic and lattice operations. That is, if $S, T \in \text{Orth}(E)$, then*

$$[S \vee T](x) = S(x) \vee T(x) \quad \text{and} \quad [S \wedge T](x) = S(x) \wedge T(x)$$

holds for all $x \in E^+$.

Proof. Let $T: E \to E$ be an orthomorphism. Then by Theorem 2.40 the modulus of T exists and satisfies $|T|(x) = |Tx|$ for each $x \in E^+$. Clearly, $|T|$ is an orthomorphism, and this shows that $\text{Orth}(E)$ is a Riesz space. The two formulas follow from the lattice identities

$$S \vee T = \tfrac{1}{2}(S + T + |S - T|) \quad \text{and} \quad S \wedge T = \tfrac{1}{2}(S + T - |S - T|) \,,$$

and the proof is finished. ∎

Every orthomorphism is an order continuous operator. This result was established via representation theorems by A. Bigard and K. Keimel [**34**], and independently by P. F. Conrad and J. E. Diem [**50**]. A direct proof was presented by W. A. J. Luxemburg and A. R. Schep [**128**].

Theorem 2.44. *Every orthomorphism on an Archimedean Riesz space is order continuous.*

Proof. Let $T: E \to E$ be a positive orthomorphism on an Archimedean Riesz space, and let $x_\alpha \downarrow 0$ in E. We can assume that $0 \le x_\alpha \le x$ holds for all α and some $x \in E$.

Now suppose that some $y \in E^+$ satisfies $0 \leq y \leq T(x_\alpha) \leq T(x)$ for all α. Since $(x_\alpha - \epsilon x)^+ \wedge (x_\alpha - \epsilon x)^- = 0$ holds for each $\epsilon > 0$, we see that

$$0 \leq (y - \epsilon Tx)^+ \wedge (x_\alpha - \epsilon x)^- \leq T(x_\alpha - \epsilon x)^+ \wedge (x_\alpha - \epsilon x)^- = 0 \,.$$

Therefore,

$$0 = (y - \epsilon Tx)^+ \wedge (x_\alpha - \epsilon x)^- \uparrow_\alpha (y - \epsilon Tx)^+ \wedge \epsilon x$$

holds, from which it follows that $(y - \epsilon Tx)^+ \wedge x = 0$ for all $\epsilon > 0$. Since E is Archimedean, the latter conclusion implies $y \wedge x = 0$. Consequently, $y = y \wedge T(x) = 0$, and so $T(x_\alpha) \downarrow 0$ holds in E, proving that the operator T is order continuous.

Finally, note that (since $\mathrm{Orth}(E)$ is a Riesz space) each orthomorphism on E is the difference of two positive orthomorphisms, and hence every orthomorphism is order continuous. ■

Clearly, the identity operator $I \colon E \to E$ is an orthomorphism. In fact, I is a weak order unit of $\mathrm{Orth}(E)$. Indeed, if $S \in \mathrm{Orth}(E)$ satisfies $S \wedge I = 0$, then $S(x) \wedge x = (S \wedge I)(x) = 0$ and so $S(x) \in B_x \cap B_x^d = \{0\}$ holds for all $x \in E^+$, i.e., $S = 0$.

In case E is Dedekind complete, $\mathrm{Orth}(E)$ has a nice characterization.

Theorem 2.45. *If E is a Dedekind complete Riesz space, then $\mathrm{Orth}(E)$ coincides with the band generated by the identity operator in $\mathcal{L}_b(E)$.*

Proof. Let B_I be the band generated by the identity operator I. By Theorem 1.47 we see that $B_I \subseteq \mathrm{Orth}(E)$.

For the reverse inclusion, let $0 \leq T \in \mathrm{Orth}(E)$. Since I is a weak order unit in the Archimedean Riesz space $\mathrm{Orth}(E)$, it follows that $T \wedge nI \uparrow T$. From $\{T \wedge nI\} \subseteq B_I$, we get $T \in B_I$. Hence, $\mathrm{Orth}(E) \subseteq B_I$ also holds, proving that $\mathrm{Orth}(E) = B_I$, as required. ■

An orthomorphism always extends to an orthomorphism on the Dedekind completion of the space.

Theorem 2.46. *If $T \colon E \to E$ is an orthomorphism on an Archimedean Riesz space, then T extends uniquely to an orthomorphism on E^δ (the Dedekind completion of E).*

Proof. It is enough to show that every positive orthomorphism on E extends to a positive orthomorphism on E^δ. So, let $T \colon E \to E$ be a positive orthomorphism. By Theorem 2.44 the operator T is order continuous, and by Theorem 1.65 the formula

$$T^*(x) = \sup\{T(y) \colon y \in E \text{ and } 0 \leq y \leq x\}, \quad 0 \leq x \in E^\delta,$$

defines an order continuous positive operator on E^δ. Since E is order dense in E^δ it should be clear that T^* is the only linear order continuous extension of T to E^δ.

Now to see that T^* is also an orthomorphism, let $x \wedge y = 0$ in E^δ. Note that if $u, v \in E$ satisfy $0 \leq u \leq x$ and $0 \leq v \leq y$, then $u \wedge v = 0$ holds in E, and so $u \wedge Tv = 0$. Thus,

$$
\begin{aligned}
x \wedge T^*y &= \left[\sup\{u \in E\colon\ 0 \leq u \leq x\}\right] \wedge \left[\sup\{Tv\colon\ v \in E \text{ and } 0 \leq v \leq y\}\right] \\
&= \sup\{u \wedge Tv\colon\ u, v \in E,\ 0 \leq u \leq x,\ \text{and } 0 \leq v \leq y\} = 0,
\end{aligned}
$$

so that T^* is an orthomorphism. ∎

For each $T \in \mathrm{Orth}(E)$ denote by T^* the (unique) orthomorphism in $\mathrm{Orth}(E^\delta)$ that by Theorem 2.46 extends T linearly to all of E^δ. This means that a mapping $T \mapsto T^*$ is defined from $\mathrm{Orth}(E)$ to $\mathrm{Orth}(E^\delta)$, which is clearly linear and one-to-one. In addition, it is readily seen that the operator $T \mapsto T^*$ is a lattice isomorphism (into). The image of $T \mapsto T^*$ is the Riesz subspace of $\mathrm{Orth}(E^\delta)$ consisting of all orthomorphisms of $\mathrm{Orth}(E^\delta)$ that leave E invariant. Thus, if we identify $\mathrm{Orth}(E)$ with its image in $\mathrm{Orth}(E^\delta)$ under the lattice isomorphism $T \mapsto T^*$, then

$$
\mathrm{Orth}(E) = \left\{T \in \mathrm{Orth}(E^\delta)\colon\ T(E) \subseteq E\right\}.
$$

Summarizing the above, we have the following basic theorem.

Theorem 2.47. *If E is an Archimedean Riesz space, then*
$$
\mathrm{Orth}(E) = \left\{T \in \mathrm{Orth}(E^\delta)\colon\ T(E) \subseteq E\right\}.
$$

If two orthomorphisms agree on some vector, then they also agree on the band generated by that vector. The details follow.

Theorem 2.48. *The kernel of an orthomorphism is a band. In particular, if two orthomorphisms agree on a set, then they also agree on the band generated by that set.*

Proof. Let $T\colon E \to E$ be an orthomorphism. Then by Theorem 2.40 we have $\mathrm{Ker}\,(T) = N_T$. Since T is order continuous, the ideal N_T is a band, and so $\mathrm{Ker}\,(T)$ is a band.

If two orthomorphisms S and T agree on some set A, then the orthomorphism $R = S - T$ is zero on A, and hence $A \subseteq \mathrm{Ker}\,(R)$. Since $\mathrm{Ker}\,(R)$ is a band, the band generated by A is included in $\mathrm{Ker}\,(R)$, and so S and T agree on that band. ∎

We continue with a few extension properties of orthomorphisms.

Theorem 2.49. *Let G be a Riesz subspace of a Dedekind complete Riesz space E, and let an operator $T: G \to E$ satisfy $0 \le T(x) \le x$ for all $x \in G^+$. Then T extends to a positive orthomorphism on E.*

In particular, if E is Dedekind complete and $x \in E^+$, then for each $|y| \le x$ there exists an orthomorphism $T: E \to E$ (uniquely determined on B_x) with $T(x) = y$.

Proof. Note that the formula $p(x) = x^+$ defines a sublinear mapping on E such that $T(x) \le p(x)$ holds for all $x \in G$. By the Hahn–Banach Theorem 1.25, the operator T has an extension to all of E (which we shall denote by T again) such that $T(x) \le p(x)$ holds for all $x \in E$. Now it easily follows that $0 \le T \le I$ holds in $\mathcal{L}_b(E)$.

For the last part let $|y| \le x$. First, assume that $0 \le y \le x$ holds. Then $G = \{\lambda x: \lambda \in \mathbb{R}\}$ is a Riesz subspace of E and $T: G \to E$, defined by $T(\lambda x) = \lambda x$, satisfies $0 \le T(z) \le z$ for all $z \in G^+$ and $T(x) = y$. By the first part, T extends to an orthomorphism to all of E. Now for the general case $|y| \le x$. Pick two orthomorphisms S and T with $S(x) = y^+$ and $T(x) = y^-$, and note that the orthomorphism $S - T$ satisfies $(S - T)(x) = y$. ∎

We have mentioned before that every positive orthomorphism is a lattice homomorphism. However, a lattice homomorphism need not be an orthomorphism. For instance, the shift operator $T: \ell_2 \to \ell_2$, defined by $T(x_1, x_2, \dots) = (0, x_1, x_2, \dots)$, is a lattice homomorphism but it is not an orthomorphism. The next result of S. S. Kutateladze [108] presents an interesting characterization of lattice homomorphisms in terms of orthomorphisms.

Theorem 2.50 (Kutateladze). *For a positive operator $T: E \to F$ between two Riesz spaces with F Dedekind complete the following statements are equivalent.*

(1) *T is a lattice homomorphism.*

(2) *For every operator $S: E \to F$ with $0 \le S \le T$, there exists a positive orthomorphism $R \in \mathrm{Orth}(F)$ satisfying $S = RT$.*

Proof. (1) $\Longrightarrow$ (2) Assume that T is a lattice homomorphism and that $S \in \mathcal{L}(E, F)$ satisfies $0 \le S \le T$. Start by observing that if $Tx = Ty$ holds, then the relation $0 \le |Sx - Tx| \le S|x - y| \le T|x - y| = |Tx - Ty| = 0$ implies $Sx = Sy$. Thus, the formula

$$R(Tx) = Sx, \quad x \in E,$$

defines a positive operator on the range $T(E)$ of T. Since $T(E)$ is a Riesz subspace of F and $0 \le R(Tx) = Sx \le Tx$ holds for each $x \in E^+$, it follows from Theorem 2.49 that R extends to a positive orthomorphism on F. Clearly, $S = RT$ holds.

(2) $\Longrightarrow$ (1) Assume that $x \wedge y = 0$ holds in E. Let S be the operator given by Theorem 1.28 that agrees with T on the ideal E_x and is zero on E_x^d. Clearly, $0 \le S \le T$. So, by our hypothesis, there exists a positive orthomorphism $R \colon F \to F$ with $S = RT$. In particular, note that R agrees with the identity operator on $T(E_x)$, and so by Theorem 2.48 the operator R also agrees with the identity operator on the ideal generated by $T(E_x)$. Thus,

$$0 \le Tx \wedge Ty = R(Tx \wedge Ty) = R(Tx) \wedge R(Ty) = Tx \wedge Sy = Tx \wedge 0 = 0 \,,$$

and so $Tx \wedge Ty = 0$. That is, T is a lattice homomorphism, and the proof is finished. ■

Now consider a majorizing vector subspace G of a Riesz space E, and let $T \colon G \to F$ be a positive operator, where F is Dedekind complete. We already know (Theorem 1.33) that the convex set $\mathcal{E}(T)$ of all positive extensions of T to all of E is not merely nonempty but also has extreme points. In case G is a Riesz subspace and T is a lattice homomorphism, we have also seen (in the proof of Theorem 2.29) that the extreme points of $\mathcal{E}(T)$ are lattice homomorphisms. We are now in the position to establish the converse of the last statement, i.e., that the lattice homomorphisms are precisely the extreme points of $\mathcal{E}(T)$. This result is due to Z. Lipecki [114] and W. A. J. Luxemburg [126].

Theorem 2.51 (Lipecki–Luxemburg). *Let G be a majorizing Riesz subspace of a Riesz space E, and let $T \colon G \to F$ be a lattice homomorphism, where F is Dedekind complete. Then an operator S of (the nonempty convex set) $\mathcal{E}(T)$ is an extreme point of $\mathcal{E}(T)$ if and only if S is a lattice homomorphism.*

Proof. The proof of Theorem 2.29 shows that the extreme points of $\mathcal{E}(T)$ are lattice homomorphisms.

For the converse assume that $S \in \mathcal{E}(T)$ is a lattice homomorphism and that $S = \alpha S_1 + (1 - \alpha) S_2$ holds for some $S_1, S_2 \in \mathcal{E}(T)$ and some $0 < \alpha < 1$. To finish the proof we have to show that $S_1 = S_2 = S$.

Since $0 \le \alpha S_1 \le S$ holds, it follows from Theorem 2.50 that there exists a positive orthomorphism $R \in \text{Orth}(F)$ with $\alpha S_1 = RS$. Also, since S_1 and S are both extensions of T, we have $S_1 x = Sx = Tx$ for each $x \in G$, and so

$$R(Sx) = \alpha S_1 x = \alpha Sx = [\alpha I](Sx)$$

holds for all $x \in G$. In other words, R and αI agree on $S(G)$. Since G majorizes E, then band generated by $S(G)$ includes $S(E)$. Therefore,

$$\alpha S_1 x = R(Sx) = \alpha S(x)$$

holds for all $x \in E$, and so $S_1 = S$. Similarly, $S_2 = S$, and the proof is finished. ∎

The range and kernel of an orthomorphism are related as follows.

Theorem 2.52. *If $T \colon E \to E$ is an orthomorphism on an Archimedean Riesz space, then $\operatorname{Ker}(T) = [T(E)]^{\mathrm{d}}$.*

Proof. Let $T \colon E \to E$ be an orthomorphism. If $x \in [T(E)]^{\mathrm{d}}$, then $x \perp T(y)$ holds for all $y \in E$, and so $T(x) \perp T(y)$ for each $y \in E$. In particular, we have $T(x) \perp T(x)$, and so $T(x) = 0$. That is, $x \in \operatorname{Ker}(T)$, and therefore $[T(E)]^{\mathrm{d}} \subseteq \operatorname{Ker}(T)$.

For the reverse inclusion, let $x \in E$. Then since $\operatorname{Ker}(T)$ is an ideal, $\operatorname{Ker}(T) \oplus [\operatorname{Ker}(T)]^{\mathrm{d}}$ is an order dense ideal (see Theorem 1.36), and so there exists a net $\{x_\alpha + y_\alpha\} \subseteq \operatorname{Ker}(T) \oplus [\operatorname{Ker}(T)]^{\mathrm{d}}$ such that $x_\alpha + y_\alpha \overset{o}{\longrightarrow} x$. By the order continuity of T, we have $T(y_\alpha) \overset{o}{\longrightarrow} T(x)$. But by Theorem 2.36 we have $T(y_\alpha) \in [\operatorname{Ker}(T)]^{\mathrm{d}}$ for each α, and so $T(x) \in [\operatorname{Ker}(T)]^{\mathrm{d}}$. Consequently, $T(E) \subseteq [\operatorname{Ker}(T)]^{\mathrm{d}}$ holds, and since $\operatorname{Ker}(T)$ is a band, it follows that $\operatorname{Ker}(T) = [\operatorname{Ker}(T)]^{\mathrm{dd}} \subseteq [T(E)]^{\mathrm{d}}$. Therefore, $\operatorname{Ker}(T) = [T(E)]^{\mathrm{d}}$ holds, as claimed. ∎

We now turn our attention to f-algebras (the letter "f" comes from the word function) and their connection with orthomorphisms. The class of f-algebras was introduced by G. Birkhoff and R. S. Pierce [**38**] as follows.

Definition 2.53 (Birkhoff–Pierce). *A Riesz space E under an associative multiplication is said to be a **Riesz algebra** whenever the multiplication makes E an algebra (with the usual properties), and in addition it satisfies the following property: If $x, y \in E^+$, then $xy \in E^+$.*

*A Riesz algebra E is said to be an f-**algebra** whenever $x \wedge y = 0$ implies $(xz) \wedge y = (zx) \wedge y = 0$ for each $z \in E^+$.*

Now let E be an f-algebra. Then for each fixed $y \in E^+$ the mappings $x \mapsto xy$ and $x \mapsto yx$ (from E to E) are clearly positive orthomorphisms. In particular, for each $y \in E$ the mappings

$$x \mapsto xy = xy^+ - xy^- \quad \text{and} \quad x \mapsto yx = y^+x - y^-x$$

are both orthomorphisms on E. In the sequel, we shall take advantage of the properties of orthomorphisms to study the algebraic structure of f-algebras.

Theorem 2.54. *In any f-algebra $x \perp y$ implies $xy = yx = 0$.*

Proof. Assume at the beginning that $x \wedge y = 0$ holds. Then $(xy) \wedge y = 0$, and so $xy = (xy) \wedge (xy) = 0$. Now if $x \perp y$ holds, then by the preceding case we have

$$xy = (x^+ - x^-)(y^+ - y^-) = x^+ y^+ - x^- y^+ - x^+ y^- + x^- y^- = 0.$$

By the symmetry of the situation $yx = 0$, and the proof is finished. ∎

An immediate consequence of the preceding theorem is the following.

Corollary 2.55. *If x is a vector in an f-algebra, then $x^2 = (x^+)^2 + (x^-)^2 \geq 0$.*

We shall show next that every Archimedean f-algebra is necessarily a commutative algebra. This remarkable result is due to I. Amemiya [19] and G. Birkhoff and R. S. Pierce [38]; see also G. Birkhoff [37, p. 405].

Theorem 2.56 (Amemiya–Birkhoff–Pierce). *Every Archimedean f-algebra is commutative.*

Proof. Fix some $y \in E$ and note that the two formulas $T(x) = xy$ and $S(x) = yx$ define orthomorphisms on E. Since $S(y) = T(y) = y^2$ holds, it follows from Theorem 2.48 that $S = T$ on E_y. On the other hand, if $x \in E_y^d$, then (by Theorem 2.54) $S(x) = T(x) = 0$. Thus, $S = T$ holds on the order dense ideal $E_y \oplus E_y^d$, and consequently (by Theorem 2.48) $S = T$ holds on E. In other words, $yx = xy$ for all $x \in E$ (and all $y \in E$). That is, E is a commutative algebra, as desired. ∎

The above elegant proof is due to A. C. Zaanen [196]. Now let E be an Archimedean f-algebra with a multiplicative unit vector e. In view of $e = e^2 \geq 0$, we see that e must be a positive vector. On the other hand, since $e \wedge x = 0$ implies $x = x \wedge x = (xe) \wedge x = 0$, it follows that e is a weak order unit, and therefore $x \wedge ne \uparrow x$ must hold for each $x \in E^+$. Surprisingly enough, B. de Pagter [158] has shown that, in fact, $\{x \wedge ne\}$ converges relatively uniformly to x. The details follow.

Theorem 2.57 (de Pagter). *Let E be an Archimedean f-algebra with a multiplicative unit e. Then for each $x \in E^+$ and each $n \in \mathbb{N}$ we have*

$$0 \leq x - x \wedge ne \leq \tfrac{1}{n} x^2.$$

Proof. Let $x \in E^+$ be fixed. From $(ne - x \wedge ne) \wedge (x - x \wedge ne) = 0$, it follows that $\left[\tfrac{1}{n} x(ne - x \wedge ne)\right] \wedge (x - x \wedge ne) = 0$. Taking into account that $y \mapsto \left(\tfrac{1}{n} x^2\right) y$ is a lattice homomorphism, we see that

$$\left[x - \left(\tfrac{1}{n} x^2\right) \wedge x\right] \wedge (x - x \wedge ne) = \left[\tfrac{1}{n} x(ne - x \wedge ne)\right] \wedge (x - x \wedge ne) = 0.$$

Now combine the last identity with the relation

$$\begin{aligned}
0 \leq x - x \wedge ne \ &\leq \ x = \left(\tfrac{1}{n} x^2\right) \wedge e + \left[x - \left(\tfrac{1}{n} x^2\right) \wedge e\right] \\
&\leq \ \tfrac{1}{n} x^2 + \left[x - \left(\tfrac{1}{n} x^2\right) \wedge e\right],
\end{aligned}$$

to obtain that $0 \leq x - x \wedge ne \leq \tfrac{1}{n} x^2$ for all n. ∎

It is an interesting property that in a Riesz space E, if we fix a vector $e > 0$, then there exists at most one algebra multiplication on E that makes E an f-algebra having e as its unit vector.

Theorem 2.58. *Let E be an Archimedean Riesz space and let $e > 0$. Then there exists at most one product on E that makes E an f-algebra having e as its multiplicative unit.*

Proof. Assume that two products $\cdot$ and $\star$ make E an f-algebra with e as a unit for both products. Then for each fixed $y \in E$ the orthomorphism $T(x) = y \cdot x - y \star x$ satisfies $T(e) = 0$. Since e is (in this case) a weak order unit, we have $B_e = E$, and so from Theorem 2.48 it follows that $T = 0$. That is, $y \cdot x = y \star y$ holds for all $x \in E$ (and all $y \in E$). ∎

At this point let us bring into the picture the orthomorphisms.

Theorem 2.59. *For every Archimedean Riesz space E, the Riesz space $\mathrm{Orth}(E)$ under composition is an Archimedean f-algebra, having the identity operator I as its multiplicative unit. Moreover, $\mathrm{Orth}(E)$ is an f-subalgebra of $\mathrm{Orth}(E^\delta)$.*

Proof. It should be clear that under composition $\mathrm{Orth}(E)$ is an Archimedean Riesz algebra with the identity operator as a multiplicative unit.

To see that $\mathrm{Orth}(E)$ is an f-algebra, let $S, T \in \mathrm{Orth}(E)$ satisfy $S \wedge T = 0$, and let $0 \leq R \in \mathrm{Orth}(E)$. If $x \in E^+$, then $Sx \wedge Tx = [S \wedge T](x) = 0$ implies $[RS \wedge T](x) = RSx \wedge Tx = 0$, and so $(RS) \wedge T = 0$. On the other hand, if $R_n = R \wedge nI$, then $R_n x \uparrow Rx$ holds for all $x \in E^+$, and so (by the order continuity of S) it follows that $SR_n \uparrow SR$ in $\mathrm{Orth}(E)$. Therefore, $(SR_n) \wedge T \uparrow (SR) \wedge T$ likewise holds in $\mathrm{Orth}(E)$. Now since $S \wedge T = 0$ and $0 \leq SR_n \leq nS$, we see that $(SR_n) \wedge T = 0$ holds for all n, and therefore $(SR) \wedge T = 0$. ∎

As an application of the preceding connection between f-algebras and orthomorphisms, we shall show that the adjoint of an orthomorphism is again an orthomorphism.

Theorem 2.60. *If $T \colon E \to E$ is an orthomorphism on a Riesz space, then its adjoint $T' \colon E^\sim \to E^\sim$ is likewise an orthomorphism.*

Moreover, the operator $T \mapsto T'$, from $\mathrm{Orth}(E)$ to $\mathrm{Orth}(E^\sim)$, is a lattice homomorphism, i.e.,

$$\left| T' \right| = |T|'$$

holds for all $T \in \mathrm{Orth}(E)$.

Proof. We can assume that $T: E \to E$ is a positive orthomorphism. From Theorems 2.57 and 2.59 we have $0 \leq T - T \wedge nI \leq \frac{1}{n}T^2$ for all n, and so

$$0 \leq T' - T' \wedge nI' \leq T' - (T \wedge nI)' \leq \frac{1}{n}(T')^2$$

also holds for all n. This easily implies that T' is an orthomorphism.

For the last part, let $T \in \mathrm{Orth}(E)$ be fixed. If $0 \leq f \in E^\sim$ and $x \in E^+$, then from Lemma 1.75 it follows that

$$\langle |T|'f, x \rangle = \langle f, |T|x \rangle = \langle f, |Tx| \rangle \leq \langle |T'|f, x \rangle.$$

Therefore, $|T|' \leq |T'|$ holds, and since $|T'| \leq |T|'$ is trivially true, we see that $|T'| = |T|'$, as required. ∎

When the order dual of a Riesz space separates its points, an order bounded operator is an orthomorphism if and only if its adjoint is an orthomorphism. This is due to A. W. Wickstead [191].

Theorem 2.61 (Wickstead). *Let E be a Riesz space such that $E^\sim$ separates the points of E. Then an order bounded operator $T: E \to E$ is an orthomorphism if and only if its adjoint $T': E^\sim \to E^\sim$ is an orthomorphism.*

Proof. By Theorem 2.60 we know that if T is an orthomorphism, then T' is an orthomorphism. For the converse, assume that T' is an orthomorphism. Then by Theorem 2.60 the operator $T'': E^{\sim\sim} \to E^{\sim\sim}$ is also an orthomorphism. Now taking into account the facts that E is a Riesz subspace of $E^{\sim\sim}$ (Theorem 1.69) and T'' agrees with T on E, it is a routine matter to verify that $T: E \to E$ is itself an orthomorphism. ∎

Now let E be an Archimedean f-algebra. Then a mapping $u \mapsto T_u$ from E to $\mathrm{Orth}(E)$ can be defined by $T_u(x) = ux$ for each $x \in E$ (and $u \in E$). It is easy to see that $u \mapsto T_u$ is a multiplication preserving operator. When E has a multiplicative unit, then $u \mapsto T_u$ is an onto f-isomorphism (i.e., a lattice isomorphism that preserves multiplication). This result is due to A. C. Zaanen [196] .

Theorem 2.62 (Zaanen). *Let E be an Archimedean f-algebra with a multiplicative unit. Then $u \mapsto T_u$ (where $T_u(x) = ux$) is an f-isomorphism from E onto $\mathrm{Orth}(E)$; and therefore, subject to this f-isomorphism, we have $\mathrm{Orth}(E) = E$. In particular, for each Archimedean Riesz space E we have $\mathrm{Orth}(\mathrm{Orth}(E)) = \mathrm{Orth}(E)$.*

Proof. Let $e > 0$ denote the multiplicative unit vector of E. To see that the mapping $u \mapsto T_u$ is onto, let $T \in \mathrm{Orth}(E)$. Then the orthomorphism $S(x) = T(x) - T(e)x$ satisfies $S(e) = 0$. Since $B_e = E$ holds (since e is also a weak order unit), it follows that $S = 0$ on E, and so $T(x) = T(e)x$ holds for all $x \in E$. That is, $T = T_u$ holds for $u = T(e)$, and so $u \mapsto T_u$ is onto.

To finish the proof note that $u \mapsto T_u$ is one-to-one, and then apply Theorem 2.15 by observing that $T_u \geq 0$ holds if and only if $u \geq 0$. ∎

Theorem 2.62 justifies the name **multiplication operator** as an alternative name for orthomorphisms used by several authors.

We list below a few important f-algebras with multiplicative units.

(1) The Riesz space $C(X)$ of all continuous real-valued functions on a topological space X (pointwise operations; multiplicative unit is the constant function one).

(2) The Riesz space $C_b(X)$ of all (uniformly) bounded continuous real-valued functions on a topological space X (pointwise operations; multiplicative unit is the constant function one).

(3) The Riesz space $\ell_\infty(X)$ of all bounded real-valued functions on a nonempty set X (pointwise operations; multiplicative unit is the constant function one).

(4) The Riesz space $\mathbb{R}^X$ of all real-valued functions on a nonempty set X (pointwise operations; multiplicative unit is the constant function one).

(5) The Riesz space $L_\infty(\mu)$ of all almost everywhere bounded real-valued functions on a measure space (X, Σ, μ) with the pointwise operations and multiplicative unit the constant function one; almost everywhere equal functions are identified.

(6) The Riesz space $\mathcal{M}$ of all measurable functions on a measure space with the pointwise operations, where almost everywhere equal functions are identified.

The rest of the section will be devoted to extension properties of orthomorphisms. To do this we shall need the concept of the universal completion of an Archimedean Riesz space.

Recall that a Riesz space is called **laterally complete** whenever every set of pairwise disjoint positive vectors has a supremum. A Riesz space that is at the same time laterally complete and Dedekind complete is referred to as a **universally complete Riesz space**. If E is an Archimedean Riesz space, then there exists a unique (up to lattice isomorphism) universally complete Riesz space E^u (called the **universal completion** of E) such that E is Riesz isomorphic to an order dense Riesz subspace of E^u. Identifying E with its copy in E^u, we have the Riesz subspace inclusion $E \subseteq E^u$ with E order dense in E^u. In particular, the Dedekind completion E^δ of E can be identified with the ideal generated by E in E^u, and so also we have the Riesz subspace inclusions $E \subseteq E^\delta \subseteq E^u$ with E order dense in E^u.

The Riesz space E^u is of the form $C^\infty(X)$ for some Hausdorff extremally disconnected compact topological space X. The symbol $C^\infty(X)$ denotes the collection of all continuous functions $f \colon X \to [-\infty, \infty]$ for which the open set $\{x \in X \colon -\infty < f(x) < \infty\}$ is dense in X. When X is extremally disconnected (i.e., when the closure of every open set of X is also open), $C^\infty(X)$ is a universally complete Riesz space under the pointwise algebraic and lattice operations. If E has a weak order unit e, then the embedding of E into $C^\infty(X)$ can be taken so that e corresponds to the constant function one on X. For details see [**132**, Section 50] and [**7**, Chapter 7].

For our purpose here it is important to observe that $C^\infty(X)$, with X extremally disconnected, under the pointwise multiplication is an Archimedean f-algebra with multiplicative unit the constant function one. Therefore, the universal completion E^u of any Archimedean Riesz space E is an f-algebra with a multiplicative unit. (Exercise 13 at the end of this section provides an alternative approach to the existence of the universal completion using orthomorphisms.)

Now consider an Archimedean Riesz space E and a positive orthomorphism $T \colon E \to E$. Since T is order continuous, it follows from Theorem 1.65 that the formula

$$T^*(x) = \sup\{T(y) \colon y \in E \text{ and } 0 \le y \le x\}, \quad 0 \le x \in E^\delta,$$

defines a positive orthomorphism on E^δ that extends the operator T; see the proof of Theorem 2.46. Considering E^δ embedded in E^u, we obtain that the operator $T^* \colon E^\delta \to E^u$ is an order continuous lattice homomorphism satisfying $T^*x \wedge y = 0$ in E^u whenever $x \wedge y = 0$ holds in E^δ. Thus, by Theorem 2.32, the mapping $T^\star \colon E^u \to E^u$ defined for each $0 \le x \in E^u$ via the formula

$$
\begin{aligned}
T^\star(x) &= \sup\{T^*(y) \colon y \in E^\delta \text{ and } 0 \le y \le x\} \\
&= \sup\{T(y) \colon y \in E \text{ and } 0 \le y \le x\},
\end{aligned}
$$

extends T^* (and hence T) uniquely to an order continuous lattice homomorphism on E^u. Clearly, this extension $T^\star \colon E^u \to E^u$ is also a positive orthomorphism on E^u.

Therefore, every orthomorphism T on E extends to a unique orthomorphism T^* on E^δ, and from there it can be extended uniquely to an orthomorphism $T^\star$ on E^u (where there, since E^u is an f-algebra with a multiplicative unit, it is a multiplication operator). In particular, the mappings $T \mapsto T^* \mapsto T^\star$ from $\mathrm{Orth}(E) \to \mathrm{Orth}(E^\delta) \to \mathrm{Orth}(E^u)$ are lattice isomorphisms (into). If we identify $\mathrm{Orth}(E)$ with its image in $\mathrm{Orth}(E^u)$

under $T \mapsto T^{\star}$, then we see that $\mathrm{Orth}(E)$ consists precisely of all ortho-morphisms on E^{u} that leave E invariant. In particular, since E^{u} is an Archimedean f-algebra with multiplicative unit, every orthomorphism on E is a **multiplication operator**. That is, if $T \in \mathrm{Orth}(E)$, then there exists some $y \in E^{\mathrm{u}}$ such that $T(x) = yx$ holds for all $x \in E$ (where the multiplication is, of course, taken in E^{u}).

Summarizing the above discussion, we have the following powerful result.

Theorem 2.63. *If E is an Archimedean Riesz space, then*

$$\mathrm{Orth}(E) = \left\{ T \in \mathrm{Orth}(E^{\mathrm{u}}) \colon \ T(E) \subseteq E \right\}.$$

Moreover, the following f-algebra inclusions hold:

$$\mathrm{Orth}(E) \subseteq \mathrm{Orth}(E^{\delta}) \subseteq \mathrm{Orth}(E^{\mathrm{u}}).$$

As a first application of the preceding theorem we have the following result.

Theorem 2.64. *Every Archimedean f-algebra E with a multiplicative unit e can be considered as an (order dense) f-subalgebra of E^{u} with the same unit e.*

Proof. Embed E (order densely) in E^{u} in such a way that e corresponds to $\mathbf{1}$, the constant function one on $E^{\mathrm{u}} = C^{\infty}(X)$. Denote by $\cdot$ the multi-plicative product of E and by $\star$ the pointwise multiplication on $C^{\infty}(X)$.

For each fixed $u \in E$, the formula $T(x) = u \cdot x$ $(x \in E)$ defines an ortho-morphism on E, and hence by Theorem 2.63, it extends to an orthomorphism on E^{u}. Consequently, by Theorem 2.62 there exists some $w \in E^{\mathrm{u}}$ such that $T(x) = w \star x$ holds for all $x \in E^{\mathrm{u}}$. In particular, $u = u \cdot e = T(e) = w \star e = w$. Thus, $u \cdot x = u \star x$ holds for all $u, x \in E$, and so $\star$ extends $\cdot$ to E^{u}. In other words, E is an order dense f-subalgebra of E^{u} with the same unit e. ∎

With the help of Theorem 2.62 we are now in the position to describe the orthomorphisms for a number of Riesz spaces. The examples below were presented first by A. C. Zaanen [196].

Example 2.65. Let $E = c_0(X)$ for some nonempty set X. In other words, a function $f \colon X \to \mathbb{R}$ belongs to $c_0(X)$ if and only if for each $\epsilon > 0$ there exists a finite subset A of X such that $|f(x)| < \epsilon$ holds for all $x \notin A$. Then $u \mapsto T_u$, where $T_u(f) = uf$ (pointwise product) for each $u \in \ell_\infty(X)$ and $f \in c_0(X)$, is an f-algebra isomorphism from $\ell_\infty(X)$ (the f-algebra of all bounded real-valued functions on X) onto $\mathrm{Orth}(E)$, and so $\mathrm{Orth}(E) = \ell_\infty(X)$ holds.

The difficult part is to show that the mapping $u \mapsto T_u$ is onto. To see this, let $T \in \mathrm{Orth}(E)$, and note that the universal completion of $c_0(X)$

is $\mathbb{R}^X$. By Theorem 2.63 the orthomorphism T extends to an orthomorphism on $\mathbb{R}^X$, and thus there exists some $u \in \mathbb{R}^X$ such that $T(f) = uf$ (pointwise product) holds for each $f \in E$. Now if some countable subset $\{x_1, x_2, \ldots\}$ of X satisfies $|u(x_n)| \geq n^2$, then the function $g \colon X \to \mathbb{R}$, defined by $g(x_n) = \frac{1}{n}$ and $g(x) = 0$ for $x \notin \{x_1, x_2, \ldots\}$, belongs to E and satisfies $|u(x_n)g(x_n)| \geq n$. This implies that $T(g) = ug \notin E$, which is a contradiction. This argument shows that $u \in \ell_\infty(X)$, and so $T = T_u$. ∎

Example 2.66. Let X be a locally compact Hausdorff topological space, and let $E = C_c(X)$, the f-algebra of all continuous real-valued functions on X with compact support. We claim that $u \mapsto T_u$, where $T_u(f) = uf$ (pointwise product) for each $u \in C(X)$ and $f \in C_c(X)$, is an f-algebra isomorphism from $C(X)$ onto $\mathrm{Orth}(E)$, and so $\mathrm{Orth}(C_c(X)) = C(X)$ holds.

Again the difficult part is to show that $u \mapsto T_u$ is onto. Start by observing that $C(X)$ is an f-algebra with multiplicative unit the constant function **1**. Hence, by Theorem 2.64, $C(X)$ can be embedded (order densely) in its universal completion in such a way that $C(X)$ is an f-subalgebra with the same multiplicative unit **1**. Since $E = C_c(X)$ is order dense in $C(X)$, the universal completion of $C(X)$ serves equally well as the universal completion of E. Thus, the following f-algebra inclusions hold: $E \subseteq C(X) \subseteq E^u$. Now let $0 \leq T \in \mathrm{Orth}(E)$. By Theorem 2.63, the operator T can be considered as an orthomorphism on E^u. Therefore, there exists some $0 \leq u \in E^u$ such that $T(f) = uf$ holds for each $f \in E$. Note that $u = T(\mathbf{1})$.

Next, observe that for every open set V, the set

$$\{f \in C_c(X) \colon f = 0 \text{ on } V\}$$

is a band of E. Thus, by Theorem 2.36, if $f \in C_c(X)$ vanishes on some open set V, then $T(f) = 0$ also holds on V. In particular, if $f, g \in C_c(X)$ satisfy $f = g$ on V, then they also satisfy $T(f) = T(g)$ on V. Now for each open set V with compact closure, choose $f_V \in C_c(X)$ such that $0 \leq f_V \leq 1$ and $f_V = 1$ on V. Then $f_V(x) \uparrow_V \mathbf{1}$ holds for each $x \in X$, and so by the order continuity of T we have $T(f_V) \uparrow_V T(\mathbf{1}) = u$ in E^u. On the other hand, in view of $T(f_V) = T(f_W)$ on $V \cap W$, it follows that $T(f_V)(x) \uparrow_V h(x)$ holds for some $h \in C(X)$. This implies $T(f_V) \uparrow_V h$ in $C(X)$, and so $T(f_V) \uparrow_V h$ also holds in E^u. Thus, $u = h \in C(X)$. This establishes that $u \mapsto T_u$ from $C(X)$ to $\mathrm{Orth}(C_c(X))$ is onto. ∎

Example 2.67. Assume that (X, Σ, μ) is a σ-finite measure space, and let $E = L_p(\mu)$ with $0 < p < \infty$. Then $u \mapsto T_u$, where $T_u(f) = uf$ (pointwise product) for each $f \in E$, is an f-algebra isomorphism from $L_\infty(\mu)$ onto $\mathrm{Orth}(E)$, and therefore $\mathrm{Orth}(L_p(\mu)) = L_\infty(\mu)$ holds.

That the mapping is onto is the only thing that needs proof. To see this, let $T \in \mathrm{Orth}(E)$. Note that E^u is the f-algebra $\mathcal{M}$ of all equivalence classes of μ-measurable functions. By Theorem 2.63, the operator T can be considered as an orthomorphism on $\mathcal{M}$, and so there exists some $u \in \mathcal{M}$ such that $T(f) = uf$ holds for all $f \in E$. From standard arguments, it easily follows that $u \in L_\infty(\mu)$. Thus, $T = T_u$ holds, and hence $u \mapsto T_u$ is onto. ∎

Our discussion concludes with a few bibliographical remarks. As we have mentioned before, there exists a vast bibliography on orthomorphisms. The theory of orthomorphisms was originated with H. Nakano [148] and was developed by A. Bigard and K. Keimel [34] and P. F. Conrad and J. E. Diem [50] by means of representation theory. Along these lines also were the works of A. Bigard [33], A. Bigard, K. Keimel, and S. Wolfenstein [35], M. Meyer [137, 138, 139], and A. W. Wickstead [191, 192]. Orthomorphisms without the use of representation theorems were investigated by A. C. Zaanen [196], S. J. Bernau [31], C. D. Aliprantis and O. Burkinshaw [13], and M. Duhoux and M. Meyer [55]. A thorough study of f-algebras and orthomorphisms is included in the Ph.D. dissertation of B. de Pagter [158], and in C. B. Huijsmans and B. de Pagter [77]. Other interesting works on orthomorphisms are those of Y. A. Abramovich [3], Y. A. Abramovich, A. I. Veksler, and A. V. Koldunov [4], G. Birkhoff and R. S. Pierce [38], R. C. Buck [44], H. O. Flösser, G. Gierz, and K. Keimel [63], W. A. J. Luxemburg [126], A. R. Schep [177], W. A. J. Luxemburg and A. R. Schep [128], B. de Pagter [160], and W. Wils [195].

Exercises

1. If $T: E \to F$ is an order bounded operator between two Archimedean Riesz spaces and T preserves disjointness, then show there exist two lattice homomorphisms $R, S: E \to F$ satisfying $T = R - S$.

2. Show that a positive operator $T: E \to E$ on a Riesz space is an orthomorphism if and only if $I + T$ is a lattice homomorphism.

3. Show that if E is Dedekind complete, then the range of every orthomorphism on E is an ideal.

4. (Luxemburg [126]) Let $T: E \to E$ be an order continuous lattice homomorphism on an Archimedean Riesz space. Show that T is an orthomorphism if and only if $0 < x \in [\operatorname{Ker}(T)]^{\mathrm{d}}$ implies $x \wedge Tx > 0$.

5. Let E be Dedekind complete. By Theorem 2.45 we know that $\operatorname{Orth}(E)$ is the band generated in $\mathcal{L}_\mathrm{b}(E)$ by the identity operator I, and so $\operatorname{Orth}(E)$ is a projection band of $\mathcal{L}_\mathrm{b}(E)$. This exercise presents a description, due to A. R. Schep [177], of the order projection of $\mathcal{L}_\mathrm{b}(E)$ onto $\operatorname{Orth}(E)$.

   Show that the projection of an arbitrary operator $0 \leq T \in \mathcal{L}_\mathrm{b}(E)$ onto $\operatorname{Orth}(E)$ is given by

   $$\inf\left\{ \sum_{i=1}^n P_i T P_i: \text{ each } P_i \text{ is an order projection and } \sum_{i=1}^n P_i = I \right\}.$$

6. Let (P) denote any one of the properties: Dedekind completeness, Dedekind σ-completeness, projection property, principal projection property, and uniform completeness.

    Show that if a Riesz space E has property (P), then $\mathrm{Orth}(E)$ also has property (P).

7. (Zaanen [**196**]) Show that if E is the Riesz space of all continuous piecewise linear functions on $[0,1]$, then $\mathrm{Orth}(E) = \{\alpha I : \ \alpha \in \mathbb{R}\}$.

8. Let E be a Dedekind complete Riesz space. Use Theorem 1.50 to present an alternative proof of the fact that $\mathrm{Orth}(E)$ under the pointwise algebraic and lattice operations is a Riesz space.

9. (de Pagter [**158**]) Show that an order bounded operator $T\colon E \to E$ on an Archimedean Riesz space is an orthomorphism if and only if it leaves invariant every uniformly closed ideal of E. [*Hint*: Use Theorem 2.57.]

10. (de Pagter [**158**]) Let E and F be two Archimedean Riesz spaces with F Dedekind complete. Show that for each $S \in \mathrm{Orth}(E)$ and $R \in \mathrm{Orth}(F)$, the operator $T \mapsto RTS$, from $\mathcal{L}_\mathrm{b}(E,F)$ to $\mathcal{L}_\mathrm{b}(E,F)$, is an orthomorphism. [*Hint*: Use Theorem 2.57.]

11. Let $T\colon E \to F$ be a positive operator between two Riesz spaces with F Dedekind complete. If R and S ate two positive orthomorphisms on F, then show that

$$(R \wedge S)T = RT \wedge ST \quad \text{and} \quad (R \vee S)T = RT \vee ST$$

hold in $\mathcal{L}_\mathrm{b}(E,F)$. Similarly, if R and S are two orthomorphisms on E, then show that

$$T(R \wedge S) = TR \wedge TS \quad \text{and} \quad T(R \vee S) = TR \vee TS.$$

12. For an f-algebra E establish the following statements:
    (a) $|uv| = |u| \cdot |v|$ holds for all $u, v \in E$.
    (b) If E is Archimedean with a multiplicative unit, then $u^2 = 0$ holds if and only if $u = 0$.
    (c) If E is Archimedean with a multiplicative unit, then $u \perp v$ holds if and only if $uv = 0$.
    [*Hint*: For each $u \in E$ consider the orthomorphism $x \mapsto ux$.]

13. This exercise presents by steps another approach to the existence of the universal completion using the notion of orthomorphisms. For our discussion let E be a fixed Archimedean Riesz space.
    (a) Using Theorems 1.65 and 2.32 show that E has at most one (up to a lattice isomorphism) universal completion.
    (b) An order bounded operator $T\colon A \to E$, where A is an order dense ideal of E, is called a **generalized orthomorphism** if $u \perp v$ in A implies $Tu \perp v$ in E. If T is in addition positive, then T is called a **generalized positive orthomorphism**.

        Show that every generalized positive orthomorphism is order continuous, and that every generalized orthomorphism can be written as a difference of two generalized positive orthomorphisms.

(c) Let $\text{Orth}^\infty(E)$ denote the collection of all generalized orthomorphisms. In $\text{Orth}^\infty(E)$ we introduce an equivalence relation by letting $S \sim T$ whenever $S = T$ holds on an order dense ideal of E. As usual, the equivalence classes of $\text{Orth}^\infty(E)$ are denoted by $\text{Orth}^\infty(E)$ again.

Show that under the pointwise algebraic and lattice operations $\text{Orth}^\infty(E)$ is an Archimedean laterally complete f-algebra with unit the identity operator. (Multiplication is, of course, the composition operation.)

(d) Show that if E is Dedekind complete, then $\text{Orth}^\infty(E)$ is a universally complete f-algebra with a multiplicative unit.

(e) Assume that E is Dedekind complete with a weak order unit e. Then E_e is an order dense ideal of E. By Theorem 2.49, for each $x \in E_e$ there exists a unique orthomorphism $T_x \colon E_e \to E_e$ with $T_x(e) = x$. Clearly, $T_x \in \text{Orth}^\infty(E)$. Thus, a mapping $x \mapsto T_x$ from E_e to $\text{Orth}^\infty(E)$ can be defined such that $T_e = I$ holds.

Show that $x \mapsto T_x$ is an order continuous lattice isomorphism whose range is order dense in $\text{Orth}^\infty(E)$. Also show (by using Theorem 2.32) that $x \mapsto T_x$ extends to a lattice isomorphism from E into $\text{Orth}^\infty(E)$ (and hence, in this case, $\text{Orth}^\infty(E)$ is the universal completion of E).

(f) Let E be a Dedekind complete Riesz space. Pick a maximal disjoint set $\{e_i \colon i \in I\}$ of nonzero positive vectors of E, and let E_i denote the band generated by e_i in E. For each $x \in E$, let x_i denote the projection of x onto E_i, i.e., $x_i = P_{e_i}(x)$. Consider E_i embedded in $\text{Orth}^\infty(E)$ as determined by (e) above.

Show that the mapping $x \mapsto \{T_{x_i}\}$, from E to the universally complete Riesz space $\prod_{i \in I} \text{Orth}^\infty(E_i)$, is a lattice isomorphism whose range is order dense (and hence $\prod_{i \in I} \text{Orth}^\infty(E_i)$ is the universal completion of E).

(g) For the general case, embed E into its Dedekind completion E^δ and then use part (f) to establish that E has a universal completion.

Chapter 3

Topological Considerations

It is well known that operator theory is intrinsically related to the topological structures associated with the spaces upon which the operators act. The theory of positive operators is no exception to this phenomenon. The various topological notions provide an invaluable insight into the properties of operators. This chapter is devoted to the basic topological concepts needed for the study of positive operators. The presentation (although concise) is quite complete. The discussion focuses on locally convex spaces, Banach spaces, and locally solid Riesz spaces.

3.1. Topological Vector Spaces

The basic concepts from the theory of topological vector spaces needed for our study will be reviewed briefly in this section. For detailed treatments of the theory of topological vector spaces the interested reader can consult the references [**57, 73, 76, 99, 169, 173, 194**]. All vector spaces are assumed to be real vectors spaces. Unless otherwise stated, all topological spaces are considered to be Hausdorff. We shall adhere to the following topological notation: If (X, τ) is a topological space and A is a subset of X, then (A, τ) will denote the set A equipped with the topology induced by τ.

The standard notation involving subsets of a vector space will be employed. For instance, if A and B are two arbitrary subsets of a vector space, then their **algebraic sum** and their **algebraic difference** are defined by

$$A + B := \{a + b \colon \ a \in A \text{ and } b \in B\},$$

and

$$A - B := \{a - b \colon \ a \in A \text{ and } b \in B\}.$$

Also if A is a subset of a vector space, then for each $\lambda \in \mathbb{R}$ we let

$$\lambda A := \{\lambda a \colon \ a \in A\} \, .$$

Let X be a vector space. Then a nonempty subset A of X is said to be:

(1) **convex**, whenever $x, y \in A$ and $0 \leq \lambda \leq 1$ imply $\lambda x + (1 - \lambda)y \in A$.

(2) **circled** (or **balanced**), whenever $x \in A$ and $|\lambda| \leq 1$ imply $\lambda x \in A$.

(3) **absorbing**, if for each $x \in X$ there exists some $\lambda > 0$ satisfying $x \in \lambda A$ (or equivalently, if there exists some $\lambda > 0$ with $\lambda x \in A$).

It can be shown by induction that a subset A of X is convex if and only if it satisfies the following property: For any arbitrary vectors $x_1, \ldots, x_n \in A$ and any arbitrary nonnegative scalars $\lambda_1, \ldots, \lambda_n$ with $\lambda_1 + \cdots + \lambda_n = 1$ we have $\lambda_1 x_1 + \cdots + \lambda_n x_n \in A$.[1] The **convex hull** co A is the smallest (with respect to inclusion) convex set that includes A. An easy argument shows that co A consists of all convex combinations of A, i.e.,

$$\text{co } A := \Big\{ \sum_{i=1}^{n} \lambda_i x_i \colon \ x_i \in A, \ \lambda_i \geq 0, \ \text{and} \ \sum_{i=1}^{n} \lambda_i = 1 \Big\} \, .$$

Similarly, it can be seen that the set

$$\Big\{ \sum_{i=1}^{n} \lambda_i x_i \colon \ x_i \in A \ \text{for each } i \ \text{and} \ \sum_{i=1}^{n} |\lambda_i| \leq 1 \Big\}$$

is the **convex circled hull** of A; i.e., the smallest convex and circled set that includes A.[2]

A topology τ on a vector space X is called a **linear topology** whenever the addition function

$$(x, y) \mapsto x + y, \quad \text{from } X \times X \text{ to } X,$$

and the scalar multiplication function

$$(\lambda, x) \mapsto \lambda x, \quad \text{from } \mathbb{R} \times X \text{ to } X,$$

are both continuous. The pair (X, τ) is then called a **topological vector space**. It should be immediate that in a topological vector space the closure of a circled set is circled, and the closure of a convex set is convex.

Now let (X, τ) be a topological vector space. Since $x \mapsto a + x$ is a homeomorphism, it easily follows that every τ-neighborhood at a is of the form $a + V$, where V is a τ-neighborhood of zero. Thus, the τ-neighborhoods of zero determine the structure of τ. The continuity of $\lambda \mapsto \lambda x$ at zero (x fixed) guarantees that each τ-neighborhood of zero is an absorbing set. On

---

[1] The vectors of the form $\lambda_1 x_1 + \cdots + \lambda_n x_n$, where $x_i \in A$, $\lambda_i \geq 0$, and $\lambda_1 + \cdots + \lambda_n = 1$, are referred to as **convex combinations** of A.

[2] The convex circled hull of A is also known as the **absolute convex hull** of A.

the other hand, the continuity of $(x, y) \mapsto x + y$ at $(0, 0)$ implies that for each τ-neighborhood V of zero there exists another τ-neighborhood W of zero with $W + W \subseteq V$ (and so, by induction, for each τ-neighborhood V of zero and each n there exists another τ-neighborhood W of zero with $W + \cdots + W \subseteq V$, where the sum to the left of the inclusion has n summands). Now it is not difficult to see that the τ-neighborhood system at zero has a base $\mathcal{B}$ such that:

(a) Each $V \in \mathcal{B}$ is circled and absorbing.

(b) For each $V \in \mathcal{B}$ there exists some $W \in \mathcal{B}$ with $W + W \subseteq V$.

Note that if $V, W \in \mathcal{B}$ satisfy $W + W \subseteq V$, then $\overline{W} \subseteq V$ holds. (Indeed, if $x \in \overline{W}$, then $(x + W) \cap W \neq \emptyset$ holds, which implies $x \in V$.) In particular, it follows that the τ-closed, circled τ-neighborhoods of zero form a base for the τ-neighborhoods at zero. In the converse direction, if $\mathcal{B}$ is a family of subsets of a vector space X such that

(i) for each $V \in \mathcal{B}$ we have $0 \in V$,

(ii) for each $V, W \in \mathcal{B}$ there exists some $U \in \mathcal{B}$ with $U \subseteq V \cap W$,

(iii) each $V \in \mathcal{B}$ is circled and absorbing, and

(iv) for each $V \in \mathcal{B}$ there exists some $W \in \mathcal{B}$ with $W + W \subseteq V$,

then there exists a unique linear topology τ on X having $\mathcal{B}$ as a base at zero. It should be observed that the linear topology τ is Hausdorff if and only if $\bigcap \{V \colon V \in \mathcal{B}\} = \{0\}$. As we have mentioned before, unless otherwise stated, this condition will be assumed throughout this book.

A subset A of a topological vector space (X, τ) is said to be:

(4) τ-**bounded**, whenever for each τ-neighborhood V of zero there exists some $\lambda > 0$ satisfying $\lambda A \subseteq V$ (or, equivalently, whenever for each τ-neighborhood V of zero there exists some $\lambda > 0$ satisfying $A \subseteq \lambda V$), and

(5) τ-**totally bounded**, whenever for each τ-neighborhood V of zero there is a finite subset Φ of A such that $A \subseteq \bigcup_{x \in \Phi}(x + V) = \Phi + V$.

Clearly, subsets, sums, and scalar multiples of τ-bounded sets are likewise τ-bounded. Also, it should be clear that:

$$\tau\text{-compactness} \implies \tau\text{-total boundedness} \implies \tau\text{-boundedness}.$$

The most useful characterizations of totally bounded sets are included in the next theorem.

Theorem 3.1. *For a subset A of a topological vector space (X, τ) the following statements are equivalent:*

(1) *A is τ-totally bounded.*

(2) *For each τ-neighborhood V of zero there exists a finite subset Φ of X satisfying $A \subseteq \Phi + V$.*

(3) *For each τ-neighborhood V of zero there exists a finite subset Φ of X satisfying $A \subseteq \Phi + V + V$.*

(4) *For each τ-neighborhood V of zero there exists a τ-totally bounded set B satisfying $A \subseteq B + V$.*

Proof. The implications $(1) \Longrightarrow (2) \Longrightarrow (3) \Longrightarrow (4)$ are obvious.

$(4) \Longrightarrow (1)$ Let V be a τ-neighborhood of zero. Fix a circled τ-neighborhood W of zero with $W + W + W + W \subseteq V$. Now pick a τ-totally bounded set B with $A \subseteq B + W$, and then select a finite subset D of B with $B \subseteq D + W$. Thus, $A \subseteq D + W + W$ holds, and we can assume that $A \cap (x + W + W) \neq \emptyset$ for each $x \in D$. Given $x \in D$, fix some $a_x \in A \cap (x + W + W)$, and note that the set $\Phi = \{a_x : x \in D\}$ is a finite subset of A. Now if $a \in A$, then there exist $x \in D$ and $v, w \in W$ with $a = x + v + w$. Therefore,

$$a = a_x + (x - a_x) + v + w \in a_x + W + W + W + W \subseteq a_x + V,$$

and so $A \subseteq \Phi + V$ holds, proving that A is a τ-totally bounded set. ∎

It is important to know that the algebraic operations on totally bounded sets produce totally bounded sets.

Theorem 3.2. *If A and B are τ-totally bounded subsets of a topological vector space (X, τ), then λA, $A + B$, and $\overline{A}$ are likewise τ-totally bounded.*

Proof. We prove the statement about the closure. Given a circled τ-neighborhood V of zero, choose a finite subset Φ of A with $A \subseteq \Phi + V$. Now if $x \in \overline{A}$, then $(x + V) \cap (\Phi + V) \neq \emptyset$, and so there exist $u, v \in V$ and $a \in \Phi$ with $x + u = a + v$. Thus, $x = a + v - u \in \Phi + V + V$, and so $\overline{A} \subseteq \Phi + V + V$ holds, and the total boundedness of $\overline{A}$ follows from Statement (3) of Theorem 3.1. ∎

Continuous linear mappings between topological vector spaces carry totally bounded sets to totally bounded sets.

Theorem 3.3. *Let $T: (X, \tau) \to (Y, \xi)$ be an operator between two topological vector spaces. If T is continuous on the τ-bounded subsets of X, then T carries τ-totally bounded sets to ξ-totally bounded sets.*

Proof. Let A be a τ-totally bounded subset of X, and let V be an arbitrary ξ-neighborhood of zero. Since $A - A$ is τ-totally bounded and by our hypothesis $T: (A - A, \tau) \to (Y, \xi)$ is continuous, there exists a τ-neighborhood W of zero satisfying $(A - A) \cap W \subseteq (A - A) \cap T^{-1}(V)$. Now pick a finite subset Φ of A with $A \subseteq \Phi + W$, and note that $T(A) \subseteq T(\Phi) + V$ holds. That is, $T(A)$ is a ξ-totally bounded set, as desired. ∎

Before proceeding further, let us mention two properties that will be used extensively in the sequel (the easy proofs of which can be furnished by the reader).

A subset A of a vector space is:

(i) circled if and only if $|\lambda| \leq |\mu|$ implies $\lambda A \subseteq \mu A$, and

(ii) convex if and only if $\lambda A + \mu A = (\lambda + \mu)A$ holds for all $\lambda, \mu \geq 0$.

The locally convex topologies are the most important linear topologies. A linear topology τ on a vector space X is called **locally convex** (and (X, τ) is called a **locally convex space**) whenever τ has a base at zero consisting of convex sets. Since in a topological vector space the closure of a convex set is also convex, it follows that a linear topology is locally convex if and only if the τ-closed, convex, and circled neighborhoods of zero form a base for the neighborhood system at zero. The role of convexity will be discussed below.

Recall that in a topological vector space the closure of the convex circled hull of a set A is called the **convex, circled, closed hull** of A. As a first sample, we shall establish that in a locally convex space the convex, circled, closed hull of a totally bounded set is also totally bounded. This important result is essentially due to S. Mazur [**136**].

Theorem 3.4 (Mazur). *Let (X, τ) be a locally convex space. If A is a τ-totally bounded subset of X, then its convex, circled, and τ-closed hull (and hence its convex τ-closed hull) is likewise τ-totally bounded.*

In particular, if X is a Banach space, then the convex closed hull of any compact subset of X is likewise a compact set.

Proof. Let A be a τ-totally bounded subset of (X, τ). Denote by A_c the convex, circled hull of A. By Theorem 3.2, it is enough to show that A_c is τ-totally bounded.

Assume at the beginning that A is a finite set, say $A = \{a_1, \ldots, a_n\}$. Then

$$A_c = \left\{ \sum_{i=1}^{n} \lambda_i a_i : \sum_{i=1}^{n} |\lambda_i| \leq 1 \right\}.$$

The set $K = \{(\lambda_1, \ldots, \lambda_n) \in \mathbb{R}^n : |\lambda_1| + + \cdots + |\lambda_n| \leq 1\}$ is closed and bounded in $\mathbb{R}^n$ and hence is compact. Since the function $f : K \to A_c$, defined by $f(\lambda_1, \ldots, \lambda_n) = \lambda_1 a_1 + \cdots + \lambda_n a_n$, is continuous and onto, A_c is a τ-compact subset of X. In particular, A_c is also a τ-totally bounded set.

For the general case, let V be a convex, circled τ-neighborhood of zero. Pick a finite subset Φ of A with $A \subseteq \Phi + V$. Now let $x \in A_c$. Then $x = \lambda_1 x_1 + \cdots + \lambda_n x_n$ holds with $|\lambda_1| + \cdots + |\lambda_n| \leq 1$ and $x_i \in A$ $(1 \leq i \leq n)$. Write $x_i = u_i + v_i$ with $u_i \in \Phi$ and $v_i \in V$, and note that

$$x = \sum_{i=1}^{n} \lambda_i x_i = \sum_{i=1}^{n} \lambda_i u_i + \sum_{i=1}^{n} \lambda_i v_i \in \Phi_c + V .$$

Thus, $A_c \subseteq \Phi_c + V$ holds. By the preceding case Φ_c is a τ-totally bounded set, and so by Condition (4) of Theorem 3.1 the set A_c must be a τ-totally bounded set. ∎

Let A be a convex and absorbing subset of a vector space X. Then the **Minkowski functional** (or the **supporting functional** or the **gauge**) p_A of A is defined by

$$p_A(x) := \inf\{\lambda > 0 \colon x \in \lambda A\}, \quad x \in X .$$

Note that p_A is indeed a function from X to $\mathbb{R}$. The Minkowski functional p_A satisfies the following properties:

(a) $p_A(x) \geq 0$ for all $x \in X$.

(b) $p_A(\lambda x) = \lambda p_A(x)$ for all $\lambda \geq 0$ and all $x \in X$.

(c) $p_A(x + y) \leq p_A(x) + p_A(y)$ for all $x, y \in X$.

(d) If A is also circled, then p_A is a seminorm. That is, in addition to satisfying (a) and (c), it also satisfies $p_A(\lambda x) = |\lambda| p_A(x)$ for all $\lambda \in \mathbb{R}$ and all $x \in X$.

Clearly, (b) and (c) express the fact that the Minkowski functional p_A is a sublinear function.

Now let (X, τ) be a locally convex space. Denote by $\mathcal{B}$ the collection of all τ-neighborhoods of zero that are τ-closed, circled, and convex; clearly $\mathcal{B}$ is a base for the τ-neighborhood system at zero. The collection of seminorms $\{P_V \colon V \in \mathcal{B}\}$ has the following properties (whose easy verifications are left for the reader):

(i) For each $V, W \in \mathcal{B}$ we have $p_{V \cap W}(x) = \max\{p_V(x), p_W(x)\}$.

(ii) For each $V \in \mathcal{B}$ and each $\lambda > 0$ we have $p_{\lambda V} = \frac{1}{\lambda} p_V$.

(iii) Each $V \in \mathcal{B}$ is the closed unit ball of p_V. That is,

$$V = \{x \in X \colon p_V(x) \leq 1\}.$$

Statements (i) and (ii) imply that $\{p_V \colon V \in \mathcal{B}\}$ is a saturated family of seminorms.[3]

---

[3] A family of seminorms that is closed under finite suprema and multiplication by positive scalars is referred to as a **saturated family of seminorms**.

In the converse direction, let $\{p_i\colon\ i \in I\}$ be a saturated family of semi-norms on a vector space. Denote by V_i the closed unit ball of p_i (i.e., $V_i = \{x\colon\ p_i(x) \leq 1\}$), and note that the family of convex sets $\{V_i\colon\ i \in I\}$ satisfies the necessary properties for determining a (unique) locally convex topology. This topology is referred to as **the locally convex topology generated by the family of seminorms** $\{p_i\colon\ i \in I\}$. Note that with respect to this topology each V_i is also closed. In general, if $\{p_i\colon\ i \in I\}$ is a family of seminorms on a vector space (not necessarily saturated), then the collection of all finite suprema of the positive multiples of $\{p_i\colon\ i \in I\}$ is a saturated family of seminorms. This saturated family generates a locally convex topology, which is referred to as the locally convex topology generated by the family of seminorms $\{p_i\colon\ i \in I\}$. It should be noted that if τ is the locally convex topology generated by $\{p_i\colon\ i \in I\}$, then $x_\alpha \xrightarrow{\ \tau\ } 0$ if and only if $p_i(x_\alpha) \to 0$ holds in $\mathbb{R}$ for each $i \in I$. Also, a subset A is τ-bounded if and only if $p_i(A)$ is a bounded subset of $\mathbb{R}$ for each $i \in I$.

Recapitulating the above discussion, we see that a linear topology on a vector space is locally convex if and only if it is generated by a family of seminorms—which can be taken to be saturated. In particular, note that a locally convex topology that is generated by a family of seminorms $\{p_i\colon\ i \in I\}$ is Hausdorff if and only if $p_i(x) = 0$ for all $i \in I$ implies $x = 0$.

Recall that the **algebraic dual** X^* of a vector space X is the vector space consisting of all linear functionals on X. If τ is a linear topology on X, then the **topological dual** X' of (X, τ) is the vector subspace of X^* consisting of all τ-continuous linear functionals on X. That is,

$$X' := \{f \in X^*\colon\ f \text{ is } \tau\text{-continuous}\}.$$

Following the standard notation, we shall designate the elements of X' by primes (x', y', etc.). The continuous linear functionals are characterized as follows. (Recall that if f is a linear functional on a vector space X, then its kernel is denoted by $\operatorname{Ker} f := \{x \in X\colon\ f(x) = 0\}$.)

Theorem 3.5. *For a linear functional f on a locally convex space (X, τ) the following statements are equivalent.*

(1) *f is τ-continuous.*

(2) *The kernel of f is a τ-closed vector subspace.*

(3) *f is bounded on a τ-neighborhood of zero.*

(4) *There is a τ-continuous seminorm p on X such that $\bigl|f(x)\bigr| \leq p(x)$ holds for all $x \in X$.*

Proof. (1) $\Longrightarrow$ (2) Note that $\operatorname{Ker} f = f^{-1}(\{0\})$. Thus, if f is τ-continuous, then $\operatorname{Ker} f$ is a τ-closed vector subspace of X.

(2) $\Longrightarrow$ (3) Assume that f is nonzero. Fix some $x \in X$ with $f(x) = 1$. Consequently, $x \notin \operatorname{Ker} f$, and since $\operatorname{Ker} f$ is τ-closed, there exists a circled τ-neighborhood V of zero with $(x+V) \cap \operatorname{Ker} f = \emptyset$. We claim that $|f(y)| \leq 1$ holds for all $y \in V$.

To see this, let $y \in V$, and suppose by way of contradiction that we have $|f(y)| > 1$. If $\alpha = \frac{1}{f(y)}$, then $|\alpha| < 1$ holds, and so $x - \alpha y \in x + V$. On the other hand,

$$f(x - \alpha y) = f(x) - \alpha f(y) = 0$$

implies $x - \alpha y \in \operatorname{Ker} f$, contrary to $(x + V) \cap \operatorname{Ker} f = \emptyset$. Therefore, f is bounded on V.

(3) $\Longrightarrow$ (4) Pick a circled, convex τ-neighborhood V of zero such that $|f(x)| \leq 1$ holds for all $x \in V$, and note that $|f(x)| \leq p_V(x)$ for all $x \in X$.

(4) $\Longrightarrow$ (1) Obvious. ∎

The continuous linear functionals on a vector subspace Y of a locally convex space X are merely the restrictions on Y of the continuous linear functionals on X.

Theorem 3.6. *If f is a continuous linear functional defined on a vector subspace of a locally convex space (X, τ), then f extends to a continuous linear functional on (X, τ).*

Proof. Let Y be a vector subspace of a locally convex space (X, τ), and let $f: Y \to \mathbb{R}$ be a τ-continuous linear functional. Pick a circled, convex τ-neighborhood V of zero such that $|f(y)| \leq 1$ holds for all $y \in V \cap Y$. Therefore, $f(y) \leq p_V(y)$ holds for all $y \in Y$. By the Hahn–Banach Theorem 1.25 there exists an extension of f to all of X (which we denote by f again) satisfying $f(x) \leq p_V(x)$ for all $x \in X$. Thus, $|f(x)| \leq p_V(x)$ holds for all $x \in X$, and so f is a τ-continuous linear functional on X. ∎

Another property of locally convex spaces is that they have an abundance of continuous linear functionals.

Theorem 3.7. *If Y is a closed vector subspace of a locally convex space (X, τ) and $a \notin Y$, then there exists a τ-continuous linear functional f on X such that $f(y) = 0$ for all $y \in Y$ and $f(a) = 1$. In particular, X' separates the points of X.*

Proof. Let Z be the vector subspace generated by Y and a, that is,

$$Z = \{y + \lambda a: \ y \in Y \ \text{and} \ \lambda \in \mathbb{R}\}.$$

Define the linear functional $f: Z \to \mathbb{R}$ by $f(y + \lambda a) = \lambda$, and note that $\operatorname{Ker} f = Y$. Since Y is closed, by Theorem 3.5 the linear functional $f: Z \to \mathbb{R}$

is continuous, and so by Theorem 3.6 it extends continuously to all of X. This extension satisfies the required properties.

To see that X' separates the points on X, let $a \neq b$. This means that $a - b$ does not belong to the closed vector subspace $\{0\}$, and so by the above there exists some $x' \in X'$ satisfying $x'(a) - x'(b) = x'(a - b) = 1 \neq 0$. ∎

Now we turn our attention to separation properties of convex sets by continuous linear functionals.

Definition 3.8. *Let A and B be two (nonempty) subsets of a vector space X. Then a linear functional x^* on X is said to*

(1) *separate A and B, whenever there exists some $c \in \mathbb{R}$ satisfying $x^*(a) \geq c$ for all $a \in A$ and $x^*(b) \leq c$ for all $b \in B$,*

(2) *strictly separate A and B, whenever there exist some $c \in \mathbb{R}$ and some $\epsilon > 0$ satisfying $x^*(a) \geq c + \epsilon$ for all $a \in A$ and $x^*(b) \leq c$ for all $b \in B$.*

The geometrical meanings of separations are shown in the figure below.

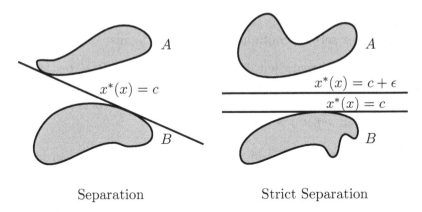

Separation   Strict Separation

One reason for the importance of locally convex topologies is that the continuous linear functionals do not merely separate the points but also separate disjoint open convex sets. The details are explained in the next few results.

Theorem 3.9. *Let C be a convex subset of a locally convex space (X, τ) such that $0 \notin C$. If C has an interior point, then there exists a nonzero τ-continuous linear functional x' on X satisfying $x'(x) \geq 0$ for all $x \in C$.*

Proof. Let $a \in C$ be an interior point of C. Then zero is an interior point of the convex set $K = a - C$. Let V be a circled, convex, τ-neighborhood of zero satisfying $V \subseteq K$. Clearly, K is an absorbing set.

Next, note that $p_K(a) \geq 1$. Indeed, if $p_K(a) < 1$ holds, then there exists some $0 \leq \lambda < 1$ with $a \in \lambda K$. Thus, for some $v \in K$ we have

$$a = \lambda v = \lambda v + (1 - \lambda)0 \in K.$$

This implies $0 \in C$, which is a contradiction. Hence, $p_K(a) \geq 1$ holds.

Now consider the vector subspace $Y = \{\lambda a \colon \lambda \in \mathbb{R}\}$, and define the nonzero linear functional $x' \colon Y \to \mathbb{R}$ by $x'(\lambda a) = \lambda p_K(a)$. Clearly, we have $x'(\lambda a) \leq p_K(\lambda a)$ for each λ. Since p_K is sublinear on X, the Hahn–Banach Theorem 1.25 guarantees that x' has a linear extension to all of X, which we denote by x' again, satisfying $x'(x) \leq p_K(x) \leq p_V(x)$ for all $x \in X$. From

$$-x'(x) = x'(-x) \leq p_K(-x) \leq p_V(-x) = p_V(x),$$

we see that $|x'(x)| \leq p_V(x)$ holds for all $x \in X$, and so, by Theorem 3.5, x' is a nonzero τ-continuous linear functional on X. Finally, note that if $c \in C$, then

$$p_K(a) - x'(c) = x'(a) - x'(c) = x'(a - c) \leq p_K(a - c) \leq 1,$$

and so $x'(c) \geq p_K(a) - 1 \geq 0$ for all $c \in C$, and the proof is finished. ∎

The next theorem, known as the **separation theorem**, deals with the separation of convex sets by continuous linear functionals.

Theorem 3.10 (The Separation Theorem). *Let A and B be two disjoint (nonempty) convex subsets of a locally convex space (X, τ). If either A or B has an interior point, then there exists a nonzero τ-continuous linear functional on X that separates A and B.*

Proof. Note that $0 \notin A - B$, and $A - B$ is a convex set with an interior point. By Theorem 3.9, there exists a nonzero τ-continuous linear functional x' on X such that $x'(x) - x'(y) \geq 0$ holds for all $x \in A$ and all $y \in B$. Put $s = \inf\{x'(x) \colon x \in A\}$ and $t = \sup\{x'(y) \colon y \in B\}$, and note that $-\infty < t \leq s < \infty$ holds. Now if $c \in \mathbb{R}$ satisfies $t \leq c \leq s$, then we have $x'(x) \geq c$ for all $x \in A$ and $x'(y) \leq c$ for all $y \in B$. Therefore, the nonzero τ-continuous linear functional x' separates the sets A and B. ∎

In general, two disjoint convex sets cannot be strictly separated by a continuous linear functional, even when both are closed. However, if one of them is closed and the other is compact, then strict separation is always possible. To establish this result, we need the following useful lemma dealing with the sum of two closed sets.

Lemma 3.11. *Let A and B be two subsets of a topological vector space (X, τ). If A is τ-closed and B is τ-compact, then $A + B$ is τ-closed.*

Proof. Assume that a net $\{a_\alpha + b_\alpha\} \subseteq A + B$, with $\{a_\alpha\} \subseteq A$ and $\{b_\alpha\} \subseteq B$, satisfies $a_\alpha + b_\alpha \xrightarrow{\tau} x$ in X. By the τ-compactness of B we can assume (by passing to a subnet) that $b_\alpha \xrightarrow{\tau} b \in B$ holds. Hence,

$$a_\alpha = (a_\alpha + b_\alpha) - b_\alpha \xrightarrow{\tau} x - b = a,$$

and since A is τ-closed we get $a \in A$. Thus, $x = a + b \in A + B$, and this shows that the set $A + B$ is τ-closed. ∎

We are now in the position to establish that two disjoint convex sets, one of which is closed and the other is compact, can always be strictly separated by a continuous linear functional.

Theorem 3.12 (The Strict Separation Theorem). *Let A and B be two nonempty disjoint convex subsets of a locally convex space (X, τ). If A is τ-closed and B is τ-compact, then A and B can be strictly separated by a τ-continuous linear functional. That is, there exist a τ-continuous linear functional x', some $c \in \mathbb{R}$, and some $\epsilon > 0$ such that*

$$x'(x) \leq c < c + \epsilon \leq x'(y)$$

holds for all $x \in B$ and all $y \in A$.

Proof. Clearly, $A - B$ is convex, $0 \notin A - B$, and (by Lemma 3.11) $A - B$ is also τ-closed. Pick a circled, convex τ-neighborhood V of zero satisfying $V \cap (A - B) = \emptyset$. Since zero is an interior point of V, Theorem 3.10 applied to the pair $(V, A - B)$ guarantees the existence of a nonzero τ-continuous linear functional x' on X and some $\epsilon \in \mathbb{R}$ such that $x'(v) \leq \epsilon$ for all $v \in V$ and $x'(y) - x'(x) \geq \epsilon$ for all $x \in B$ and $y \in A$. Since V is circled and absorbing and x' is nonzero, it is easy to see that $\epsilon > 0$ must hold. Now put $c = \max\{x'(x) : x \in B\}$, and note that $x'(x) \leq c < c + \epsilon \leq x'(y)$ holds for all $x \in B$ and $y \in A$. ∎

For a locally convex space (X, τ) the sets of the form $\{x \in X : x'(x) \geq c\}$, where x' is a τ-continuous linear functional on X and $c \in \mathbb{R}$, are referred to as the τ-**closed half-spaces**.

Theorem 3.13. *In a locally convex space (X, τ) the τ-closure of any convex set is the intersection of all τ-closed half-spaces that include it.*

In particular, if two locally convex topologies on a vector space have the same topological dual, then they also have the same closed convex sets.

Proof. Let A be a τ-closed convex subset of (X, τ). If $a \notin A$, then $\{a\}$ is a τ-compact convex set that is disjoint from A. Thus by Theorem 3.12 there exists some $x' \in X'$ and some $c \in \mathbb{R}$ such that $x'(x) \geq c$ for all $x \in A$ and $x'(a) < c$. Now from this the desired conclusion follows. ∎

Recall that a point e of a convex set A is said to be an **extreme point** of A whenever $e = \lambda x + (1 - \lambda)y$ with $0 < \lambda < 1$ and $x, y \in A$ imply $x = y = e$. The extreme points of convex sets play an important role in analysis, and for this reason conditions that guarantee their existence are useful.

The next theorem, due to M. Krein and D. Milman [**102**], gives a topological condition under which a convex set has an extreme point. This result has a wide range of applications.

Theorem 3.14 (Krein–Milman). *If a convex set C of a vector space X is compact for some locally convex topology τ on X, then C has an extreme point. Moreover, C is the τ-closed convex hull of its extreme points.*

Proof. The proof below is due to J. L. Kelley [**96**]. We establish first that C has an extreme point. To do this, start by saying that a nonempty subset A of C is an *extremal set* whenever $\lambda x + (1 - \lambda)y \in A$, $0 < \lambda < 1$, and $x, y \in C$ imply $x, y \in A$. Let $\mathcal{A}$ be the collection of all τ-closed extremal subsets of C, ordered by inclusion. That is,

$$\mathcal{A} = \{A \subseteq C : A \text{ is extremal and } \tau\text{-closed}\}.$$

Since $C \in \mathcal{A}$, we have $\mathcal{A} \neq \emptyset$. Now if $\{A_i : i \in I\}$ is an arbitrary chain of $\mathcal{A}$, then $\{A_i : i \in I\}$ has the finite intersection property, and since C is τ-compact, it follows that $\bigcap\{A_i : i \in I\} \neq \emptyset$. An easy argument shows that $\bigcap\{A_i : i \in I\} \in \mathcal{A}$. That is, every chain of $\mathcal{A}$ has a lower bound in $\mathcal{A}$. Consequently, by Zorn's lemma, $\mathcal{A}$ has a minimal element, say M. Assume that $a, b \in M$ and $a \neq b$. Then by Theorem 3.7 there exists some $x' \in X'$ with $x'(a) \neq x'(b)$. Put $m = \min\{x'(x) : x \in M\}$ and consider the set $B = \{x \in M : x'(x) = m\}$. Clearly, B is a nonempty proper subset of M which is τ-closed and convex. On the other hand, it is easy to see that B is an extremal subset of C, contradicting the minimality property of M. Hence, M consists of one point, which must be an extreme point of C.

Now let $\mathcal{E}$ denote the (nonempty) set of all extreme points of C. Assume that there exists some $a \in C$ with $a \notin \overline{\operatorname{co}}(\mathcal{E})$. By Theorem 3.12 there exists some $x' \in X'$ such that for some $c \in \mathbb{R}$ and some $\epsilon > 0$ we have $x'(a) < c$ and $x'(x) \geq c + \epsilon$ for all $x \in \overline{\operatorname{co}}(\mathcal{E})$. Put $s = \min\{x'(x) : x \in C\}$, and then define the nonempty τ-closed convex set $K = \{x \in C : x'(x) = s\}$. Clearly, $K \cap \mathcal{E} = \emptyset$ holds, and by the first part, K must have an extreme point. However, since every extreme point of K is also an extreme point of C (why?), this is a contradiction. Hence, $C = \overline{\operatorname{co}}(\mathcal{E})$ holds, and the proof is finished. ∎

We leave now the separation properties and turn our attention to the important concept of duality. A **dual system** $\langle X, X' \rangle$ is a pair of vector spaces X and X' together with a bilinear form $(x, x') \mapsto \langle x, x' \rangle$ (i.e., with a function $\langle \cdot, \cdot \rangle \colon X \times X' \to \mathbb{R}$ which is linear in each variable separately) such that:

(1) If $\langle x, x' \rangle = 0$ holds for all $x \in X$, then $x' = 0$.

(2) If $\langle x, x' \rangle = 0$ holds for all $x' \in X'$, then $x = 0$.

If (X, τ) is a locally convex space, then $\langle X, X' \rangle$ is a dual system under the bilinear form $\langle x, x' \rangle = x'(x)$. (To see this, use Theorem 3.7.) As we shall see, this example is representative of the typical dual system.

Now let $\langle X, X' \rangle$ be a dual system. Then note that for each $x' \in X'$ the formula $f(x) = \langle x, x' \rangle$, $x \in X$, defines a linear functional on X. Moreover, the mapping $x' \mapsto \langle \cdot, x' \rangle$, from X' to X^*, is one-to-one and linear. In other words, identifying x' with the linear functional $\langle \cdot, x' \rangle$, we see that each $x' \in X'$ can be viewed as a linear functional on X, and in addition the following subspace inclusions hold:

$$X' \subseteq X^* \subseteq \mathbb{R}^X .$$

The **weak topology** $\sigma(X', X)$ on X' is the locally convex topology induced by the product topology of $\mathbb{R}^X$ on X'. The product topology of $\mathbb{R}^X$ is the locally convex topology generated by the family of seminorms $\{p_x \colon x \in X\}$, where $p_x(f) = |f(x)|$ holds for all $f \in \mathbb{R}^X$. Thus $\sigma(X', X)$ is generated by the family of seminorms $\{p_x \colon x \in X\}$, where $p_x(x') = |\langle x, x' \rangle|$ holds for all $x' \in X'$. In particular, it should be noted that the collection of sets

$$\{x' \in X' \colon |\langle x_i, x' \rangle| \le 1 \text{ for } i = 1, \ldots, n\}$$

form a base at zero for $\sigma(X', X)$. Clearly, a net $\{x'_\alpha\} \subseteq X'$ satisfies $x'_\alpha \xrightarrow{\sigma(X', X)} 0$ if and only if $\langle x, x'_\alpha \rangle \to 0$ holds in $\mathbb{R}$ for each $x \in X$; and this justifies the alternative name **the topology of pointwise convergence on X** used for $\sigma(X', X)$.

Similarly, for each $x \in X$ the formula $x(x') = \langle x, x' \rangle$ defines a linear functional on X', which we identify with x. Thus, the following vector subspace inclusions hold:

$$X \subseteq (X')^* \subseteq \mathbb{R}^{X'} .$$

The *weak topology* $\sigma(X, X')$ (or **the topology of pointwise convergence on X'**) is the locally convex topology induces on X by the product topology of $\mathbb{R}^{X'}$. Note that $\sigma(X, X')$ is generated by the family of seminorms $\{p_{x'} \colon x' \in X'\}$, where $p_{x'}(x) = |\langle x, x' \rangle|$. In particular, a net $\{x_\alpha\} \subseteq X$ satisfies $x_\alpha \xrightarrow{\sigma(X, X')} 0$ if and only if $\langle x_\alpha, x' \rangle \to 0$ holds in $\mathbb{R}$ for each $x' \in X'$. Also, it should be noted that the collection of sets of the form $\{x \in X \colon |\langle x, x'_i \rangle| \le 1 \text{ for } i = 1, \ldots, n\}$ is a base at zero for the topology $\sigma(X, X')$.

The next result characterizes the linear functionals that belong to the vector space generated by a finite collection of linear functionals.

Lemma 3.15. *Let $f, f_1, \ldots, f_n$ be linear functionals on a vector space X. Then f is a linear combination of $f_1, \ldots, f_n$ (i.e., there exist real numbers $\lambda_1, \ldots, \lambda_n$ such that $f = \lambda_1 f_1 + \cdots + \lambda_n f_n$) if and only if $\bigcap_{i=1}^{n} \operatorname{Ker} f_i \subseteq \operatorname{Ker} f$.*

Proof. Clearly, if $f = \sum_{i=1}^{n} \lambda_i f_i$, then $\bigcap_{i=1}^{n} \operatorname{Ker} f_i \subseteq \operatorname{Ker} f$ holds. For the converse assume $\bigcap_{i=1}^{n} \operatorname{Ker} f_i \subseteq \operatorname{Ker} f$.

Define the operator $T \colon X \to \mathbb{R}^n$ by $T(x) = \big(f_1(x), \ldots, f_n(x)\big)$ for each $x \in X$. Note that if $T(x) = T(z)$, then $f_i(x - z) = 0$ holds for all $1 \leq i \leq n$, and from our hypothesis it follows that $f(x) = f(z)$. This means that the formula $\phi(Tx) = f(x)$, $x \in X$, defines a linear functional on the vector subspace $T(X)$ of $\mathbb{R}^n$, and hence ϕ can be extended linearly to all of $\mathbb{R}^n$. Thus, there exist real numbers $\lambda_1, \ldots, \lambda_n$ such that

$$\phi(y_1, \ldots, y_n) = \lambda_1 y_1 + \cdots + \lambda_n y_n$$

holds for all $(y_1, \ldots, y_n) \in \mathbb{R}^n$. In particular, we have

$$f(x) = \phi(Tx) = \sum_{i=1}^{n} \lambda_i f_i(x)$$

for all $x \in X$, as desired. ∎

The topological dual of $\big(X, \sigma(X, X')\big)$ is precisely X'. The details follow.

Theorem 3.16. *Let $\langle X, X' \rangle$ be a dual system. Then the topological dual of X with $\sigma(X, X')$ is precisely X'. That is, a linear functional f on X is $\sigma(X, X')$-continuous if and only if there exists a (unique) $x' \in X'$ satisfying*

$$f(x) = \langle x, x' \rangle$$

for all $x \in X$. Similarly, we have $\big(X', \sigma(X', X)\big)' = X$.

Proof. Let $f \in X^*$. If some $x' \in X'$ satisfies $f(x) = \langle x, x' \rangle$ for all $x \in X$, then f is clearly a $\sigma(X, X')$-continuous linear functional.

For the converse assume that f is a $\sigma(X, X')$-continuous linear functional. By Theorem 3.5 there exist $x_1', \ldots, x_n' \in X'$ such that f is bounded on the $\sigma(X, X')$-neighborhood of zero $\{x \in X \colon |x_i'(x)| \leq 1 \text{ for } 1 \leq i \leq n\}$. Now an easy argument shows that $\operatorname{Ker} x_1' \cap \cdots \cap \operatorname{Ker} x_n' \subseteq \operatorname{Ker} f$ holds. Thus, by Lemma 3.15, f must be a linear combination of $x_1', \ldots, x_n'$, and so $f \in X'$. ∎

If (X, τ) is a Hausdorff locally convex space, then (by Theorem 3.7) the topological dual X' of (X, τ) separates the points of X, and so $\langle X, X' \rangle$ is a dual system under the duality $\langle x, x' \rangle = x'(x)$. Theorem 3.16 simply tells us that these are the only type of dual systems.

Fix a dual system $\langle X, X' \rangle$. If A is a subset of X, then its **polar** A° is defined by

$$A^\circ := \{ x' \in X' : \ |\langle x, x' \rangle| \le 1 \text{ for all } x \in A \} .$$

Clearly, A° is a convex, circled, and $\sigma(X', X)$-closed subset of X'. Similarly, if B is a subset of X', then its **polar** is defined by

$$B^\circ := \{ x \in X : \ |\langle x, x' \rangle| \le 1 \text{ for all } x' \in B \} .$$

Obviously, B° is a convex, circled, and $\sigma(X, X')$-closed subset of X. If A is a subset of X (or X'), then the set $A^{\circ\circ} := (A^\circ)^\circ$ is called the **bipolar** of A. The following are two elementary properties of polars.

(a) If $A \subseteq B$, then $B^\circ \subseteq A^\circ$.

(b) For any set A we have $A \subseteq A^{\circ\circ}$.

The next result, known as the *bipolar theorem*, is quite important.

Theorem 3.17 (The Bipolar Theorem). *Let $\langle X, X' \rangle$ be a dual system, and let A be a nonempty subset of X. Then the bipolar $A^{\circ\circ}$ is the convex, circled, $\sigma(X, X')$-closed hull of A, i.e., $A^{\circ\circ}$ is the smallest convex, circled, and $\sigma(X, X')$-closed set that includes A.*

Similarly, if A is an arbitrary subset of X', then $A^{\circ\circ}$ is the convex, circled, $\sigma(X', X)$-closed hull of A.

Proof. Let C be the convex, circled, $\sigma(X, X')$-closed hull of A. Clearly, $A \subseteq C \subseteq A^{\circ\circ}$ holds. Assume by way of contradiction that $A^{\circ\circ} \ne C$. Thus, there exists some $a \in A^{\circ\circ}$ with $a \notin C$. Now applying Theorem 3.12 to the pair of convex sets $(C, \{a\})$ and taking into account Theorem 3.16, we see that there exist $x' \in X'$ and $c \in \mathbb{R}$ satisfying $\langle x, x' \rangle \le c$ for all $x \in C$ and $\langle a, x' \rangle > c$. In view of $0 \in C$, we can assume that $c = 1$. Now since C is circled, it follows that $|\langle x, x' \rangle| \le 1$ holds for all $x \in C$, and so $x' \in A^\circ$. In particular, since $a \in A^{\circ\circ}$, we have $|\langle a, x' \rangle| \le 1$, contrary to $\langle a, x' \rangle > 1$. Hence, $C = A^{\circ\circ}$ holds, and the proof is finished. ∎

More properties of polars are included in the next result.

Theorem 3.18. *Let $\langle X, X' \rangle$ be a dual system, and let $\{ A_i : \ i \in I \}$ be a family of (nonempty) subsets of X. Then we have:*

(1) $\left(\bigcup_{i \in I} A_i \right)^\circ = \bigcap_{i \in I} A_i^\circ$.

(2) *If each A_i is convex, circled, and $\sigma(X, X')$-closed, then the set $\left(\bigcap_{i \in I} A_i \right)^\circ$ is the convex, circled, $\sigma(X', X)$-closed hull of $\bigcup_{i \in I} A_i^\circ$.*

Similar statements hold true if $\{ A_i : \ i \in I \}$ is a family of subsets of X'.

Proof. (1) Straightforward.

(2) By the bipolar theorem we have $A_i^{\circ\circ} = A_i$ for each i, and so by (1)

$$\left(\bigcup_{i\in I} A_i^{\circ}\right)^{\circ\circ} = \left[\left(\bigcup_{i\in I} A_i^{\circ}\right)^{\circ}\right]^{\circ} = \left[\bigcap_{i\in I} A_i^{\circ\circ}\right]^{\circ} = \left[\bigcap_{i\in I} A_i\right]^{\circ},$$

which (by the bipolar theorem again) establishes our claim. ∎

In term of polars the topological dual of a locally convex space is described as follows.

Theorem 3.19. *If (X,τ) is a locally convex space, then*

$$X' = \bigcup\{V^{\circ}: \ V \ \text{is a } \tau\text{-neighborhood of zero}\},$$

where the polars are taken in the algebraic dual X^*.

Proof. If $f \in X'$, then $V = f^{-1}([-1,1])$ is a τ-neighborhood of zero and $f \in V^{\circ}$ holds. On the other hand, if $f \in X^*$ satisfies $f \in V^{\circ}$ for some τ-neighborhood V of zero, then f is bounded on V, and so (by Theorem 3.5) $f \in X'$. ∎

Let (X,τ) be a Hausdorff locally convex space. Then X always will be considered in duality with X' under $\langle x, x'\rangle = x'(x)$. From the preceding result it should be immediate that for an arbitrary τ-neighborhood of zero its polars in X' and X^* coincide. As we shall see next, the polar of a neighborhood is always $\sigma(X', X)$-compact. This important result is due to L. Alaoglu [5].

Theorem 3.20 (Alaoglu). *If (X,τ) is a locally convex space and V is a τ-neighborhood of zero, then its polar V° is a $\sigma(X', X)$-compact set.*

Proof. Let V be a τ-neighborhood of zero. We have $V^{\circ} \subseteq X' \subseteq \mathbb{R}^X$, and $\sigma(X', X)$ is the product topology of $\mathbb{R}^X$ restricted to X'. In particular, V° is $\sigma(X', X)$-compact if and only if V° is compact in $\mathbb{R}^X$, and this in turn (by Tychonoff's classical theorem) is the case if and only if V° is closed and bounded in $\mathbb{R}^X$.

Clearly, V° is bounded in $\mathbb{R}^X$. On the other hand, if a net $\{x'_\alpha\} \subseteq V^{\circ}$ satisfies $x'_\alpha \to f$ in $\mathbb{R}^X$ (i.e., if $x'_\alpha(x) \to f(x)$ holds in $\mathbb{R}$ for each $x \in X$), then f is a linear functional on X and $|f(x)| \leq 1$ holds for all $x \in V$. By Theorem 3.5, $f \in X'$, and thus $f \in V^{\circ}$, so that V° is also a closed subset of $\mathbb{R}^X$. By the above, V° is $\sigma(X', X)$-compact. ∎

Let (X,τ) be a locally convex space. A subset A of X' is said to be a τ-**equicontinuous set**, whenever for each $\epsilon > 0$ there exists a τ-neighborhood V of zero such that $|x'(x)| \leq \epsilon$ holds for all $x \in V$ and all $x' \in A$. The equicontinuous sets are characterized as follows.

Theorem 3.21. *Let (X, τ) be a locally convex space. Then a subset A of X' is an equicontinuous set if and only if there exists a τ-neighborhood V of zero such that $A \subseteq V^\circ$. In particular, every equicontinuous subset of X' is relatively $\sigma(X', X)$-compact.*

Proof. Assume that $A \subseteq V^\circ$ holds for some τ-neighborhood V of zero. Then for each $x \in \epsilon V$ and each $x' \in A$ we have $|x'(x)| \leq \epsilon$, which shows that A is an equicontinuous set. Conversely, if A is an equicontinuous set, then there exists a τ-neighborhood V of zero satisfying $|x'(x)| \leq 1$ for all $x \in V$ and all $x' \in A$, and so $A \subseteq V^\circ$ holds. The last part follows from Theorem 3.20. ■

We continue now with the introduction of the $\mathfrak{S}$-topologies. Let $\langle X, X' \rangle$ be a dual system, and let $\mathfrak{S}$ be a collection of $\sigma(X, X')$-bounded subsets of X. (To keep the topologies Hausdorff, it will be tacitly assumed that $\bigcup\{A: A \in \mathfrak{S}\}$ always spans X.) Then for each $A \in \mathfrak{S}$ the formula

$$\rho_A(x') = \sup\{|\langle x, x' \rangle|: x \in A\}, \quad x' \in X',$$

defines a seminorm on X', and so $\mathfrak{S}$ generates a locally convex topology on X' via the family of seminorms $\{\rho_A: A \in \mathfrak{S}\}$. Note that the closed unit ball of ρ_A is precisely the polar of A, i.e., $A^\circ = \{x' \in X': \rho_A(x') \leq 1\}$. This topology is referred to as the $\mathfrak{S}$-topology. Since a net $\{x'_\alpha\} \subseteq X'$ satisfies $x'_\alpha \xrightarrow{\mathfrak{S}} 0$ if and only if $\{x'_\alpha\}$ converges uniformly to zero on the sets of $\mathfrak{S}$, the $\mathfrak{S}$-topology is also known as the **topology of uniform convergence on the sets of $\mathfrak{S}$**. A moment's thought reveals that the sets of the form

$$\epsilon\left(A_1^\circ \cap \cdots \cap A_n^\circ\right),$$

where $A_1, \ldots, A_n \in \mathfrak{S}$ and $\epsilon > 0$, form a base at zero for the $\mathfrak{S}$-topology.

A collection $\mathfrak{S}$ of subsets of X is called **full** whenever it satisfies the following properties:

(a) If $A \in \mathfrak{S}$, then $\lambda A \in \mathfrak{S}$ holds for all $\lambda > 0$.

(b) If $A, B \in \mathfrak{S}$, then there exists some $C \in \mathfrak{S}$ with $A \cup B \subseteq C$.

(c) $\bigcup\{A: A \in \mathfrak{S}\}$ spans X.

Note that if $\langle X, X' \rangle$ is a dual system and $\mathfrak{S}$ is a full collection of $\sigma(X, X')$-bounded subsets of X, then the polars of the sets of $\mathfrak{S}$ form a base at zero for the $\mathfrak{S}$-topology on X'. Clearly, a completely symmetric situation occurs when $\mathfrak{S}$ is a collection of $\sigma(X', X)$-bounded subsets of X'.

It is interesting to know that every locally convex topology on a vector space is an appropriate $\mathfrak{S}$-topology.

Theorem 3.22. *Let (X, τ) be a locally convex space. Then τ is the $\mathfrak{S}$-topology of uniform convergence on the τ-equicontinuous subsets of X'.*

Proof. By Theorem 3.16 the topological dual of $(X, \sigma(X, X'))$ is also X', and so (by Theorem 3.13) τ and $\sigma(X, X')$ have the same convex closed sets. In particular, if V is a convex, circled, τ-closed neighborhood of zero, then (by the bipolar theorem) we have $V = V^{\circ\circ}$. Therefore, if

$$\mathfrak{S} = \{A \subseteq X' \colon \ A \text{ is } \tau\text{-equicontinuous}\},$$

then according to Theorem 3.21 this $\mathfrak{S}$-topology coincides with τ. ∎

Consider a dual system $\langle X, X'\rangle$. Then a locally convex topology τ on X is said to be **consistent** (or **compatible**) with the dual system whenever the topological dual of (X, τ) is precisely X'. The consistent topologies on X' are defined analogously. In order to characterize the consistent topologies we need a lemma.

Lemma 3.23. *If $A_1, \ldots, A_n$ are convex, circled, and compact subsets in a topological vector space, then the convex circled hull of $A_1 \cup \cdots \cup A_n$ is a compact set.*

Proof. Let $A_1, \ldots, A_n$ be convex, circled, and compact subsets in a topological vector space (X, τ), and let B denote the convex circled hull of $A_1 \cup \cdots \cup A_n$. Then an easy argument shows that

$$B = \left\{\sum_{i=1}^{n} \lambda_i x_i \colon \ x_i \in A_i \text{ for each } i = 1, \ldots, n \text{ and } \sum_{i=1}^{n} |\lambda_i| \leq 1\right\}.$$

Put $K = \{\lambda = (\lambda_1, \ldots, \lambda_n) \in \mathbb{R}^n \colon |\lambda_1| + \cdots + |\lambda_n| \leq 1\}$, and note that K is a compact subset of $\mathbb{R}^n$. Now consider each A_i and B with the topology τ, and note that the function $f \colon K \times A_1 \times \cdots \times A_n \to B$, defined by $f(\lambda, x_1, \ldots, x_n) = \lambda_1 x_1 + \cdots + \lambda_n x_n$, is continuous and onto. In particular, this implies that B is a compact set, as desired. ∎

G. W. Mackey [133] and R. Arens [27] characterized the consistent topologies as follows.

Theorem 3.24 (Mackey–Arens). *Let $\langle X, X'\rangle$ be a dual system. Then a locally convex topology τ on X is consistent with the dual system if and only if τ is a $\mathfrak{S}$-topology for some collection $\mathfrak{S}$ of convex, circled, and $\sigma(X', X)$-compact sets that cover X'.*

Proof. Let τ be a locally convex topology on X consistent with the dual system $\langle X, X'\rangle$. By Theorem 3.22 the topology τ is the $\mathfrak{S}$-topology for the collection $\mathfrak{S} = \{V^{\circ} \colon V \text{ is a } \tau\text{-neighborhood of zero}\}$. Clearly, each V° is convex, circled, and (by Theorem 3.20) $\sigma(X', X)$-compact. Moreover, $\bigcup\{V^{\circ} \colon V^{\circ} \in \mathfrak{S}\} = X'$.

For the converse assume that $\mathfrak{S}$ is a collection of convex, circled, and $\sigma(X', X)$-compact subsets of X' with $\bigcup\{A \colon A \in \mathfrak{S}\} = X'$. Denote by Y

the topological dual of X with the $\mathfrak{S}$-topology, i.e., $Y = (X, \mathfrak{S})'$, and note that $X' \subseteq Y$ holds. Now consider the dual system $\langle X, X^* \rangle$, and let

$$\mathcal{B} = \left\{ V \subseteq X \colon \exists \epsilon > 0 \text{ and } A_1, \ldots, A_n \in \mathcal{S} \text{ with } V = \epsilon\left(A_1^\circ \cap \cdots \cap A_n^\circ\right) \right\}.$$

Since $\mathcal{B}$ is a base at zero for the $\mathfrak{S}$-topology, it follows from Theorem 3.19 that $Y = \bigcup\{V^\circ \colon V \in \mathcal{B}\}$, where the polars are taken in X^*. Now let $A_1, \ldots, A_n \in \mathfrak{S}$. If C denotes the convex circled hull of $A_1 \cup \cdots \cup A_n$ in X', then by Lemma 3.23 the set C is $\sigma(X', X)$-compact and hence $\sigma(X^*, X)$-compact. Therefore, by the bipolar theorem we have

$$(A_1 \cup \cdots \cup A_n)^{\circ\circ} = (A_1^\circ \cap \cdots \cap A_n^\circ)^\circ = C,$$

where the polars are taken with respect to the dual system $\langle X, X^* \rangle$. This implies that $V^\circ \subseteq X'$ holds for all $V \in \mathcal{B}$, and so $Y \subseteq X'$ also holds. Consequently, $Y = X'$, and the proof is finished. ∎

An immediate consequence of Theorem 3.24 is that in a dual system $\langle X, X' \rangle$ there always exists a largest consistent locally convex topology on X, namely, the $\mathfrak{S}$-topology, where $\mathfrak{S}$ is the collection of all convex, circled and $\sigma(X', X)$-compact subsets of X'. This largest topology is denoted by $\tau(X, X')$ and is called the **Mackey topology** of X. The Mackey topology $\tau(X', X)$ on X' is defined in a similar manner. Taking into account Theorem 3.24 once more, it is easy to see that a locally convex topology τ on X is consistent with respect to the dual system $\langle X, X' \rangle$ if and only if

$$\sigma(X, X') \subseteq \tau \subseteq \tau(X, X').$$

Similarly, a locally convex topology τ on X' is consistent with $\langle X, X' \rangle$ if and only if $\sigma(X', X) \subseteq \tau \subseteq \tau(X', X)$.

It is surprising to learn that all consistent topologies have the same bounded sets. This important result is due to G. W. Mackey [**133**].

Theorem 3.25 (Mackey). *Let $\langle X, X' \rangle$ be a dual system. Then all consistent locally convex topologies on X have the same bounded subsets of X.*

Proof. It is enough to establish that every $\sigma(X, X')$-bounded subset of X is $\tau(X, X')$-bounded. To this end, let A be a $\sigma(X, X')$-bounded subset of X, and let B be a convex, circled, and $\sigma(X', X)$-compact subset of X'. To complete the proof we have to demonstrate the existence of some $\lambda > 0$ satisfying $\lambda A \subseteq B^\circ$. The existence of such λ will be based upon the following claim:

> *There exists some $x' \in B$, some $k \in \mathbb{N}$, and some $\sigma(X', X)$-neighborhood V of zero such that*
>
> $$B \cap (x' + V) \subseteq kA^\circ. \qquad (\star)$$

To see this, assume by way of contradiction that the above claim is false. Then there exists a sequence $\{x'_n\} \subseteq B$ and a sequence $\{V_n\}$ of open $\sigma(X', X)$-neighborhoods of zero satisfying

$$x'_{n+1} + \overline{V}_{n+1} \subseteq (x'_n + V_n) \cap [X' \setminus (n+1)A^\circ]$$

for all n. The existence of the two sequences can be demonstrated by induction as follows. Start with $n = 0$ by choosing an arbitrary $x'_0 \in B$ and an arbitrary open $\sigma(X', X)$-neighborhood V_0 of zero. If $x'_n \in B$ and V_n have been chosen, then from $B \cap (x'_n + V_n) \not\subseteq (n+1)A^\circ$, it follows that there exists some $x'_{n+1} \in B \cap (x'_n + V_n) \cap [X' \setminus (n+1)A^\circ]$. Since $(x'_n + V_n) \cap [X' \setminus (n+1)A^\circ]$ is a $\sigma(X', X)$-open set, there exists an open $\sigma(X', X)$-neighborhood V_{n+1} of zero with $x'_{n+1} + \overline{V}_{n+1} \subseteq (x'_n + V_n) \cap [X' \setminus (n+1)A^\circ]$, where the bar denotes $\sigma(X', X)$-closure, and the induction is complete.

Next note that $\{B \cap (x'_n + \overline{V}_n)\}$ is a decreasing sequence of $\sigma(X', X)$-closed nonempty subsets of B, and so by the $\sigma(X', X)$-compactness of B we must have $\bigcap_{n=1}^\infty B \cap (x'_n + \overline{V}_n) \neq \emptyset$. However, if $x' \in \bigcap_{n=1}^\infty B \cap (x'_n + \overline{V}_n)$, then $x' \notin nA^\circ$ for all n, contradicting the fact that A° is absorbing. Thus, $(\star)$ holds.

Now write$(\star)$ in the form $(B - x') \cap V \subseteq kA^\circ - x'$, and then pick some $0 < \epsilon < 1$ with $\epsilon(B - x') \subseteq V$. Since zero belongs to the convex set $B - x'$, we also have $\epsilon(B - x') \subseteq B - x'$, and so $\epsilon(B - x') \subseteq (B - x') \cap V \subseteq kA^\circ - x'$. Therefore, $\epsilon B \subseteq kA^\circ + (\epsilon - 1)x'$. Since A° is absorbing, there exists some $m \in \mathbb{N}$ with $(\epsilon - 1)x' \in mA^\circ$, and so by the convexity of A° we get

$$\epsilon B \subseteq kA^\circ + (\epsilon - 1)x' \subseteq kA^\circ + mA^\circ = (k + m)A^\circ.$$

Thus, $\lambda B \subseteq A^\circ$ holds for some $\lambda > 0$. Taking polars we obtain

$$A \subseteq A^{\circ\circ} \subseteq (\lambda B)^\circ = \tfrac{1}{\lambda} B^\circ,$$

so that $\lambda A \subseteq B^\circ$ holds, as required. ∎

When X is a Banach space its norm topology coincides with the Mackey topology $\tau(X, X')$. Thus, for a Banach space X, Theorem 3.25 simply states that a subset of X is norm bounded if and only if it is weakly bounded. This property is known as the **principle of uniform boundedness**.

Recall that the **adjoint** (or **transpose**) of an operator $T \colon X \to Y$ between two vector spaces is the operator $T^* \colon Y^* \to X^*$ defined via the duality identity

$$\langle x, T^* y^* \rangle = \langle Tx, y^* \rangle, \quad x \in X, \ y^* \in Y^*.$$

If $\langle X, X' \rangle$ and $\langle Y, Y' \rangle$ is a pair of dual systems, then an operator $T \colon X \to Y$ is said to be **weakly continuous** whenever $T \colon (X, \sigma(X, X')) \to (Y, \sigma(Y, Y'))$ is continuous. For this to happen it is necessary and sufficient for the algebraic adjoint T^* to carry Y' onto a vector subspace of X'.

Theorem 3.26. *Let $\langle X, X' \rangle$ and $\langle Y, Y' \rangle$ be a pair of dual systems, and let $T \colon X \to Y$ be an operator. Then T is weakly continuous if and only if*

$$T^*(Y') \subseteq X'.$$

Proof. Assume first that T is weakly continuous, and let $y' \in Y'$. From $T^* y'(x) = y'(Tx)$, $x \in X$, it is easy to see that $T^* y'$ is $\sigma(X, X')$-continuous, and so (by Theorem 3.16) we have $T^* y' \in X'$. Hence, $T^*(Y') \subseteq X'$.

For the converse, suppose that $T^*(Y') \subseteq X'$. Let $y'_1, \ldots, y'_n \in Y'$, and consider the $\sigma(Y, Y')$-neighborhood of zero

$$V = \{ y \in Y \colon \ |\langle y, y'_i \rangle| \le 1 \text{ for } i = 1, \ldots, n \}.$$

Since $T^* y'_1, \ldots, T^* y'_n \in X'$, the set

$$W = \{ x \in X \colon \ |\langle x, T^* y'_i \rangle| \le 1 \text{ for } i = 1, \ldots, n \}$$

is a $\sigma(X, X')$-neighborhood of zero. From $\langle Tx, y'_i \rangle = \langle x, T^* y'_i \rangle$, it follows that $T(W) \subseteq V$. This shows that T is weakly continuous. ∎

Consider a pair of dual systems $\langle X, X' \rangle$ and $\langle Y, Y' \rangle$, and let $T \colon X \to Y$ be a weakly continuous operator. By the preceding theorem we know that T^* carries Y' into X'. In this case, the adjoint operator T^* restricted to Y' will be denoted by T'. Thus, if $T \colon X \to Y$ is weakly continuous, then $T' \colon Y' \to X'$ satisfies the duality identity

$$\langle x, T' y' \rangle = \langle Tx, y' \rangle, \quad x \in X, \ y' \in Y'.$$

In connection with weakly continuous operators, the following remarkable theorem of A. Grothendieck [73] describes an important duality property of totally bounded sets.

Theorem 3.27 (Grothendieck). *Let $\langle X, X' \rangle$ and $\langle Y, Y' \rangle$ be a pair of dual systems. Let $\mathfrak{S}$ be a full collection of $\sigma(X, X')$-bounded subsets of X, and let $\mathfrak{S}'$ be another full collection of $\sigma(Y', Y)$-bounded subsets of Y'. Then for a weakly continuous operator $T \colon X \to Y$ the following statements are equivalent:*

(1) *$T(A)$ is $\mathfrak{S}'$-totally bounded for each $A \in \mathfrak{S}$.*

(2) *$T'(B)$ is $\mathfrak{S}$-totally bounded for each $B \in \mathfrak{S}'$.*

Proof. By the symmetry of the situation it is enough to show that (1) implies (2). Therefore, assume that (1) is true.

Fix $B \in \mathfrak{S}'$ and let $A \in \mathfrak{S}$. By hypothesis $T(A)$ is $\mathfrak{S}'$-totally bounded. Therefore, there exists a finite subset $\Phi = \{ y_1, \ldots, y_n \} \subseteq Y$ such that

$$T(A) \subseteq \Phi + \tfrac{1}{3} B^\circ. \tag{$\star$}$$

Since B is $\sigma(Y', Y)$-bounded, the set $\{(\langle y_1, y'\rangle, \ldots, \langle y_n, y'\rangle) \colon y' \in B\}$ is a bounded subset of $\mathbb{R}^n$ and hence is totally bounded. Pick a finite subset C of B such that for each $y' \in B$ there exists some $c' \in C$ with $|\langle y_i, y' - c'\rangle| \leq \frac{1}{3}$ for each $i = 1, \ldots, n$.

Now let $b' \in B$. Fix $c' \in C$ so that $|\langle y_i, b' - c'\rangle| \leq \frac{1}{3}$ holds for all $i = 1, \ldots, n$. If $x \in A$, then by $(\star)$ there exists some $1 \leq j \leq n$ with $T(x) - y_j \in \frac{1}{3}B^\circ$, and so

$$\begin{aligned} |\langle x, T'(b' - c')\rangle| &= |\langle Tx, b' - c'\rangle| \\ &= |\langle T(x) - y_j, b'\rangle| + |\langle T(x) - y_j, c'\rangle| + |\langle y_j, b' - c'\rangle| \\ &\leq \tfrac{1}{3} + \tfrac{1}{3} + \tfrac{1}{3} = 1. \end{aligned}$$

This implies $T'(b') - T'(c') \in A^\circ$, and so $T'(B) \subseteq T'(C) + A^\circ$ holds. Since $\{A^\circ \colon A \in \mathfrak{S}\}$ is a base at zero for the $\mathfrak{S}$-topology, the latter shows that $T'(B)$ is an $\mathfrak{S}$-totally bounded set, and the proof is finished. ∎

We continue with the introduction of the strong topologies. Consider a dual system $\langle X, X'\rangle$. Then the **strong topology** $\beta(X', X)$ on X' is the $\mathfrak{S}$-topology when $\mathfrak{S}$ is the collection of all $\sigma(X, X')$-bounded subsets of X. Similarly, the strong topology $\beta(X, X')$ on X is the $\mathfrak{S}$-topology when $\mathfrak{S}$ is the collection of all $\sigma(X', X)$-bounded subsets of X'.

If (X, τ) is a locally convex space, then its **second dual** X'' is the topological dual of $(X', \beta(X', X))$. The elements of X'' will be denoted by double primes, for instance, x'', y'', etc. If no specific topology on X'' is considered, then X'' will always be assumed to be equipped with $\beta(X'', X')$. There is a natural embedding $x \mapsto \hat{x}$ from X into X'', defined by

$$\hat{x}(x') = x'(x) \quad \text{for all } x' \in X' \text{ and all } x \in X.$$

Clearly, $x \mapsto \hat{x}$ is linear and (by Theorem 3.7) it is also one-to-one. Identifying each x with its image $\hat{x}$, we can consider X as a vector subspace of X'. In conjunction with this identification, it should be noted that the locally convex topology $\sigma(X'', X')$ induces $\sigma(X, X')$ on X.

For a Banach space X, the strong topology $\beta(X, X')$ is the norm topology of X, and likewise $\beta(X', X)$ is the norm topology of X'. Also, it should be noted that the natural embedding $x \mapsto \hat{x}$ is, in this case, an isometry.

Exercises

1. Prove that a convex set A is circled if and only if $x \in A$ implies $-x \in A$.

2. For a finite collection $A_1, \ldots, A_n$ of subsets of a vector space establish the following:

(a) If each A_i is convex, then

$$\text{co}\left(\bigcup_{i=1}^{n} A_i\right) = \left\{\sum_{i=1}^{n} \lambda_i x_i : \lambda_i \geq 0, \ x_i \in A_i, \text{ and } \sum_{i=1}^{n} \lambda_i = 1\right\}.$$

(b) If each A_i is convex and circled, then the convex circled hull of $\bigcup_{i=1}^{n} A_i$ is precisely the set

$$\left\{\sum_{i=1}^{n} \lambda_i x_i : x_i \in A_i \text{ and } \sum_{i=1}^{n} |\lambda_i| \leq 1\right\}.$$

3. Show that a subset A of a topological vector space (X, τ) is τ-bounded if and only if for each sequence $\{x_n\} \subseteq A$ and each sequence $\{\lambda_n\} \subseteq \mathbb{R}$ with $\lambda_n \to 0$ we have $\lambda_n x_n \xrightarrow{\tau} 0$.

4. Give an example of two disjoint closed convex sets in a locally convex space that cannot be strictly separated.

5. Let τ be the locally convex topology on a vector space X generated by a family of seminorms $\{p_i : \ i \in I\}$. Show that:
   (a) A subset A of X is τ-bounded if and only if $p_i(A)$ is a bounded subset of $\mathbb{R}$ for each $i \in I$.
   (b) A net $\{x_\alpha\}$ of X satisfies $x_\alpha \xrightarrow{\tau} 0$ if and only if $p_i(x_\alpha) \to 0$ holds in $\mathbb{R}$ for each $i \in I$.

6. Let (X, τ) be a locally convex space. If p is a τ-continuous seminorm on X, then show that for each $a \in X$ there exists a τ-continuous linear functional x' on X satisfying:
   (a) $|x'(x)| \leq p(x)$ for all $x \in X$, and
   (b) $x'(a) = p(a)$.

7. If X is a Banach space, then show that:
   (a) The Mackey topology $\tau(X, X')$ coincides with the norm topology of X.
   (b) The strong topology $\beta(X', X)$ coincides with the norm topology of X'.

8. Let $\langle X, X' \rangle$ be a dual system and let $\mathfrak{S}$ be a collection of $\sigma(X', X)$-bounded subsets of X'. Then show that the $\mathfrak{S}$-topology on X does not change if we replace $\mathfrak{S}$ by any one of the following collections of subsets of X':
   (a) The finite unions of the sets of $\mathfrak{S}$.
   (b) The subsets of the sets of $\mathfrak{S}$.
   (c) The sets of the form λA, $\lambda \in \mathbb{R}$ and $A \in \mathfrak{S}$.
   (d) The $\sigma(X', X)$-closures of the sets in $\mathfrak{S}$.
   (e) The circled hulls of the sets in $\mathfrak{S}$.
   (f) The convex, circled, $\sigma(X', X)$-closed hulls of the sets of $\mathfrak{S}$.
   (g) The sets of the form λA, where $\lambda \in \mathbb{R}$ and A is a subset of the convex circled $\sigma(X', X)$-closed hull of some finite union of sets in $\mathfrak{S}$.

9. Show that a sequence $\{x_n\}$ in a locally convex space (X, τ) satisfies $x_n \xrightarrow{\tau} 0$ if and only if $x'_n(x_n) \to 0$ holds in $\mathbb{R}$ for each τ-equicontinuous sequence $\{x'_n\}$ of X'.

10. For a locally convex space (X, τ) establish the following statements.
   (a) The topology $\sigma(X'', X')$ induces $\sigma(X, X')$ on X.
   (b) X is $\sigma(X'', X')$-dense in X''.
   (c) A subset B of X is $\sigma(X, X')$-compact if and only if B is bounded and $\sigma(X, X')$-complete.

11. Let $\langle X, X' \rangle$ and $\langle Y, Y' \rangle$ be a pair of dual systems. Then for an operator $T \colon X \to Y$ show that the following statements are equivalent.
   (a) $T \colon (X, \sigma(X, X')) \to (Y, \sigma(Y, Y'))$ is continuous.
   (b) $T \colon (X, \tau(X, X')) \to (Y, \tau(Y, Y'))$ is continuous.

12. This exercise discusses quotient topological vector spaces. Let Y be a vector subspace of topological vector spaces (X, τ), and let $Q \colon X \to X/Y$ denote the canonical projection of X onto X/Y.
   (a) Show that $\{Q(V) \colon V \text{ is a } \tau\text{-neighborhood of zero}\}$ is a base at zero for a linear topology τ_Q on X/Y (called the **quotient topology** on X/Y).
   (b) Show that $Q \colon (X, \tau) \to (X/Y, \tau_Q)$ is an open mapping.
   (c) Show that τ_Q is a Hausdorff topology if and only if Y is a τ-closed vector subspace.
   (d) If τ is locally convex, then show that τ_Q is likewise locally convex.
   (e) What is the topological dual of $(X/Y, \tau_Q)$?

3.2. Weak Topologies on Banach Spaces

A vector space equipped with a norm is called a normed vector space or simply a normed space. If a normed space is complete with respect to the metric generated by its norm, then it is called a **Banach space**. The normed spaces are special examples of topological vector spaces. In this section we shall review (with proofs) the most important properties of the weak topologies on normed spaces.

Theorem 3.28. *On a finite dimensional vector space, the Euclidean topology is the only Hausdorff linear topology that the space admits. In particular, on a finite dimensional vector space any two norms are equivalent.*

Proof. We can assume that the finite dimensional vector space is $\mathbb{R}^n$. Let $\| \cdot \|$ denote the Euclidean norm on $\mathbb{R}^n$, i.e., $\|x\| = (x_1^2 + \cdots + x_n^2)^{\frac{1}{2}}$, and let τ be a linear Hausdorff topology on $\mathbb{R}^n$. We must show that the identity operator $I \colon (\mathbb{R}^n, \| \cdot \|) \to (\mathbb{R}^n, \tau)$ is a homeomorphism.

Since convergence with respect to the Euclidean norm is equivalent to pointwise convergence, we see that $I \colon (\mathbb{R}^n, \| \cdot \|) \to (\mathbb{R}^n, \tau)$ is a continuous operator. Consequently, to complete the proof, it is enough to establish that $B = \{x \in \mathbb{R}^n \colon \|x\| < 1\}$ is a τ-neighborhood of zero.

To this end, let $S = \{x \in \mathbb{R}^n \colon \|x\| = 1\}$. Since S is closed and bounded, it is a norm compact subset of $\mathbb{R}^n$, and so by the continuity of $I \colon (\mathbb{R}^n, \|\cdot\|) \to (\mathbb{R}^n, \tau)$ the set S is also τ-compact. In particular, S is τ-closed. Since $0 \notin S$, there exists a circled τ-neighborhood V of zero with $V \cap S = \emptyset$. We claim that $V \subseteq B$ holds. Indeed, if some $x \in V$ satisfies $\|x\| \geq 1$, then we have $\frac{x}{\|x\|} \in V \cap S$, which is impossible. Thus, $V \subseteq B$ holds, proving that B is a τ-neighborhood of zero, as required. ∎

An immediate consequence of the preceding result is the following.

Corollary 3.29. *Every finite dimensional vector subspace of a Hausdorff topological vector space is closed.*

Now some notation is in order. Let X be a normed vector space. Then the norm dual of X will be denoted (as usual) by X' and its second norm dual by X''. Recall that X' is a Banach space under the norm

$$\|x'\| := \sup\{|x'(x)| \colon \|x\| = 1\}.$$

The closed unit balls of X, X', and X'' will be denoted by U, U', and U'', respectively. That is,

$$\begin{aligned} U &:= \{x \in X \colon \|x\| \leq 1\}, \\ U' &:= \{x' \in X' \colon \|x'\| \leq 1\}, \\ U'' &:= \{x'' \in X'' \colon \|x''\| \leq 1\}. \end{aligned}$$

With the vector space X we associate the pair of dual systems $\langle X, X' \rangle$ and $\langle X', X'' \rangle$, whose dualities are given by the bilinear forms

$$\langle x, x' \rangle = x'(x) \quad \text{and} \quad \langle x', x'' \rangle = x''(x').$$

The closed unit balls U and U' are in duality with respect to $\langle X, X' \rangle$, that is, $U^\circ = U'$ and $(U')^\circ = U$ hold. Similarly, U' and U'' are in duality with respect to the dual system $\langle X', X'' \rangle$. Note that (by Alaoglu's Theorem 3.20) U' is $\sigma(X', X)$-compact and U'' is $\sigma(X'', X')$-compact.

Definition 3.30. *Let X be a normed vector space. The weak topology $\sigma(X, X')$ on X will be denoted by w, i.e., $w := \sigma(X, X')$. Similarly, the topology $\sigma(X', X)$ on X' is called the **weak* topology** and will be denoted by w^*, i.e, $w^* := \sigma(X', X)$.*

Consequently, in a normed space X the symbol $x_\alpha \xrightarrow{w} x$ means that $x'(x_\alpha) \to x'(x)$ holds in $\mathbb{R}$ for each $x' \in X'$. Similarly, $x'_\alpha \xrightarrow{w^*} x'$ means that $x'_\alpha(x) \to x'(x)$ holds in $\mathbb{R}$ for each $x \in X$. Also, note that $x'_\alpha \xrightarrow{w} x'$ in X' means that $x''(x'_\alpha) \to x''(x')$ for each $x'' \in X''$.

The next result is a Banach space version of Theorem 3.6 and is often quite useful.

Lemma 3.31. *Let X be a normed space. If f is a continuous linear functional on a vector subspace of X, then f has a continuous extension to all of X that preserves its original norm.*

In particular, for each $x \in X$ there exists a continuous linear functional x' on X such that $\|x'\| = 1$ and $x'(x) = \|x\|$.

Proof. Let Y be a vector subspace of a normed space X, and let $f : Y \to \mathbb{R}$ be a continuous linear functional. Denote by α the norm of f, i.e.,

$$\alpha = \sup\{|f(y)| : \ y \in Y \ \text{and} \ \|y\| = 1\}.$$

Then the mapping $p : X \to \mathbb{R}$, defined by $p(x) = \alpha\|x\|$ for each $x \in X$, is a sublinear mapping satisfying $f(y) \leq p(y)$ for all $y \in Y$. By the Hahn–Banach Theorem 1.25 there exists an extension of f to all of X (which we denote by f again) satisfying $f(x) \leq p(x)$ for all $x \in X$. This implies $|f(x)| \leq \alpha\|x\|$ for all $x \in X$, which shows that f is continuous, and moreover we have $\|f\| = \alpha$.

To see the last claim, let $x \in X$ and put $Y = \{\lambda x : \ \lambda \in \mathbb{R}\}$. Now consider the continuous linear functional $x' : Y \to \mathbb{R}$, where $x'(\lambda x) = \lambda\|x\|$. Observe that (when $x \neq 0$) we have $\|x'\| = 1$, and then apply the preceding conclusion. ∎

Every vector $x \in X$ gives rise to a continuous linear functional $\hat{x}$ on X' via the formula $\hat{x}(x') = x'(x)$ for all $x' \in X'$. An easy application of the preceding lemma shows that the natural embedding $x \mapsto \hat{x}$, of a normed vector space X into its second dual X'', is a linear isometry. Thus, X can be considered as a vector subspace of X''. In case X is a Banach space and $x \mapsto \hat{x}$ is onto (i.e., for each $x'' \in X''$ there exists some $x \in X$ satisfying $x''(x') = x'(x)$ for all $x' \in X'$), then X is referred to as a **reflexive Banach space**.

Now let X be normed space. Consider the dual system $\langle X', X'' \rangle$, and view X as a vector subspace of X''. Note that U is a convex circled subset of X'', and so the $\sigma(X'', X')$-closure of U in X'' must be its convex circled and $\sigma(X'', X')$-closed hull. On the other hand, the identity $U'' = (U')^\circ = U^{\circ\circ}$, coupled with the bipolar (Theorem 3.17), shows that U'' is also the convex, circled and $\sigma(X'', X')$-closed hull of U in X''. Thus, the following important density theorem of H. H. Goldstein [71] is true.

Theorem 3.32 (Goldstein). *If X is a normed space, then the closed unit ball U of X is $\sigma(X'', X')$-dense in U'' (and hence X is also $\sigma(X'', X')$-dense in X''). In particular, a Banach space X is reflexive if and only if its closed unit ball is weakly compact.*

The next three results deal with metrizability properties of the weak topologies. We first show that the weak topology is seldom metrizable.

Theorem 3.33. *For a Banach space X the following statements are equivalent.*

   (1) *The vector space X is finite dimensional.*

   (2) *The weak topology is metrizable.*

   (3) *The weak* topology is metrizable.*

Proof. (1) $\Longrightarrow$ (2) If X is finite dimensional, then the weak topology w of X coincides with the Euclidean topology (Theorem 3.28), and hence w is metrizable.

(2) $\Longrightarrow$ (1) Let $\{x_1', x_2', \ldots\}$ be a countable subset of X' such that the w-neighborhoods $V_n = \{x \in X\colon |x_i'(x)| \leq 1$ for $i = 1, \ldots, n\}$ $(n \in \mathbb{N})$ form a base at zero for w. Let Y_n denote the vector subspace generated by $\{x_1', \ldots, x_n'\}$ in X'.

Now if $x' \in X'$, then the set $W = \{x \in X\colon |x'(x)| \leq 1\}$ is a w-neighborhood of zero, and so $V_n \subseteq W$ must hold for some n. In particular, it follows that $\operatorname{Ker} x_1' \cap \cdots \cap \operatorname{Ker} x_n' \subseteq \operatorname{Ker} x'$, and so (by Lemma 3.15) $x' \in Y_n$. Therefore, $X' = \bigcup_{n=1}^{\infty} Y_n$. Since by Corollary 3.29 each Y_n is a closed set, it follows from Baire's category theorem that some Y_n must have an interior point, and consequently $X' = Y_n$ must be true for some n. That is, X' is finite dimensional. Since X is a vector subspace of X'', we see that X is also finite dimensional.

(1) $\Longleftrightarrow$ (3) Repeat the above arguments in a dual fashion. ∎

In the sequel we shall make use of the following result:

   • *Let K and Ω be two topological spaces with K compact and Ω Hausdorff. If a map $f\colon K \to \Omega$ is onto, one-to-one, and continuous, then f is a homeomorphism.*

To see this, let C be a closed subset of K. Then C is a compact subset of K, and so $f(C)$ is a compact (and hence a closed) subset of Ω. Now the identity $(f^{-1})^{-1}(C) = f(C)$ implies that f^{-1} is also continuous.

The metrizability of the weak* topology on the norm bounded subsets of X' is equivalent to the separability of X. (Recall that a topological space is said to be **separable** whenever there exists an at most countable subset that is dense.)

Theorem 3.34. *A normed space X is separable if and only if the closed unit ball of X' is weak* metrizable.*

Proof. Assume that X is separable.   Fix a countable dense subset $\{x_1, x_2, \ldots\}$ of X, and define

$$d(x', y') = \sum_{n=1}^{\infty} 2^{-n} \frac{|(x' - y')(x_n)|}{1 + |(x' - y')(x_n)|}$$

for each $x', y' \in X'$. A straightforward verification shows that d is a metric on X', and we claim that the topology generated by d agrees with w^* on U'. To see this, it is enough to show that (in view of the w^*-compactness of U') that the identity mapping $I\colon (U', w^*) \to (U', d)$ is continuous. Indeed, if $x'_\alpha \xrightarrow{w^*} x'$ holds in U', then the inequality

$$d(x'_\alpha, x') \leq \sum_{n=1}^{k} 2^{-n} \frac{|(x'_\alpha - x')(x_n)|}{1 + |(x'_\alpha - x')(x_n)|} + \frac{1}{2^k},$$

implies $\limsup d(x'_\alpha, x') \leq 2^{-k}$ for each k, and so $\lim d(x'_\alpha, x') = 0$.

For the converse, assume that w^* induces a metrizable topology on U'. Choose a sequence $\{x_n\}$ of X such that the w^*-neighborhoods of zero

$$V_n = \{x' \in X' \colon |x'(x_i)| \leq 1 \text{ for all } i = 1, \ldots, n\}$$

satisfy $\bigcap_{n=1}^{\infty} V_n \cap U' = \{0\}$. Let Y be the norm closed vector subspace generated in X by $\{x_n\}$. Clearly, Y is separable, and we claim that $Y = X$. Indeed, if there exists some $x \notin Y$, then by Theorem 3.7 there exists some $x' \in U'$ such that $x'(y) = 0$ for all $y \in Y$ and $x'(x) \neq 0$. In particular, we have $x' \in V_n \cap U'$ for all n, a contradiction, and the proof is finished. ∎

The dual of the preceding result is also true. That is, the metrizability of the weak topology on the norm bounded subsets of X is equivalent to the separability of X'.

Theorem 3.35. *The norm dual X' of a normed vector space X is separable if and only if the closed unit ball of X is weakly metrizable.*

Proof. Assume first that X' is separable. Then by Theorem 3.34 the topology $\sigma(X'', X')$ is metrizable on U''. Since w is the restriction of $\sigma(X'', X')$ on U, it follows that w is metrizable on U.

For the converse, assume that w is metrizable on U. Choose a sequence $\{x'_n\}$ of X' such that the sequence of sets $\{V_n\}$, where

$$V_n = \{x \in U \colon |x'_i(x)| \leq 1 \text{ for each } i = 1, \ldots, n\},$$

is a w-base at zero on U. Let Y be the norm closed vector subspace generated by $\{x'_n\}$ in X'. We claim that $Y = X'$. Indeed, if this is not the case, then by Theorem 3.7 there exists some $x'' \in U''$ with $x'' = 0$ on Y and $x'' \neq 0$. By Theorem 3.32 there exists a net $\{x_\alpha\} \subseteq U$ such that $x_\alpha \xrightarrow{\sigma(X'', X')} x''$. Now, since for each fixed n we have $x'_n(x_\alpha) = x_\alpha(x'_n) \to x''(x'_n) = 0$, it is easy to see that $x_\alpha \xrightarrow{w} 0$ holds in X and so $x_\alpha \xrightarrow{\sigma(X'',X')} 0$. Thus, $x'' = 0$, which is a contradiction. Hence, $Y = X'$ holds, and so X' is separable. ∎

We have seen so far that the weak topology on an infinite dimensional Banach space is not metrizable. Thus, in order to describe various topological properties involving the weak topology, one needs to employ nets. For instance, a set is weakly compact if and only if every net of the set has a weakly convergent subnet to some point of the set. Therefore, it is remarkable and surprising to learn that a set is weakly compact if and only if every sequence of the set has a weakly convergent subsequence to some vector of the set. This important (and extremely useful) result is known as the Eberlein–Šmulian theorem, and its proof is the subject of our next discussion.

Lemma 3.36. *If Y is a vector subspace of a normed space X, then $\sigma(Y, Y')$ and $\sigma(X, X')$ induce the same topology on every subset of Y.*

Proof. Every $y' \in Y'$ defines a seminorm on Y by letting $p(y) = |y'(y)|$ for all $y \in Y$. The collection of all such seminorms generates $\sigma(Y, Y')$. Similarly, $\sigma(X, X')$ is generated by the seminorms of the form $p(x) = |x'(x)|$ for each $x \in X$ ($x' \in X'$). By Lemma 3.31 we know that Y' consists precisely of the restrictions of the members of X' to Y. In particular, this implies that the seminorms generating $\sigma(Y, Y')$ and the seminorms generating $\sigma(X, X')$ agree on Y, and so they induce the same topology on every subset of Y. ∎

The following simple property of separable normed spaces will be needed.

Lemma 3.37. *If X is a separable normed space, then X' admits a countable set that separates the points of X.*

Proof. Pick a countable dense set $\{x_1, x_2, \ldots\}$ of X, and then use Lemma 3.31 to select a sequence $\{x_n'\}$ of X' satisfying $\|x_n'\| = 1$ and $x_n'(x_n) = \|x_n\|$ for each n. Now if $x_n'(x) = 0$ holds for each n, then the inequalities

$$\|x\| \le \|x - x_n\| + \|x_n\| = \|x - x_n\| + x_n'(x_n - x) \le 2\|x - x_n\|$$

easily imply $x = 0$, so that $\{x_n'\}$ separates the points of X. ∎

In a separable normed space, the weak topology induces a metrizable topology on the weakly compact sets.

Theorem 3.38. *Let X be a normed space. If X' admits a countable set that separates the points of X (in particular, if X is separable), then the weak topology on every weakly compact subset of X is metrizable.*

Proof. Let $\{x_1', x_2', \ldots\} \subseteq U'$ be a countable set that separates the points of X, and let A be a weakly compact subset of X. Since, by Theorem 3.25, the set A is norm bounded we have

$$d(x,y) = \sum_{n=1}^{\infty} 2^{-n} |x_n'(x-y)|$$

is well defined for each $x, y \in A$. Clearly, d is a metric on A, and moreover it is easy to see that the identity mapping $I\colon (A, w) \to (A, d)$ is continuous. Since A is weakly compact, this implies that I is a homeomorphism, and the conclusion follows. ∎

The proof of the Eberlein–Šmulian theorem will be based upon the following ingenious lemma that is essentially due to R. J. Whitley [190].

Lemma 3.39 (Whitley). *Let A be a norm bounded subset of a normed space X, and let x'' be a $\sigma(X'', X')$-closure point of A in X''. Then there exists a sequence $\{x_n\}$ of A (depending upon x'') such that:*

(a) *$\{x_n\}$ has at most one weak accumulation point in X.*

(b) *If $\{x_n\}$ has a weak accumulation point, then x'' belongs to X and is the weak accumulation point of $\{x_n\}$.*

Proof. Let x'' be a $\sigma(X'', X')$-closure point of A in X''. We shall construct by induction a strictly increasing sequence $\{k_n\}$ of natural numbers and sequences $\{x_n\} \subseteq A$ and $\{x_n'\} \subseteq X'$ such that:

(1) $\|x_n'\| = 1$ for each n.

(2) If y'' belongs to the vector subspace generated by

$$\{x'', x'' - x_1, \ldots, x'' - x_n\}$$

in X'', then $\max\{|y''(x_i')|\colon i = 1, \ldots, k_{n+1}\} \geq \frac{1}{2}\|y''\|$.

(3) For each n we have $\max\{|(x'' - x_n)(x_i')|\colon i = 1, \ldots, k_n\} < \frac{1}{n}$.

To see this, fix some $x_1' \in X'$ with $\|x_1'\| = 1$, and let $k_1 = 1$. Since x'' is a $\sigma(X'', X')$-closure point of A, there is some $x_1 \in A$ with $|(x'' - x_1)(x_1')| < 1$. For the basic step of the induction, assume that $\{k_1, \ldots, k_n\}$, $\{x_1, \ldots, x_n\}$, and $\{x_1', \ldots, x_{k_n}'\}$ have been constructed. Let Y be the vector subspace of X'' generated by $\{x, x'' - x_1, \ldots, x'' - x_n\}$. Since $S = \{y'' \in Y\colon \|y''\| = 1\}$ is a norm compact set (see Theorem 3.28), there exists $y_{k_n+1}'', \ldots, y_{k_{n+1}}'' \in S$ such that whenever $y'' \in Y$ satisfies $\|y''\| = 1$, then $\|y'' - y_i''\| < \frac{1}{4}$ holds for some $k_n + 1 \leq i \leq k_{n+1}$. Now for each $k_n + 1 \leq i \leq k_{n+1}$ pick some $x_i' \in X'$ with $\|x_i'\| = 1$ and $|y_i''(x_i')| > \frac{3}{4}$, and note that

$$\max\{|y''(x_i')|\colon i = k_n + 1, \ldots, k_{n+1}\} \geq \frac{1}{2}\|y''\|$$

holds for all $y'' \in Y$. Since x'' is a $\sigma(X'', X')$-closure point of A, there exists some $x_{n+1} \in A$ satisfying $\left|(x'' - x_{n+1})(x_i')\right| < \frac{1}{n+1}$ for each $i = 1, \ldots, k_{n+1}$. It is a routine matter to verify that the above constructed sequences satisfy (1)–(3).

Next, we claim that the sequence $\{x_n\}$ satisfies the properties of the lemma. Let $x \in X$ be a w-accumulation point of $\{x_n\}$. It is enough to show that $x = x''$.

To this end, note first that by Theorem 3.13 the vector x belongs to the norm closure of the vector subspace of X generated by $\{x_1, x_2, \ldots\}$. Thus, if Z denotes the vector subspace generated by $\{x'', x'' - x_1, x'' - x_2, \ldots\}$ in X'', then $x'' - x$ belongs to the norm closure of Z. Since by (2) every $y'' \in Z$ satisfies $\sup\{|y''(x_i')|: \ i = 1, 2, \ldots\} \geq \frac{1}{2}\|y''\|$, we see that the same inequality holds true for every point in the closure of Z. In particular, we have

$$\sup\{\left|(x'' - x)(x_i')\right|: \ i = 1, 2, \ldots\} \geq \tfrac{1}{2}\|x'' - x\|. \qquad (\star)$$

Now let i be fixed. Given $\epsilon > 0$ and $p > i$, choose some $n > p$ such that $\left|x_i'(x_n - x)\right| < \epsilon$. (Such an $n \in \mathbb{N}$ exists since x is a w-accumulation point of $\{x_n\}$. Taking into account that $i < n \leq k_n$ holds, it follows from property (3) that

$$\left|(x'' - x)(x_i')\right| \leq \left|(x'' - x_n)(x_i')\right| + \left|(x_n - x)(x_i')\right| < \tfrac{1}{n} + \epsilon < \tfrac{1}{p} + \epsilon.$$

Since $\epsilon > 0$ and $p > i$ are arbitrary, $(x'' - x)(x_i') = 0$ holds for all i. Now a glance at $(\star)$ shows that $x'' = x$ must hold, and the proof is finished. ∎

We are now ready to prove that a set is weakly relatively compact if and only if every sequence of the set has a weakly convergent subsequence. This important result is due to W. F. Eberlein [58] and V. L. Šmulian [179].

Theorem 3.40 (Eberlein–Šmulian). *A subset A of a normed space X is weakly relatively compact (resp. weakly compact) if and only if every sequence of A has a subsequence that converges to some vector of X (resp. to some vector of A).*

Proof. We prove the result for the weakly relatively compact sets, and leave the identical arguments for the weakly compact sets to the reader. Assume first that A is weakly relatively compact. Taking the weak closure of A, we can assume that A is weakly compact. Let $\{x_n\}$ be a sequence of A, and denote by Y the norm closed vector subspace of X generated by $\{x_n\}$. Clearly, Y is separable, $\{x_n\} \subseteq A \cap Y$ holds, and (by Lemma 3.36) $A \cap Y$ is $\sigma(Y, Y')$-compact. Now, by Theorem 3.38, $\sigma(Y, Y')$ is metrizable on $A \cap Y$, and so $\{x_n\}$ has a $\sigma(Y, Y')$-convergent subsequence in Y. The same subsequence converges (by Lemma 3.36) weakly in X.

For the converse, assume that every sequence of A has a weakly conver-
gent subsequence in X. By Alaoglu's Theorem 3.20 the $\sigma(X'', X')$-closure $\overline{A}$
of A in X'' is $\sigma(X'', X')$-compact. Now an easy application of Lemma 3.39
shows that $\overline{A} \subseteq X$ holds, and since $\sigma(X'', X')$ induces $\sigma(X, X')$ on X, it
follows that A is weakly relatively compact.   ∎

Our next goal is to establish (with the help of the Eberlein–Šmulian
theorem) that in a Banach space the convex circled hull of a weakly relatively
compact set is itself weakly relatively compact.

To do this, we need a result that is also of some independent interest.
Let X be a Banach space, and let f be a linear functional on X'. We
already know (see Theorem 3.16) that if f is weak* continuous, then it can
be represented by a unique point of X. It is surprising to learn that the
weak* continuity of f on U' is sufficient to guarantee such a representation
for f. This was established first by S. Banach [**30**] for the separable case.
(The interested reader can find various extensions to locally convex spaces
in H. H. Schaefer [**173**, Section 6, p. 147] and J. Horváth [**76**, Section 11,
p. 247].)

Theorem 3.41 (Banach). *If X is a Banach space and f is a linear functional on X', then the following statements are equivalent.*

(1) *f is weak* continuous on X'.*

(2) *f is weak* continuous on the closed unit ball of X'.*

Proof. (1) $\Longrightarrow$ (2) Obvious.

(2) $\Longrightarrow$ (1) Observe first that since (U', w^*) is a compact topological
space, $f(U')$ must be a bounded subset of $\mathbb{R}$, and so $f \in X''$. Consider the
dual system $\langle X', X'' \rangle$ and view X as a subset of X''.

Let $\epsilon > 0$. By the w^*-continuity of f at zero on U', there exists a finite
subset $D = \{x_1, \ldots, x_n\}$ of X such that $x' \in D^\circ \cap U'$ implies $|f(x')| \leq \epsilon$,
that is

$$f \in \epsilon(D^\circ \cap U')^\circ = \epsilon[D^\circ \cap (U'')^\circ]^\circ = \epsilon[(D \cup U'')^\circ]^\circ$$
$$= \epsilon(D \cup U'')^{\circ\circ}. \tag{†}$$

Now note that by the bipolar theorem, we have

$$D^{\circ\circ} = \left\{ \sum_{i=1}^{n} \lambda_i x_i : \sum_{i=1}^{n} |\lambda_i| \leq 1 \right\},$$

and so $D^{\circ\circ}$ is a subset of X. By Lemma 3.11 the set $D^{\circ\circ} + U''$ is $\sigma(X'', X')$-
closed. Since $D \cup U'' \subseteq D^{\circ\circ} + U''$ holds and since $D^{\circ\circ} + U''$ is clearly convex
and circled, it follows from the bipolar theorem that $(D \cup U'')^{\circ\circ} \subseteq D^{\circ\circ} + U''$.
From (†) we see that $f \in \epsilon(D^{\circ\circ} + U'')$, and so there exists some $x \in D^{\circ\circ} \subseteq X$

such that $\|f - \epsilon x\| \leq \epsilon$. This means that f belongs to the norm closure of X in X'' which (since X is a Banach space) equals X. Therefore, $f \in X$, and the proof is finished. ∎

We now have the background to prove that the convex circled hull of a weakly relatively compact set in a Banach space is itself weakly relatively compact. This important result is due to M. Krein and V. Šmulian [**104**].

Theorem 3.42 (Krein–Šmulian). *The convex circled hull (and hence the convex hull) of a weakly relatively compact subset of a Banach space is weakly relatively compact.*

Proof. Let A be a weakly relatively compact subset of a Banach space X. Replacing A by its weak closure, we can assume that A is weakly compact. In particular, note that A is norm bounded, and so there exists some $M > 0$ such that $\|x\| \leq M$ holds for all $x \in A$. Now let $\{u_n\}$ be a sequence in the convex circled hull of A. To complete the proof, it is enough to establish (by Theorem 3.40) that $\{u_n\}$ has a weakly convergent subsequence in X.

To this end, note first that each u_n is a linear combination of a finite subset of A. Thus, there exists an at most countable subset A_0 of A such that each u_n is a linear combination of a finite subset of A_0. If Y denotes the closed vector subspace of X generated by A_0, then Y is a separable Banach space, $\{u_n\} \subseteq Y$, and (by Lemma 3.36) the set $A \cap Y$ is $\sigma(Y, Y')$-compact. Thus, using Lemma 3.36 again, we see that by replacing X with Y (if necessary), we can assume without loss of generality that X is a separable Banach space. In doing so, we gain that w^* is metrizable on U'; see Theorem 3.34.

Now consider the compact topological space (A, w) and the Banach space $C(A)$ of all continuous real-valued functions (with the sup norm). Note that for each $x' \in X'$ the restriction Rx' of x' to A belongs to $C(A)$. Thus, the restriction process defines an operator $R\colon X' \to C(A)$. By the Riesz representation theorem (see, for instance [**8**, Section 38]) we know that $C'(A)$ consists of all regular Borel measures on A.

Let $\mu \in C'(A)$ be fixed. Then the formula

$$f(x') = \langle Rx', \mu \rangle = \int_A R(x')\, d\mu$$

defines a linear functional on X', and we claim that f is w^*-continuous on U'. To see this, let $x'_n \xrightarrow{w^*} x'$ in U'. Since $|x'_n(x)| \leq M$ holds for all $x \in A$, it follows from the Lebesgue dominated convergence theorem that

$$f(x'_n) = \int_A R(x'_n)\, d\mu \longrightarrow \int_A R(x')\, d\mu = f(x'),$$

and hence f is w^*-continuous on U'. Therefore, by Theorem 3.41 there exists a unique $x \in X$ satisfying $\langle Rx', \mu \rangle = \langle x', x \rangle$ for all $x' \in X'$. This shows that $R'\mu = x$ holds, and hence $R' \colon C'(A) \to X$.

From $\langle Rx', \mu \rangle = \langle x', R'\mu \rangle$ and $R' \colon C'(A) \to X$, it follows that R' is continuous for $\sigma(C'(A), C(A))$ and $\sigma(X, X')$. Let $B = \{\mu \in C'(A) \colon \|\mu\| \leq 1\}$. Then (by Alaoglu's Theorem 3.20) B is w^*-compact, and hence $R'(B)$ is a weakly compact subset of X. Since B is convex and circled, $R'(B)$ is also convex and circled. Now for each $x \in A$, let δ_x denote the Dirac measure supported at x, i.e., $\langle f, \delta_x \rangle = f(x)$ holds for all $f \in C(A)$. Clearly, $\delta_x \in B$ and $R'(\delta_x) = x$ for each $x \in A$. Thus, $A \subseteq R'(B)$ holds. In particular, we have $\{u_n\} \subseteq R'(B)$, and so by Theorem 3.40 the sequence $\{u_n\}$ has a convergent subsequence in X, as desired. ∎

If a sequence $\{x_n\}$ in a normed space satisfies $x_n \xrightarrow{w} x$, then it is easy to see that the set $\{x, x_1, x_2, \ldots\}$ is weakly compact, and so the set $\{x_1, x_2, \ldots\}$ is weakly relatively compact. Therefore, the following consequence of the preceding theorem should be immediate.

Corollary 3.43. *If $\{x_n\}$ is a weakly convergent sequence in a Banach space, then the set*

$$\left\{ \sum_{n=1}^{\infty} \lambda_n x_n \colon \sum_{n=1}^{\infty} |\lambda_n| \leq 1 \right\}$$

is weakly relatively compact.

Finally, we close the section we another useful characterization of the weakly compact sets.

Theorem 3.44 (Grothendieck). *A subset A of a Banach space X is weakly relatively compact if and only if for each $\epsilon > 0$ there exists a weakly compact subset W of X satisfying*

$$A \subseteq W + \epsilon U.$$

Proof. If A is weakly relatively compact, then it should be clear that it satisfies the desired condition. For the converse, assume that A satisfies the stated condition.

Fix $\epsilon > 0$ and then pick a weakly compact subset W of X such that $A \subseteq W + \epsilon U \subseteq W + \epsilon U''$. Clearly, A is a norm bounded set, and so its w^*-closure $\overline{A}$ in X'' is w^*-compact. On the other hand, it follows from Lemma 3.11 that $\overline{A} \subseteq W + \epsilon U''$. Thus, $\overline{A} \subseteq X + \epsilon U''$ holds for each $\epsilon > 0$. Since X is norm closed in X'', it follows that $\overline{A} \subseteq X$. Taking into consideration that $\sigma(X'', X')$ induces $\sigma(X, X')$ on X, we conclude that A is a weakly relatively compact subset of X, as desired. ∎

Exercises

1. Let (X, τ) be a Hausdorff topological vector space, and let Y and Z be two τ-closed vector subspaces.
   (a) Give an example where $Y + Z$ is not a τ-closed vector subspace.
   (b) If Y is finite dimensional, then show that $Y + Z$ is a τ-closed vector subspace.
   [*Hint:* For (b) consider the canonical projection $Q: X \to X/Z$. Since Z is τ-closed X/Z is a Hausdorff topological vector space; see Exercise 12 of Section 3.1. By Corollary 3.29 we know that $Q(Y)$ is a closed vector subspace of X/Z. Now note that $Y + Z = Q^{-1}(Q(Y))$.]

2. If the range of an operator $T: X \to Y$ between two topological vector spaces is finite dimensional, then show that T is continuous if and only if $\mathrm{Ker}\,(T)$ is a closed vector subspace of X. [*Hint:* Put $Z = \mathrm{Ker}(T)$, and consider the diagram

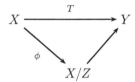

   where ϕ is the canonical projection of X onto X/Z.]

3. Show that in a normed space the weak topology coincides with the norm topology if and only if the vector space is finite dimensional.

4. Show that in a normed space the norm is weakly lower semicontinuous, i.e., show that $x_\alpha \xrightarrow{\;w\;} x$ implies $\|x\| \le \liminf \|x_\alpha\|$.

5. Show that if the dual of a normed space X is separable, then X is also separable. [*Hint:* If $\{x_1', x_2', \ldots\}$ is a countable dense subset of X', then for each n pick some $x_n \in X$ with $\|x_n\| = 1$ and $|x_n'(x_n)| \ge \frac{1}{2}\|x_n'\|$. Now consider the norm closed vector subspace of X generated by $\{x_1, x_2, \ldots\}$.]

6. Show that a normed space X is reflexive if and only if the weak and weak* topologies on X' coincide.

7. Show that if a normed space X is separable, then every norm bounded sequence of X' has a w^*-convergent subsequence.

8. If $x_n \xrightarrow{\;w\;} 0$ holds in a normed space, then show that there exist integers $0 = k_0 < k_1 < \cdots$ and a sequence $\{\alpha_n\} \subseteq [0, 1]$ such that:
   (a) $\sum_{i=k_n+1}^{k_{n+1}} \alpha_i = 1$ for $n = 0, 1, 2, \ldots$.
   (b) If $y_n = \sum_{i=k_n+1}^{k_{n+1}} \alpha_i x_i$, then $\sum_{n=0}^{\infty} \|y_n\| < \infty$.
   [*Hint:* For each fixed k note that the zero vector is in the weak closure of the set $\mathrm{co}\,\{x_n: \ n > k\}$. Consequently, given any $\epsilon > 0$ there exist (by Theorem 3.13) $\alpha_{k+1}, \ldots, \alpha_{k+p} \in [0, 1]$ with $\alpha_{k+1} + \cdots + \alpha_{k+p} = 1$ and $\|\alpha_{k+1}x_{k+1} + \cdots + \alpha_{k+p}x_{k+p}\| < \epsilon$.]

9. Let Ω be a Hausdorff compact topological space, and consider the Banach space $C(\Omega)$ with the sup norm. Show that a sequence $\{f_n\} \subseteq C(\Omega)$ satisfies $f_n \xrightarrow{\;w\;} 0$ if and only if

(a) $\{f_n\}$ is norm bounded, and

(b) $f_n(\omega) \to 0$ holds in $\mathbb{R}$ for each $\omega \in \Omega$.

10. The function $r_n\colon [0,1] \to \mathbb{R}$ defined by $r_n(t) = \operatorname{sgn} \sin(2^n \pi t)$ is known as the nth **Rademacher function**.

(a) Draw the graph of r_n.

(b) Show that $r_n \xrightarrow{w} 0$ holds in $L_1[0,1]$.

(c) Does $r_n \xrightarrow{w} 0$ hold in $L_p[0,1]$ for $1 < p < \infty$?

11. If $f_n(t) = \sin(nt)$, then show that $f_n \xrightarrow{w} 0$ in $L_p[0,1]$ for each $1 \le p < \infty$.

12. Show that a Banach space X is reflexive if and only if every closed vector subspace of X is reflexive.

13. Let A be a compact subset of a normed space. Show that there exists a null sequence whose convex closed hull includes A.

14. Let A be a (nonempty) norm bounded subset of c_0, the Banach space of all null real sequences with the sup norm. Put

$$s_n = \sup\{|a_n|\colon a = (a_1, a_2, \ldots) \in A\}, \quad n = 1, 2, \ldots .$$

Show that A is norm totally bounded if and only if $s_n \to 0$. [*Hint:* Assume that $s_n \to 0$ and let $\epsilon > 0$. Pick some k with $s_n < \epsilon$ for all $n > k$, and note that

$$A \subseteq [-s_1, s_1] \times \cdots \times [-s_k, s_k] \times \{0\} \times \{0\} \times \cdots + \epsilon U.$$

Now apply Theorem 3.1.]

15. Let A be a (nonempty) norm bounded subset of ℓ_p ($1 \le p < \infty$). Put

$$s_n = \sup\left\{\sum_{i=n}^{\infty} |a_i|^p\colon a = (a_1, a_2, \ldots) \in A\right\}, \quad n = 1, 2, \ldots .$$

Show that A is norm totally bounded if and only if $s_n \to 0$.

16. For a continuous operator $T\colon X \to Y$ between two Banach spaces establish the following statements.

(a) If T is onto, then T' is one-to-one.

(b) If T' is onto, then T is one-to-one.

(c) If T is one-to-one and has a closed range, then T' is onto.

(d) The range of T is dense in Y if and only if T' is one-to-one.

17. Let $T\colon X \to Y$ be an operator between two normed spaces. Recall that the norm of T is defined by

$$\|T\| = \sup\{\|Tx\|\colon \|x\| = 1\}.$$

Show that:

(a) $\|T\| = \sup\{\|Tx\|\colon \|x\| \le 1\}$.

(b) T is continuous if and only if $\|T\| < \infty$.

(c) If T is continuous, then its adjoint $T'\colon Y' \to X'$ is also continuous and $\|T'\| = \|T\|$ holds.

(d) If T is an isometry (i.e., if $\|Tx\| = \|x\|$ holds for all $x \in X$), then $T''\colon X'' \to Y''$ is likewise an isometry.

[*Hint*: For (d) let $x'' \in X''$. If $x' \in X'$ satisfies $\|x'\| = 1$, then show that there exists some $y' \in Y'$ with $\|y'\| = 1$ and $T'y' = x'$. Now note that

$$\langle x', x'' \rangle = \langle T'y', x'' \rangle = \langle y', T''x'' \rangle \leq \|T''x''\| \leq \|x''\|.]$$

3.3. Locally Convex-Solid Riesz Spaces

The locally convex-solid topologies bind the algebraic and lattice structures of Riesz spaces together. They were introduced by G. T. Roberts [**168**] and they were studied systematically for the first time by I. Namioka [**153**] and H. H. Schaefer [**171**]. The reader can find extensive treatments of locally solid topologies in the authors' book [**7**] and in the book by D. H. Fremlin [**64**]. Recall that a subset A of a Riesz space is called **solid** whenever $|x| \leq |y|$ and $y \in A$ imply $x \in A$. Clearly, every solid set is circled.

Let E be a Riesz space. A seminorm p on E is said to be a **lattice** (or a **Riesz**) **seminorm** whenever $|x| \leq |y|$ in E implies $p(x) \leq p(y)$. Note that a seminorm p on E is a lattice seminorm if and only if its closed unit ball, i.e., the set $U_p := \{x \in E : p(x) \leq 1\}$, is a solid set. Indeed, observe first that if p is a lattice seminorm, then U_p is a solid set. On the other hand, if U_p is a solid set and $|x| \leq |y|$ holds in E, then we have

$$\left| \tfrac{1}{p(y)+\epsilon} x \right| \leq \left| \tfrac{1}{p(y)+\epsilon} y \right| \quad \text{and} \quad \tfrac{1}{p(y)+\epsilon} y \in U_p,$$

and so $\tfrac{1}{p(y)+\epsilon} x \in U_p$ for all $\epsilon > 0$. Thus, $p(x) \leq p(y) + \epsilon$ holds for all $\epsilon > 0$, and so $p(x) \leq p(y)$, proving that p is a lattice seminorm. A locally convex topology τ on E that is generated by a family of lattice seminorms is referred to as a **locally convex-solid topology** (and (E, τ) is called a **locally convex-solid Riesz space**). By the above discussion the following result should now be immediate.

Theorem 3.45. *A locally convex topology on a Riesz space is locally convex-solid if and only if it has a base at zero consisting of convex and solid sets.*

In a Riesz space E, the functions

(1) $x \mapsto x^+$ from E to E,

(2) $x \mapsto x^-$ from E to E,

(3) $x \mapsto |x|$ from E to E,

(4) $(x, y) \mapsto x \vee y$ from $E \times E$ to E, and

(5) $(x, y) \mapsto x \wedge y$ from $E \times E$ to E,

are referred to as the **lattice operations** of E. From the lattice inequalities of Theorem 1.9, it should be immediate that every locally convex-solid topology on E makes the lattice operations continuous (in fact, uniformly continuous) functions.

At this point it should be remarked that the lattice operations in a locally convex-solid Riesz space are seldom weakly continuous. For instance, consider $E = L_1[0, 1]$ with the locally convex-solid topology generated by the L_1-norm, and note that $E' = L_\infty[0, 1]$. Now if $x_n(t) = \sin(\pi nt)$, then (by the Riemann–Lebesgue lemma) we have $x_n \xrightarrow{w} 0$. On the other hand, since $\int_0^1 |\sin(\pi nt)|\, dt = \frac{2}{\pi}$ holds for all n, we infer that $\{|x_n|\}$ does not converge weakly to zero. Thus, $x \mapsto |x|$ is not weakly continuous on $L_1[0, 1]$.

Some immediate consequences of the continuity of the lattice operations are included in the next result.

Theorem 3.46. *In a (Hausdorff) locally convex-solid Riesz space (E, τ) the following statements hold:*

    (1) *The positive cone E^+ is τ-closed.*

    (2) *If a net $\{x_\alpha\}$ of E satisfies $x_\alpha \uparrow$ and $x_\alpha \xrightarrow{\tau} x$, then $x = \sup\{x_\alpha\}$.*

    (3) *Every band of E is τ-closed.*

    (4) *The τ-closure of a Riesz subspace of E is also a Riesz subspace.*

Proof. (1) Since $x \mapsto x^-$ is continuous and $E^+ = \{x \in E: \ x^- = 0\}$ holds, we see that E^+ is τ-closed.

(2) Since $x_\beta - x_\alpha \xrightarrow{\tau} x - x_\alpha$ and $x_\beta - x_\alpha \in E^+$ holds for all $\beta \succeq \alpha$, it follows from (1) that $x - x_\alpha \in E^+$ for each α. Thus, x is an upper bound of the net $\{x_\alpha\}$. On the other hand, if $x_\alpha \leq y$ holds for all α, then $0 \leq y - x_\alpha \xrightarrow{\tau} y - x$ implies $y - x \geq 0$, and so $y \geq x$. In other words, $x_\alpha \uparrow x$ holds in E.

(3) Let A be a subset of E. From $A^d = \{x \in E: \ |x| \wedge |y| = 0 \text{ for all } y \in A\}$, it is easy to see that A^d is τ-closed. Now if B is a band of E, then $B = B^{dd}$ holds, and so B is τ-closed.

(4) This follows from the continuity of the lattice operations. ■

A few elementary properties of locally convex-solid Riesz spaces are included in the next result.

Theorem 3.47. *If τ is a locally convex-solid topology on a Riesz space E, then the following statements hold:*

    (1) *The order bounded subsets of E are τ-bounded.*

    (2) *The τ-closure of a solid subset of E is likewise solid (and hence the τ-closure of an ideal is also an ideal).*

Proof. (1) Let $[x, y]$ be an order interval. If V is a solid τ-neighborhood of zero, then pick some $\lambda > 0$ with $\lambda(|x| + |y|) \in V$, and note that $\lambda[x, y] \subseteq V$.

(2) Let A be a solid set. Assume that $|x| \leq |y|$ holds with $y \in \overline{A}$, the τ-closure of the set A. Pick a net $\{y_\alpha\} \subseteq A$ with $y_\alpha \xrightarrow{\tau} y$, and then put $x_\alpha = (x \wedge |y_\alpha|) \vee (-|y_\alpha|)$ for each α. Clearly, $|x_\alpha| \leq |y_\alpha|$ holds for each α, and so $\{x_\alpha\} \subseteq A$. On the other hand, the continuity of the lattice operations implies that $x_\alpha \xrightarrow{\tau} x$, and so $x \in \overline{A}$. Thus, $\overline{A}$ is a solid set. ∎

An immediate consequence of the preceding result is that in a locally convex-solid Riesz space (E, τ) the convex, solid, and τ-closed neighborhoods of zero is a base at zero. In turn, this implies that if A is a τ-bounded subset of E, then its solid hull is likewise τ-bounded. Recall that the **solid hull** is the smallest solid set including A and is exactly the set

$$\mathrm{Sol}\,(A) := \{x \in E \colon \exists\, y \in A \text{ with } |x| \leq |y|\}.$$

It should be noted that the convex hull of a solid set is also solid. However, the solid hull of a convex set need not be convex.

It is useful to know that when the lattice operations are applied to totally bounded sets, then they also produce totally bounded sets.

Theorem 3.48. *Let (E, τ) be a locally convex-solid Riesz space. If A and B are two τ-totally bounded subsets of E, then the sets*

$$A^+, \quad A^-, \quad |A|, \quad A \vee B, \quad \text{and } A \wedge B$$

are likewise τ-totally bounded.

Proof. We prove first that $A \vee B$ is τ-totally bounded. To this end, let V be a solid τ-neighborhood of zero. Pick two finite subsets Φ_1 and Φ_2 with $A \subseteq \Phi_1 + V$ and $B \subseteq \Phi_2 + V$. If $a \in A$ and $b \in B$, then pick $x \in \Phi_1$ and $y \in \Phi_2$ with $a - x \in V$ and $b - y \in V$, and then use the inequality $|a \vee b - x \vee y| \leq |a - x| + |b - y|$ to obtain $|a \vee b - x \vee y| \in V + V$. The solidness of $V + V$ implies that $a \vee b - x \vee y \in V + V$, and so $A \vee B \subseteq \Phi_1 \vee \Phi_2 + V + V$, which (by Theorem 3.1) shows that $A \vee B$ is a τ-totally bounded set.

Next, note that we have the relations

$$A^+ = A \vee \{0\}, \quad A^- = (-A) \vee \{0\}, \quad |A| \subseteq A \vee (-A),$$

and

$$A \wedge B \subseteq A + B - A \vee B.$$

Now use the preceding case to complete the proof. ∎

The topological dual of a locally-convex solid Riesz space is always a Riesz space and, in fact, it is an ideal in its order dual.

Theorem 3.49. *The topological dual E' of a locally convex-solid Riesz space (E, τ) is an ideal in its order dual $E^\sim$ (and hence E' is a Dedekind complete Riesz space in its own right). Moreover, for each $f, g \in E'$ and $x \in E^+$ we have*

$$[f \vee g](x) = \sup\{f(y) + g(z) \colon \ y, z \in E^+ \ \text{and} \ y + z = x\}$$

and

$$[f \wedge g](x) = \inf\{f(y) + g(z) \colon \ y, z \in E^+ \ \text{and} \ y + z = x\}.$$

Proof. By Theorem 3.47, every order bounded set is τ-bounded. Thus, each $f \in E'$ carries order bounded sets to bounded subsets of $\mathbb{R}$, and hence E' is a vector subspace of $E^\sim$.

To see that E' is an ideal of $E^\sim$, assume that $|g| \leq |f|$ holds in $E^\sim$ with $f \in E'$. By Theorem 3.5 there exists a τ-continuous lattice seminorm p on E satisfying $|f(x)| \leq p(x)$ for each $x \in E$. Now applying Theorem 1.18, we see that

$$|g(x)| \leq |g|(|x|) \leq |f|(|x|) = \sup\{|f(y)| \colon \ |y| \leq |x|\} \leq p(x)$$

also holds for all $x \in E$. Therefore, g is likewise τ-continuous. For the lattice operations use Theorem 1.18. ∎

Let E be a Riesz space. Then every $f \in E^\sim$ defines a lattice seminorm p_f on E via the formula

$$p_f(x) = |f|(|x|), \quad x \in E.$$

In particular, every nonempty subset A of $E^\sim$ generates a (not necessarily Hausdorff) locally convex-solid topology on E via the family of lattice seminorms $\{p_f \colon \ f \in A\}$. This topology is called the **absolute weak topology** generated by A on E and is denoted by $|\sigma|(E, A)$. (Of course, if A separates the points of E, then $|\sigma|(E, A)$ is a Hausdorff topology.) As we shall see next, the topological dual of $(E, |\sigma|(E, A))$ is precisely the ideal generated by A in $E^\sim$. This result is due to S. Kaplan [**93**, Note II].

Theorem 3.50 (Kaplan). *Let E be a Riesz space and let A be a nonempty subset of $E^\sim$. Then the topological dual of $(E, |\sigma|(E, A))$ is precisely the ideal generated by A in $E^\sim$.*

Proof. Let E' be the topological dual of $(E, |\sigma|(E, A))$. Theorem 3.49 shows that E' is an ideal of $E^\sim$. From the inequality $|f(x)| \leq |f|(|x|)$, it follows that $A \subseteq E'$, and so the ideal generated by A is included in E'.

To see that E' is also included in the ideal generated by A in $E^\sim$, let $f \in E'$. Then there exist linear functionals $f_1, \ldots, f_n \in A$ and some $\lambda > 0$

such that $|f(x)| \leq \lambda \sum_{i=1}^{n} |f_i|(|x|)$ holds for all $x \in E$. Therefore, if $x \in E^+$, then it follows from Theorem 1.18 that

$$|f|(x) = \sup\{|f(y)|: \ |y| \leq x\} \leq \lambda \sum_{i=1}^{n} |f_i|(x).$$

Hence, $|f| \leq \lambda \sum_{i=1}^{n} |f_i|$ holds, and so f belongs to the ideal generated by A in $E^\sim$, as desired. ∎

We now introduce the concept of a Riesz dual system.

Definition 3.51. *Let E be a Riesz space, and let E' be an ideal of $E^\sim$ separating the points of E. Then the pair $\langle E, E' \rangle$, under its natural duality $\langle x, x' \rangle := x'(x)$, will be referred to as a **Riesz dual system**.*

Consider a Riesz dual system $\langle E, E' \rangle$. Note that the polars of solid sets are solid. To see this, let $A \subseteq E$ be a solid set, and let $|y'| \leq |x'|$ hold in E' with $x' \in A^\circ$. Then for each $x \in A$, it follows from Theorem 1.18 that

$$\left| \langle x, y' \rangle \right| \leq \langle |x|, |y'| \rangle \leq \langle |x|, |x'| \rangle = \sup\{|\langle y, x' \rangle|: \ |y| \leq |x|\} \leq 1,$$

and so $y' \in A^\circ$. (If A is a solid subset of E', then use Theorem 1.23 instead of Theorem 1.18.) Also, arguing as in the proof of Theorem 1.69, we see that the mapping $x \mapsto \hat{x}$, where $\hat{x}(x') = x'(x)$ for each $x' \in E'$, is a lattice isomorphism from E into $(E')_n^\sim$. Thus, E can be considered as a Riesz subspace of $(E')_n^\sim$ (and hence as a Riesz subspace of $(E')^\sim$).

In case (E, τ) is a Hausdorff locally convex-solid Riesz space, E' is an ideal of $E^\sim$ separating the points of E , and so $\langle E, E' \rangle$ under the duality $\langle x, x' \rangle = x'(x)$ is a Riesz dual system. In fact, this example is representative of the arbitrary Riesz dual system. The following discussion will clarify the situation.

Let $\langle E, E' \rangle$ be a Riesz dual system. Then, according to Theorem 3.50, the topological dual of $(E, |\sigma|(E, E'))$ is precisely E', and so $|\sigma|(E, E')$ is consistent with $\langle E, E' \rangle$. That is,

$$\sigma(E, E') \subseteq |\sigma|(E, E') \subseteq \tau(E, E')$$

holds. On the other hand, if τ is a consistent locally convex-solid topology on E, then it is easy to see that $x_\alpha \overset{\tau}{\longrightarrow} 0$ implies $x_\alpha \overset{|\sigma|(E, E')}{\longrightarrow} 0$, and so $|\sigma|(E, E') \subseteq \tau$ holds. In other words, $|\sigma|(E, E')$ is the smallest locally convex-solid topology on E which is consistent with $\langle E, E' \rangle$.

Now if $\mathcal{B}$ is a base at zero consisting of solid convex τ-neighborhoods of zero, then (by Theorem 3.22) τ is the $\mathfrak{S}$-topology for $\mathfrak{S} = \{V^\circ: V \in \mathcal{B}\}$. This shows that every consistent locally convex-solid topology on E is an $\mathfrak{S}$-topology for some collection of solid, convex, and $\sigma(E', E)$-compact subsets of E'. Thus, if we define the **absolute Mackey topology** $|\tau|(E', E)$ as the

$\mathfrak{S}$-topology for the collection of all solid, convex and $\sigma(E', E)$-compact sets, then $|\tau|(E, E')$ is consistent with $\langle E, E' \rangle$. Moreover, a locally convex-solid topology τ on E is consistent if and only if $|\sigma|(E, E') \subseteq \tau \subseteq |\tau|(E, E')$ holds. Therefore, the "spectrum" of the locally convex-solid topologies on E consistent with $\langle E, E' \rangle$ is slightly narrower than that of the consistent locally convex topologies on E.

Since the absolute weak topology $|\sigma|(E, E')$ is a consistent locally convex-solid topology on E, it follows from Theorem 3.25 that a subset of E is $\sigma(E, E')$-bounded if and only if its solid hull is $\sigma(E, E')$-bounded. From this observation we easily infer that the strong topology $\beta(E', E)$ is the $\mathfrak{S}$-topology for the collection of all $\sigma(E, E')$-bounded solid subsets of E. Therefore, $\beta(E', E)$ is a locally convex-solid topology on E'. By Theorem 3.49 the topological dual E'' of $\big(E', \beta(E', E)\big)$ is also an ideal of $(E')^\sim$, and so $\langle E', E'' \rangle$ is also a Riesz dual system.

The next result is a Dini-type theorem. It shows that monotone convergence plus pointwise convergence imply topological convergence.

Theorem 3.52. *Assume that a net $\{x_\alpha\}$ in a locally convex-solid Riesz space (E, τ) satisfies $x_\alpha \downarrow 0$. Then $x_\alpha \xrightarrow{\tau} 0$ if and only if $x_\alpha \xrightarrow{\sigma(E,E')} 0$.*

Proof. Let $x_\alpha \downarrow 0$ in E, and assume that $x_\alpha \xrightarrow{\sigma(E,E')} 0$. It is enough to show that $x_\alpha \xrightarrow{\tau} 0$ holds.

To this end, let V be a solid τ-neighborhood of zero. Note that zero belongs to the $\sigma(E, E')$-closure of $\mathrm{co}\,\{x_\alpha\}$. By Theorem 3.13, zero must belong to the τ-closure of $\mathrm{co}\,\{x_\alpha\}$, and so $V \cap \mathrm{co}\,\{x_\alpha\} \neq \emptyset$. Pick indices $\alpha_1, \ldots, \alpha_n$ and positive constants $\lambda_1, \ldots, \lambda_n$ with $\lambda_1 + \cdots + \lambda_n = 1$ and $\lambda_1 x_{\alpha_1} + \cdots + \lambda_n x_{\alpha_n} \in V$. Now if $\alpha \succeq \alpha_i$ for each $i = 1, \ldots, n$, then it follows from

$$0 \leq x_\alpha = \sum_{i=1}^{n} \lambda_i x_\alpha \leq \sum_{i=1}^{n} \lambda_i x_{\alpha_i} \in V$$

that $x_\alpha \in V$. This shows that $x_\alpha \xrightarrow{\tau} 0$ holds, as desired. ∎

Topologies for which order convergence implies topological convergence are very useful. They are known as order continuous topologies.

Definition 3.53. *A linear topology τ on a Riesz space is said to be **order continuous** whenever $x_\alpha \downarrow 0$ implies $x_\alpha \xrightarrow{\tau} 0$.*

A lattice seminorm p on a Riesz space is said to be **order continuous** if $x_\alpha \downarrow 0$ implies $p(x_\alpha) \downarrow 0$. Note that a locally convex-solid topology on a Riesz space is order continuous if and only if it is generated by a family of order continuous lattice seminorms.

Theorem 3.54. *For a Riesz dual system* $\langle E, E' \rangle$ *the following statements are equivalent.*

(1) *Every consistent locally convex-solid topology on* E *is order continuous.*

(2) $\sigma(E, E')$ *is order continuous.*

(3) $E' \subseteq E_n^\sim$ *holds.*

(4) *Every band of* E' *is* $\sigma(E', E)$-*closed.*

(5) E *is an order dense Riesz subspace of* $(E')_n^\sim$.

Proof. (1) $\iff$ (2) By Theorem 3.50 the locally convex-solid topology $|\sigma|(E, E')$ is consistent with $\langle E, E' \rangle$. The equivalence of (1) and (2) is now an easy consequence of Theorem 3.52.

(2) $\implies$ (3) Obvious.

(3) $\implies$ (4) Let B be a band of E', and let $\{f_\alpha\} \subseteq B$ satisfy $f_\alpha \xrightarrow{\sigma(E', E)} f$ in E'. Write $f = g + h$ with $f \in B$ and $h \in B^d$. We have to show that $h = 0$, and for this it is enough to establish that $C_h = \{0\}$. So, let $0 \leq x \in C_h$.

Note that if $\phi_\alpha = f_\alpha - g$, then $\{\phi_\alpha\} \subseteq B$ and $\phi_\alpha \xrightarrow{\sigma(E', E)} h$. By Theorem 1.67 we have $C_h \subseteq N_{\phi_\alpha}$ for each α. Thus, for each $y \in C_h$ we also have $h(y) = \lim \phi_\alpha(y) = 0$, and so $|h|(x) = \sup\{|h(y)|: |y| \leq x\} = 0$, and hence $x = 0$, as desired.

(4) $\implies$ (5) Let $0 < \phi \in (E')_n^\sim$. We have to establish the existence of some $x \in E$ satisfying $0 < x \leq \phi$. The proof is along the lines of that of Theorem 1.70.

Fix some $0 < f \in C_\phi$. Since (by hypothesis) the band B_f^d is $\sigma(E', E)$-closed and $f \notin B_f^d$, there exists (by Theorems 3.7 and 3.16) some $u \in E$ with $f(u) \neq 0$ and $g(u) = 0$ for all $g \in B_f^d$. An easy application of Theorem 1.23 shows that we can assume $u > 0$. Note that $u \wedge \phi > 0$. (Indeed, if $u \wedge \phi = 0$ holds, then by Theorem 1.67 we have $C_\phi \subseteq N_u$, and so $f(u) = 0$, contrary to the choice of u.) Thus, replacing ϕ by $\phi \wedge u$, we can assume that $0 < \phi \leq u$ holds in $(E')_n^\sim$. Next pick some $0 < \epsilon < 1$ with $\psi = (\phi - \epsilon u)^+ > 0$. Fix some $0 < h \in C_\psi$, and then (as above) choose some $0 < v \in E$ with $h(v) > 0$ and $g(v) = 0$ for all $g \perp h$. Put $x = v \wedge \epsilon u \in E$, and we claim that $0 < x \leq \phi$ holds in $(E')_n^\sim$.

We show first that $x > 0$ holds. Indeed, if $x = v \wedge \epsilon u = 0$, then $u \wedge v = 0$, and so $\psi \wedge v = 0$. The latter (by Theorem 1.67) implies $C_\psi \subseteq N_v$, and so $h(v) = v(h) = 0$, a contradiction. Thus, $x > 0$ holds.

Finally, we prove that $x \leq \phi$. To see this, assume by way of contradiction that $\omega = (x - \phi)^+ > 0$. Pick some $0 < g \in C_\omega$. From

$$0 \leq \omega = (x - \phi)^+ \leq (\epsilon u - \phi)^+ = (\phi - \epsilon u)^- \perp (\phi - \epsilon u)^+ = \psi$$

and Theorem 1.67, we see that $C_\omega \perp C_\psi$. In particular, we have $g \perp h$, and so

$$0 < \omega(g) = (x - \phi)^+(g) \leq v(g) = g(v) = 0,$$

which is a contradiction. Thus, $0 < x \leq \phi$ holds, as desired.

$(5) \Longrightarrow (2)$ Let $x_\alpha \downarrow 0$ in E. Then $x_\alpha \downarrow 0$ holds in $(E')_n^\sim$, and so $f(x_\alpha) \downarrow 0$ holds for each $0 \leq f \in E'$. Thus, $x_\alpha \xrightarrow{\ \sigma(E, E')\ }$ holds, and the proof of the theorem is finished. ∎

Statement (4) of the preceding theorem is essentially due to W. A. J. Luxemburg and A. C. Zaanen [**130**, Note XI].

If $\langle E, E' \rangle$ is a Riesz dual system, then the **absolute weak topology** $|\sigma|(E', E)$ on E' is the locally convex-solid topology generated by the family of lattice seminorms $\{p_x \colon x \in E\}$, where

$$p_x(x') = |x'|(|x|), \quad x' \in E'.$$

It is easy to see that each p_x is order continuous, and so $|\sigma|(E', E)$ is always order continuous. From the identities

$$|x'|(|x|) = \sup\{|x'(y)| \colon |y| \leq |x|\} = \sup\{|y'(x)| \colon |y'| \leq |x'|\},$$

the following result should be clear.

Theorem 3.55. *If $\langle E, E' \rangle$ is a Riesz dual system, then:*

(1) *The topology $|\sigma|(E, E')$ on E is the $\mathfrak{S}$-topology of uniform convergence on the order intervals of E'.*

(2) *The topology $|\sigma|(E', E)$ on E' is the $\mathfrak{S}$-topology of uniform convergence on the order intervals of E.*

In general, the topology $|\sigma|(E', E)$ is not consistent with $\langle E, E' \rangle$. In order to characterize the consistency of $|\sigma|(E', E)$ we need a lemma.

Lemma 3.56. *Let $\langle E, E' \rangle$ be a Riesz dual system. For each $x \in E^+$ put*

$$[0, x] = \{y \in E \colon 0 \leq y \leq x\} \quad \text{and} \quad [\![0, x]\!] = \{x'' \in E'' \colon 0 \leq x'' \leq x\}.$$

Then the interval $[0, x]$ is $\sigma(E'', E')$-dense in the $\sigma(E'', E')$-compact interval $[\![0, x]\!]$.

Proof. Clearly, $[\![0, x]\!]$ is closed and bounded for $\sigma(E'', E')$, and hence it is $\sigma(E'', E')$-compact. To see that $[0, x]$ is $\sigma(E'', E')$-dense in $[\![0, x]\!]$, let A be the $\sigma(E'', E')$-closure of $[0, x]$ in E''. Clearly, A is a convex $\sigma(E'', E')$-compact subset of $[\![0, x]\!]$. If there exists some $x'' \in [\![0, x]\!]$ with $x'' \notin A$, then by Theorems 3.12 and 3.16 there exist $x' \in E'$ and $c \in \mathbb{R}$ satisfying $x'(y) \leq c$ for all $y \in [0, x]$ and $x''(x') > c$. In particular, from Theorem 1.18 it follows that

$$x''(x') \leq x''\big((x')^+\big) \leq (x')^+(x) = \sup\{x'(y)\colon\ y \in [0, x]\} \leq c,$$

which is impossible. Thus, $[0, x]$ is $\sigma(E'', E')$-dense in $[\![0, x]\!]$. ∎

The consistency of $|\sigma|(E', E)$ is characterized as follows.

Theorem 3.57. *If $\langle E, E' \rangle$ is a Riesz dual system, then the following statements are equivalent.*

(1) *$|\sigma|(E', E)$ is consistent with $\langle E, E' \rangle$.*

(2) *E is Dedekind complete and $\sigma(E, E')$ is order continuous.*

(3) *E is an ideal of $(E')^\sim_{\mathrm{n}}$.*

(4) *E is an ideal of E''.*

(5) *Each order interval of E is $\sigma(E, E')$-compact.*

Proof. (1) $\Longrightarrow$ (2) Consider E as a Riesz subspace of $(E')^\sim$, and note that $\langle E', (E')^\sim \rangle$ is a Riesz dual system. By Theorem 3.50 the topological dual of $\big(E', |\sigma|(E', E)\big)$ is the ideal generated by E in $(E')^\sim$, which by our hypothesis must coincide with E, and the desired conclusion follows.

(2) $\Longrightarrow$ (3) By Theorem 3.54, E is an order dense Riesz subspace of $(E')^\sim_{\mathrm{n}}$. Thus, by Theorem 2.31, E is an ideal of $(E')^\sim_{\mathrm{n}}$.

(3) $\Longrightarrow$ (4) Recall that E'' is the topological dual of $\big(E', \beta(E', E)\big)$. Since $\beta(E', E)$ is a locally convex-solid topology, E'' is an ideal of $(E')^\sim$, and from this we see that E is an ideal of E''.

(4) $\Longrightarrow$ (1) Clearly, E is also an ideal of $(E')^\sim$, and the conclusion follows from Theorem 3.50 applied to the Riesz dual system $\langle E', (E')^\sim \rangle$.

(4) $\Longleftrightarrow$ (5) We use the notation of Lemma 3.56. Note that E is an ideal of E'' if and only if $[0, x] = [\![0, x]\!]$ holds for each $x \in E^+$. On the other hand, since the topologies $\sigma(E'', E')$ and $\sigma(E, E')$ agree on E, it follows from Lemma 3.56 that this is the case if and only if every order interval of E is $\sigma(E, E')$-compact. ∎

For Banach lattices statement (5) of the preceding theorem is attributed by M. Nakamura [**146**, p. 106] to T. Ogasawara.

The Riesz dual systems have an asymmetry reminiscent to that of the dual systems $\langle X, X' \rangle$ with X a Banach space. In the Banach space case, the strong topology $\beta(X', X)$ is not in general consistent with $\langle X, X' \rangle$. In a Riesz dual system $\langle E, E' \rangle$, the locally convex-solid topology $|\sigma|(E', E)$ is not in general consistent with $\langle E, E' \rangle$. However, if a Riesz dual system satisfies any one of the conditions of Theorem 3.57, then a complete symmetry occurs.

Consider a Riesz dual system $\langle E, E' \rangle$. We have already seen that the polars of solid sets are likewise solid sets. In case A is an ideal of E, then it is not difficult to see that

$$A^\circ = \left\{ x' \in E' : \ x'(x) = 0 \ \text{for all} \ x \in A \right\}.$$

From this, it follows that A° is a band of E'. Similarly, if A is an ideal of E', then (by Theorem 1.23) we see that

$$A^\circ = \left\{ x \in E : \ x'(x) = 0 \ \text{for all} \ x' \in A \right\},$$

and so A° is an ideal of E.

The next two results deal with order projections in Riesz dual systems.

Theorem 3.58. *Let $\langle E, E' \rangle$ be a Riesz dual system. If A and B are two bands of E satisfying $E = A \oplus B$, then $E' = A^\circ \oplus B^\circ$ holds.*

In particular, if P denotes the order projection of E onto A, then P is weakly continuous and its adjoint P' is the order projection of E' onto B°.

Proof. We have mentioned before that both A° and B° are bands, and clearly $A^\circ \cap B^\circ = \{0\}$. If P denotes the order projection of E onto A, then $x' \circ P \in B^\circ$ and $x' - x' \circ P \in A^\circ$ hold for each $x' \in E'$. From $x' = (x' - x' \circ P) + x' \circ P$, we see that $E' = A^\circ \oplus B^\circ$.

Now from $E = A \oplus B$ and $E' = A^\circ \oplus B^\circ$, it is easy to see that $P'x' = 0$ for all $x' \in A^\circ$ and $P'x' = x'$ for all $x' \in B^\circ$. Therefore, P' is the order projection of E' onto the band B°. ∎

The following is a "dual" of the preceding result.

Theorem 3.59. *Let $\langle E, E' \rangle$ be a Riesz dual system with E Dedekind complete and $\sigma(E, E')$ order continuous. If A and B are two bands of E' satisfying $E' = A \oplus B$, then $E = A^\circ \oplus B^\circ$.*

In particular, if P is an order projection on E', then there exists an order projection Q on E such that $Q' = P$.

Proof. By Theorem 3.57 we know that E is an ideal of $(E')^\sim$, and so $\langle E', E \rangle$ is a Riesz dual system. From Theorem 3.58 we see that $E = A^\circ \oplus B^\circ$. Thus, if P is the order projection of E' onto A and Q is the order projection of E onto B°, then $Q' = P$ holds. ∎

We continue with an important density property of E in $(E')^\sim_n$.

Theorem 3.60. *If $\langle E, E' \rangle$ is a Riesz dual system, then the Riesz space E is $|\sigma|\left((E')^\sim_n, E'\right)$-dense in $(E')^\sim_n$.*

Proof. Note that $\langle E', (E')_n^\sim \rangle$ is a Riesz dual system. Since E' is Dedekind complete and $\sigma(E', (E')_n^\sim)$ is order continuous, it follows from Theorem 3.57 that $|\sigma|((E')_n^\sim, E')$ is consistent with $\langle E', (E')_n^\sim \rangle$.

Denote by $\overline{E}$ the $|\sigma|((E')_n^\sim, E')$-closure of E in $(E')_n^\sim$. If $\overline{E} \neq (E')_n^\sim$ holds, then according to Theorems 3.7 and 3.16 there exists some nonzero $x' \in E'$ satisfying $x(x') = x'(x) = 0$ for all $x \in E$. However, the latter means that $x' = 0$, which is a contradiction. Hence, $\overline{E} = (E')_n^\sim$ holds, as asserted. ∎

If an order interval $[0, x]$ is compact for some locally convex topology, then by the classical Krein–Milman Theorem 3.14 we know that the order interval $[0, x]$ is the closed convex hull of its extreme points. Remarkably, if τ is a locally convex solid topology on a Riesz space E with the principal projection property, then (although the order intervals of E need not be compact) for every $x \in E^+$ the order interval $[0, x]$ is always the τ-closed convex hull of its extreme points. This useful result (which is essentially due to H. H. Schaefer [**174**, p. 142]) is stated next.

Theorem 3.61 (Schaefer). *Let (E, τ) be a locally convex-solid Riesz space with the principal projection property. Then for each $x \in E^+$ the order interval $[0, x]$ is the τ-closed convex hull of its extreme points—or, equivalently, $[0, x]$ is the τ-closed convex hull of the components of x.*

Proof. Let $x \in E^+$ be fixed. By Theorem 1.49 we know that the extreme points of $[0, x]$ are precisely the components of x. Let A be the τ-closed convex hull of the extreme points of $[0, x]$, and assume by way of contradiction that $A \neq [0, x]$.

Pick $0 < z < x$ with $z \notin A$. Since $\{z\}$ is convex and τ-compact, A is convex and τ-closed, and $\{z\} \cap A = \emptyset$, it follows from Theorem 3.12 that there exist $c \in \mathbb{R}$ and $f \in E'$ such that $f(y) \leq c$ holds for all $y \in A$ and $f(z) > c$. Now by Theorem 1.50 we have

$$f^+(x) = \sup\{f(y): \ y \wedge (x - y) = 0\},$$

and so, since the components of x belong to A, we infer that $f^+(x) \leq c$. In particular, it follows that

$$c < f(z) \leq f^+(z) \leq f^+(x) \leq c,$$

which is absurd. So, $A = [0, x]$ must be true and the proof is finished. ∎

Exercises

1. Show that the convex hull of a solid set is likewise a solid set. Is the solid hull of a convex set necessarily convex? [*Hint*: Use Theorem 1.13.]

2. Let E be a Riesz space such that $E^\sim$ separates its points. Then show that for an ideal E' of $E^\sim$ the following statements are equivalent.
   (a) $\langle E, E' \rangle$ is a Riesz dual system.
   (b) E' is $\sigma(E^\sim, E)$-dense in $E^\sim$.
   (c) If $x \in E$ satisfies $x'(x) \geq 0$ for all $0 \leq x' \in E'$, then $x \geq 0$.

3. If $\langle E, E' \rangle$ is a Riesz dual system, then show that E^+ is $\sigma(E, E')$-closed and E'_+ is $\sigma(E', E)$-closed.

4. Let $\langle E, E' \rangle$ be a Riesz dual system with E Dedekind complete and with $\sigma(E, E')$ order continuous. Show that the order intervals of E are complete with respect to any consistent locally convex-solid topology on E.

5. Show that an order continuous locally convex-solid topology on a Riesz space E has a unique order continuous locally convex-solid extension to the Dedekind completion of E.

6. (Andô) Show that a locally convex-solid topology τ on a Riesz space is order continuous if and only if every τ-closed ideal is a band.

7. Let $\langle E, E' \rangle$ be a Riesz dual system. Show that $\sigma(E, E')$ is order continuous if and only if N_f is a band for each $f \in E'$.

8. Let $\langle E, E' \rangle$ and $\langle F, F' \rangle$ be two Riesz dual systems, and let $T \colon E \to F$ be a weakly continuous operator. Then show the following:
   (a) $T[0, x]$ is $|\sigma|(F, F')$-totally bounded for each $x \in E^+$ if and only if $T'[0, x']$ is $|\sigma|(E', E)$-totally bounded for each $x' \in F'_+$.
   (b) If τ is a consistent locally convex-solid topology on F, then $T[0, x]$ is τ-totally bounded for each $x \in E^+$ if and only if $T'(A)$ is $|\sigma|(E', E)$-totally bounded for every τ-equicontinuous subset A of F'.
   [*Hint*: For (a) use Theorems 3.27 and 3.55, and for (b) use Theorems 3.22, 3.27, and 3.55.]

9. Let $\langle E, E' \rangle$ be a Riesz dual system and let A be a projection band of E. Show that $(A^\circ)^d = (A^d)^\circ$.

10. Let $\langle E, E' \rangle$ be a Riesz dual system, and let τ be a locally convex topology on E consistent with $\langle E, E' \rangle$. Show that τ is locally convex-solid if and only if the solid hull of every τ-equicontinuous subset of E' is also τ-equicontinuous.

11. Let $\langle E, E' \rangle$ and $\langle F, F' \rangle$ be two Riesz dual systems. If $T \colon E \to F$ is a weakly continuous positive operator, then show that
$$T \colon \big(E, |\sigma|(E, E')\big) \to \big(F, |\sigma|(F, F')\big)$$
is continuous.

12. Let (E, τ) be a locally convex-solid Riesz space, and let A be a τ-closed ideal of E. Then show the following:
   (a) The quotient Riesz space E/A with the quotient topology (see Exercise 12 of section 3.1) is a locally convex-solid Riesz space.
   (b) If τ is order continuous, then the quotient topology is likewise order continuous.

Chapter 4

Banach Lattices

It is well known that most classical Banach spaces are, in fact, Banach lattices on which positive operators appear naturally. This chapter is devoted to the study of Banach lattices with special emphasis on Banach lattices with order continuous norms. The classes of *AL*- and *AM*-spaces are investigated thoroughly. In addition, some interesting connections between weak compactness and the lattice structure of a Banach lattice are presented. One section deals exclusively with embeddings of the classical sequence spaces into Banach spaces. As we shall see, a number of important properties of Banach spaces and Banach lattices are reflected upon the embeddability (or nonembeddability) of c_0, ℓ_1, and ℓ_∞ into the spaces. Also, Banach lattices of operators are considered and some useful approximation properties of positive operators are obtained.

4.1. Banach Lattices with Order Continuous Norms

Recall that a norm $\|\cdot\|$ on a Riesz space is said to be a **lattice norm** whenever $|x| \leq |y|$ implies $\|x\| \leq \|y\|$. A Riesz space equipped with a lattice norm is known as a **normed Riesz space**. If a normed Riesz space is also norm complete, then it is referred to as a **Banach lattice**. Clearly, the normed Riesz spaces are special examples of locally convex-solid Riesz spaces.

It should be obvious that in a normed Riesz space $\|x\| = \||x|\|$ holds for all x. Also, in view of Theorem 1.9, in a normed Riesz space the following inequalities hold:

$$\|x^+ - y^+\| \leq \|x - y\| \quad \text{and} \quad \||x| - |y|\| \leq \|x - y\|.$$

In particular, these inequalities show that the lattice operations of a normed Riesz space are uniformly continuous functions.

Theorem 4.1. *The norm dual of a normed Riesz space is a Banach lattice.*

Proof. Let E be a normed Riesz space. By Theorem 3.49 we know that E' is an ideal of $E^{\sim}$, and hence E' is a Riesz space in its own right. Since E' is also a Banach space, it remains to be shown that the norm of E' is a lattice norm. To this end, let $|x'| \le |y'|$ in E'. From the inequalities

$$\begin{aligned}|x'(x)| &\le |x'|(|x|) \le |y'|(|x|) \\ &= \sup\{|y'(y)|:\ |y| \le |x|\} \le \|y'\| \cdot \|x\|,\end{aligned}$$

it easily follows that $\|x'\| \le \|y'\|$, and the proof is finished. ∎

Let E be a normed Riesz space. If $0 \le x' \in E'$ holds, then from the inequality $|x'(x)| \le x'(|x|)$, it is easy to see that

$$\|x'\| = \sup\{x'(x):\ 0 \le x \in E \ \text{and} \ \|x\| = 1\}.$$

Similarly, if $x \in E^+$, then

$$\|x\| = \sup\{x'(x):\ 0 \le x' \in E' \ \text{and} \ \|x'\| = 1\}.$$

In addition, since E' is an ideal of $E^{\sim}$ and separates the points of E, it follows that the natural embedding of E into E'' is a lattice isometry; see Theorem 1.69 and the discussion after it. In other words, E is a Riesz subspace of the Banach lattice E''. In particular, the norm closure of E in E'' is the norm completion of E. On the other hand, the continuity of the lattice operations imply that the norm closure of E in E'' is also a Riesz subspace, and so we have proven the following result.

Theorem 4.2. *The norm completion of a normed Riesz space E is a Banach lattice including E as a Riesz subspace.*

Positive operators between Banach lattices are necessarily continuous. This important result was first established for integral operators by S. Banach [**30**]. For positive linear functionals, the next theorem appeared in the first edition of G. Birkhoff's book [**37**], and was re-proved my M. G. Krein and M. A. Rutman [**103**]. In a more general setting the result was proven by I. A. Bahtin, M. A. Krasnoselskii, and V. Y. Stecenko [**29**].

Theorem 4.3. *Every positive operator from a Banach lattice to a normed Riesz space is continuous.*

Proof. Let $T\colon E \to F$ be a positive operator from a Banach lattice E to a normed Riesz space. Assume by way of contradiction that T is not norm bounded. Then there exists a sequence $\{x_n\}$ of E satisfying $\|x_n\| = 1$ and $\|Tx_n\| \ge n^3$ for each n. In view of $|Tx_n| \le T|x_n|$, we can assume that $x_n \ge 0$

holds for each n. From $\sum_{n=1}^{\infty} \frac{\|x_n\|}{n^2} < \infty$ and the norm completeness of E, it follows that the series $\sum_{n=1}^{\infty} \frac{x_n}{n^2}$ is norm convergent in E. Let $x = \sum_{n=1}^{\infty} \frac{x_n}{n^2}$. Clearly, $0 \leq \frac{x_n}{n^2} \leq x$ holds for all n, and so

$$n \leq \left\| T\left(\tfrac{x_n}{n^2}\right) \right\| \leq \|Tx\| < \infty$$

also holds for each n, which is impossible. Thus, T must be normed bounded and hence continuous. ∎

A striking consequence of the preceding result is that a given Riesz space admits essentially at most one lattice norm under which it is a Banach lattice. This is due to C. Goffman [70].[1]

Corollary 4.4 (Goffman). *All lattice norms that a make a Riesz space a Banach lattice are equivalent.*

Proof. Let a Riesz space E be a Banach lattice under the two lattice norms $\| \cdot \|_1$ and $\| \cdot \|_2$. Then, according to Theorem 4.3, the identity operator $I \colon (E, \| \cdot \|_1) \to (E, \| \cdot \|_2)$ is a homeomorphism, and this guarantees that the norms $\| \cdot \|_1$ and $\| \cdot \|_2$ are equivalent. ∎

Since every order bounded linear functional on a Riesz space can be written as a difference of two positive linear functionals, the following result of G. Birkhoff [37, 1940 Edition] should also be immediate from Theorem 4.3.

Corollary 4.5 (Garrett Birkhoff). *The norm dual of a Banach lattice E coincides with its order dual, i.e., $E' = E^{\sim}$.*

Now let $\{X_n\}$ be a sequence of Banach spaces. Then recall that the L_p-**sum** of the sequence $\{X_n\}$ $(1 \leq p < \infty)$ is the Banach space

$$(X_1 \oplus X_2 \oplus \cdots)_p$$

$$:= \left\{ x = (x_1, x_2, \ldots) \colon x_n \in X_n \text{ and } \|x\| := \left[\sum_{n=1}^{\infty} \|x_n\|^p \right]^{\frac{1}{p}} < \infty \right\}.$$

As usual we also define

$$(X_1 \oplus X_2 \oplus \cdots)_{\infty} := \left\{ x = (x_1, x_2, \ldots) \colon x_n \in X_n \text{ and } \|x\| := \sup\{\|x_n\|\} < \infty \right\}$$

and

$$(X_1 \oplus X_2 \oplus \cdots)_0 := \left\{ x = (x_1, x_2, \ldots) \colon x_n \in X_n \text{ and } \lim_{n \to \infty} \|x_n\| = 0 \right\}.$$

If each X_n is a Banach lattice, then clearly under the pointwise ordering each $(X_1 \oplus X_2 \oplus \cdots)_p$ is likewise a Banach lattice.

---

[1] Recall that two norms $\| \cdot \|_1$ and $\| \cdot \|_2$ on a vector space X are said to be **equivalent** if there exist constants $K, M > 0$ satisfying $K\|x\|_1 \leq \|x\|_2 \leq M\|x\|_1$ for all $x \in X$.

Theorem 4.6. *Let $\{X_n\}$ be a sequence of Banach spaces, and assume that $1 < p, q < \infty$ satisfy $\frac{1}{p} + \frac{1}{q} = 1$. Then we have*

$$\left(X_1 \oplus X_2 \oplus \cdots\right)_p' = \left(X_1' \oplus X_2' \oplus \cdots\right)_q,$$

where the equality holds subject to the duality

$$\langle x, x'\rangle = \sum_{n=1}^{\infty} x_n'(x_n)$$

for $x = (x_1, x_2, \ldots) \in \left(X_1 \oplus X_2 \oplus \cdots\right)_p$ and $x' = (x_1', x_2', \ldots) \in \left(X_1' \oplus X_2' \oplus \cdots\right)_q$.
    Similarly, we have

$$\left(X_1 \oplus X_2 \oplus \cdots\right)_1' = \left(X_1' \oplus X_2' \oplus \cdots\right)_\infty$$

and

$$\left(X_1 \oplus X_2 \oplus \cdots\right)_0' = \left(X_1' \oplus X_2' \oplus \cdots\right)_1.$$

Proof. We shall establish the result when $1 < p, q < \infty$ satisfy $\frac{1}{p} + \frac{1}{q} = 1$, and leave the other cases for the reader. To this end, it is enough to show and the mapping $x' \mapsto \phi_{x'}$ where $\phi_{x'}(x_1, x_2, \ldots) = \sum_{n=1}^{\infty} x_n'(x_n)$, is an isometry from $\left(X_1' \oplus X_2' \oplus \cdots\right)_q$ onto $\left(X_1 \oplus X_2 \oplus \cdots\right)_p'$.

Fix $x' = (x_1', x_2', \ldots) \in \left(X_1' \oplus X_2' \oplus \cdots\right)_q$. Then for each $x = (x_1, x_2, \ldots)$ in $\left(X_1 \oplus X_2 \oplus \cdots\right)_p$ we have $|x_n'(x_n)| \leq \|x_n'\| \cdot \|x_n\|$, and so

$$
\begin{aligned}
\left| \sum_{n=1}^{k} x_n'(x_n) \right| &\leq \sum_{n=1}^{k} \|x_n'\| \cdot \|x_n\| \\
&\leq \left[\sum_{n=1}^{\infty} \|x_n'\|^q \right]^{\frac{1}{q}} \cdot \left[\sum_{n=1}^{\infty} \|x_n\|^p \right]^{\frac{1}{p}} \\
&= \|x'\|_q \cdot \|x\|_p
\end{aligned}
$$

holds for all k. Thus, the formula $\phi_{x'}(x) = \sum_{n=1}^{\infty} x_n'(x_n)$ defines a continuous linear functional on $\left(X_1 \oplus X_2 \oplus \cdots\right)_p$ satisfying

$$\|\phi_{x'}\| \leq \|x'\|_q \qquad\qquad (\star)$$

for all $x' \in \left(X_1' \oplus X_2' \oplus \cdots\right)_q$. Clearly, $x' \mapsto \phi_{x'}$ is a linear operator.
    Now let $\phi \in \left(X_1 \oplus X_2 \oplus \cdots\right)_p'$. If $x_n' \colon X_n \to \mathbb{R}$ is defined by

$$x_n'(x_n) = \phi(0, \ldots, 0, x_n, 0, 0, \ldots),$$

then $x_n' \in X_n'$, and moreover $\phi(x_1, x_2, \ldots) = \sum_{n=1}^{\infty} x_n'(x_n)$ holds for all $(x_1, x_2, \ldots) \in \left(X_1 \oplus X_2 \oplus \cdots\right)_p$. Fix $0 < \epsilon < 1$. For each n pick some

$y_n \in X_n$ with $\|y_n\| = 1$ and $x'_n(y_n) \geq \epsilon \|x'_n\|$. Put $z_n = \|x'_n\|^{q-1} y_n$, and note that for each k we have

$$\epsilon \sum_{n=1}^{k} \|x'_n\|^q = \sum_{n=1}^{k} \|x'_n\|^{q-1} \epsilon \|x'_n\| \leq \sum_{n=1}^{k} \|x'_n\|^{q-1} x'_n(y_n)$$

$$= \phi(z_1, \ldots, z_k, 0, 0, \ldots) \leq \|\phi\| \cdot \left[\sum_{n=1}^{k} \|z_n\|^p \right]^{\frac{1}{p}}$$

$$= \|\phi\| \cdot \left[\sum_{n=1}^{k} \|x'_n\|^q \right]^{\frac{1}{p}}.$$

Therefore, $\epsilon \left(\sum_{n=1}^{k} \|x'_n\|^q \right)^{\frac{1}{q}} \leq \|\phi\|$ holds for all k, and each $0 < \epsilon < 1$. This implies that $x' = (x'_1, x'_2, \ldots) \in \left(X'_1 \oplus X'_2 \oplus \cdots \right)_q$, $\phi = \phi_{x'}$, and $\|x'\|_q \leq \|\phi_{x'}\|$. By $(\star)$ we see that $\|\phi_{x'}\| = \|x'\|_q$, and thus $x' \mapsto \phi_{x'}$ is an onto linear isometry. Finally, it should be noted that if each X_n is a Banach lattice, then $x' \mapsto \phi_{x'}$ is an onto lattice isometry. ∎

We now turn our attention to Banach lattices with order continuous norms.

Definition 4.7. *A lattice seminorm p on a Riesz space is said to be **order continuous** whenever $x_\alpha \downarrow 0$ implies $p(x_\alpha) \downarrow 0$.*

*If the above condition holds for sequences, i.e., $x_n \downarrow 0$ implies $p(x_n) \downarrow 0$, then p is said to be σ-**order continuous**.*

Clearly, a lattice seminorm p is order continuous if and only if $A \downarrow 0$ implies $\{p(a): a \in A\} \downarrow 0$. A lattice norm is, of course, order continuous if and only if it generates an order continuous locally convex-solid topology. In addition, note that a lattice seminorm p is order continuous if and only if $0 \leq x_\alpha \uparrow x$ implies $p(x - x_\alpha) \downarrow 0$.

To characterize Banach lattices with order continuous norms, we need the following simple result.

Lemma 4.8. *If $0 \leq x_\alpha \uparrow \leq x$ holds in an Archimedean Riesz space E, then the set $D = \{y \in E: x_\alpha \leq y \text{ for all } \alpha\}$ is directed downward and $y - x_\alpha \downarrow_{y,\alpha} 0$.*

Proof. Clearly, D is directed downward. Let $0 \leq u \leq y - x_\alpha$ holds for all α and all $y \in D$. Then $x_\alpha \leq y - u$ also holds for all α, and so $y - u \in D$ for all $y \in D$. By induction, $y - nu \in D$ for all n and all $y \in D$. In particular, we have $0 \leq nu \leq x$ for each n. Since E is Archimedean, it follows that $u = 0$, and so $y - x_\alpha \downarrow_{y,\alpha} 0$. ∎

A few elementary characterizations of Banach lattices with order continuous norms are included in the next theorem. They are essentially due to H. Nakano [**152**, p. 321].

Theorem 4.9 (Nakano). *For a Banach lattice E the following statements are equivalent.*

(1) *E has order continuous norm.*

(2) *If $0 \le x_n \uparrow \le x$ holds in E, then $\{x_n\}$ is a norm Cauchy sequence.*

(3) *E is Dedekind σ-complete and $x_n \downarrow 0$ in E implies $\|x_n\| \downarrow 0$.*

(4) *E is an ideal of E''.*

(5) *Each order interval of E is weakly compact.*

Proof. (1) $\Longrightarrow$ (2) Let $0 \le x_\alpha \uparrow \le x$ hold in E, and let $\epsilon > 0$. By Lemma 4.8 there exists a net $\{y_\lambda\} \subseteq E$ with $y_\lambda - x_\alpha \downarrow 0$. Thus, there exists λ_0 and α_0 such that $\|y_\lambda - x_\alpha\| < \epsilon$ holds for all $\lambda \succeq \lambda_0$ and $\alpha \succeq \alpha_0$. From the inequality

$$\|x_\alpha - x_\beta\| \le \|x_\alpha - y_{\lambda_0}\| + \|x_\beta - y_{\lambda_0}\| \,,$$

we see that $\|x_\alpha - x_\beta\| < 2\epsilon$ holds for all $\alpha, \beta \succeq \alpha_0$. Hence, $\{x_\alpha\}$ is a norm Cauchy net. (If y is its norm limit, then $x_\alpha \uparrow y$, and so E is also Dedekind complete.)

(2) $\Longrightarrow$ (3) It follows immediately from part (2) of Theorem 3.46.

(3) $\Longrightarrow$ (1) Let $x_\alpha \downarrow 0$. If $\{x_\alpha\}$ is not a norm Cauchy net, then there exist some $\epsilon > 0$ and a sequence $\{\alpha_n\}$ of indices with $\alpha_n \uparrow$, and $\|x_{\alpha_n} - x_{\alpha_{n+1}}\| > \epsilon$ for all n. Since E is Dedekind σ-complete, there exists some $x \in E$ with $x_{\alpha_n} \downarrow x$. Now from our hypothesis, we see that $\{x_{\alpha_n}\}$ is a norm Cauchy sequence, which contradicts $\|x_{\alpha_n} - x_{\alpha_{n+1}}\| > \epsilon$. Thus, $\{x_\alpha\}$ is a norm Cauchy net, and so $\{x_\alpha\}$ is norm convergent to some $y \in E$. By part (2) of Theorem 3.46 we see that $y = 0$, and so $\|x_\alpha\| \downarrow 0$ holds.

The other equivalences easily follow from Theorems 3.57 and 3.54. ∎

Corollary 4.10. *Every Banach lattice with order continuous norm is Dedekind complete.*

When does the norm completion of a normed Riesz space have order continuous norm? The next theorem of W. A. J. Luxemburg [**125**, Note XVI] provides the answer.

Theorem 4.11 (Luxemburg). *For a normed Riesz space E the following statements are equivalent.*

(1) *The norm completion of E has order continuous norm.*

(2) *If $0 \le x_n \uparrow \le x$ holds in E, then $\{x_n\}$ is a norm Cauchy sequence.*

Proof. (1) $\Longrightarrow$ (2) This follows immediately from Theorem 4.9.

(2) $\Longrightarrow$ (1) Let $0 \leq x_n \downarrow$ hold in the norm completion of E. By statement (2) of Theorem 4.9, it is enough to show that $\{x_n\}$ is a norm Cauchy sequence. To this end, let $\epsilon > 0$. For each n pick some $y_n \in E^+$ with $\|x_n - y_n\| < \epsilon 2^{-n}$. Put $z_n = \bigwedge_{i=1}^n y_i$, and note that $0 \leq z_n \downarrow$ holds in E.

Now, by our hypothesis, there exists some k with $\|z_n - z_m\| < \epsilon$ for all $n, m \geq k$. From

$$x_n - z_n = \bigvee_{i=1}^n (x_n - y_i) \leq \bigvee_{i=1}^n (x_i - y_i) \leq \sum_{i=1}^n |x_i - y_i|,$$

and

$$-(x_n - z_n) = z_n - x_n = \bigwedge_{i=1}^n (y_i - x_n) \leq y_n - x_n \leq \sum_{i=1}^n |x_i - y_i|,$$

we see that

$$|x_n - z_n| \leq \sum_{i=1}^n |x_i - y_i|.$$

Therefore, $\|x_n - z_n\| \leq \sum_{i=1}^n \|x_i - y_i\| < \sum_{i=1}^n \epsilon 2^{-i} < \epsilon$ holds for all n. In particular, for $n, m \geq k$ we have

$$\|x_n - x_m\| \leq \|x_n - z_n\| + \|z_n - z_m\| + \|z_m - x_m\| < \epsilon + \epsilon + \epsilon = 3\epsilon,$$

and so $\{x_n\}$ is a norm Cauchy sequence, as desired. ∎

Examples of Banach lattices with order continuous norms are provided by the classical $L_p(\mu)$-spaces, where $1 \leq p < \infty$. From statement (4) of Theorem 4.9, it is easy to see that every reflexive Banach lattice has order continuous norm. The Banach lattices $C[0,1]$, $L_\infty[0,1]$, and ℓ_∞ (all with the sup norm) are examples of Banach lattices without order continuous norms.

Recall that a sequence $\{x_n\}$ in a Riesz space is said to be **disjoint** whenever $|x_n| \wedge |x_m| = 0$ holds for $n \neq m$. The next theorem of the authors describes an important approximation property of increasing order bounded sequences in terms of disjoint sequences.

Theorem 4.12 (Aliprantis–Burkinshaw). *If $0 \leq x_n \uparrow \leq x$ holds in a Riesz space, then for each $k \in \mathbb{N}$ there exist disjoint sequences $\{y_n^1\}, \ldots, \{y_n^k\}$ of $[0, x]$ such that for each n we have*

$$y_n^1 + \cdots + y_n^k \leq x_{n+1} - x_n \leq y_n^1 + \cdots + y_n^k + \tfrac{2}{k+3} x.$$

Proof. The proof uses the *k-disjointness technique* introduced by D. H. Fremlin [64]. Start by saying that a sequence $\{u_n\}$ in a Riesz space is *k*-**disjoint** whenever for every set I of k natural numbers, we have $\inf\{|u_i|: i \in I\} = 0$. In this terminology, a disjoint sequence is a 2-disjoint sequence.

Now let $0 \leq x_n \uparrow \leq x$ hold in a Riesz space. The proof will be based upon two steps.

Step 1: *For each k there exists a $(k+1)$-disjoint sequence $\{u_n\}$ of $[0, x]$ satisfying $u_n \leq x_{n+1} - x_n \leq u_n + \frac{1}{k+1}x$ for all n.*

To see this, put $u_n = \left(x_{n+1} - x_n - \frac{1}{k+1}x\right)^+$. Clearly,

$$u_n \leq x_{n+1} - x_n = u_n + (x_{n+1} - x_n) \wedge \frac{1}{k+1}x \leq u_n + \frac{1}{k+1}x$$

holds for all n, and we claim that $\{u_n\}$ is a $(k+1)$-disjoint sequence. Indeed, if I is an arbitrary set of $k+1$ natural numbers, then the vector $u = \inf\{u_i \colon i \in I\}$ satisfies

$$
\begin{aligned}
0 &= u \wedge \sum_{i \in I}\left(x_{i+1} - x_i - \tfrac{1}{k+1}x\right)^- = u \wedge \sum_{i \in I}\left(\tfrac{1}{k+1}x + x_i - x_{i+1} + u_i\right) \\
&= u \wedge \left[x - \sum_{i \in I}(x_{i+1} - x_i) + \sum_{i \in I} u_i\right] \\
&\geq u \wedge \sum_{i \in I} u_i = u \geq 0,
\end{aligned}
$$

so that $u = 0$. Therefore, $\{u_n\}$ is a $(k+1)$-disjoint sequence.

Step 2: *If $\{u_n\} \subseteq [0, x]$ is a $(k+1)$-disjoint sequence $(k > 1)$ and $0 < \epsilon < 1$, then there exist a k-disjoint sequence $\{w_n\} \subseteq [0, x]$ and a disjoint sequence $\{y_n\} \subseteq [0, x]$ such that for each n we have*

$$y_n + w_n \leq u_n \leq y_n + w_n + \epsilon x.$$

To see this, put $u_0 = 0$ and $y_n = \left(u_n - \frac{1}{\epsilon}\sum_{i=0}^{n-1} u_i - \epsilon x\right)^+$ for all $n \geq 1$. Then for $m > n \geq 1$ the inequalities

$$
\begin{aligned}
0 \leq \epsilon y_m &= \left[\epsilon u_n - \sum_{i=0}^{m-1} u_i - \epsilon^2 x\right]^+ \leq \left[\epsilon x - \sum_{i=0}^{n-1} u_i - u_n\right]^+ \\
&\leq \left[\epsilon x + \frac{1}{\epsilon}\sum_{i=0}^{n-1} u_i - u_n\right]^+ = \left[u_n - \frac{1}{\epsilon}\sum_{i=0}^{n-1} u_i - \epsilon x\right]^+ \perp y_n
\end{aligned}
$$

imply $y_m \wedge y_n = 0$, and so $\{y_n\}$ is a disjoint sequence.

Next, for each $n = 1, 2, \ldots$ let $w_n = u_n \wedge \left(\frac{1}{\epsilon}\sum_{i=0}^{n-1} u_i\right)$. We claim that $\{w_n\}$ is a k-disjoint sequence. To see this, let $1 \leq n_1 < \cdots < n_k$, and note that

$$0 \leq \bigwedge_{i=1}^{k} w_{n_i} \leq u_{n_1} \wedge \cdots \wedge u_{n_k} \wedge \left[\tfrac{1}{\epsilon} \sum_{i=0}^{n_1-1} u_i \right]$$

$$\leq \tfrac{1}{\epsilon} \sum_{i=0}^{n_1-1} u_{n_1} \wedge \cdots \wedge u_{n_k} \wedge u_i = 0 \,,$$

where each term of the last sum is zero by virtue of the $(k+1)$-disjointness of $\{u_n\}$. Now note that

$$y_n + w_n \leq u_n = y_n + u_n \wedge \left[\tfrac{1}{\epsilon} \sum_{i=0}^{n-1} u_i + \epsilon x \right] \leq y_n + w_n + \epsilon x$$

holds for all n, and this completes the proof of Step 2.

Now let $k \geq 1$ be fixed, and put $\epsilon = \frac{1}{(k+1)(k+3)}$. By Step 1 there exists a $(k+1)$-disjoint sequence $\{u_n^1\}$ of $[0, x]$ satisfying

$$u_n^1 \leq x_{n+1} - x_n \leq u_n^1 + \tfrac{1}{k+1} x$$

for all n. If $k = 1$, then $\{u_n^1\}$ is a disjoint sequence, and the proof of the theorem is finished. If $k > 1$ holds, then apply Step 2 to $\{u_n^1\}$ to get the k-disjoint sequence $\{u_n^2\}$ of $[0, x]$ and the disjoint sequence $\{y_n^1\}$ of $[0, x]$ satisfying

$$y_n^1 + u_n^2 \leq x_{n+1} - x_n \leq y_n^1 + u_n^2 + \epsilon x + \tfrac{1}{k+1} x$$

for all n. Repeat applying Step 2 until after $(k-1)$-times we have the inequalities

$$y_n^1 + \cdots + y_n^k \leq x_{n+1} - x_n \leq y_n^1 + \cdots + y_n^k + (k-1)\epsilon x + \tfrac{1}{k+1} x$$

for all $n \geq 1$, where each $\{y_n^i\}$ is a disjoint sequence of $[0, x]$. To complete the proof, note that $(k-1)\epsilon + \frac{1}{k+1} = \frac{2}{k+3}$ holds. ∎

As a first application of the preceding theorem, we have the following result of D. H. Fremlin [**64**, p. 56].

Theorem 4.13 (Fremlin). *For a lattice seminorm p on a Riesz space E and a vector $x \in E^+$, the following statements are equivalent:*

(1) $0 \leq x_n \uparrow \leq x$ *in E implies that $\{x_n\}$ is a p-Cauchy sequence.*

(2) *If $\{x_n\}$ is a disjoint sequence of $[0, x]$, then $\lim p(x_n) = 0$.*

Proof. (1) $\Longrightarrow$ (2) If $\{x_n\}$ is a disjoint sequence of $[0, x]$, then

$$0 \leq y_n = \sum_{i=1}^{n} x_i = \bigvee_{i=1}^{n} x_i \uparrow \leq x$$

holds, and so $p(x_n) = p(y_n - y_{n-1}) \to 0$.

(2) $\implies$ (1) Assume $0 \le x_n \uparrow \le x$ in E. If $\{x_n\}$ is not p-Cauchy sequence, then passing to a subsequence, we can assume that there exist some $\epsilon > 0$ satisfying $p(x_{n+1} - x_n) > 2\epsilon$ for all n. Fix some positive integer k with $\frac{2}{k+3} p(x) < \epsilon$, and then use Theorem 4.12 to select the disjoint sequences $\{y_n^1\}, \ldots, \{y_n^k\}$ of $[0, x]$ such that

$$0 \le x_{n+1} - x_n \le y_n^1 + \cdots + y_n^k + \tfrac{2}{k+3} x$$

for all n. Now by our hypothesis we have $\lim_{n \to \infty} p(y_n^i) = 0$ for all i, and thus $p(x_{n+1} - x_n) < 2\epsilon$ must hold for all n sufficiently large, which is impossible. Hence, $\{x_n\}$ is a p-Cauchy sequence, and the proof is finished. ∎

Taking into account Theorem 4.9, an easy application of Theorem 4.13 yields the following important characterization (due to D. H. Fremlin [64, p. 56] and P. Meyer-Nieberg [141]) of the order continuity of the norm in a Banach lattice in terms of disjoint sequences.

Theorem 4.14 (Fremlin–Meyer-Nieberg). *A Banach lattice has order continuous norm if and only if every order bounded disjoint sequence is norm convergent to zero.*

A positive linear functional f on a Riesz space is said to be **strictly positive** whenever $x > 0$ implies $f(x) > 0$. Riesz spaces admitting strictly positive linear functionals are useful in many contexts, and they have been studied by many authors. In [144] L. C. Moore, Jr., studied the existence of strictly positive linear functionals in connection with strictly monotone lattice norms.

A Banach lattice with order continuous norm and a weak order unit always admits a strictly positive linear functional. The details are included in the next theorem.

Theorem 4.15. *If a Banach lattice E has order continuous norm, then for each $x > 0$ there exists a positive linear functional on E that is strictly positive on the order interval $[0, x]$.*

Proof. Let E be a Banach lattice with order continuous norm, and let $0 < x \in E$. For each $0 < f \in E'$, put $N_f = \{y \in E: f(|y|) = 0\}$ and $C_f = N_f^{\mathrm{d}}$. Since f is order continuous, N_f is a band of E, and moreover f is strictly positive on C_f.

Now put $A = \bigcup \{C_f: 0 < f \in E'\}$. Since $0 \le f \le g$ implies $C_f \subseteq C_g$, it is easy to see that A is an ideal of E, and we claim that A is order dense in E. To see this, let $0 \le y \in A^{\mathrm{d}}$. That is, $y \perp C_f$ holds for all $0 < f \in E'$, and so $y \in C_f^{\mathrm{d}} = N_f^{\mathrm{dd}} = N_f$ for all $0 < f \in E'$. Therefore, $f(y) = 0$ holds for all $0 \le f \in E'$, from which it follows that $y = 0$. Hence, $A^{\mathrm{d}} = \{0\}$, and so by

Theorem 1.36 the ideal A is order dense in E. Since E has order continuous norm, this implies that A is also norm dense in E.

Next, we claim that A is, in fact, a norm closed ideal. To see this, pick a sequence $\{y_n\} \subseteq A$ with $\lim \|y_n - y\| = 0$. Then for each n there exists some $0 < f_n \in E'$ with $\|f_n\| = 1$ and $y_n \in C_{f_n}$. Put $f = \sum_{n=1}^{\infty} 2^{-n} f_n$, and note that $0 < f \in E'$. Also, $N_f \subseteq N_{f_n}$ holds for all n. This implies that $C_{f_n} \subseteq C_f$ for all n, and so $\{y_n\} \subseteq C_f$. Since C_f is a band, it follows from Theorem 3.46 that C_f is norm closed, and hence $y \in C_f \subseteq A$. Thus, A is norm closed, and since A is norm dense in E, we see that $A = E$.

Finally, note that there exists some $0 < g \in E'$ so that $[0, x] \subseteq C_g$ holds. Clearly, g is strictly positive on $[0, x]$, and the proof is finished. ■

If f is a strictly positive linear functional on a normed Riesz space E, then the formula $p(x) = f(|x|)$ defines a lattice norm on E. A connection between this norm and the original on E is described in the next result.

Lemma 4.16. *Let E be a Banach lattice with order continuous norm, and let $0 \le f \in E'$ be strictly positive on an order interval $[0, x]$. Then for each $\epsilon > 0$ there exists some $\delta > 0$ such that $y \in [0, x]$ and $f(y) < \delta$ imply $\|y\| < \epsilon$.*

Proof. Assume by way of contradiction that the claim is false. Then there exist some $\epsilon > 0$ and a sequence $\{x_n\} \subseteq [0, x]$ satisfying $f(x_n) < 2^{-n}$ and $\|x_n\| > \epsilon$ for each n.

Now let $y_n = \bigvee_{i=n}^{\infty} x_i$. Clearly, $\{y_n\} \subseteq [0, x]$ and $y_n \downarrow$. Since f is order continuous, it follows that

$$f(y_n) \le \sum_{i=n}^{\infty} f(x_i) \le \sum_{i=n}^{\infty} 2^{-i} = 2^{1-n},$$

and so $\lim f(y_n) = 0$. From this, and the strict positivity of f on $[0, x]$, we see that $y_n \downarrow 0$ must hold in E. On the other hand, $\|y_n\| \ge \|x_n\| > \epsilon$ for each n contradicts the order continuity of the norm, and the proof is finished. ■

In a Banach lattice with order continuous norm, there is an important relationship between the norm and the absolute weak topologies.

Theorem 4.17. *In a Banach lattice E with order continuous norm, the norm topology and $|\sigma|(E, E')$ agree on each order interval of E.*

In particular, in this case, $|\sigma|(E, E')$ and the norm topology have the same order bounded totally bounded sets.

Proof. Since $|\sigma|(E, E')$ is coarser than the norm topology, we have only to show that the norm topology is coarser than $|\sigma|(E, E')$ on the order intervals of E.

To this end, let V be an open ball with radius $\epsilon > 0$ centered at some $u \in [-x, x]$. By Theorem 4.15 there exists a positive linear functional f on E that is strictly positive on $[0, x]$, and by Lemma 4.16 there exists some $\delta > 0$ such that $y \in [-x, x]$ and $f(|y|) < \delta$ imply $\|y\| < \epsilon$. In particular, if $W = \{z \in E: \ f(|z - u|) < \delta\}$, then W is a $|\sigma|(E, E')$-neighborhood of u, and moreover $[-x, x] \cap W \subseteq [-x, x] \cap V$ holds. This shows that the norm topology is coarser that $|\sigma|(E, E')$ on $[-x, x]$, as desired.   ∎

It is useful to know that in a Banach lattice E every order bounded disjoint sequence converges weakly to zero. Indeed, if $\{x_n\}$ is a disjoint sequence of E satisfying $|x_n| \leq x$ for all n and $f \in E'$, then

$$\sum_{n=1}^{k} |f(x_n)| \leq \sum_{n=1}^{k} |f|(|x_n|) = |f|\left(\sum_{n=1}^{k} |x_n|\right) = |f|\left(\bigvee_{n=1}^{k} |x_n|\right) \leq |f|(x)$$

holds for each k, and so $\sum_{n=1}^{\infty} f(x_n)$ converges in $\mathbb{R}$. In particular, we have $f(x_n) \to 0$, and hence $x_n \xrightarrow{w} 0$ holds in E.

The next two theorems, due to P. G. Dodds and D. H. Fremlin [54], characterize the order continuity of the norm in a Banach lattice in terms of an approximation property.

Theorem 4.18 (Dodds–Fremlin). *A Banach lattice E has order continuous norm if and only if for each $\epsilon > 0$ and each $x \in E^+$ there exists some $0 \leq y' \in E'$ satisfying*

$$\left(|x'| - y'\right)^+(x) < \epsilon$$

for all $x' \in E'$ with $\|x'\| \leq 1$.

Proof. Assume first that E has order continuous norm. Fix $\epsilon > 0$ and $x \in E^+$. By Theorem 4.15 there exists some $0 \leq y' \in E'$ that is strictly positive on $[0, x]$. Also, by Lemma 4.16 there exists some $\delta > 0$ such that $y \in [0, x]$ and $y'(y) \leq \delta$ imply $\|y\| < \epsilon$. Now let $\alpha = \frac{\|x\|}{\delta}$, and let $0 \leq y \leq x$ and $0 \leq x' \in U'$ be arbitrary. Then we claim that $(x' - \alpha y')(y) < \epsilon$ holds. Indeed, if we have $(x' - \alpha y')(y) \geq \epsilon$, then we also have

$$\|x\| \geq \|y\| \geq x'(y) \geq \epsilon + \alpha y'(y) \geq \epsilon . \tag{$\star$}$$

In particular, it follows from $\alpha y'(y) \leq \|x\|$ that $y'(y) \leq \delta$ holds, and so $\|y\| < \epsilon$. However, from $(\star)$ we see that $\|y\| \geq \epsilon$, a contradiction. Thus, $(x' - \alpha y')(y) < \epsilon$, and so

$$\left(x' - \alpha y'\right)^+(x) = \sup\{(x' - \alpha y')(y): \ 0 \leq y \leq x\} \leq \epsilon$$

holds for each $0 \leq x' \in U'$, as desired.

For the converse, assume that the condition is satisfied, and let $\{x_n\}$ be a disjoint sequence satisfying $0 \leq x_n \leq x$ for each n. By Theorem 4.14 it is enough to show that $\|x_n\| \to 0$. To this end, let $\epsilon > 0$. Pick some

$0 \leq y' \in E'$ such that $(|x'| - y')^+(x) < \epsilon$ holds for all $x' \in U'$. Now for each $0 \leq x' \in U'$ we have

$$x'(x_n) = [x' \wedge y'](x_n) + (x' - y')^+(x_n) \leq y'(x_n) + \epsilon \,,$$

and so

$$\|x_n\| = \sup\{x'(x_n) \colon 0 \leq x' \in U'\} \leq y'(x_n) + \epsilon \,.$$

Since $x_n \xrightarrow{w} 0$ is true, we see that $\limsup \|x_n\| \leq \epsilon$ holds for all $\epsilon > 0$, and thus $\|x_n\| \to 0$, as required. ∎

The dual companion of the preceding theorem is the following.

Theorem 4.19 (Dodds–Fremlin). *The norm dual E' of a Banach lattice E has order continuous norm if and only if for each $\epsilon > 0$ and each $0 \leq x' \in E'$ there exists some $y \in E^+$ satisfying*

$$x'\big((|x| - y)^+\big) < \epsilon$$

for all $x \in E$ with $\|x\| \leq 1$.

Proof. Assume that E' has order continuous norm, and let $\epsilon > 0$ and $0 \leq x' \in E'$. By Theorem 4.18 there exists some $0 \leq y'' \in E''$ satisfying $(|x''| - y'')^+(x') < \epsilon$ for all $x'' \in U''$. Since $E'' = (E')_n^\sim$ holds, it follows from Theorem 3.60 that E is $|\sigma|(E'', E')$-dense in E''. In particular, there exists some $0 \leq y \in E$ satisfying $|y'' - y|(x') < \epsilon$. Therefore,

$$\big(|x''| - y\big)^+(x') \leq \big(|x''| - y''\big)^+(x') + \big(y'' - y\big)^+(x') < \epsilon + \epsilon = 2\epsilon$$

holds for all $x'' \in U''$.

For the converse repeat the arguments of the second part of the proof of Theorem 4.18. ∎

We now turn our attention to the study of *AL*- and *AM*-spaces. These spaces play a significant role in analysis.

Definition 4.20. *A Banach lattice E is said to be:*

(1) *An **abstract L_p-space** for some $1 \leq p < \infty$, whenever its norm is p-**additive** in the sense that*

$$\|x + y\|^p = \|x\|^p + \|y\|^p$$

holds for all $x, y \in E^+$ with $x \wedge y = 0$.

(2) *An **abstract M-space**, whenever its norm is an M-**norm**, i.e., if $x \wedge y = 0$ in E implies*

$$\|x \vee y\| = \max\{\|x\|, \|y\|\} \,.$$

An abstract L_1-space is referred to as an *AL*-**space** and an abstract M-space is known as an *AM*-**space**. This standard terminology will be employed here. The *AL*-spaces were introduced by G. Birkhoff [**36**], and the *AM*-spaces were studied systematically for the first time by S. Kakutani [**82**] (although they appeared in a fragmented way in the work of S. Banach [**30**]).

It should be noted that every abstract L_p-space E has order continuous norm. Indeed, if $\{x_n\} \subseteq [0, x]$ is a disjoint sequence, then from the inequality

$$\sum_{n=1}^{k} \|x_n\|^p = \left\| \sum_{n=1}^{k} x_n \right\|^p = \left\| \bigvee_{n=1}^{k} x_n \right\|^p \leq \|x\|^p,$$

it follows that $\sum_{n=1}^{\infty} \|x_n\|^p < \infty$, and so $\|x_n\| \to 0$ holds. By Theorem 4.14 the norm of E is order continuous.

The next result indicates the importance of *AM*-spaces.

Theorem 4.21. *Let E be a Banach lattice, and let $x \in E$. Then the principal ideal E_x generated by x in E under the norm $\|\cdot\|_\infty$, defined by*

$$\|y\|_\infty = \inf\{\lambda > 0: \ |y| \leq \lambda|x|\}, \quad y \in E_x,$$

is an AM-space, whose closed unit ball is the order interval $[-|x|, |x|]$.

Proof. It is a routine matter to verify that $\|\cdot\|_\infty$ is a lattice norm on E_x having $[-|x|, |x|]$ as its closed unit ball.

Now assume that a sequence $\{x_n\}$ of E_x is $\|\cdot\|_\infty$-Cauchy. By passing to a subsequence, we can assume that

$$|x_{n+p} - x_n| \leq 2^{-n}|x| \tag{$\star$}$$

holds for all n and p. Hence, $\{x_n\}$ is a norm Cauchy sequence of E. If y is its norm limit, then by letting $p \to \infty$ in $(\star)$, we see that $|y - x_n| \leq 2^{-n}|x|$ holds for all n. The latter shows that $y \in E_x$ and that $\|y - x_n\|_\infty \to 0$. Therefore, $(E_x, \|\cdot\|_\infty)$ is a Banach lattice.

Finally, let us show that $\|\cdot\|_\infty$ is an M-norm. To this end, let $u \wedge v = 0$ in E_x. Put $m = \max\{\|u\|_\infty, \|v\|_\infty\}$, and note that $m \leq \|u + v\|_\infty = \|u \vee v\|_\infty$ holds. On the other hand, the inequalities

$$0 \leq u \vee v \leq \big[\|u\|_\infty|x|\big] \vee \big[\|v\|_\infty|x|\big] \leq m|x| \vee m|x| = m|x|$$

show that $\|u \vee v\|_\infty \leq m$ also holds, and the proof is finished. ∎

Recall that a vector $e > 0$ in a Riesz space is said to be an **order unit** whenever for each x there exists some $\lambda > 0$ with $|x| \leq \lambda e$. Now if a Banach lattice E has an order unit $e > 0$, then $E_e = E$ holds, and so by Corollary 4.4 the norm

$$\|x\|_\infty = \inf\{\lambda > 0: \ |x| \leq \lambda e\}$$

is equivalent to the original norm of E. In other words, if a Banach lattice E has an order unit, then E can be renormed in such a way that it becomes an AM-space having the order interval $[-e, e]$ as its closed unit ball. In the sequel, unless otherwise stated, by the phrase E **is an** AM**-space with unit** we shall mean a Banach lattice whose norm is the $\|\cdot\|_\infty$ norm for some order unit $e > 0$.

We continue our discussion with a useful approximation lemma.

Lemma 4.22. *Let E be a Banach lattice. If $x' \wedge y' = 0$ holds in E', then for each $\epsilon > 0$ there exist $x, y \in U$ with $x \wedge y = 0$ and*
$$\|x'\| \le x'(x) + \epsilon \quad and \quad \|y'\| \le y'(y) + \epsilon.$$

Proof. Let $x', y' \in E'$ satisfy $x' \wedge y' = 0$, and let $\epsilon > 0$. Pick $u, v \in U$ with $\|x'\| \le x'(u) + \epsilon$ and $\|y'\| \le y'(v) + \epsilon$. From $[x' \wedge y'](u) = 0$, it follows that there exist $u_1, u_2 \in E^+$ with $u = u_1 + u_2$ and $x'(u_1) + y'(u_2) < \epsilon$. Similarly, there exists $v_1, v_2 \in E^+$ with $v = v_1 + v_2$ and $x'(v_1) + y'(v_2) < \epsilon$. Put $x = u_2 - v_1 \wedge u_2$ and $y = v_1 - v_1 \wedge u_2$, and note that $x, y \in U$ and $x \wedge y = 0$ both hold. On the other hand, observe that
$$\begin{aligned} x'(x) &= x'(u_2) - x'(v_1 \wedge u_2) \ge x'(u_2) - \epsilon = x'(u) - x'(u_1) - \epsilon \\ &\ge (\|x'\| - \epsilon) - \epsilon - \epsilon = \|x'\| - 3\epsilon. \end{aligned}$$
Similarly, $y'(y) \ge \|y'\| - 3\epsilon$ holds, and the conclusion follows. ∎

We now come to an important duality property between AL- and AM-spaces.

Theorem 4.23. *A Banach lattice E is an AL-space (resp. an AM-space) if and only if E' is an AM-space (resp. an AL-space).*

Proof. We show first that if E is an AL-space, then E' is an AM-space. To this end, assume that E is an AL-space, and let $x' \wedge y' = 0$ in E'. Put $m = \max\{\|x'\|, \|y'\|\}$, and note that $m \le \|x' + y'\|$ holds trivially. Now let $\epsilon > 0$. Choose some $x \in E^+$ with $\|x\| = 1$ and $\|x' + y'\| \le (x' + y')(x) + \epsilon$. Since $[x' \wedge y'](x) = 0$, there exist $u, v \in E^+$ with $u + v = x$ and $x'(u) + y'(v) < \epsilon$. From $(v - v \wedge u) \wedge (u - v \wedge u) = 0$, $0 \le u + v - 2(u \wedge v) \le x$, and the fact that E is an AL-space, it follows that
$$\|v - v \wedge u\| + \|u - v \wedge u\| = \|u + v - 2(v \wedge u)\| \le \|x\| \le 1,$$
and consequently
$$\begin{aligned} \|x' + y'\| &\le x'(x) + y'(x) + \epsilon = x'(v) + y'(u) + x'(u) + y'(v) + \epsilon \\ &\le x'(v) + y'(u) + 2\epsilon \le x'(v - v \wedge u) + y'(u - v \wedge u) + 3\epsilon \\ &\le m[\|v - v \wedge u\| + \|u - v \wedge u\|] + 3\epsilon \le m + 3\epsilon. \end{aligned}$$

Since $\epsilon > 0$ is arbitrary, $\|x' + y'\| \leq m$ also holds, and hence

$$\|x' \vee y'\| = \|x' + y'\| = \max\{\|x'\|, \|y'\|\}.$$

Next we show that the norm dual of an AM-space is an AL-space. To this end, let E be an AM-space, let $x' \wedge y' = 0$ in E', and let $\epsilon > 0$. By Lemma 4.22 there exist $x, y \in U$ with

$$x \wedge y = 0, \quad \|x'\| \leq x'(x) + \epsilon, \quad \text{and} \quad \|y'\| \leq y'(y) + \epsilon.$$

Since E is an AM-space, we have $\|x + y\| = \max\{\|x\|, \|y\|\} \leq 1$, and so

$$\|x'\| + \|y'\| \leq x'(x + y) + y'(x + y) + 2\epsilon \leq \|x' + y'\| \cdot \|x + y\| + 2\epsilon$$
$$\leq \|x' + y'\| + 2\epsilon \leq \|x'\| + \|y'\| + 2\epsilon$$

holds for all $\epsilon > 0$. Therefore, $\|x' + y'\| = \|x'\| + \|y'\|$. Thus, E' is an AL-space.

To complete the proof, note that if E' is an AL-space, then E'' is an AM-space, and hence the closed Riesz subspace E of E'' is likewise an AM-space. A similar observation is true when E' is an AM-space. ∎

It is a routine matter to verify that an AM-space E with unit e enjoys also the following useful property: *If $x' \in E'$, then $\|x'\| = |x'|(e)$.* This observation will be used in the proof of the next theorem which deals with pointwise limits of sequences of linear functionals.

Theorem 4.24. *Let E be a Banach lattice with order continuous norm, and let $x \in E^+$. Then for each norm bounded sequence $\{x'_n\}$ in E' there exist a subsequence $\{y'_n\}$ of $\{x'_n\}$ and some $x' \in E'$ such that $x'(y) = \lim y'_n(y)$ holds for each $0 \leq y \leq x$.*

Proof. Without loss of generality we can assume that $\{x'_n\} \subseteq E'_+$. By Theorem 4.21, the ideal E_x generated by x in E is an AM-space under the norm $\|y\|_\infty = \inf\{\lambda > 0 : |y| \leq \lambda x\}$. Let $J : E_x \to E$ denote the natural inclusion, i.e., $Jy = y$ for all $y \in E_x$. Then J is interval preserving and so, by Theorem 2.19, $J' : E' \to E'_x$ is a lattice homomorphism.

Now let $\epsilon > 0$. Since E has order continuous norm, there exists by Theorem 4.18 some $0 \leq y' \in E'$ such that $(x'_n - y')^+(x) < \epsilon$ holds for all n. Therefore,

$$\|(J'x'_n - J'y')^+\| = [J'(x'_n - y')]^+(x) = [J'(x'_n - y')^+](x)$$
$$= (x'_n - y')^+(Jx) = (x'_n - y')^+(x) < \epsilon$$

holds for all n. Using the identity $u = (u - v)^+ + u \wedge v$, we see that

$$\{J'x'_1, J'x'_2, \ldots\} \subseteq [0, J'y'] + \epsilon B, \qquad (\star)$$

where B is the closed unit ball of E'_x. By Theorem 4.23 we know that E'_x is an AL-space, and hence it has order continuous norm. Thus, $[0, J'y']$ is weakly compact (Theorem 4.9), and so from $(\star)$ and Theorem 3.44 it follows that $\{J'x'_n\}$ is a weakly relatively compact subset of E'_x. Pick a subsequence $\{y'_n\}$ of $\{x'_n\}$ and some $f \in E'_x$ such that $J'y'_n \xrightarrow{w} f$ holds in E'_x. Now if $x' \in E'$ is any w^*-accumulation point of $\{y'_n\}$, then note that

$$\lim_{n \to \infty} y'_n(y) = \lim_{n \to \infty} [J'y'_n](y) = x'(y)$$

holds for all $y \in [0, x]$. ∎

Let $\{x_n\}$ be a sequence in a Banach space X. Then recall that $\{x_n\}$ is said to be a **weak Cauchy sequence** whenever $\lim x'(x_n)$ exists in $\mathbb{R}$ for each $x' \in X'$. Note that (by the uniform boundedness principle) for this to happen, it is necessary and sufficient that $x_n \xrightarrow{w^*} x''$ holds in X'' for some $x'' \in X''$.

The next theorem deals with weak Cauchy sequences in Banach lattices.

Theorem 4.25. *Let E be a Banach lattice such that both E and E' have order continuous norms. Then every norm bounded sequence of E has a weak Cauchy subsequence.*

Proof. Let $\{x_n\}$ be an arbitrary norm bounded sequence of E and put $x = \sum_{n=1}^{\infty} 2^{-n}|x_n|$. By Theorem 4.15 there exists some $0 \le f \in E'$ which is strictly positive on $[0, x]$. Since E' has order continuous norm, there exist (by Theorem 4.24) a subsequence $\{y_n\}$ of $\{x_n\}$ and some $x'' \in E''$ such that $y_n(y') = y'(y_n) \to x''(y')$ holds for each $y' \in [0, f]$.

Now let A be the ideal generated by f in E', and let B be the band generated by f. Since E' has order continuous norm, B is the norm closure of A (see Theorem 1.38), and thus $y_n(y') \to x''(y')$ holds for all $y' \in B$.

Next, we claim that $g(x_n) = 0$ holds for all $g \in B^{\mathrm{d}}$. To see this, let $g \in B^{\mathrm{d}}$. Then, by Theorem 1.67, we see that $C_f \subseteq N_g$. On the other hand, since f is strictly positive on $[0, x]$, we have $E_x \cap N_f = \{0\}$. Therefore,

$$E_x \subseteq N_f^{\mathrm{d}} = C_f \subseteq N_g,$$

from which it follows that $g(x_n) = 0$ holds for all n and all $g \in B^{\mathrm{d}}$.

In view of $E' = B \oplus B^{\mathrm{d}}$, we see that $\lim y'(y_n)$ exists in $\mathbb{R}$ for each $y' \in E'$. In other words, $\{y_n\}$ is a weak Cauchy subsequence of $\{x_n\}$, and the proof is finished. ∎

Examples of abstract L_p-spaces are provided by the Banach lattices $L_p(\mu)$ for $1 \le p < \infty$. Our next objective is to show that, in fact, these are the only type of abstract L_p-spaces. To do this, we need to recall a few things about Boolean algebras.

Recall that a complemented distributive lattice is referred to as **Boolean algebra**. Two Boolean algebras $\mathcal{A}$ and $\mathcal{B}$ are said to be **isomorphic** whenever there is a one-to-one mapping from $\mathcal{A}$ onto $\mathcal{B}$ preserving the lattice operations as well as the complements. If Ω is a topological space, then the collection of all subsets of Ω that are simultaneously closed and open (called **clopen** sets) is a Boolean algebra under the usual operations of union, intersection, and complementation. Remarkably, every Boolean algebra $\mathcal{A}$ is isomorphic to some Boolean algebra of clopen sets (called a representation of $\mathcal{A}$).

We are interested in the representations of Dedekind complete Boolean algebras. A Boolean algebra $\mathcal{A}$ is said to be **Dedekind complete** whenever every nonempty subset of $\mathcal{A}$ has a supremum. Also, recall that a topological space Ω is said to be **extremally disconnected** if the closure of every open set is itself an open set. The following classical theorem of M. H. Stone [**180**] presents a link between Dedekind complete Boolean algebras and Hausdorff compact extremally topological spaces. (Exercise 20 at the end of the section gives an indication of how one can prove this important representation theorem.)

Theorem 4.26 (Stone). *A Boolean algebra is Dedekind complete if and only if it is isomorphic to the Boolean algebra of all clopen subsets of a (unique up to homeomorphism) Hausdorff, compact and extremally disconnected topological space.*

Let $\mathcal{A}$ be a Dedekind complete Boolean algebra. Then the Boolean algebra of all clopen subsets of the unique Hausdorff, compact and extremally disconnected topological space that is isomorphic to $\mathcal{A}$ is known as the **Stone space** of $\mathcal{A}$.

Now consider a Hausdorff, compact and extremally disconnected topological space Ω, and denote by Σ the Boolean algebra of all clopen subsets of Ω. Since every set in Σ is necessarily compact, it is easy to see that every finitely additive measure on Σ is automatically σ-additive, i.e., it is a measure. This observation will be used in the proof of the next theorem that characterizes the abstract L_p-spaces. For AL-spaces the result is due to S. Kakutani [**81**]. For $1 < p < \infty$ the representation theorem was established by H. F. Bohnenblust [**40**] and H. Nakano [**151**] under some additional conditions.

Theorem 4.27 (Kakutani–Bohnenblust–Nakano). *A Banach lattice E is an abstract L_p-space for some $1 \leq p < \infty$ if and only if E is lattice isometric to some concrete $L_p(\mu)$-space.*

Proof. The "only if" part needs proof. To this end, let E be an abstract L_p-space for some $1 \leq p < \infty$. Assume at the beginning that E has also a weak order unit $e > 0$. Let $\mathcal{B}$ denote the Dedekind complete Boolean algebra of all components of e, i.e., $\mathcal{B} = \{x \in E \colon x \wedge (e - x) = 0\}$ (see Theorem 1.49). Also let E_0 denote the Riesz subspace of E consisting of all e-step functions of E. That is, s belongs to E_0 if and only if there exist pairwise disjoint components $e_1, \ldots, e_n$ of e with $\sum_{i=1}^n e_i = e$ and real constants $\alpha_1, \ldots, \alpha_n$ satisfying $s = \sum_{i=1}^n \alpha_i e_i$. By Theorem 2.8 we know that E_0 is order dense in E. Since E has order continuous norm, E_0 is also norm dense in E. By Theorem 4.26 there exists a Hausdorff, compact and extremally disconnected topological space Ω such that $\mathcal{B}$ is isomorphic to the Boolean algebra Σ of all clopen subsets of Ω. Let $x \mapsto S_x$ be this isomorphism. Next, define a set function $\mu \colon \Sigma \to [0, \infty)$ by $\mu(S_x) = \|x\|^p$, and note that the p-additivity of the norm implies that μ is a finitely additive measure, and hence μ is a measure on Σ. Thus, we can apply the Carathéodory extension procedure and obtain the σ-algebra of all μ-measurable subsets of Ω; see [**8**, Chapter 3]. Let L_0 be the Riesz subspace of $L_p(\mu)$ consisting of all step functions of the measure space (Ω, Σ, μ). Clearly, L_0 is order (and hence norm) dense in $L_p(\mu)$.

Now consider the mapping $T \colon E_0 \to L_0$ defined by

$$T\left(\sum_{i=1}^n \alpha_i e_i\right) = \sum_{i=1}^n \alpha_i \chi_{S_{e_i}} .$$

It is not difficult to see that T is well defined, i.e, $T(s)$ depends only upon s and not on its particular representation as an e-step function. Also, it is easy to see that T is an onto lattice isometry. To see that T is onto, let S be a μ-measurable subset of Ω. If S is a σ-set, then there exists a sequence $\{x_n\}$ of $\mathcal{B}$ with $x_n \uparrow$ and $S = \bigcup_{n=1}^\infty S_{x_n}$. It follows that $x_n \uparrow x$ holds in $\mathcal{B}$ (and in E), and so by the order continuity of the norm, we have

$$\mu^*(S) = \lim_{n \to \infty} \mu(S_{x_n}) = \lim_{n \to \infty} \|x_n\|^p = \|x\|^p = \mu(S_x) .$$

Therefore, the vector $x \in \mathcal{B}$ satisfies $S \subseteq S_x$ and $\mu^*(S) = \mu(S_x)$. Now for the general case, pick a sequence $\{y_n\} \subseteq \mathcal{B}$ with $S \subseteq S_{y_{n+1}} \subseteq S_{y_n}$ for all n and $\mu^*(S) = \lim \mu(S_{y_n})$, and then note that the vector $y = \inf\{y_n\} \in \mathcal{B}$ satisfies $S \subseteq S_y$ and $\mu^*(S) = \mu(S_y)$. Therefore, $\chi_S = \chi_{S_y}$ holds, and this shows that $T \colon E_0 \to L_0$ is onto. Finally, since E_0 and L_0 are both order and norm dense in E and $L_p(\mu)$, respectively, it is easy to see that T extends to a lattice isometry from E onto $L_p(\mu)$.

If E does not have a weak unit, consider a maximal disjoint family $\{e_i \colon i \in I\}$ of nonzero positive vectors of E. Let $B_i = B_{e_i}$, the principal band generated by e_i in E. By the preceding case, for each $i \in I$ there exists a measure space $(\Omega_i, \Sigma_i, \mu_i)$ and an onto lattice isometry $T_i \colon B_i \to L_p(\mu_i)$.

We can assume that $\Omega_i \cap \Omega_j = \emptyset$ for $i \neq j$. Put

$$\Omega = \bigcup_{i \in I} \Omega_i, \quad \Sigma = \{S \subseteq \Omega : \ S \cap \Omega_i \in \Sigma_i \text{ for all } i \in I\},$$

and

$$\mu(S) = \sum_{i \in I} \mu_i(S \cap \Omega_i) \text{ for all } S \in \Sigma,$$

and observe that (Ω, Σ, μ) is a measure space. Next, for each $x \in E$ put $x_i = P_{B_i}(x)$, and note that $|x| = \sup\{|x_i| : \ i \in I\}$ holds. By the order continuity and p-additivity of the norm, we see that $\|x\|^p = \sum_{i \in I} \|x_i\|^p$ holds, from which it easily follows that the mapping $T \colon E \to L_p(\mu)$, defined by $T(x) = \sum_{i \in I} T_i(x_i)$ is a lattice isometry from E onto $L_p(\mu)$. The proof of the theorem is now complete. ∎

Now let E be an AL-space. Then from Theorem 4.27 it should be clear that E has an additive norm. That is, $\|x + y\| = \|x\| + \|y\|$ holds for all $x, y \in E^+$. This, coupled with Lemma 1.10, shows that the formula

$$e(x) = \|x^+\| - \|x^-\|, \quad x \in E,$$

defines a positive linear functional on E. On the other hand, the inequality

$$|x'|(|x|) \leq \|x'\| \cdot \|x\| = \|x'\| e(|x|)$$

shows that e is an order unit of E' and that $U' = [-e, e]$ holds. In other words, we have shown that:

- *The norm dual of an AL-space is an AM-space with unit.*

For the representations of AM-spaces, we shall need the following result.

Theorem 4.28. *Let E be an AM-space, and let $0 \leq x' \in E'$ satisfy $\|x'\| = 1$. Then x' is an extreme point of U'_+ if and only if x' is a lattice homomorphism (from E to $\mathbb{R}$).*

Proof. Assume first that x' is an extreme point of U'_+. If $0 < y' < x'$ holds, then

$$x' = \|y'\| \cdot \frac{y'}{\|y'\|} + \|x' - y'\| \cdot \frac{x' - y'}{\|x' - y'\|},$$

and since E' is an AL-space, we have $\|y'\| + \|x' - y'\| = \|x'\| = 1$, and so $y' = \|y'\| x'$. Thus, if $|y'| \leq x'$ holds, then there exists some λ with $|\lambda| \leq 1$ and $y' = \lambda x'$. Now let $x \in E$. Then by Theorem 1.23 there exists some $y' \in E'$ with $|y'| \leq x'$ and $x'(|x|) = |y'(x)|$, and so from

$$|x'(x)| \leq x'(|x|) = |y'(x)| = |\lambda x'(x)| \leq |x'(x)| \leq x'(|x|),$$

we get $|x'(x)| = x'(|x|)$. That is, x' is a lattice homomorphism.

For the converse assume that x' is a lattice homomorphism. Note first that if $0 \leq y' \leq x'$ holds and $x \in \operatorname{Ker} x'$, then it follows from

$$|y'(x)| \leq y'(|x|) \leq x'(|x|) = |x'(x)| = 0$$

that $x \in \operatorname{Ker} y'$. That is, $\operatorname{Ker} x' \subseteq \operatorname{Ker} y'$, and so by Lemma 3.15 there exists some $0 \leq \lambda \leq 1$ with $y' = \lambda x'$. Now assume that $x' = \alpha y' + (1 - \alpha)z'$ holds for some $y', z' \in U'_+$ and some $0 < \alpha < 1$. Clearly, $\|y'\| = \|z'\| = 1$. On the other hand, from $0 \leq \alpha y' \leq x'$ and $0 \leq (1 - \alpha)z' \leq x'$ (and the above discussion), there exist $\beta, \gamma > 0$ with $y' = \beta x'$ and $z' = \gamma x'$. It follows that $\beta = \gamma = 1$, so that x' is an extreme point of U'_+. ∎

If Ω is a Hausdorff compact topological space, then $C(\Omega)$ with the sup norm, i.e., with the norm $\|f\|_\infty = \sup\{|f(\omega)|\colon \omega \in \Omega\}$, is an AM-space having unit the constant function one. It is remarkable that the $C(\Omega)$ Banach lattices (with Ω Hausdorff and compact) are the only type of AM-spaces with unit. This important result was proved by S. Kakutani [82] and was extended by H. F. Bohnenblust and S. Kakutani [41]. The same result was established independently by M. Krein and S. Krein [101].

Theorem 4.29 (Kakutani–Bohnenblust and M. Krein–S. Krein). *A Banach lattice E is an AM-space with unit if and only if is lattice isometric to some $C(\Omega)$ for a (unique up to homeomorphism) Hausdorff compact topological space Ω.*

In particular, a Banach lattice is an AM-space if and only if it is lattice isometric to a closed Riesz subspace of some $C(\Omega)$-space.

Proof. Assume first that $E = C(\Omega)$ holds for some Hausdorff compact topological space Ω. Let us examine next the role of Ω in connection with the lattice structure of E. Denote by e the constant function one on Ω.

Consider the set

$$K = \left\{x' \in U'_+\colon \ x' \text{ is an extreme point of } U'_+ \text{ with } \|x'\| = x'(e) = 1\right\}. \ (\star)$$

By Theorem 4.28 we know that

$$K = \left\{x' \in U'_+\colon \ x' \text{ is a lattice homomorphism with } \|x'\| = x'(e) = 1\right\}, \ (\star\star)$$

from which it easily follows that K is w^*-closed, and hence w^*-compact. On the other hand, if $x' \in K$, then by Theorem 2.33 there exists a unique $\omega \in \Omega$ satisfying $x' = \delta_\omega$. Thus, a mapping $\omega \mapsto \delta_\omega$, from Ω onto (K, w^*), is established, which is easily seen to be one-to-one and continuous (and hence a homeomorphism). In other words, Ω can be interpreted as playing the role of all nonzero extreme points of U'_+ with the w^*-topology. In particular, if $C(\Omega)$ is lattice isometric to some $C(\Omega_1)$, then Ω and Ω_1 must be homeomorphic.

Now let E be an AM-space with unit $e > 0$. By the preceding discussion, for E to be lattice isometric to some $C(\Omega)$ there is only one choice for Ω; namely, Ω must be homeomorphic to K as defined by $(\star)$. So, consider K as in $(\star)$ (which by Theorem 3.14 is nonempty), and note that $(\star\star)$ holds. Thus, (K, w^*) is a compact topological space. Consider the mapping $T\colon E \to C(K)$ defined by

$$[Tx](x') = x'(x), \quad x \in E \text{ and } x' \in K.$$

Since the set of extreme points of U'_+ is $K \cup \{0\}$, it follows from Theorem 3.14 that T is a norm preserving lattice isomorphism from E to $C(K)$. Also, $[Te](x') = x'(e) = 1$ holds for all $x' \in K$, and clearly $T(E)$ separates the points of K. Since $T(E)$ is closed, it follows from the classical Stone–Weirstrass theorem (see for instance [**8**, Theorem 11.3, p. 88]) that T is onto. Thus, E is lattice isometric to $C(K)$ as desired.

Finally, to establish the last claim of the theorem, note that if E is an AM-space, then E' is an AL-space, and so E'' is an AM-space with unit including E as a closed Riesz subspace. ∎

The next two results are consequences of the preceding theorem and describe some important properties of AM-spaces. The first one is due to U. Krengel [**105**].

Theorem 4.30 (Krengel). *If A is a nonempty norm totally bounded subset of an AM-space E, then the set D of all finite suprema of A is likewise norm totally bounded. In particular, $\sup A$ exists in E and $\sup A \in \overline{D}$ holds.*

Proof. By Theorem 4.29 we can assume that E is a closed Riesz subspace of some $C(\Omega)$-space with Ω Hausdorff and compact. Thus, A can be considered as a norm totally bounded subset of $C(\Omega)$, which according to the classical Ascoli–Arzelà theorem means that A is norm bounded and equicontinuous; see for instance [**8**, Theorem 9.10, p. 75].

Now let $g \in D$. Pick $f_1, \ldots, f_n \in A$ with $g = \bigvee_{i=1}^n f_i$. Since A is equicontinuous on Ω, given $\epsilon > 0$ and $\omega \in \Omega$, there exists a neighborhood V of ω such that $|f(t) - f(\omega)| < \epsilon$ holds for all $t \in V$ and all $f \in A$. From the inequality

$$\big| g(t) - g(\omega) \big| \leq \max\{|f_i(t) - f_i(\omega)| \colon \ i = 1, \ldots, n\},$$

it follows that $|g(t) - g(\omega)| < \epsilon$ holds for all $t \in V$ (and all $g \in D$). That is, D is also equicontinuous on Ω, and since D is clearly norm bounded, it follows (from the Ascoli–Arzelà theorem again) that D is likewise norm totally bounded.

Finally, let $\mathcal{A}$ denote the set of all finite subsets of A directed by the inclusion $\subseteq$. For each $\alpha \in \mathcal{A}$ put $g_\alpha = \sup \alpha$, and note that the net $\{g_\alpha\} \subseteq D$

satisfies $g_\alpha \uparrow$. By the norm compactness of $\overline{D}$ there exists a subnet of $\{g_\alpha\}$ that converges in norm to some $g \in \overline{D}$, and from this it easily follows that $\sup A = \sup D = \sup\{g_\alpha\} = g$. ∎

A Banach lattice E is said to have **weakly sequentially continuous lattice operations** whenever $x_n \xrightarrow{w} 0$ in E implies $|x_n| \xrightarrow{w^*} 0$ in E.

In case $E = C(\Omega)$ for some Hausdorff compact topological space Ω, an easy application of the Riesz representation theorem shows that a sequence $\{f_n\} \subseteq E$ satisfies $f_n \xrightarrow{w} 0$ in E if and only if $\{f_n\}$ is norm bounded and $f_n(\omega) \to 0$ holds for each $\omega \in \Omega$. Therefore, $f_n \xrightarrow{w} 0$ in $C(\Omega)$ implies $|f_n| \xrightarrow{w} 0$, and so by Theorem 4.29 every AM-space has weakly sequentially continuous lattice operations. Thus (from Theorem 3.40), the following result should be immediate.

Theorem 4.31. *In an AM-space the lattice operations are weakly sequentially continuous.*

In particular, if A is a weakly relatively compact subset of an AM-space, then $|A|$, A^+, and A^- are likewise weakly relatively compact subsets.

The interested reader will find more on Banach lattices in the books by H. H. Schaefer [**174**], H. E. Lacey [**110**], and J. Lindenstrauss and L. Tzafriri [**112, 113**]. Also, the reader will benefit by reading the survey articles [**46**] and [**47**].

Exercises

1. If $\{X_n\}$ is a sequence of Banach spaces, then show that:
   (a) $(X_1 \oplus X_2 \oplus \cdots)_1' = (X_1' \oplus X_2' \oplus \cdots)_\infty$.
   (b) $(X_1 \oplus X_2 \oplus \cdots)_0' = (X_1' \oplus X_2' \oplus \cdots)_1$.

2. Show that for each $1 \le p < \infty$ the Banach lattice $(\ell_p \oplus \ell_p \oplus \cdots)_p$ is lattice isometric to ℓ_p.

3. This exercise shows that a positive linear functional on a normed Riesz space need not be continuous. Let E be the Riesz space of all real sequences that are eventually zero. Show that:
   (a) Under the sup norm E is a normed Riesz space but not a Banach lattice.
   (b) The norm completion of E is c_0.
   (c) The formula

$$f(x_1, x_2, \ldots) = \sum_{n=1}^{\infty} n x_n$$

   defines a positive linear functional on E that fails to be continuous.

4. Show that every separable Banach lattice admits a strictly positive linear functional. [*Hint:* Pick a countable w^*-dense subset $\{x_1', x_2', \ldots\}$ of U_+', and consider the positive linear functional $x' = \sum_{n=1}^{\infty} 2^{-n} x_n'$.]

5. Recall that a Banach lattice is said to have σ-**order continuous norm** whenever $x_n \downarrow 0$ implies $\|x_n\| \downarrow 0$.
   (a) Give an example of a Banach lattice whose norm is σ-order continuous but not order continuous.
   (b) Show that a Banach lattice E has σ-order continuous norm if and only if $E_c^{\sim} = E'$.

6. (Aliprantis–Burkinshaw–Kranz [18]) Show that a Dedekind σ-complete Banach lattice E has order continuous norm if and only if $0 \leq T_n \uparrow T$ in $\mathcal{L}_b(E)$ implies $T_n^2 \uparrow T^2$. [*Hint:* Use Theorem 1.79.]

7. Let A be an ideal of a Banach lattice E, and let p be a lattice norm on A such that the natural embedding $J: (A, p) \to E$ (i.e., $Jx = x$ for all $x \in A$) is continuous. If $\hat{A}$ denotes the norm completion of (A, p), then show that the unique continuous extension of J to all of $\hat{A}$ is an interval preserving lattice homomorphism. [*Hint:* Denote by J again the extension, and note that the extension is clearly a lattice homomorphism. To see that J is interval preserving, assume that $0 \leq x \in \hat{A}$ and $y \in E^+$ satisfy $0 \leq y \leq Jx$. Pick a sequence $\{x_n\} \subseteq A^+$ with $p(x - x_n) \to 0$, and note that

$$\|y \wedge x_n - y\| = \|y \wedge Jx_n - y \wedge Jx\| \leq \|Jx_n - Jx\| \to 0. \qquad (\star)$$

On the other hand, the inequality $|y \wedge x_n - y \wedge x_m| \leq |x_n - x_m|$ shows that $\{y \wedge x_n\}$ is a p-Cauchy sequence of A^+, and so for some $0 \leq z \in \hat{A}$ we have $p(y \wedge x_n - z) \to 0$. From $0 \leq y \wedge x_n \leq x_n$, it follows that $0 \leq z \leq x$. Finally, from $(\star)$ and $\|y \wedge x_n - Jz\| = \|J(y \wedge x_n) - Jz\| \to 0$, we see that $y = Jz$ holds.]

8. Let E be a Riesz space. Then for a subset A of E define

$$A^{\vee} := \left\{ x \in E: \ \exists x_1, \ldots, x_n \in A \ \text{with} \ x = \bigvee_{i=1}^{n} x_i \right\},$$

and

$$A^{\wedge} := \left\{ x \in E: \ \exists x_1, \ldots, x_n \in A \ \text{with} \ x = \bigwedge_{i=1}^{n} x_i \right\},$$

Also, we let $A^{\vee\wedge} := (A^{\vee})^{\wedge}$ and $A^{\wedge\vee} := (A^{\wedge})^{\vee}$. Show that:
   (a) $A^{\vee\wedge} = A^{\wedge\vee}$.
   (b) If A is a vector subspace, then $A^{\vee\wedge}$ is the Riesz subspace generated by A (i.e., $A^{\vee\wedge}$ is the smallest Riesz subspace including A) and

$$A^{\vee\wedge} = A^{\vee} - A^{\vee} = A^{\wedge} - A^{\wedge}.$$

9. Show that every separable vector subspace of a Banach lattice is included in a separable Banach sublattice. [*Hint:* Let X be a separable vector subspace of a Banach lattice E. If C is a countable dense subset of X,

then $C^{\wedge\vee}$ (for notation see the previous exercise) is countable, and by the continuity of the lattice operations $C^{\wedge\vee}$ is norm dense in $\overline{X^{\wedge\vee}}$.]

10. This exercise adds several equivalent statements to Theorem 4.13. They are due to O. Burkinshaw [48], D. H. Fremlin [64], P. G. Dodds and D. H. Fremlin [54], and P. Meyer-Nieberg [142].

    For a lattice seminorm p on a Riesz space E and a vector $x \in E^+$ show that the following statements are equivalent.
    (a) If $\{x_n\}$ is a disjoint sequence of $[0, x]$, then $\lim p(x_n) = 0$.
    (b) If $0 \leq x_n \uparrow \leq x$ holds in E, then $\{x_n\}$ is a p-Cauchy sequence.
    (c) For each $\epsilon > 0$ there exists some $0 \leq g \in E^\sim$ such that whenever $f \in E^\sim$ satisfies $|f(y)| \leq p(y)$ for all $y \in E$, then we have

    $$(|f| - g)^+(x) < \epsilon.$$

    (d) For each $\epsilon > 0$ there exist $0 \leq g \in E^\sim$ and $\delta > 0$ such that $y \in [0, x]$ and $g(y) < \delta$ imply $p(y) < \epsilon$.
    (e) If $\{x_n\}$ is a sequence of $[0, x]$ with $x_n \xrightarrow{\sigma(E, E^\sim)} 0$, then $\lim p(x_n) = 0$.
    (f) If $\{f_n\}$ is a disjoint sequence of $E^\sim$ satisfying $|f_n(y)| \leq p(y)$ for all $y \in E$ and all n, then $\lim |f_n|(x) = 0$.
    (g) If $\{x_n\}$ is a disjoint sequence of $[0, x]$ and $\{f_n\}$ is a disjoint sequence of $E^\sim$ satisfying $|f_n(y)| \leq p(y)$ for all $y \in E$ and all n, then $\lim f_n(x_n) = 0$.

11. Show that a Banach lattice E has order continuous norm if and only if $|\sigma|(E', E)$ is consistent with the Riesz dual system $\langle E, E' \rangle$.

12. If E is an Archimedean uniformly complete Riesz space, then show that every principal ideal E_x of E is an AM-space under the norm

    $$\|y\|_\infty = \min\{\lambda \geq 0: \; |y| \leq \lambda|x|\}.$$

13. Let A be a normed closed ideal of a normed Riesz space E. Show that:
    (a) The quotient Riesz space E/A under the quotient norm

    $$\|\dot{x}\| = \inf\{\|y\|: \; y \in \dot{x}\}$$

    is a normed Riesz space.
    (b) If the norm of E is order continuous, then the quotient norm on E/A also is order continuous.
    (c) If E is an AM-space (resp. an AL-space), then E/A is likewise an AM-space (resp. an AL-space).

14. If E is a normed Riesz space, then show that:
    (a) The band generated by E in E'' is precisely $(E')^\sim_n$.
    (b) $(E')^\sim_n = E^{\uparrow\downarrow\uparrow\downarrow}$ (see Section 2.1 for the notation).

15. Show that for each $0 < \epsilon < 1$ there exists a sequence $\{f_n\} \subseteq C[0, 1]$ with the following properties:
    (a) $0 \leq f_n \leq 1$.
    (b) $f_n \downarrow 0$ in $C[0, 1]$.
    (c) $\int_0^1 f_n(x)\, dx \geq 1 - \epsilon$ holds for all n.
    [Hint: Use the Cantor set of Lebesgue measure ϵ.]

16. Let Ω be a Hausdorff compact topological space. Then show that $C(\Omega)$ is a Dedekind complete Riesz space if and only if Ω is extremally disconnected (i.e., if and only if the closure of every open set is also open).

17. Show that a Hausdorff compact topological space Ω is metrizable if and only if $C(\Omega)$ is a separable Banach lattice.

18. (Bohnenblust [**40**]) Let $c_0(\Omega)$ denote the vector space of all real-valued functions f on a nonempty set Ω such that for each $\epsilon > 0$ there exists a finite subset Φ of Ω (depending upon f and ϵ) satisfying $|f(\omega)| < \epsilon$ for all $\omega \notin \Phi$.
   (a) Show that under the sup norm $c_0(\Omega)$ is an AM-space with order continuous norm.
   (b) Show that $c_0(\Omega)$ is an AM-space with unit if and only if Ω is finite.
   (c) Show that if E is an AM-space with order continuous norm, then there exists a nonempty set Ω so that E is lattice isometric to $c_0(\Omega)$.

19. Let E be a Banach lattice. Show that E and E' both have order continuous norms if and only if E is an order dense ideal of E''.

20. This exercise presents by steps the Stone representation theorem for Boolean algebras [**180**].

Let $\mathcal{B}$ be a Boolean algebra. A nonempty subset J of $\mathcal{B}$ is said to be an **ideal** whenever
   (i) for each $x, y \in J$ we have $x \vee y \in J$, and
   (ii) $x \leq y$ and $y \in J$ imply $x \in J$.
If $J \neq \mathcal{B}$ holds, then J is called a **proper ideal**. A proper ideal J of $\mathcal{B}$ is said to be a **maximal ideal** whenever the only ideal that includes J properly is $\mathcal{B}$ itself. Finally, a proper ideal J is said to be a **prime ideal** if $x \wedge y \in J$ implies either $x \in J$ or $y \in J$.
   (a) Show that a proper ideal of $\mathcal{B}$ is a prime ideal if and only if it is a maximal ideal.
   (b) If an ideal J of $\mathcal{B}$ is a maximal ideal with respect to the property of not including an element x, then show that J is a prime ideal.
   (c) If J is an ideal of $\mathcal{B}$ and $x \notin J$, then show that there exists a prime ideal I with $J \subseteq I$ and $x \notin I$.
   (d) Let Ω denote the collection of all proper prime ideals of $\mathcal{B}$. For each $x \in \mathcal{B}$ put $\Omega_x = \{\omega \in \Omega: x \notin \omega\}$. Show that $\{\Omega_x: x \in \mathcal{B}\}$ forms a base for a Hausdorff compact topology τ on Ω called the **hull-kernel topology**. (The topological space (Ω, τ) is referred to as the **Stone space** of $\mathcal{B}$.)
   (e) Show that a subset A of Ω is clopen with respect to τ if and only if there exists some $x \in \mathcal{B}$ with $A = \Omega_x$.
   (f) Show that $x \mapsto \Omega_x$ is an isomorphism from $\mathcal{B}$ onto the Boolean algebra of all clopen subsets of (Ω, τ).
   (g) Show that (Ω, τ) is extremally disconnected if and only if $\mathcal{B}$ is a Dedekind complete Boolean algebra.
   (h) Show that if Ω is a Hausdorff, compact, extremally disconnected topological space, then there exists a unique (up to isomorphism)

Dedekind complete Boolean algebra whose Stone space is homeo-morphic to Ω.

4.2. Weak Compactness in Banach Lattices

On an infinite dimensional normed Riesz space, the weak topology is not locally solid and the lattice operations are seldom weakly sequentially continuous. For these reasons, relationships between weak compactness and the lattice structure are very subtle. However, there are some important connections between weak compactness and the order structure of a Banach lattice, and they will be discussed in this section. Our approach will be based upon two fundamental properties of the Banach lattice ℓ_1.

Recall that a Banach space is said to have the **Schur property** whenever every weak convergent sequence is norm convergent, i.e., whenever $x_n \xrightarrow{w} 0$ implies $\|x_n\| \to 0$. Note that (by Theorem 3.40) a Banach space has the Schur property if and only if every weakly compact set is norm compact.

S. Banach [**30**] has proved that ℓ_1 has the Schur property. This will be a basic result for the study of weak compactness in Banach lattices.

Theorem 4.32 (Banach). *The Banach lattice ℓ_1 has the Schur property, i.e., $x_n \xrightarrow{w} 0$ in ℓ_1 implies $\|x_n\|_1 \to 0$.*

Proof. Let $x_n = (x_1^n, x_2^n, \ldots) \xrightarrow{w} 0$ in ℓ_1, and assume by way of contradiction that $\|x_n\|_1 \nrightarrow 0$. Then, by passing to a subsequence, we can assume without loss of generality that for some $\epsilon > 0$ we have

$$\|x_n\|_1 = \sum_{i=1}^{\infty} |x_i^n| > 5\epsilon$$

for all n. Keep in mind that (in view of $\ell_1' = \ell_\infty$) we have $x_i^n \xrightarrow[n \to \infty]{} 0$ for each i.

Now we claim that there exist two strictly increasing sequences $\{k_n\}$ and $\{m_n\}$ of natural numbers such that for each n we have:

(a) $\sum_{i=1}^{k_n} |x_i^{m_n}| < \epsilon$.

(b) $\sum_{i=k_n+1}^{k_{n+1}} |x_i^{m_n}| > 3\epsilon$.

(c) $\sum_{i>k_{n+1}} |x_i^{m_n}| < \epsilon$.

To see this, we shall use induction. Start with $k_1 = 1$, and then choose an integer $m_1 \geq 1$ with $|x_1^{m_1}| < \epsilon$ (this is possible since $x_1^n \xrightarrow[n \to \infty]{} 0$). Now for the inductive argument: Assume $k_1 < \cdots < k_n$ and $m_1 < \cdots < m_n$ satisfy (a), (b), and (c). Since $x_{m_n} \in \ell_1$, there exists some $k_{n+1} > k_n$ such that

$\sum_{i>k_{n+1}} |x_i^{m_n}| < \epsilon$. From $\lim_{m\to\infty} \sum_{i=1}^{k_{n+1}} |x_i^m| = 0$, it follows that for some $m_{n+1} > m_n$ we have $\sum_{i=1}^{k_{n+1}} |x_i^{m_{n+1}}| < \epsilon$. On the other hand, note that

$$\sum_{i=k_n+1}^{k_{n+1}} |x_i^{m_n}| = \sum_{i=1}^{\infty} |x_i^{m_n}| - \sum_{i=1}^{k_n} |x_i^{m_n}| - \sum_{i>k_{n+1}} |x_i^{m_n}| > 5\epsilon - \epsilon - \epsilon = 3\epsilon,$$

and the induction is finished.

Next put $y_1 = 1$, and for each $k_n < i \le k_{n+1}$ let $y_i \in \{-1, 1\}$ be such that $y_i x_i^{m_n} = |x_i^{m_n}|$. Then $y = (y_1, y_2, \ldots) \in \ell_\infty = \ell_1'$, and

$$|\langle x_{m_n}, y \rangle| = \left| \sum_{i=1}^{\infty} y_i x_i^{m_n} \right| \ge \sum_{i=k_n+1}^{k_{n+1}} |x_i^{m_n}| - \sum_{i=1}^{k_n} |x_i^{m_n}| - \sum_{i>k_{n+1}} |x_i^{m_n}|$$

$$> 3\epsilon - \epsilon - \epsilon = \epsilon$$

for all n. However, this contradicts $x_{m_n} \xrightarrow{w} 0$, and so $\lim \|x_n\|_1 = 0$ holds, as desired. ∎

The next simple characterization of the norm totally bounded subsets of ℓ_1 will be needed for our discussion. Keep in mind that (by the previous theorem) a subset of ℓ_1 is norm compact if and only if it is weakly compact.

Theorem 4.33. *For a norm bounded subset A of ℓ_1 the following statements are equivalent:*

(1) *A is norm totally bounded.*

(2) *For each $\epsilon > 0$ there exists some n such that $\sum_{i=n}^{\infty} |x_i| < \epsilon$ holds for each $x = (x_1, x_2, \ldots) \in A$.*

Proof. (1) $\Longrightarrow$ (2) Let $\epsilon > 0$. Pick a finite subset Φ of A with $A \subseteq \Phi + \epsilon U$. Since Φ is a finite set, it is easy to see that there exists some n satisfying $\sum_{i=n}^{\infty} |y_i| < \epsilon$ for each $y = (y_1, y_2, \ldots) \in \Phi$. Now if $x = (x_1, x_2, \ldots) \in A$, then pick $y \in \Phi$ and $u \in U$ with $x = y + \epsilon u$, and note that

$$\sum_{i=n}^{\infty} |x_i| \le \sum_{i=n}^{\infty} |y_i| + \epsilon \sum_{i=n}^{\infty} |u_i| < \epsilon + \epsilon = 2\epsilon.$$

(2) $\Longrightarrow$ (1) Put $s_k = \sup\{|x_k|: x = (x_1, x_2, \ldots) \in A\}$. Since A is norm bounded, each s_k is a nonnegative real number. Now let $\epsilon > 0$. Pick some n so that (2) holds, and consider the set

$$B = [-s_1, s_1] \times \cdots \times [-s_n, s_n] \times \{0\} \times \{0\} \times \cdots .$$

Clearly, B is a closed and bounded subset lying in a finite dimensional vector subspace of ℓ_1, and hence B is a compact subset of ℓ_1. On the other hand, $A \subseteq B + \epsilon U$ holds, and this easily implies that A is a norm totally bounded subset of ℓ_1 (see also Theorem 3.1). ∎

We start our discussion on the weak compactness in Banach lattices with an important relationship between weakly compact sets and disjoint sequences.

Theorem 4.34. *If W is a weakly relatively compact subset of a Banach lattice, then every disjoint sequence in the solid hull of W converges weakly to zero.*

Proof. Let W be a relatively weakly compact subset of a Banach lattice E, and let $\{x_n\} \subseteq E$ be a disjoint sequence lying in the solid hull of W. Pick a sequence $\{y_n\} \subseteq W$ satisfying $|x_n| \leq |y_n|$ for all n, and let $\epsilon > 0$ and $0 \leq x' \in E'$ be fixed.

Consider each x_n as an element of E'', and denote by P_n the order projection of E' onto the carrier C_{x_n} of x_n. From $x_n \perp x_m$ and Theorem 1.67, we see that $P_n x' \perp P_m x'$ holds for $n \neq m$. Also, from Theorem 1.23 we have

$$
\begin{aligned}
\left|x'(x_n)\right| &\leq x'(|x_n|) = [P_n x'](|x_n|) \leq [P_n x'](|y_n|) \\
&= \max\{y'(y_n) \colon |y'| \leq P_n x'\},
\end{aligned}
$$

and so for each n there exists some $y'_n \in E'$ with $|y'_n| \leq P_n x'$ and

$$
\left|x'(x_n)\right| \leq y'_n(y_n). \tag{$\star$}
$$

Next, note that for each $x \in E$ and each k we have

$$
\sum_{i=1}^{k} |y'_i(x)| \leq \left[\sum_{i=1}^{k} P_i x'\right](|x|) \leq x'(|x|),
$$

and so $(y'_1(x), y'_2(x), \ldots) \in \ell_1$. Now define the operator $T \colon E \to \ell_1$ by

$$
Tx = (y'_1(x), y'_2(x), \ldots).
$$

The inequality $\|Tx\|_1 = \sum_{n=1}^{\infty} |y'_n(x)| \leq x'(|x|) \leq \|x'\| \cdot \|x\|$ guarantees that T is continuous. In particular, $T(W)$ is a weakly relatively compact subset of ℓ_1.

Since (by Theorem 4.32) weak and norm convergence of sequences in ℓ_1 coincide, we see that $T(W)$ is a norm totally bounded subset of ℓ_1. By Theorem 4.33, there exists some k such that $\sum_{i=k}^{\infty} |y'_i(x)| < \epsilon$ holds for all $x \in W$. In particular, since $\{y_n\} \subseteq W$ holds, it follows from $(\star)$ that

$$
\left|x'(x_n)\right| \leq y'_n(y_n) \leq \sum_{i=k}^{\infty} |y'_i(y_n)| < \epsilon
$$

for all $n \geq k$. Therefore, $\lim x'(x_n) = 0$. That is, $x_n \xrightarrow{w} 0$ holds in E, as claimed. ∎

In the sequel, the following technique for constructing disjoint sequences will be needed.

Lemma 4.35. *Let E be a Riesz space, and let $\{x_n\}$ be a sequence of E^+. If some $x \in E^+$ satisfies $2^{-n}x_n \leq x$ for all n, then the sequence $\{u_n\}$, defined by*

$$u_n = \left[x_{n+1} - 4^n \sum_{i=1}^{n} x_i - 2^{-n}x \right]^+,$$

is a disjoint sequence.

Proof. If $m > n \geq 1$, then note that

$$0 \leq 4^{-m}u_m = \left[4^{-m}x_{m+1} - \sum_{i=1}^{m} x_i - 2^{-3m}x \right]^+ \leq \left[4^{-m}x_{m+1} - x_{n+1} \right]^+$$

$$\leq \left[2^{-n} \cdot 2^{-m-1}x_{m+1} - x_{n+1} \right]^+ \leq \left[2^{-n}x + 4^n \sum_{i=1}^{n} x_i - x_{n+1} \right]^+$$

$$= \left[x_{n+1} - 4^n \sum_{i=1}^{n} x_i - 2^{-n}x \right]^- \perp u_n.$$

This implies $u_n \perp u_m$. ∎

The next theorem is a powerful tool. It describes an important approximation property of continuous operators in terms of disjoint sequences.

Theorem 4.36. *Let $T: E \to X$ be a continuous operator from a Banach lattice E to a Banach space X, let A be a norm bounded solid subset of E, and let ρ be a norm continuous seminorm on X. If $\lim \rho(Tx_n) = 0$ holds for each disjoint sequence $\{x_n\}$ in A, then for each $\epsilon > 0$ there exists some $u \in E^+$ lying in the ideal generated by A such that*

$$\rho\big(T[(|x| - u)^+]\big) < \epsilon$$

holds for all $x \in A$.

Proof. Suppose that the claim is false. Then there exists some $\epsilon > 0$ such that for each $u \geq 0$ in the ideal generated by A we have $\rho\big((T(|x| - u)^+\big) \geq \epsilon$ for at least one $x \in A$. In particular, there exists a sequence $\{x_n\} \subseteq A$ such that for each n we have

$$\rho\Big(T\Big(|x_{n+1}| - 4^n \sum_{i=1}^{n} |x_i|\Big)^+\Big) \geq \epsilon. \tag{$\star$}$$

Now put $y = \sum_{n=1}^{\infty} 2^{-n}|x_n|$. Also, let $w_n = \big(|x_{n+1}| - 4^n \sum_{i=1}^{n} |x_i|\big)^+$ and $v_n = \big(|x_{n+1}| - 4^n \sum_{i=1}^{n} |x_i| - 2^{-n}y\big)^+$. By Lemma 4.35, the sequence $\{v_n\}$

is disjoint. Also, since A is solid and $0 \leq v_n \leq |x_{n+1}|$ holds, we see that $\{v_n\} \subseteq A$, and so by our hypothesis

$$\lim_{n \to \infty} \rho(T v_n) = 0.$$

On the other hand, we have $0 \leq w_n - v_n \leq 2^{-n} y$, and so $\|w_n - v_n\| \leq 2^{-n} \|y\|$. In particular, it follows that $\lim \rho(T(w_n - v_n)) = 0$. From

$$\rho(T w_n) \leq \rho(T(w_n - v_n)) + \rho(T v_n),$$

we see that $\lim \rho(T w_n) = 0$. However, this contradicts $(\star)$, and the proof is finished. ∎

The next theorem, showing that the weakly compact sets enjoy a useful lattice approximation property, is the single most important result in this section.

Theorem 4.37. *Let W be a weakly relatively compact subset of a Banach lattice E. Then for each $\epsilon > 0$ and each $0 \leq x' \in E'$ there exists some $u \geq 0$ lying in the ideal generated by W such that*

$$x'(|x| - u)^+ < \epsilon$$

holds for all x in the convex solid hull of W.

Proof. Let A denote the solid hull of a weakly relatively compact subset W of a Banach lattice E, and let ρ be the norm continuous seminorm defined by $\rho(x) = x'(|x|)$, where $x' \geq 0$ is fixed. If $I \colon E \to E$ is the identity operator, then Theorem 4.34 shows that $\lim \rho(I x_n) = 0$ holds for every disjoint sequence $\{x_n\} \subseteq A$. Thus, by Theorem 4.36, there exists some $u \geq 0$ in the ideal generated by W satisfying

$$\rho(I(|x| - u)^+) = x'(|x| - u)^+ < \epsilon \tag{†}$$

for all $x \in W$. Clearly, (†) also holds for each $x \in A$. On the other hand, if x belongs to the convex solid hull of W, then there exist $x_1, \ldots, x_n \in A$ and positive constants $\alpha_1, \ldots, \alpha_n$ with $\alpha_1 + \cdots + \alpha_n = 1$ and $x = \alpha_1 x_1 + \cdots + \alpha_n x_n$ (see Exercise 1 of Section 3.3). Therefore,

$$x'(|x| - u)^+ \leq x'\left(\sum_{i=1}^{n} \alpha_i (|x_i| - u)^+\right) = \sum_{i=1}^{n} \alpha_i x'(|x_i| - u)^+ < \sum_{i=1}^{n} \alpha_i \epsilon = \epsilon$$

holds, and the proof is finished. ∎

By Dini's classical theorem, it is easy to see that order convergence in the dual of a Banach lattice E implies uniform convergence on the weakly compact subsets of E^+. We are now in a position to prove something stronger, namely, that order convergence in the dual implies uniform convergence on the convex solid hull of any arbitrary weakly compact set.

Theorem 4.38. *For a Banach lattice E the following statements hold:*

(1) *If $x'_\alpha \downarrow 0$ in E', then the net $\{x'_\alpha\}$ converges uniformly to zero on the convex solid hull of any weakly relatively compact subset of E.*

(2) *Every order bounded disjoint sequence of E' converges uniformly to zero on the convex solid hull of any weakly relatively compact subset of E.*

Proof. (1) Let $x'_\alpha \downarrow 0$ in E', let A be the convex solid hull of a weakly relatively compact subset of E, and let $\epsilon > 0$. If β is a fixed index, then there exists (by Theorem 4.37) some $u \in E^+$ such that $x'_\beta(|x| - u)^+ < \epsilon$ holds for all $x \in A$. In view of $x'_\alpha(u) \downarrow 0$, there exists some $\alpha_0 \succeq \beta$ satisfying $x'_\alpha(u) < \epsilon$ for all $\alpha \succeq \alpha_0$. Thus, for $\alpha \succeq \alpha_0$ and $x \in A$ we have

$$
\begin{aligned}
\left|x'_\alpha(x)\right| &\leq x'_\alpha(|x|) = x'_\alpha(|x| \wedge u) + x'_\alpha(|x| - u)^+ \\
&\leq x'_\alpha(u) + x'_\beta(|x| - u)^+ < \epsilon + \epsilon = 2\epsilon,
\end{aligned}
$$

and this shows that $\{x'_\alpha\}$ converges uniformly to zero on A.

(2) Let $\{x'_n\}$ be a disjoint sequence of E' satisfying $|x'_n| \leq x'$ for each n. From $\sum_{i=1}^n |x'_i| \uparrow \leq x'$, we see that $\sum_{i=1}^n |x'_i| \uparrow y'$ holds in E'. Thus, by (1) the sequence $\{\sum_{i=1}^n |x'_i|\}$ converges uniformly to y' on the convex solid hull of any weakly relatively compact subset of E. The latter easily implies that $\{|x'_n|\}$ (and hence $\{x'_n\}$) converges uniformly to zero on the convex solid hull of every weakly relatively compact subset of E. ∎

When is the solid hull of a weakly relatively compact set weakly relatively compact?

The next theorem provides some answers.

Theorem 4.39 (Abramovich–Wickstead). *For a Banach lattice E we have the following.*

(1) *E is an ideal of E'' if and only if every weakly relatively compact subset of E^+ has a weakly relatively compact solid hull.*

(2) *If E is a band of E'', then every weakly relatively compact subset of E has a weakly relatively compact solid hull.*

Proof. (1) Assume first that E is an ideal of E'' (i.e., assume that E has order continuous norm), and let $A \subseteq E^+$ be a weakly relatively compact set. Let $x'' \in E''$ be in the w^*-closure of $\mathrm{Sol}\,(A)$. Pick a net $\{x_\alpha\} \subseteq \mathrm{Sol}\,(A)$ with $x_\alpha \xrightarrow{w^*} x''$. For each α choose some $y_\alpha \in A$ with $-y_\alpha \leq x_\alpha \leq y_\alpha$. By passing to a subnet, we can assume that $y_\alpha \xrightarrow{w} y$ holds in E. This implies that $-y \leq x'' \leq y$ in E'', and so $x'' \in E$. Therefore, $\mathrm{Sol}\,(A)$ is a weakly relatively compact subset of E.

For the converse, assume that every weakly relatively compact subset of E^+ has a weakly relatively compact solid hull. Note that for each $x \in E^+$ we have $\operatorname{Sol}(\{x\}) = [-x, x]$, and so the order intervals of E are weakly compact. By Theorem 4.9, E is an ideal of E''.

(2) Assume that E is a band of $E'' = (E')^\sim$. By Theorem 3.60, we know that E is $|\sigma|(E'', E')$-dense in $(E')^\sim$. Since every band of E'' is $|\sigma|(E'', E')$-closed (see Theorem 3.46), we conclude that $E = (E')^\sim_n$ holds.

Now let A be the convex solid hull of a weakly relatively compact subset of E, and let $\overline{A}$ denote its w^*-closure in E''. We must show that $\overline{A} \subseteq E$ holds. In view of $E = (E')^\sim_n$, it is enough to show that every vector of $\overline{A}$ is order continuous on E'.

To this end, let $x'' \in \overline{A}$, let $x'_\alpha \downarrow 0$ in E', and fix $\epsilon > 0$. By Theorem 4.38 the net $\{x'_\alpha\}$ converges uniformly to zero on A, and so there exists some α_0 satisfying $|x'_\alpha(x)| < \epsilon$ for all $\alpha \succeq \alpha_0$ and all $x \in A$. Since $x'' \in \overline{A}$, for each α there exists some $y_\alpha \in A$ with $|(x'' - y_\alpha)(x'_\alpha)| < \epsilon$. Then for $\alpha \succeq \alpha_0$ we have

$$\left|x''(x'_\alpha)\right| \leq \left|(x'' - y_\alpha)(x'_\alpha)\right| + \left|x'_\alpha(y_\alpha)\right| < \epsilon + \epsilon = 2\epsilon .$$

Therefore, $x'' \in (E')^\sim_n$, and the proof is finished. ∎

In the preceding theorem, the first part was proved by A. W. Wickstead [193] and the second part by Y. A. Abramovich [2].

Recall that (by Theorem 4.9) a Banach lattice is an ideal in its double dual if and only if it has order continuous norm. In the next section (Theorem 4.60), we shall characterize the Banach lattices that are bands in their double duals. Also, it should be noted that in a Banach lattice with order continuous norm the solid hull of a weakly relatively compact set need not be weakly relatively compact and that the converse of part (2) of Theorem 4.39 is false; see Exercises 11 and 12 at the end of this section.

We now turn our attention to weak* convergence. The next technical result is due to D. H. Fremlin [64] and is similar to Theorem 4.36.

Theorem 4.40 (Fremlin). *Let E be a normed Riesz space, and let A be a norm bounded subset of E'. Then for a vector $x \in E^+$ the following statements are equivalent.*

(1) *Every disjoint sequence of $[0, x]$ converges uniformly to zero on A.*

(2) *For each $\epsilon > 0$ there exists some $0 \leq y' \in E'$ lying in the ideal generated by A such that*

$$\left(|x'| - y'\right)^+(x) < \epsilon$$

holds for all $x' \in A$.

Proof. (1) $\Longrightarrow$ (2) For each $u \in E$, let

$$\rho(u) = \sup\{|x'|(|u|) \colon \ x' \in A\} = \sup\{x'(y) \colon \ x' \in A \text{ and } |y| \le |u|\} \,.$$

Since A is norm bounded, $\rho(u) \in \mathbb{R}$ holds for each $u \in E$, and clearly ρ is a lattice seminorm on E. On the other hand, if $\{x_n\}$ is a disjoint sequence of $[0, x]$, then $\lim \rho(x_n) = 0$ holds. To see this, let $\delta > 0$. For each n choose $x'_n \in A$ and $|y_n| \le x_n$ with $\rho(x_n) < \delta + x'_n(y_n)$. Since $\{y_n^+\}$ and $\{y_n^-\}$ are both disjoint sequences of $[0, x]$, it follows from our hypothesis that $\lim x'_n(y_n) = 0$, and so $\limsup \rho(x_n) \le \delta$ holds for all $\delta > 0$. Therefore, $\lim \rho(x_n) = 0$. Hence, by Theorem 4.13, we have the following property:

If $0 \le y_n \uparrow \le x$ holds in E, then $\{y_n\}$ is a ρ-Cauchy sequence. $(\star)$

Now pick some $r > 0$ such that $\|x'\| < r$ holds for all $x' \in A$, and assume by way of contradiction that (2) is false. Then there exists some $\epsilon > 0$ such that for each y' in the ideal generated by A there exists some $x' \in A$ such that $(|x'| - y')^+(x) > 2\epsilon$. In particular, there exists a sequence $\{x'_n\} \subseteq A$ satisfying $\left(|x'_{n+1}| - 2^n \sum_{i=1}^{n} |x'_i|\right)^+(x) > 2\epsilon$. For each n pick some $y_n \in [0, x]$ with $\left(|x'_{n+1}| - 2^n \sum_{i=1}^{n} |x'_i|\right)^+(y_n) > 2\epsilon$, and note that

$$|x'_i|(y_n) \le 2^{-n} r \|x\| \quad \text{for } i = 1, \dots, n \tag{\dagger}$$

and

$$|x'_{n+1}|(y_n) > 2\epsilon \quad \text{for all } n \,. \tag{$\star\star$}$$

Next, for each k and n put $v_{n,k} = \bigvee_{i=n}^{n+k} y_i$, and note that $0 \le v_{n,k} \uparrow_k \le x$ holds. By $(\star)$, for each n there exists some k_n satisfying

$$\rho(v_{n,k} - v_{n,k_n}) < 2^{-n}\epsilon \quad \text{for all } k \ge k_n \,.$$

Since $k \ge m \ge n$ implies $(y_m - v_{n,k_n})^+ \le (v_{n,k} - v_{n,k_n})^+$, it follows from the preceding inequality that for $m \ge n$ we have

$$|x'_{n+1}|\left((y_m - v_{n,k_n})^+\right) \le \rho\left((y_m - v_{n,k_n})^+\right) < 2^{-n}\epsilon \,. \tag{$\star\star\star$}$$

Now let $w_n = \bigwedge_{i=1}^{n} v_{i,k_i}$. From $(y_n - w_n)^+ \le \sum_{i=1}^{n}(y_n - v_{i,k_i})^+$, $(\star\star)$, and $(\star\star\star)$, we see that

$$|x'_{n+1}|(w_n) \ge |x'_{n+1}|(y_n) - |x'_{n+1}|\left((y_n - w_n)^+\right)$$

$$\ge 2\epsilon - \sum_{i=1}^{n}|x'_{n+1}|\left((y_n - v_{i,k_i})^+\right) \ge 2\epsilon - \sum_{i=1}^{n}2^{-i}\epsilon > \epsilon \,.$$

Also, from $(\dagger)$ we have

$$|x'_{n+1}|(w_{n+1}) \le |x'_{n+1}|(v_{n+1,k_{n+1}}) \le \sum_{i=n+1}^{\infty}|x'_{n+1}|(y_i)$$

$$\le \sum_{i=n+1}^{\infty} 2^{-i} r \|x\| = 2^{-n} r \|x\| \,.$$

Therefore,

$$\rho(w_n - w_{n+1}) \geq |x'_{n+1}|(w_n - w_{n+1}) \geq \epsilon - 2^{-n} r\|x\|$$

holds for each n, which implies that $\{w_1 - w_n\}$ is not a ρ-Cauchy sequence. However, in view of $0 \leq w_1 - w_n \uparrow \leq x$, the latter contradicts $(\star)$, and this contradiction establishes the validity of (2).

$(2) \Longrightarrow (1)$ Let $\{x_n\}$ be a disjoint sequence of $[0, x]$, and fix $\epsilon > 0$. Pick some $0 \leq y' \in E'$ so that (2) holds. Since $x_n \xrightarrow{w} 0$ (why?), there exists some k such that $y'(x_n) < \epsilon$ holds for all $n \geq k$. Therefore, for $x' \in A$ and $n \geq k$ we have

$$
\begin{aligned}
|x'(x_n)| &\leq |x'|(x_n) = (|x'| \wedge y')(x_n) + (|x'| - y')^+(x_n) \\
&\leq y'(x_n) + (|x'| - y')^+(x) < \epsilon + \epsilon = 2\epsilon,
\end{aligned}
$$

and this shows that $\{x_n\}$ converges uniformly to zero on A. ∎

With the help of the preceding theorem, we are now in a position to present A. Grothendieck's [72] classical characterizations of the weakly compact subsets in the dual of an AM-space with unit.

Theorem 4.41 (Grothendieck). *Let E be an AM-space with unit. Then for a norm bounded subset A of E', the following statements are equivalent.*

(1) *A is weakly relatively compact.*

(2) *Every norm bounded disjoint sequence of E converges uniformly to zero on A.*

(3) *For each $\epsilon > 0$ there exists some $y' \geq 0$ satisfying*

$$\|(|x'| - y')^+\| < \epsilon$$

for all $x' \in A$.

Proof. $(1) \Longrightarrow (2)$ Let e be the unit of E, and let $\{x_n\}$ be a norm bounded disjoint sequence of E. Since $[-e, e]$ is the closed unit ball of E, we see that $\{x_n\}$ is an order bounded disjoint sequence of E''. By part (2) of Theorem 4.38, the sequence $\{x_n\}$ converges uniformly to zero on A.

$(2) \Longrightarrow (3)$ Clearly, every disjoint sequence of $[0, e]$ converges uniformly to zero on A. Therefore, by Theorem 4.40, for each $\epsilon > 0$ there exists some $y' \geq 0$ in the ideal generated by A satisfying

$$\|(|x'| - y')^+\| = (|x'| - y')^+(e) < \epsilon$$

for all $x' \in A$.

$(3) \Longrightarrow (1)$ Let $\epsilon > 0$. Pick some $0 \leq y' \in E'$ satisfying $\|(|x'| - y')^+\| < \epsilon$ for each $x' \in A$. From $|x'| = |x'| \wedge y' + (|x'| - y')^+ \in [0, y'] + \epsilon U'$, we see that

$$A \subseteq [-y', y'] + \epsilon U'.$$

Now note that E' (as an AL-space) has order continuous norm. This implies that $[-y', y']$ is a weakly compact subset of E', and so from $A \subseteq [-y', y']+\epsilon U'$ and Theorem 3.44 we see that A is weakly relatively compact. ∎

It is interesting to know that the w^*-convergent sequences satisfy the equivalent conditions of Theorem 4.40. This useful result is due to O. Burkinshaw [48].

Theorem 4.42 (Burkinshaw). *Let E be a Dedekind σ-complete normed Riesz space. If a sequence $\{x'_n\}$ of E' is w^*-convergent, then for each $x \in E^+$ and each $\epsilon > 0$ there exists some $0 \le y' \in E'$ in the ideal generated by $\{x'_n\}$ such that for each n we have*

$$\left(|x'_n| - y'\right)^+(x) < \epsilon.$$

Proof. For the discussion below, the symbol $\sum_{n=1}^{\infty} x_n$ stands for the supremum $\sup\{\sum_{i=1}^{n} x_i\colon n = 1, 2, \dots\}$. The proof is based upon the following property:

- *If $\{u_n\} \subseteq E^+$ satisfies $\sum_{n=1}^{\infty} u_n \le u$, then for each $0 \le f \in E'$ and each $\epsilon > 0$, there exists a subsequence $\{v_n\}$ of $\{u_n\}$ (depending upon f and ϵ) such that the vector $v = \sum_{n=1}^{\infty} v_n$ satisfies $f(v) < \epsilon$.*

The proof of the above property goes as follows: Fix a countable partition $\{\mathbb{N}_n\}$ of the set of natural numbers $\mathbb{N}$, with each $\mathbb{N}_n$ infinite, and let $w_n = \sum_{i\in\mathbb{N}_n} u_i$. Since $\mathbb{N}_i \cap \mathbb{N}_j = \emptyset$ holds for $i \ne j$, it follows that $0 \le \sum_{i=1}^{n} w_i \le \sum_{i=1}^{\infty} u_i \le u$ for all n. Thus, $\sum_{i=1}^{\infty} f(w_i) \le f(u) < \infty$, and hence $f(w_i) < \epsilon$ holds for some i. For this index i write $\mathbb{N}_i = \{k_1, k_2, \dots\}$ with $k_n \uparrow$, put $v_n = u_{k_n}$, and note that the vector $v = \sum_{n=1}^{\infty} v_n = w_i$ satisfies $f(v) < \epsilon$.

According to Theorem 4.40, we need to show that each disjoint sequence of $[0, x]$ converges uniformly to zero on $\{x'_n\}$. Without loss of generality we can suppose that $x'_n \xrightarrow{w^*} 0$.

To establish this, assume by way of contradiction that there exist some $x \in E^+$ and a disjoint sequence $\{x_n\}$ of $[0, x]$ which does not converge uniformly to zero on $\{x'_n\}$. Thus, we assume that there exists an $\epsilon > 0$ such that for each n there exists some $m > n$ satisfying $|x'_i(x_m)| > 3\epsilon$ for at least one i. We claim that there exist a subsequence $\{w_n\}$ of $\{x_n\}$ and a subsequence $\{g_n\}$ of $\{x'_n\}$ satisfying

$$\left|g_n(w_n)\right| > 3\epsilon$$

for all n.

The existence of $\{w_n\}$ and $\{g_n\}$ can be proved by induction as follows: Put $k_1 = 1$, and then choose some m_1 with $|x'_{m_1}(x_{k_1})| > 3\epsilon$. Next, assume that $k_1 < \dots < k_n$ and $m_1 < \dots < m_n$ have been selected so that

$|x'_{m_i}(x_{k_i})| > 3\epsilon$ holds for each $i = 1, \ldots, n$. Since $\lim f(x_k) = 0$ holds in $\mathbb{R}$ for each $f \in E'$, there exists some $j > k_n$ satisfying $|x'_i(x_k)| < \epsilon$ for all $k > j$ and each $i = 1, \ldots, m_n$. Then there exists some $k_{n+1} > j > k_n$ and some m_{n+1} with $|x'_{m_{n+1}}(x_{k_{n+1}})| > 3\epsilon$. Clearly, $m_{n+1} > m_n$, and the induction is complete. Now put $w_n = x_{k_n}$ and $g_n = x'_{m_n}$. Note that $g_n \xrightarrow{w^*} 0$ and $\sum_{n=1}^{\infty} w_n \leq x$.

Next, we shall construct subsequences $\{x_1^k, x_2^k, \ldots\}$, $k = 1, 2, \ldots$, of $\{w_n\}$ and positive integers $n_1 < n_2 < \cdots$ such that for each $k \geq 2$ we have

(1) $\{x_n^k\}$ is a subsequence of $\{x_n^{k-1}\}$,

(2) The vector w_{n_k} is a member of $\{x_n^{k-1}\}$,

(3) $\sum_{i=1}^{k-1} |g_{n_k}(w_{n_i})| < \epsilon$, and

(4) $|g_{n_k}|(u_k) < \epsilon$, where $u_k = \sum_{n=1}^{\infty} x_n^k$.

We start by letting $x_n^0 = w_n$ for each n. For this construction we use induction on k. For $k = 1$, apply property ($\bullet$) at the beginning of the proof to $\{w_n\}$; extract a subsequence $\{x_n^1\}$ of $\{w_n\}$ with $u_1 = \sum_{n=1}^{\infty} x_n^1$ satisfying $|g_1|(u_1) < \epsilon$, and put $n_1 = 1$. Now assume that $\{x_n^k\}$ and n_k have been selected satisfying properties (1)–(4). Since $\lim_{n \to \infty} g_n(w_i) = 0$ holds for each $i = 1, \ldots, n_k$, there exists some $n_{k+1} > n_k$ with $\sum_{i=1}^{k} |g_{n_{k+1}}(w_i)| < \epsilon$ and with $w_{n_{k+1}}$ a member of $\{x_n^k\}$. By property ($\bullet$) at the beginning of the proof, there exists a subsequence $\{x_n^{k+1}\}$ of $\{x_n^k\}$ with $|g_{n_{k+1}}|(u_{k+1}) < \epsilon$, and the induction is complete.

Now let $w = \sum_{i=1}^{\infty} w_{n_i}$. Then for each k we have

$$0 \leq w = \sum_{i=1}^{k} w_{n_i} + \sum_{i=k+1}^{\infty} w_{n_i} \leq \sum_{i=1}^{k} w_{n_i} + u_k,$$

and hence $|g_{n_k}|(w - \sum_{i=1}^{k} w_{n_i}) \leq |g_{n_k}|(u_k) < \epsilon$ holds for all k. Thus,

$$|g_{n_k}(w)| \geq |g_{n_k}(w_{n_k})| - \left| g_{n_k}\left(\sum_{i=1}^{k-1} w_{n_i} \right) \right| - \left| g_{n_k}\left(w - \sum_{i=1}^{k} w_{n_i} \right) \right|$$
$$> 3\epsilon - \epsilon - \epsilon = \epsilon > 0$$

holds for all $k \geq 2$. However, this contradicts $\lim_{k \to \infty} g_{n_k}(w) = 0$, and the proof is finished. ∎

We now come to a useful property of w^*-convergent sequences.

Theorem 4.43. *Let E be a Dedekind σ-complete normed Riesz space, and let A denote the ideal generated by E in E''. Then a sequence $\{x'_n\} \subseteq E'$ satisfies $x'_n \xrightarrow{w^*} 0$ if and only if $x'_n \xrightarrow{\sigma(E',A)} 0$.*

Proof. Clearly, $x'_n \xrightarrow{\sigma(E',A)} 0$ implies $x'_n \xrightarrow{w^*} 0$. For the converse, assume $x'_n \xrightarrow{w^*} 0$. Let $0 \leq x'' \in A$ and $\epsilon > 0$ be fixed. Pick some $x \in E$ with $0 \leq x'' \leq x$, and then use Theorem 4.42 to select some $0 \leq y' \in E'$ satisfying

$$\left(|x'_n| - y' \right)^+ (x) < \epsilon$$

for all n. From Theorem 3.60 we know that $[0,x] \cap E$ is $|\sigma|(E'',E')$-dense in $[0,x'']$, and so there exists some $u \in [0,x] \cap E$ such that $y'(|x''-u|) < \epsilon$. Taking into account that $|x''-u| \leq x$ holds, we see that

$$\left| x'_n(x''-u) \right| \leq \left(|x'_n| - y' \right)^+ (|x''-u|) + y'(|x''-u|) < 2\epsilon.$$

In view of $\lim x'_n(u) = 0$, we get $\limsup |x'_n(x'')| \leq 2\epsilon$. Since $\epsilon > 0$ is arbitrary, the latter implies $\lim x'_n(x'') = 0$, and so $x'_n \xrightarrow{\sigma(E',A)} 0$. ∎

A Banach space X is said to be a **Grothendieck space** whenever $x'_n \xrightarrow{w^*} 0$ in X' implies $x'_n \xrightarrow{w} 0$ in X' (i.e., whenever weak* and weak convergence of sequences in X' coincide).

Clearly, every reflexive Banach space is a Grothendieck space. Another class of Grorthendieck spaces is described in the next result, which is essentially due to A. Grothendieck [**72**].

Theorem 4.44 (Grothendieck). *Every Dedekind σ-complete AM-space with unit is a Grothendieck space.*

Proof. Let E be an AM-space with unit e. If $0 \leq x'' \in E''$, then for each $0 \leq x' \in E'$ we have $x''(x') \leq \|x''\| \cdot \|x'\| = \|x''\| e(x')$, and so $0 \leq x'' \leq \|x''\| e$ holds in E''. This implies that the ideal generated by E in E'' is precisely E''. The rest of the proof follows from Theorem 4.43. ∎

Corollary 4.45. *Every $L_\infty(\mu)$-space is a Grothendieck space.*

An order projection on the dual of a Banach lattice is always continuous, and hence weakly continuous. However, in general, it is not w^*-continuous. Therefore, it is an interesting fact that an order projection on the dual of a Dedekind σ-complete Banach lattice is always sequentially w^*-continuous. This is due to H. H. Schaefer [**172**].

Theorem 4.46 (Schaefer). *If E is a Dedekind σ-complete normed Riesz space, then every order projection on E' is sequentially w^*-continuous.*

Proof. Let $x'_n \xrightarrow{w^*} 0$ in E', let P be an order projection on E', let $x \in E^+$, and let A be the ideal generated by E in E''. Denote by $P' \colon E'' \to E''$ the adjoint order projection of P. Since $0 \leq P'x \leq x$, we see that $P'x \in A$. Therefore, by Theorem 4.43 we have

$$[Px'_n](x) = x'_n(P'x) \longrightarrow 0,$$

proving that $Px'_n \xrightarrow{w^*} 0$ holds in E'. ∎

An immediate consequence of the preceding result is the following.

Corollary 4.47. *If E is a Dedekind σ-complete Banach lattice, then every band of E' is w^*-sequentially complete.*

Exercises

1. Show that $L_1[0,1]$ does not have the Schur property.

2. Let E be a normed Riesz space, let A be a norm bounded subset of E', and let $x \in E^+$. Then show that every disjoint sequence of $[0,x]$ converges uniformly to zero on A if and only if every disjoint sequence in the solid hull of A converges uniformly to zero on $[0,x]$.

3. Let E be a Banach lattice, and let A denote the ideal generated by E in E''. For a norm bounded subset W of E' show that the following two statements are equivalent:
   (a) W is relatively $\sigma(E', A)$-compact.
   (b) Every order bounded disjoint sequence of E converges uniformly to zero on W.

4. Let A be a norm bounded subset of a Banach lattice E such that its convex solid hull is weakly sequentially complete. Then show that the following two statements are equivalent:
   (a) A is relatively weakly compact.
   (b) If $x'_n \downarrow 0$ holds in E', then $\{x'_n\}$ converges uniformly to zero on A.

5. Let E be a Dedekind σ-complete Banach lattice, and let A be the ideal generated by E in E''. If $W \subseteq E'$ is sequentially w^*-compact (i.e., if every sequence of W has a w^*-convergent subsequence), then show that W is relatively $\sigma(E', A)$-compact.

6. Let E be a Dedekind σ-complete normed Riesz space. Assume that $P, P_1, P_2, \ldots$ are order projections on E' satisfying $P_n \uparrow P$. If $x'_n \xrightarrow{w^*} x'$ holds in E', then show that $P_n x'_n \xrightarrow{w^*} Px'$ also holds in E'.

7. Let $\{x_n\}$ be a weak Cauchy sequence in a normed Riesz space E. Then show that for each $0 \le x' \in E'$ and each $\epsilon > 0$ there exists some $x \in E^+$ lying in the ideal generated by $\{x_n\}$ satisfying $x'\big((|x_n| - x)^+\big) < \epsilon$ for all n.

8. Generalize Theorem 4.46 as follows: If $\langle E, E' \rangle$ is a Riesz dual system with E Dedekind σ-complete, then show that $f_n \xrightarrow{\sigma(E',E)} 0$ in E' implies $Pf_n \xrightarrow{\sigma(E',E)} 0$ for every order projection P on E'.

9. Let $E = L_p[0,1]$ $(1 \le p < \infty)$, and let $\{f_n\} \subseteq E$. Then show that $f_n \xrightarrow{w} 0$ holds in E if and only if $\lim\limits_{n\to\infty} \int_0^x f_n(t)\,dt = 0$ for each $x \in [0,1]$.

10. Let $E = L_p[0,1]$ $(1 \le p < \infty)$, $\{f_n\} \subseteq E$ and $f \in E$. Then show that $f_n \xrightarrow{w} f$ and $\|f_n\|_p \to \|f\|_p$ hold if and only if $\lim \|f_n - f\|_p = 0$.

11. (Meyer-Nieberg [**140**]) By part (1) of Theorem 4.39 we know that a Banach lattice E has order continuous norm if and only if every weakly relatively compact subset of E^+ has a weakly relatively compact solid hull. This exercise presents an example of a weakly relatively compact

subset of a Banach lattice with order continuous norm whose solid hull is not weakly relatively compact.

Let r_n denote the nth Rademacher function on $[0, 1]$. That is, we let $r_n(t) = \text{sgn} \sin(2^n \pi t)$. Then $|r_n| = 1$ and $r_n \xrightarrow{w} 0$ holds in $L_1[0, 1]$ (why?). Also, let $E = \left(L_1[0, 1] \oplus L_1[0, 1] \oplus \cdots\right)_0$.

(a) Show that E has order continuous norm.

(b) If $x_n = (r_n, \ldots, r_n, 0, 0, \ldots)$ (where the r_n occupy the first n positions), then show that $x_n \xrightarrow{w} 0$ holds in E (and so $\{x_1, x_2, \ldots\}$ is a weakly relatively compact subset of E).

(c) Show that $\{|x_n|\}$ does not have any weakly convergent subsequence in E (and so the solid hull of $\{x_1, x_2, \ldots\}$ is not a weakly relatively compact subset of E).

12. According to part (2) of Theorem 4.39 if a Banach lattice is a band in its double dual, then every weakly relatively compact set has a weakly relatively compact solid hull. Show that the converse of this statement is false even for Banach lattices with order continuous norms. [*Hint:* Consider the Banach lattice c_0, and let W be a weakly relatively compact subset of c_0. Since c_0 is an *AM*-space, it follows (from Theorem 4.31) that $|W|$ is also a weakly relatively compact set. Now use the identity $\text{Sol}(W) = \text{Sol}(|W|)$ and part (1) of Theorem 4.39 to conclude that $\text{Sol}(W)$ is likewise weakly relatively compact. Now note that c_0 is not a band in its double dual.]

13. Assume that E is a Dedekind σ-complete Riesz space. Fix $0 \leq f \in E_c^{\sim}$, and consider the function $d_f \colon E \times E \to \mathbb{R}$ defined by

$$d_f(x, y) = f(|x - y|).$$

Show that:

(a) (C_f, d_f) is a metric space.

(b) For each $0 \leq u \in C_f$ the metric space $([0, u], d_f)$ is complete. [*Hint:* For (2) assume that $\{x_n\} \subseteq [0, u]$ satisfies $f(|x_{n+1} - x_n|) < 2^{-n}$ for all n. Let $x = \limsup x_n \left(= \bigwedge_{n=1}^{\infty} \bigvee_{k=n}^{\infty} x_k\right)$. Put $y_n = \bigvee_{k=n}^{\infty} x_k$ and note that $y_n \downarrow x$. From

$$\left| \bigvee_{k=n}^{n+m} x_k - x_n \right| \leq \bigvee_{k=n}^{n+m} |x_k - x_n| \leq \sum_{i=n}^{n+m} |x_{i+1} - x_i|$$

and the σ-order continuity of f, it follows that $f(|y_n - x_n|) \leq 2^{1-n}$. Now note that

$$0 \leq f(|x - x_n|) \leq f(|x - y_n|) + f(|y_n - x_n|) \to 0.]$$

14. This exercise presents an elementary proof (due to M. Nakamura [**146**]) of the following important consequence of Theorem 4.46:

- *If E is a Dedekind σ-complete Banach lattice and if a sequence $\{f_n\}$ in $E_n^{\sim}$ satisfies $f_n \xrightarrow{w^*} f$ in E', then $f \in E_n^{\sim}$.*

To see this, assume $x_\alpha \downarrow 0$ in E. Prove that $\lim f(x_\alpha) = 0$ by following the steps below. Let $\epsilon > 0$ be fixed.

(a) The sequence $\{f_n\}$ is norm bounded, and consequently the formula $g = \sum_{n=1}^{\infty} 2^{-n}|f_n|$ defines an order continuous positive linear functional. This follows easily from the uniform boundedness principle.

(b) The null ideal N_g of g is a projection band.

Since N_g is a band, it is enough to show that $C_g = N_g^{\mathrm{d}}$ is a projection band. Let $u \in E^+$. Pick a sequence $\{v_n\} \subseteq [0, u] \cap C_g$ with $v_n \uparrow$ and $\lim g(v_n) = \sup g([0, u] \cap C_g)$. If $v_n \uparrow v$ holds in E, then $g(v) = \sup g([0, u] \cap C_g)$ holds. Now an easy argument shows that $v = \sup[0, u] \cap C_g$ in E, proving that C_g is a projection band.

(c) We can assume that $0 \le x_\alpha \le x$ holds for all α. Denote by y the projection of x onto C_g, and consider the complete metric space $([0, y], d_g)$ determined by g as in the previous exercise. Then each f_n restricted to $([0, y], d_g)$ is continuous. Note that for each $u, v \in [0, y]$ we have

$$|f_n(u) - f_n(v)| \le |f_n|(|u - v|) \le 2^n g(|u - v|) = 2^n d_g(u, v).$$

(d) If $K_n = \{u \in [0, y] \colon |f_n(u) - f_k(u)| \le \epsilon \text{ for all } k \ge n\}$, then some K_n has an interior point.

By (c), each K_n is closed. Also, $[0, y] = \bigcup_{n=1}^{\infty} K_n$. Now apply Baire's category theorem.

(e) Assume that $B(u, r) = \{v \in [0, y] \colon g(|u - v|) < r\} \subseteq K_m$ holds for some m. Then $z \in [0, y]$ and $g(z) < r$ imply $|f(z)| \le |f_m(z)| + 2\epsilon$.

Put $v = u \wedge (y - z)$, $w = z + v$, and note that $v, w \in B(u, r)$. Therefore, for each $n \ge m$ we have

$$
\begin{aligned}
\big|f_n(z)\big| - \big|f_m(z)\big| &\le \big|f_n(z) - f_m(z)\big| \\
&\le \big|f_n(v) - f_m(v)\big| + \big|f_n(w) - f_m(w)\big| \le 2\epsilon,
\end{aligned}
$$

and the desired inequality follows from $f_n(z) \to f(z)$.

(f) There exists some α_0 satisfying $|f(x_\alpha)| \le 3\epsilon$ for all $\alpha \succeq \alpha_0$.

Replacing each x_α by its projection onto C_g, we can assume that $0 \le x_\alpha \le y$ holds for each α (why?). Since g is order continuous, there exists some α_1 satisfying $g(x_\alpha) < r$ for all $\alpha \succeq \alpha_1$. Also, since f_m is order continuous, there exists some $\alpha_0 \succeq \alpha_1$ satisfying $|f_m(x_\alpha)| < \epsilon$ for all $\alpha \succeq \alpha_0$. By (e) we see that $|f(x_\alpha)| \le 3\epsilon$ holds for all $\alpha \succeq \alpha_0$.

15. Let E be an *AL-* or *AM-*space. If $x_n \xrightarrow{w} 0$ holds in E, then show that $\{x_n\}$ converges uniformly to zero on every weakly compact subset of E'.

4.3. Embedding Banach Spaces

An operator $T \colon X \to Y$ between two Banach spaces is said to be an **embedding** whenever there exist two positive constants K and M satisfying

$$K\|x\| \le \|T(x)\| \le M\|x\|$$

for all $x \in X$. In this case, $T(X)$ is, of course, a closed vector subspace of Y that can be identified with X. (The Banach space $T(X)$ is also called a **copy** of X in Y.) A Banach space X is said to be **embeddable** into another Banach space Y (or simply that X **embeds** into Y) whenever there exists an embedding from X to Y.

When an embedding $T \colon E \to F$ between two Banach lattices is also a lattice homomorphism, then T is called a **lattice embedding**. In this case $T(E)$ is a closed Riesz subspace of F which can be identified with E. A Banach lattice E is said to be **lattice embeddable** into another Banach lattice F whenever there exists a lattice embedding from E into F.

All Banach spaces are assumed to be real vector spaces. As usual, c_0, ℓ_1, and ℓ_∞ denote the Banach lattices of all sequences converging to zero, all absolutely summable sequences, and all bounded sequences, respectively. Also, e_n will denote the sequence of real numbers whose n^{th} term is one and the rest are zero, i.e.,

$$e_n := (0, 0, \ldots, 0, 1, 0, 0, \ldots) \, .$$

For our discussion we shall need the concept of a basic sequence. A sequence $\{x_n\}$ in a Banach space is said to be a **Schauder basis** (or simply a **basis**) whenever for each $x \in X$ there exists a unique sequence $\{\alpha_n\}$ of scalars satisfying $x = \sum_{n=1}^{\infty} \alpha_n x_n$ (where, as usual the convergence of the series is assumed to be in the norm). If a sequence $\{x_n\}$ in a Banach space is a basis for the closed vector subspace it generates, then $\{x_n\}$ is referred to as a **basic sequence**.

The sequences that are basic are characterized as follows.

Theorem 4.48. *A sequence $\{x_n\}$ in a Banach space is a basic sequence if and only if*

- (a) $x_n \neq 0$ *holds for all n, and*
- (b) *there exists some constant $M > 0$ such that for any $m > n$ and any choice of scalars $\alpha_1, \ldots, \alpha_n, \ldots, \alpha_m$ we have*

$$\Big\| \sum_{i=1}^{n} \alpha_i x_i \Big\| \leq M \Big\| \sum_{i=1}^{m} \alpha_i x_i \Big\| \, .$$

Proof. Let X be a Banach space. Assume first that $\{x_n\}$ is a basic sequence of X. Without loss of generality we can also assume that the closed vector subspace generated by $\{x_n\}$ is X itself. Clearly, $x_n \neq 0$ holds for all n. On the other hand, if for each $x = \sum_{n=1}^{\infty} \alpha_n x_n$ we put

$$|||x||| = \sup \Big\{ \Big\| \sum_{i=1}^{n} \alpha_i x_i \Big\| \colon \; n = 1, 2, \ldots \Big\} \, ,$$

then it is a routine matter to verify that $||| \cdot |||$ is a norm on X under which X is a Banach space. Clearly, $\|x\| \leq |||x|||$ holds for each x, and so by the open mapping theorem there exists some $M > 0$ such that $|||x||| \leq M\|x\|$ also holds for all $x \in X$. In particular, note that for $m > n$ we have

$$\left\| \sum_{i=1}^{n} \alpha_i x_i \right\| \leq \left\|\left\| \sum_{i=1}^{m} \alpha_i x_i \right\|\right\| \leq M \left\| \sum_{i=1}^{m} \alpha_i x_i \right\|.$$

For the converse, assume that (a) and (b) hold. Then it is easy to see that each $x \in X$ can be written in at most one way in the form $\sum_{n=1}^{\infty} \alpha_n x_n$. On the other hand, using (b) it is easy to see that the vector subspace

$$\left\{ x \in X \colon \exists \, \{\alpha_n\} \text{ with } x = \sum_{n=1}^{\infty} \alpha_n x_n \right\}$$

is closed, and hence it must coincide with the closed vector subspace generated by $\{x_n\}$. That is, $\{x_n\}$ is a basic sequence, as desired. ∎

Let $\{x_n\}$ be a basic sequence in a Banach space X. Then a sequence $\{y_n\}$ in another Banach space Y is said to be **equivalent** to $\{x_n\}$ whenever there exist two positive constants K and M such that for every choice of scalars $\alpha_1, \ldots, \alpha_n$ we have

$$K \left\| \sum_{i=1}^{n} \alpha_i x_i \right\| \leq \left\| \sum_{i=1}^{n} \alpha_i y_i \right\| \leq M \left\| \sum_{i=1}^{n} \alpha_i x_i \right\|.$$

From Theorem 4.48 it is easy to see that $\{y_n\}$ is also a basic sequence. Moreover, in this case, an easy application of the closed graph theorem shows that the formula

$$T\left(\sum_{n=1}^{\infty} \alpha_n x_n \right) = \sum_{n=1}^{\infty} \alpha_n y_n$$

defines an invertible continuous operator from the closed vector subspace generated by $\{x_n\}$ onto the closed vector subspace generated by $\{y_n\}$.

It is not difficult to verify that $\{e_n\}$ is a basis for c_0 (and for each ℓ_p, with $1 \leq p < \infty$). Therefore, saying that c_0 is embeddable in a Banach space X is the same thing as saying that X has a sequence that is equivalent to the basis $\{e_n\}$ of c_0. Similarly, ℓ_p $(1 \leq p < \infty)$ is embeddable in a Banach space X if and only if X has a sequence equivalent to the (standard) basis $\{e_n\}$ of ℓ_p.

The embeddability of c_0 into a Banach space has been characterized by C. Bessaga and A. Pelczynski [**32**] as follows.

Theorem 4.49 (Bessaga–Pelczynski). *For a Banach space X the following statements are equivalent:*

(1) c_0 is embeddable in X.

(2) There is a sequence $\{x_n\}$ of X such that the series $\sum_{n=1}^{\infty} x_n$ does not converge in norm and for each $x' \in X'$ we have $\sum_{n=1}^{\infty} |x'(x_n)| < \infty$.

(3) There exists a sequence $\{x_n\}$ of X and two positive constants K and M such that $x_n \xrightarrow{w} 0$, $\|x_n\| \geq K$ for all n, and for every choice of scalars $\alpha_1, \ldots, \alpha_n$ we have

$$\left\| \sum_{i=1}^{n} \alpha_i x_i \right\| \leq M \max_{1 \leq i \leq n} |\alpha_i|.$$

(4) There exist a sequence $\{x_n\}$ of X (where each x_n also can be taken to be a unit vector) and two constants $K, M > 0$ such that for every n and every choice of scalars $\alpha_1, \ldots, \alpha_n$ we have

$$K \max_{1 \leq i \leq n} |\alpha_i| \leq \left\| \sum_{i=1}^{n} \alpha_i x_i \right\| \leq M \max_{1 \leq i \leq n} |\alpha_i|.$$

Proof. $(1) \implies (2)$ Let $T \colon c_0 \to X$ be an embedding, and let $x_n = T(e_n)$. Since $\sum_{n=1}^{\infty} e_n$ is not norm convergent in c_0, we see that $\sum_{n=1}^{\infty} x_n$ does not converge in norm. On the other hand, if $T' \colon X' \to \ell_1$ denotes the adjoint of T, then for each $x' \in X'$ we have

$$\sum_{n=1}^{\infty} |x'(x_n)| = \sum_{n=1}^{\infty} |x'(Te_n)| = \sum_{n=1}^{\infty} |T'x'(e_n)| = \|T'x'\|_1 < \infty.$$

$(2) \implies (3)$ Let $\{x_n\}$ be a sequence of X satisfying the properties of (2). Since the series $\sum_{n=1}^{\infty} x_n$ is not norm convergent, it is easy to see that there exist some $\epsilon > 0$ and a strictly increasing sequence $\{k_n\}$ of natural numbers satisfying $\left\| \sum_{i=k_{2n}}^{k_{2n+1}} x_i \right\| \geq \epsilon$ for all n. Put $y_n = x_{k_{2n}} + \cdots + x_{k_{2n+1}}$, and note that $\|y_n\| \geq \epsilon$. Moreover, if $x' \in X'$, then

$$\sum_{n=1}^{\infty} |x'(y_n)| \leq \sum_{n=1}^{\infty} |x'(x_n)| < \infty$$

holds. It now follows that $x'(y_n) \to 0$, and so $y_n \xrightarrow{w} 0$.

Now, an easy application of the uniform boundedness principle shows that there exists some $M > 0$ satisfying $\sum_{n=1}^{\infty} |x'(y_n)| \leq M \|x'\|$ for all $x' \in X'$. Thus, if $\alpha_1, \ldots, \alpha_n$ are arbitrary scalars, then

$$\left\| \sum_{i=1}^{n} \alpha_i y_i \right\| = \sup \left\{ \left| \sum_{i=1}^{n} \alpha_i x'(y_i) \right| \colon \|x'\| \leq 1 \right\}$$

$$\leq \sup \left\{ \sum_{i=1}^{n} |x'(y_i)| \colon \|x'\| \leq 1 \right\} \cdot \max_{1 \leq i \leq n} |\alpha_i|$$

$$\leq M \max_{1 \leq i \leq n} |\alpha_i|,$$

and so the sequence $\{y_n\}$ satisfies the properties of (3).

(3) $\implies$ (4) Let $\{x_n\}$ satisfy the properties of (3). Replacing x_n by $x_n/\|x_n\|$, we can assume without loss of generality that $\|x_n\| = 1$ holds for all n. Let $\epsilon_n = \frac{1}{2^n}$.

We claim that there exists a subsequence $\{x_{k_n}\}$ of $\{x_n\}$ such that for each scalar α and each y in the vector subspace generated by $\{x_{k_1}, \ldots, x_{k_n}\}$ we have

$$\|y\| \leq (1 + 2\epsilon_n)\|y + \alpha x_{k_{n+1}}\|. \tag{$\star$}$$

The proof is by induction. Start with $k_1 = 1$, and assume that $x_{k_1}, \ldots, x_{k_n}$ have been chosen to satisfy $(\star)$. Let Y denote the vector subspace generated by $\{x_{k_1}, \ldots, x_{k_n}\}$. Pick unit vectors $y_1, \ldots, y_m \in Y$ such that for each unit vector $y \in Y$ there exists some $1 \leq i \leq m$ with $\|y - y_i\| < \frac{\epsilon_n}{2}$. Also, for each $1 \leq i \leq m$ fix some $y_i' \in X'$ with $\|y_i'\| = 1$ and $y_i'(y_i) = 1$. Since $x_n \xrightarrow{w} 0$ holds, there exists some $k_{n+1} > k_n$ with $|y_i'(x_{k_{n+1}})| < \frac{\epsilon_n}{4}$ for each $i = 1, \ldots, m$. Then we claim that

$$(1 + 2\epsilon_n)\|y + \alpha x_{k_{n+1}}\| \geq 1 \tag{$\star\star$}$$

holds for each $y \in Y$ with $\|y\| = 1$ and all scalars α. To see this, let α be a scalar, and let y be a unit vector of Y. For $|\alpha| \geq 2$ we have

$$(1 + 2\epsilon_n)\|y + \alpha x_{k_{n+1}}\| \geq (1 + 2\epsilon_n)\left(\|\alpha x_{k_{n+1}}\| - \|y\|\right) \geq 1,$$

and on the other hand, if $|\alpha| < 2$, then pick $1 \leq i \leq m$ with $\|y - y_i\| < \frac{\epsilon_n}{2}$, and note that

$$\begin{aligned}
(1 + 2\epsilon_n)\|y + \alpha x_{k_{n+1}}\| &\geq (1 + 2\epsilon_n)\left[\|y_i + \alpha x_{k_{n+1}}\| - \frac{\epsilon_n}{2}\right] \\
&\geq (1 + 2\epsilon_n)\left[|y_i'(y_i + \alpha x_{k_{n+1}})| - \frac{\epsilon_n}{2}\right] \\
&\geq (1 + 2\epsilon_n)\left[1 - |\alpha y_i'(x_{k_{n+1}})| - \frac{\epsilon_n}{2}\right] \\
&\geq (1 + 2\epsilon_n)\left[1 - \frac{\epsilon_n}{2} - \frac{\epsilon_n}{2}\right] \geq 1.
\end{aligned}$$

For $y = 0$ the inequality $(\star)$ is trivial. For nonzero $y \in Y$, the validity of $(\star)$ follows from $(\star\star)$ by replacing y with $y/\|y\|$ and α with $\alpha/\|y\|$.

Now for simplicity put $z_n = x_{k_n}$. Using $(\star)$, we see that for any choice of scalars $\alpha_1, \ldots, \alpha_n, \ldots, a_m$ we have

$$\begin{aligned}
\left\|\sum_{i=1}^n \alpha_i z_i\right\| &\leq (1 + 2\epsilon_n)\cdots(1 + 2\epsilon_m)\left\|\sum_{i=1}^m \alpha_i z_i\right\| \\
&\leq e^{2(\epsilon_n + \cdots + \epsilon_m)}\left\|\sum_{i=1}^m \alpha_i z_i\right\| \leq e^2 \left\|\sum_{i=1}^m \alpha_i z_i\right\|.
\end{aligned}$$

Thus, if $\alpha_1, \ldots, \alpha_n$ are arbitrary scalars, then from

$$|\alpha_i| = \|\alpha_i z_i\| = \left\|\sum_{k=1}^{i} \alpha_k z_k - \sum_{k=1}^{i-1} \alpha_k z_k\right\| \leq 2e^2 \left\|\sum_{i=1}^{n} \alpha_i z_i\right\|,$$

it follows that

$$\tfrac{1}{2} e^{-2} \max_{1 \leq i \leq n} |\alpha_i| \leq \left\|\sum_{i=1}^{n} \alpha_i z_i\right\| \leq M \max_{1 \leq i \leq n} |\alpha_i|,$$

and so the sequence $\{z_n\}$ satisfies (4).

(4) $\implies$ (1) If $(\alpha_1, \alpha_2, \ldots) \in c_0$, then for $m > n$ we have

$$\left\|\sum_{i=n}^{m} \alpha_i x_i\right\| \leq M \max_{n \leq i \leq m} |\alpha_i|.$$

This implies that $\sum_{n=1}^{\infty} \alpha_n x_n$ is norm convergent in X. In addition, if $T \colon c_0 \to X$ is defined by $T(\alpha_1, \alpha_2, \ldots) = \sum_{n=1}^{\infty} \alpha_n x_n$, then it is easy to see that

$$K\|(\alpha_1, \alpha_2, \ldots)\|_\infty \leq \|T(\alpha_1, \alpha_2, \ldots)\| \leq M\|(\alpha_1, \alpha_2, \ldots)\|_\infty$$

holds for all $(\alpha_1, \alpha_2, \ldots) \in c_0$. This shows that c_0 is embeddable in X. Finally, replacing each x_n by $x_n/\|x_n\|$, it is not difficult to see that we can assume that each x_n is a unit vector, and the proof is finished. ∎

It should be noted that statement (4) of the preceding theorem merely says that the sequence $\{x_n\}$ is equivalent to the standard basis of c_0.

The lattice embeddability of c_0 is characterized as follows.

Theorem 4.50. *The Banach lattice c_0 is lattice embeddable in a Banach lattice E if and only if there exists a disjoint sequence $\{x_n\}$ of E^+ such that*

(a) *$\{x_n\}$ does not converge in norm to zero, and*

(b) *the sequence of partial sums of $\{x_n\}$ is norm bounded, i.e., there exists some $M > 0$ satisfying $\left\|\sum_{i=1}^{n} x_i\right\| \leq M$ for all n.*

Proof. If $T \colon c_0 \to E$ is a lattice embedding, then the vectors $x_n = T(e_n)$ satisfy the desired properties.

For the converse, let $\{x_n\}$ be a disjoint sequence of E^+ satisfying (a) and (b). By passing to a subsequence, we can assume that $\|x_n\| \geq K$ holds for all n and some $K > 0$. Now if $\alpha_1, \ldots, \alpha_n$ are arbitrary constants, then the inequalities

$$
\begin{aligned}
K|\alpha_i| \;&\leq\; |\alpha_i|\cdot\|x_i\| \;\leq\; \left\|\sum_{i=1}^{n}|\alpha_i|x_i\right\| \;=\; \left\|\sum_{i=1}^{n}\alpha_i x_i\right\| \\
&\leq\; \max_{1\leq i\leq n}|\alpha_i|\cdot\left\|\sum_{i=1}^{n}x_i\right\| \\
&\leq\; M \max_{1\leq i\leq n}|\alpha_i|
\end{aligned}
$$

imply

$$
K \max_{1\leq i\leq n}|\alpha_i| \;\leq\; \left\|\sum_{i=1}^{n}\alpha_i x_i\right\| \;\leq\; M \max_{1\leq i\leq n}|\alpha_i|.
$$

Therefore, if we define $T\colon c_0 \to E$ by $T(\alpha_1,\alpha_2,\ldots) = \sum_{n=1}^{\infty}\alpha_n x_n$, then T is an embedding of c_0 into E. Moreover, we have $\left|\sum_{n=1}^{k}\alpha_n x_n\right| = \sum_{n=1}^{k}|\alpha_n|x_n$, and by taking norm limits we see that

$$
\left|T(\alpha_1,\alpha_2,\ldots)\right| = \left|\sum_{n=1}^{\infty}\alpha_n x_n\right| = \sum_{n=1}^{\infty}|\alpha_n|x_n = T(|\alpha_1|,|\alpha_2|,\ldots).
$$

Therefore, T is a lattice embedding, as desired. ∎

Next, we shall consider lattice embeddings of ℓ_∞. The first result of this kind characterizes the lattice embeddings of ℓ_∞ into Dedekind σ-complete Banach lattices. It is due to G. Ya. Lozanovsky and A. A. Mekler [124] and P. Meyer-Nieberg [141].

Theorem 4.51 (Lozanovsky–Mekler–Meyer-Nieberg). *For an arbitrary Dedekind σ-complete Banach lattice E the following statements are equivalent.*

(1) *ℓ_∞ is lattice embeddable in E.*

(2) *E does not have order continuous norm.*

(3) *There exists an order bounded disjoint sequence of E^+ that does not converge in norm to zero.*

Proof. (1) $\Longrightarrow$ (2) To see this, combine the fact that ℓ_∞ does not have order continuous norm with the following simple statement: If a Banach lattice has order continuous norm, then its Banach sublattices also have order continuous norms.

(2) $\Longrightarrow$ (3) This follows immediately from Theorem 4.14.

(3) $\Longrightarrow$ (1) Let $\{x_n\}$ be an order bounded disjoint sequence of E^+ satisfying $\|x_n\| \geq K$ for all n and some $K > 0$. Assume that $0 \leq x_n \leq x$ holds

for all n. Now for each fixed $0 \le \alpha = (\alpha_1, \alpha_2, \ldots) \in \ell_\infty$ we have

$$0 \le \sum_{i=1}^{n} \alpha_i x_i \uparrow \le \|\alpha\|_\infty x \,,$$

and so by the Dedekind σ-completeness of E the sequence of partial sums of $\{\alpha_n x_n\}$ has a supremum in E. Let $\sum_{i=1}^{n} \alpha_i x_i \uparrow T(\alpha_1, \alpha_2, \ldots)$. Clearly, $T \colon \ell_\infty^+ \to E^+$ is additive, and so by Theorem 1.10 the mapping T extends uniquely to a positive operator from ℓ_∞ to E. Since $\alpha \wedge \beta = 0$ implies

$$
\begin{aligned}
T(\alpha) \wedge T(\beta) &= \Big[\sup \Big\{ \sum_{i=1}^{n} \alpha_i x_i \Big\} \Big] \wedge \Big[\sup \Big\{ \sum_{j=1}^{n} \beta_j x_j \Big\} \Big] \\
&= \sup \Big\{ \sum_{i=1}^{n} (\alpha_i \wedge \beta_i) x_i \Big\} = 0 \,,
\end{aligned}
$$

it follows that $T \colon \ell_\infty \to E$ is also a lattice homomorphism.

Finally, note that if $\alpha = (\alpha_1, \alpha_2, \ldots) \in \ell_\infty$, then

$$K|\alpha_n| \le |\alpha_n| \cdot \|x_n\| \le \|T(|\alpha|)\| \le \|x\| \cdot \|\alpha\|_\infty$$

holds for all n. Hence,

$$K\|\alpha\|_\infty \le \|T(\alpha)\| \le \|x\| \cdot \|\alpha\|_\infty$$

holds for each $\alpha \in \ell_\infty$, and so $T \colon \ell_\infty \to E$ is a lattice embedding. ∎

A useful consequence of the preceding result is the following.

Corollary 4.52. *Every separable Dedekind σ-complete Banach lattice has order continuous norm.*

Proof. If ℓ_∞ is embeddable in a Banach lattice E, then its copy in E must be nonseparable, which is a contradiction. Hence, ℓ_∞ is not lattice embeddable in E, and so, by Theorem 4.51, the Banach lattice E must have order continuous norm. ∎

The Dedekind σ-completeness is essential for the above result. For instance, $C[0,1]$ with the sup norm is separable, but the sup norm on $C[0,1]$ is not order continuous.

The lattice embeddability of ℓ_∞ is connected with an important topological property of Banach spaces known as property (u). This property was introduced by A. Pelczynski in [**161**].

Definition 4.53 (Pelczynski). *A weak Cauchy sequence $\{x_n\}$ in a Banach space X is said to satisfy **property (u)** whenever there exists a sequence $\{y_n\}$ of X such that*

(a) $\sum_{n=1}^{\infty} |x'(y_n)| < \infty$ *holds for all $x' \in X'$; and*

(b) $x_n - \sum_{i=1}^n y_i \xrightarrow{w} 0$.

If every weak Cauchy sequence in a Banach space X satisfies property (u), then X itself is said to have **property (u)**.

A. Pelczynski [**161**] has shown that property (u) is "hereditary," that is, it is inherited by closed vector subspaces.

Theorem 4.54 (Pelczynski). *If a Banach space X has property (u), then every closed subspace of X also has property (u).*

Proof. Let Y be a closed vector subspace of a Banach space X with property (u), and let $\{y_n\}$ be a weak Cauchy sequence of Y. Clearly, $\{y_n\}$ is a weak Cauchy sequence of X, and so there exists a sequence $\{x_n\}$ of X such that

(a) $\sum_{n=1}^\infty |x'(x_n)| < \infty$ holds for all $x' \in X'$, and

(b) $u_n = y_n - \sum_{i=1}^n x_i \xrightarrow{w} 0$ in X.

According to Exercise 8 of Section 3.2, there exist $0 = k_0 < k_1 < \cdots$ and a sequence $\{\alpha_n\} \subseteq [0,1]$ with $\sum_{i=k_n+1}^{k_{n+1}} \alpha_i = 1$ for each n, and with the sequence $v_n = \sum_{i=k_n+1}^{k_{n+1}} \alpha_i u_i$ $(n = 1, 2, \ldots)$ satisfying $\sum_{n=1}^\infty \|v_n\| < \infty$. Put $w_n = \sum_{i=k_n+1}^{k_{n+1}} \alpha_i y_i$, and $z_1 = w_1$ and $z_{n+1} = w_{n+1} - w_n$. Clearly, $\{z_n\}$ is a sequence of Y, and we claim that

(1) $\sum_{n=1}^\infty |y'(z_n)| < \infty$ holds for all $y' \in Y'$, and

(2) $y_n - \sum_{i=1}^n z_i \xrightarrow{w} 0$ in Y.

To see (1), note first that for each n there exist appropriate constants c_i^n $(i = k_n + 1, \ldots, k_{n+2})$ in $[0,1]$ satisfying

$$z_{n+1} = w_{n+1} - w_n = v_{n+1} - v_n + \sum_{i=k_n+1}^{k_{n+2}} c_i^n x_i.$$

Thus, if $y' \in X'$, then

$$\sum_{n=1}^\infty |y'(z_n)| \leq 2\left[\|y'\| \cdot \sum_{n=1}^\infty \|v_n\| + \sum_{n=1}^\infty |y'(x_n)| \right] < \infty.$$

To see (2), let $y' \in Y'$, and let $\epsilon > 0$. Pick some k with $|y'(y_n - y_m)| < \epsilon$ for all $n, m > k$, and note that

$$\left| y'\left(y_n - \sum_{i=1}^n z_i\right) \right| = |y'(y_n - w_n)| = \left| y'\left(\sum_{i=k_n+1}^{k_{n+1}} \alpha_i(y_n - y_i) \right) \right|$$

$$\leq \sum_{i=k_n+1}^{k_{n+1}} \alpha_i |y'(y_n - y_i)| < \epsilon$$

holds for all $n > k$, and the proof is finished. ∎

An important example of a Banach space without property (u) is $C[0,1]$.

Example 4.55. $C[0,1]$ *and* ℓ_∞ *do not have property* (u).

Since $C[0,1]$ is lattice embeddable in ℓ_∞ (for instance, if $\{r_1, r_2, \ldots\}$ is an enumeration of the rational numbers of the interval $[0,1]$, then the mapping $f \mapsto (f(r_1), f(r_2), \ldots)$ is a lattice isometry from $C[0,1]$ into ℓ_∞), it is enough by Theorem 4.54 to show that $C[0,1]$ does not have property (u).

Let $\{x_n\}$ be a bounded sequence of $C[0,1]$. By the Riesz representation theorem it is easy to see that $\{x_n\}$ is weakly Cauchy if and only if $\{x_n\}$ converges pointwise to some real-valued function defined on $[0,1]$. Also (by the Riesz representation theorem again), $\sum_{n=1}^{\infty} |x'(x_n)| < \infty$ for each x' in the dual of $C[0,1]$ is equivalent to $\sum_{n=1}^{\infty} |x_n(t)| < \infty$ for all $t \in [0,1]$. Thus, $C[0,1]$ has property (u) if and only if for every function $f \colon [0,1] \to \mathbb{R}$ which is the pointwise limit of a bounded sequence of continuous functions there exists another sequence $\{f_n\}$ of $C[0,1]$ such that for each $t \in [0,1]$ we have

(a) $\sum_{n=1}^{\infty} |f_n(t)| < \infty$, and
(b) $f(t) = \sum_{n=1}^{\infty} f_n(t)$.

Now assume that a function $f \colon [0,1] \to \mathbb{R}$ satisfies (a) and (b) above. Put $g(t) = -\sum_{n=1}^{\infty} f_n^-(t)$ and $h(t) = -\sum_{n=1}^{\infty} f_n^+(t)$. Clearly, $f = g - h$ holds. Also, a routine argument shows that g and h are both upper semicontinuous functions.[2] Thus, by the above discussion, if $C[0,1]$ has property (u), then every real-valued function on $[0,1]$, which is the pointwise limit of a bounded sequence of continuous functions, must be written as a difference of two upper semicontinuous functions.

However, W. Sierpiński [**178**] has constructed a real-valued function on $\mathbb{R}$ which is the pointwise limit of a bounded sequence of continuous functions and which cannot be written as a difference of two upper semicontinuous functions. This real-valued function can also be assumed to have domain $[0,1]$. For details of Sierpiński's construction we refer the reader to [**178**].

Therefore, there exists a weak Cauchy sequence in $C[0,1]$ that does not satisfy property (u), and this shows that the Banach lattice $C[0,1]$ does not have property (u). ∎

In Dedekind σ-complete Banach lattices property (u) characterizes the order continuity of the norm. The details follow.

---

[2] Recall that a function $f \colon [0,1] \to \mathbb{R}$ is said to be **upper semicontinuous** whenever $f^{-1}((-\infty, a))$ is open in $[0,1]$ for all $a \in \mathbb{R}$, or, equivalently, whenever $t_n \to t$ in $[0,1]$ implies $\limsup f(t_n) \leq f(t)$.

Theorem 4.56. *For a Dedekind σ-complete Banach lattice E the following statements are equivalent.*

(1) *E has order continuous norm.*

(2) *E has property (u).*

(3) *ℓ_∞ is not embeddable in E.*

(4) *ℓ_∞ is not lattice embeddable in E.*

Proof. (1) $\Longrightarrow$ (2) Assume that $\{x_n\}$ is a weak Cauchy sequence of E. Then $x_n \xrightarrow{w^*} x''$ holds in E''. Consider the element $x = \sum_{n=1}^\infty 2^{-n}|x_n| \in E^+$, and let B denote the band generated by x in E''. From Theorem 4.46, it follows that $x'' \in B$.

Now let $v_n = (x'')^+ \wedge nx$ and $u_n = (x'')^- \wedge nx$, $n = 0, 1, 2, \ldots$. Since $v_n, u_n \in [0, nx]$ and E is an ideal of E'', we see that $\{v_n\}$ and $\{u_n\}$ are both sequences of E^+. In addition, since $v_n \uparrow (x'')^+$ and $u_n \uparrow (x'')^-$, we see that

$$v_n \xrightarrow{w^*} (x'')^+ \quad \text{and} \quad u_n \xrightarrow{w^*} (x'')^-$$

holds in E'', and so $x_n - (v_n - u_n) \xrightarrow{w} 0$ holds in E. Next, for each n put $y_n = (v_n - v_{n-1}) - (u_n - u_{n-1}) \in E$, and note that $\sum_{i=1}^n y_i = v_n - u_n$. Therefore, $x_n - \sum_{i=1}^n y_i \xrightarrow{w} 0$ holds in E. On the other hand, if $x' \in E'$, then for each k we have

$$\sum_{n=1}^k |x'(y_n)| \leq \sum_{n=1}^k |x'|(v_n - v_{n-1}) + \sum_{n=1}^k |x'|(u_n - u_{n-1})$$
$$\leq |x'|(v_k) + |x'|(u_k) \leq |x''|(|x'|) < \infty,$$

and so $\sum_{n=1}^\infty |x'(y_n)| < \infty$ holds for each $x' \in E'$. Thus, E has property (u).

(2) $\Longrightarrow$ (3) If ℓ_∞ is embeddable in E, then (by Theorem 4.54) ℓ_∞ must have property (u), contrary to Example 4.55.

(3) $\Longrightarrow$ (4) Obvious.

(4) $\Longrightarrow$ (1) This follows immediately from Theorem 4.51. ∎

Remark. In the preceding theorem, the equivalence of (2) and (3) was first obtained by G. Ya. Lozanovsky [**122**, **123**], while (1) $\Longleftrightarrow$ (4) was established by G. Ya. Lozanovsky and A. A. Mekler [**124**]. The equivalence (1) $\Longleftrightarrow$ (4) was also proven by P. Meyer-Nieberg [**141**]. Results similar to the spirit of Theorem 4.56 were also obtained by H. P. Lotz [**117**]. The implication (1) $\Longleftrightarrow$ (2) also was established by L. Tzafriri [**187**] and P. Meyer-Nieberg [**140**].

Theorem 4.56 has an immediate striking consequence.

Corollary 4.57. *In the class of all Dedekind σ-complete Banach lattices, the order continuity of the norm is a Banach space property.*

That is, two Dedekind σ-complete Banach lattices that are linearly homeomorphic either both have or both do not have order continuous norms. In contrast, this conclusion is no longer true in the general class of Banach lattices. For instance, let c be the Banach lattice of all convergent sequences. If for each $x = (x_1, x_2, \ldots) \in c$ we put $x_\infty = \lim x_n$, then

$$(x_1, x_2, \ldots) \longmapsto (x_\infty, x_1 - x_\infty, x_2 - x_\infty, \ldots)$$

is a linear homeomorphism from c onto c_0. However, c_0 has order continuous norm while the norm of c fails to be order continuous.

We now turn our attention to the important class of KB-spaces.

Definition 4.58. *A Banach lattice E is said to be a **Kantorovich–Banach space** (or briefly a **KB-space**) whenever every increasing norm bounded sequence of E^+ is norm convergent.*

Note that a Banach lattice is a KB-space if and only if $0 \leq x_\alpha \uparrow$ and $\sup\{\|x_\alpha\|\} < \infty$ imply that the net $\{x_\alpha\}$ is norm convergent. (This easily follows from the fact that if a net $\{x_\alpha\}$ with $0 \leq x_\alpha \uparrow$ is not Cauchy, then there exist some $\epsilon > 0$ and a sequence $\{\alpha_n\}$ of indices satisfying $\alpha_n \uparrow$ and $\|x_{\alpha_{n+1}} - x_{\alpha_n}\| > \epsilon$ for all n.) In particular, it follows that every KB-space has order continuous norm.

Now let $\{x_n\}$ be a sequence in a Banach lattice E satisfying $0 \leq x_n \uparrow$ and $\sup\{\|x_n\|\} < \infty$. Then $0 \leq x_n \uparrow x''$ holds in E'' for some $x'' \in E''$. In case E is reflexive, x'' belongs to E, and the order continuity of the norm implies that $\{x_n\}$ is norm convergent. On the other hand, if E is an AL-space, then E'' is also an AL-space, and so E'' has order continuous norm, from which it follows that $\{x_n\}$ is norm convergent in this case too. Therefore, reflexive Banach lattices and AL-spaces are examples of KB-spaces.

Theorem 4.59. *The norm dual E' of a Banach lattice E is a KB-space if and only if E' has order continuous norm.*

Proof. Assume that E' has order continuous norm and that $0 \leq x_n' \uparrow$ holds in E' with $\sup\{\|x_n'\|\} < \infty$. Then $x'(x) = \lim x_n'(x)$ exists in $\mathbb{R}$ for each $x \in E^+$, and moreover this formula defines a positive linear functional on E. Since $x_n' \uparrow x'$ holds in E', we see that $\{x_n'\}$ is norm convergent. ∎

A Banach space X is said to be **weakly sequentially complete** whenever every weak Cauchy sequence of X converges weakly to some vector of X. Clearly, a Banach space X is weakly sequentially complete if and only if $\{x_n\} \subseteq X$ and $x_n \xrightarrow{w^*} x''$ in X'' imply $x'' \in X$.

It is not difficult to see that in c_0 the elements $u_n = (1, \ldots, 1, 0, 0, \ldots)$, where the 1's occupy the first n positions, form a weak Cauchy sequence that fails to converge weakly in c_0. Therefore, c_0 does not embed in any weakly sequentially complete Banach space. Remarkably, the converse is true for Banach lattices. The details are included in the next important theorem that characterizes the KB-spaces. The theorem is a combination of results by many mathematicians.

Theorem 4.60. *For a Banach lattice E the following statements are equivalent.*

(1) *E is a KB-space.*

(2) *E is a band of E''.*

(3) *$E = (E')_n^\sim$.*

(4) *E is weakly sequentially complete.*

(5) *c_0 is not embeddable in E.*

(6) *c_0 is not lattice embeddable in E.*

Proof. $(1) \Longrightarrow (2)$ Since E has order continuous norm, it follows from Theorem 4.9 that E is an ideal of E''. To see that E is a band of E'', let $0 \le x_\alpha \uparrow x''$ hold in E'' with $\{x_\alpha\} \subseteq E$. From $\|x_\alpha\| \le \|x''\| < \infty$, it follows that $\{x_\alpha\}$ is a norm convergent net. If x is its norm limit, then $x_\alpha \uparrow x$ holds in E. Since E is an ideal of E'', we also have $x_\alpha \uparrow x$ in E''. Thus, $x'' = x \in E$ holds, and so E is a band of E''.

$(2) \Longrightarrow (3)$ By Theorem 3.60 we know that E is $|\sigma|(E'', E')$-dense in $(E')_n^\sim$, and by Theorem 3.46 the band E is $|\sigma|(E'', E')$-closed. Consequently, $E = (E')_n^\sim$ holds.

$(3) \Longrightarrow (4)$ Clearly, E is a band of E''. Thus, by Corollary 4.47, the Banach lattice E is w^*-sequentially complete in E'', which means that E is weakly sequentially complete.

$(4) \Longrightarrow (5)$ Let $u_n = (1, \ldots, 1, 0, 0, \ldots)$, where the 1's occupy the first n positions. Then $\{u_n\}$ is a weak Cauchy sequence of c_0. Thus, in case $T \colon c_0 \to E$ is an embedding, $\{T(u_n)\}$ is a weak Cauchy sequence of E, and so by our hypothesis $\{T(u_n)\}$ is weakly convergent in E. Since $T(c_0)$ is a closed vector subspace of E, we see that $\{T(u_n)\}$ is weakly convergent in $T(c_0)$, and consequently $\{u_n\}$ is weakly convergent in c_0. On the other hand, if $u_n \xrightarrow{w} u$ holds in c_0, then clearly $u = (1, 1, 1, \ldots)$, which is not in c_0. This contradiction shows that c_0 is not embeddable in E.

$(5) \Longrightarrow (6)$ Obvious.

(6) $\implies$ (1) Note first that E is Dedekind σ-complete. To see this, assume by way of contradiction that E is not Dedekind σ-complete. Then there exists a sequence $\{y_n\}$ of E satisfying $0 \leq y_n \uparrow \leq x$ for which $\sup\{y_n\}$ does not exist in E. In particular, $\{y_n\}$ is not a norm Cauchy sequence. Therefore, by Theorem 4.13, there exists a disjoint sequence $\{u_n\} \subseteq [0, x]$ that is not norm convergent to zero. Since $0 \leq \sum_{i=1}^{n} u_i \leq x$ holds for all n, it follows from Theorem 4.50 that c_0 is lattice embeddable in E, which is impossible. Thus, E is Dedekind σ-complete.

Since c_0 is a Banach sublattice of ℓ_∞, it follows that ℓ_∞ cannot be lattice embeddable in E. Thus, by Theorem 4.51, the norm of E is order continuous. In particular, E is an ideal of E''.

Now let $0 \leq x_n \uparrow$ in E satisfy $\sup\{\|x_n\|\} < \infty$. To finish the proof, it is enough to show that $\{x_n\}$ is a norm Cauchy sequence. To this end, assume by way of contradiction that this is not the case. Then, by passing to a subsequence, we can assume that for some $\epsilon > 0$ we have $\|x_{n+1} - x_n\| > 2\epsilon$ for all n. Let $0 \leq x_n \uparrow x''$ hold in E'', and then fix some positive integer k with $\frac{2}{k+3}\|x''\| < \epsilon$. By Theorem 4.12 there exist k disjoint sequences $\{y_n^1\}, \dots \{y_n^k\}$ of $[0, x'']$ satisfying

$$y_n^1 + \cdots + y_n^k \leq x_{n+1} - x_n \leq y_n^1 + \cdots + y_n^k + \frac{2}{k+3}x''$$

for all n. Since E is an ideal of E'', we see that $\{y_n^i\} \subseteq E$ holds for each $i = 1, \dots, k$. On the other hand, we have

$$2\epsilon < \|x_{n+1} - x_n\| \leq \|y_n^1\| + \cdots + \|y_n^k\| + \epsilon,$$

and so $\|y_n^1\| + \cdots + \|y_n^k\| > \epsilon$ holds for all n. It follows that for some $1 \leq i \leq k$ the disjoint sequence $\{y_n^i\} \subseteq E^+$ is not norm convergent to zero. In view of $0 \leq \sum_{j=1}^{n} y_j^i \leq x''$, it follows that $\{y_n^i\}$ also has a norm bounded sequence of partial sums. Therefore, by Theorem 4.50, c_0 is lattice embeddable in E, which is a contradiction, and the proof is finished. ∎

Remark. As mentioned before, the preceding theorem is due to many authors. The equivalence of (1)–(4) was established first by T. Ogasawara (see [146]), and later they were reproved to be equivalent by M. Nakamura [146, 147]. The equivalence of statements (1) and (5) was proved by G. Ya. Lozanovsky [122, 123]. Statement (6) was added to the list by P. Meyer-Nieberg [140, 141], where he also proved the other equivalences. Finally, L. Tzafriri [187] proved a result closely related to Theorem 4.60.

It is remarkable that c_0 is embeddable in a Banach lattice if and only if it is lattice embeddable. The next theorem is an immediate consequence of the preceding result and shows that c_0 is quite often embeddable in Banach lattices.

Theorem 4.61. *For a Banach lattice E the following are equivalent.*

(1) E is not a KB-space.

(2) c_0 is embeddable in E.

(3) c_0 is lattice embeddable in E.

The next result characterizes the embeddability of c_0 in the completion of a normed Riesz space.

Lemma 4.62. *The Banach lattice c_0 is embeddable in the norm completion of a normed Riesz space E if and only if there exist a sequence $\{x_n\} \subseteq E^+$ and two positive constants K and M satisfying*

$$K \max_{1 \leq i \leq n} |\alpha_i| \leq \left\| \sum_{i=1}^{n} \alpha_i x_i \right\| \leq M \max_{1 \leq i \leq n} |\alpha_i|$$

for every choice of scalars $\alpha_1, \dots, \alpha_n$.

Proof. The "only if" part follows from Theorem 4.49. For the "if" part assume that c_0 is embeddable in the norm completion $\widehat{E}$ of E. Since $\widehat{E}$ is a Banach lattice (Theorem 4.2), it follows from Theorem 4.61 that c_0 is lattice embeddable in $\widehat{E}$. Thus, there exist a sequence $\{u_n\} \subseteq \widehat{E}^+$ and two positive constants K and M_1 satisfying

$$2K \max_{1 \leq i \leq n} |\alpha_i| \leq \left\| \sum_{i=1}^{n} \alpha_i u_i \right\| \leq M_1 \max_{1 \leq i \leq n} |\alpha_i|$$

for every choice of scalars $\alpha_1, \dots, \alpha_n$.

Now for each n pick some $x_n \in E^+$ with $\|u_n - x_n\| < 2^{-n} K$. If $\alpha_1, \dots, \alpha_n$ are arbitrary scalars, then note that

$$\left\| \sum_{i=1}^{n} \alpha_i x_i \right\| \geq \left\| \sum_{i=1}^{n} \alpha_i u_i \right\| - \left\| \sum_{i=1}^{n} \alpha_i (x_i - u_i) \right\| \geq K \max_{1 \leq i \leq n} |\alpha_i|$$

and

$$\left\| \sum_{i=1}^{n} \alpha_i x_i \right\| \leq \left\| \sum_{i=1}^{n} \alpha_i (x_i - u_i) \right\| + \left\| \sum_{i=1}^{n} \alpha_i u_i \right\|$$
$$\leq (K + M_1) \max_{1 \leq i \leq n} |\alpha_i|,$$

and the proof is finished. ∎

A continuous operator $T \colon X \to Y$ between two Banach spaces is said to **factor through a Banach space** Z whenever there exist continuous operators

$$X \xrightarrow{S} Z \xrightarrow{R} Y$$

such that $T = RS$ holds. (The operators R and S are called **factors** of T.)

The next interesting result of N. Ghoussoub and W. B. Johnson [69] deals with factorization of operators through KB-spaces.

Theorem 4.63 (Ghoussoub–Johnson). *Let $T: E \to X$ be a continuous operator from a Banach lattice E into a Banach space X. If c_0 does not embed in X, then T admits a factorization through a KB-space F*

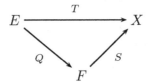

with the factor Q being a lattice homomorphism.

Proof. The formula

$$\rho(x) = \sup\{\|Ty\|\colon \ |y| \le |x|\}, \quad x \in E,$$

defines a lattice seminorm on E. (To see the triangle inequality use Theorem 1.13.) If N is the null ideal of ρ, i.e., $N = \{x \in E\colon \rho(x) = 0\}$, then E/N is a normed Riesz space under the norm

$$\||\dot{x}\|| = \rho(x).$$

The norm completion F of E/N is a Banach lattice (Theorem 4.2).

Clearly,

$$\|Tx\| \le \||\dot{x}\|| \le \|T\| \cdot \|x\|$$

holds for all $x \in X$. In particular, this implies that the formula $S(\dot{x}) = T(x)$ defines a continuous operator from E/N into X. Denote again by S the unique linear extension of S to all of F. Also, if $Q: E \to F$ is defined by $Q(x) = \dot{x}$, then Q is a lattice homomorphism, and moreover we have the factorization

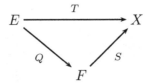

To finish the proof, it is enough to show that F is a KB-space.

If c_0 is embeddable in F, then by Lemma 4.62 there exist a sequence $\{u_n\}$ of E^+ and two constants $K, M > 0$ satisfying

$$2K \max_{1 \le i \le n} |\alpha_i| \le \left\|\left|\sum_{i=1}^{n} \alpha_i \dot{u}_i\right\|\right| \le M \max_{1 \le i \le n} |\alpha_i|$$

for every choice of scalars $\alpha_1, \ldots, \alpha_n$. Clearly, $\rho(u_n) = \||\dot{u}_n\|| \ge 2K$. Thus, for each n there exists some $|v_n| \le u_n$ with $\|Tv_n\| = \|S\dot{v}_n\| > K$. Since

$\dot{u}_n \xrightarrow{w} 0$ holds in F, it follows from $|\dot{v}_n| \leq \dot{u}_n$ that $\dot{v}_n \xrightarrow{w} 0$ in F, and hence $S\dot{v}_n \xrightarrow{w} 0$ holds in X. On the other hand, for every choice of scalars $\alpha_1, \ldots, \alpha_n$ we have

$$\left\| \sum_{i=1}^n \alpha_i S\dot{v}_i \right\| \leq \|S\| \cdot \left\| \left| \sum_{i=1}^n \alpha_i \dot{v}_i \right| \right\| \leq \|S\| \cdot \left\| \left| \sum_{i=1}^n |\alpha_i| \cdot |\dot{v}_i| \right| \right\|$$

$$\leq \|S\| \cdot \left\| \left| \sum_{i=1}^n |\alpha_i| \dot{u}_i \right| \right\| \leq M \|S\| \max_{1 \leq i \leq n} |\alpha_i|.$$

Therefore, the sequence $\{S\dot{v}_n\}$ of X satisfies statement (3) of Theorem 4.49, and so c_0 is embeddable in X, which is a contradiction. Consequently, F is a KB-space, as desired. ∎

Recall that a closed vector subspace Y of a Banach space X is said to be **complemented** (or that Y has a complement in X) whenever there exists another closed vector subspace Z of X such that $X = Y \oplus Z$. The closed vector subspace Z is referred to as a **complement** of Y (and, of course, Y is a complement of Z). By the closed graph theorem, it is easy to see that the projection of X onto Y along Z is continuous. In fact, an easy argument shows that a closed vector subspace Y of a Banach space X is complemented if and only if there exists a continuous projection on X whose range is Y. Also, recall that a Banach space Y is said to **embed complementably** into another Banach space X whenever there exists an embedding $T \colon Y \to X$ so that $T(Y)$ is complemented in X.

Regarding embeddings of Banach spaces into KB-spaces, we have the following remarkable result of W. B. Johnson and L. Tzafriri [80].

Theorem 4.64 (Johnson–Tzafriri). *If a Banach space X embeds complementably into a Banach lattice and c_0 does not embed in X, then X also embeds complementably in a KB-space.*

Proof. Let Y be a complemented closed vector subspace of a Banach lattice E, let $T \colon X \to Y$ be an invertible continuous operator from X onto Y, and let $P \colon E \to E$ be a continuous projection with range Y. Also, let $J \colon Y \to E$ denote the natural inclusion. Clearly, c_0 does not embed in Y, and so, by Theorem 4.63, the operator $P \colon E \to Y$ factors through a KB-space F. In other words, we have the following scheme of operators

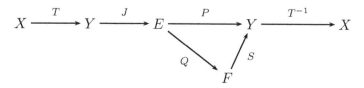

with F a KB-space. We claim that $QJT\colon X \to F$ is an embedding whose image is complemented in F. This claim will be established by steps.

Step 1. QJT *is one-to-one.*

Indeed, if $QJT(x) = 0$, then note that

$$x = [T^{-1}PJT](x) = [T^{-1}SQJT](x) = 0 \,.$$

Step 2. $QJT(X)$ *is a closed vector subspace of* F, *and so* $QJT\colon X \to F$ *is an embedding.*

If $QJTx_n \to y$ holds in F, then we have

$$x_n = [T^{-1}SQJT](x_n) \to T^{-1}Sy = z \in X \,.$$

Now note that $y = QJTz$ holds.

Step 3. $F = QJT(X) \oplus S^{-1}(\{0\})$ *holds, proving that* $QJT(X)$ *is a complemented closed vector subspace of* F.

Let $y \in F$. Pick some $x \in X$ with $Sy = Tx$. Then $z = y - QJTx \in F$ satisfies $S(z) = 0$, and so $z \in S^{-1}(\{0\})$. Now note that $y = QJTx + z$ holds.

To see that F is the direct sum of $QJT(X)$ and $S^{-1}(\{0\})$, let $QJTx+y=0$ with $y \in S^{-1}(\{0\})$. Then

$$x = T^{-1}SQJTx = T^{-1}S(QJTx + y) = 0 \,,$$

and so $y = 0$. This completes the proof of the theorem. ■

The next result is a dual of Theorem 4.63 and is also due to N. Ghoussoub and W. B. Johnson [**69**].

Theorem 4.65 (Ghoussoub–Johnson). *Let* $T\colon X \to E$ *be a continuous operator from a Banach space* X *into a Banach lattice* E. *If* c_0 *does not embed in* X', *then* T *admits a factorization through a Banach lattice* F

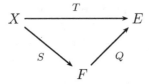

such that F' *is a* KB-space *and* Q *is an interval preserving lattice homomorphism.*

Proof. We denote by U the closed unit ball of X. Let A be the ideal generated by $T(X)$ in E, and let C be the convex solid hull of $T(U)$. Clearly,

C is a norm bounded subset of E, and moreover $C \subseteq A$ holds. Then the Minkowski functional ρ of C on A, i.e.,

$$\rho(x) = \inf\{\lambda \colon x \in \lambda C\}, \quad x \in A,$$

defines a lattice norm on A. Let F be the norm completion of (A, ρ). By Theorem 4.2, F is a Banach lattice. Since $\|x\| \le \|T\|\rho(x)$ holds for all $x \in A$ (why?), it follows that the natural inclusion $J \colon (A, \rho) \to E$ is continuous. Thus (according to Exercise 7 of Section 4.1), J extends to a unique interval preserving lattice homomorphism Q from F to E. Also, if $S \colon X \to F$ is defined by $S(x) = T(x)$, then (in view of $S(U) = T(U) \subseteq C$) the operator S is continuous, and moreover we have the factorization

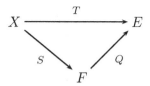

We show next that F' is a KB-space. For this, it is enough to show that F' has order continuous norm. To this end, let $\{f_n\}$ be a disjoint sequence of F' such that $0 \le f_n \le f$ holds for all n and some $f \in F'$. If ρ' denotes the norm of F', then (in view of Theorem 4.14) it suffices to establish that $\lim \rho'(f_n) = 0$. Form the inequality

$$\begin{aligned} \rho'(f_n) &= \sup\{|f_n(y)| \colon \rho(y) \le 1\} = \sup\{|f_n(y)| \colon y \in C\} \\ &\le \sup\{f_n(|Tx|) \colon x \in U\}, \end{aligned}$$

we see that for each n there exists some $x_n \in U$ with $\rho'(f_n) - \frac{1}{n} < f_n(|Tx_n|)$. Also, by Theorem 1.23, for each n there exists some $g_n \in F'$ with $|g_n| \le f_n$ and $f_n(|Tx_n|) = g_n(Tx_n) = g_n(Sx_n)$. Thus,

$$\rho'(f_n) < \tfrac{1}{n} + [S'g_n](x_n) \le \tfrac{1}{n} + \|S'g_n\|. \tag{$\star$}$$

Next, note that for each $x'' \in X''$, we have

$$\sum_{n=1}^{\infty} |x''(S'g_n)| = \sum_{n=1}^{\infty} |S''x''(g_n)| \le \sum_{n=1}^{\infty} |S''x''|(|g_n|) \le |S''x''|(f) < \infty.$$

Since c_0 does not embed in X', it follows from condition (2) of Theorem 4.49 that $\sum_{n=1}^{\infty} S'g_n$ is norm convergent in X'. In particular, we have $\lim \|S'g_n\| = 0$, and so from $(\star)$ we see that $\lim \rho'(f_n) = 0$, as desired. ∎

We now turn our attention to embeddings of ℓ_1. Before doing so, let us mention a few important properties of ℓ_1. Clearly, ℓ_1 is a KB-space, and so ℓ_1 is a band in its double dual. In fact, the next theorem shows that $\ell_\infty' = \ell_1 \oplus (c_0)^\circ$.

Theorem 4.66. *If ℓ_1^{d} denotes the disjoint complement of ℓ_1 in ℓ'_∞, then for an element $f \in \ell'_\infty$ the following statements are equivalent:*

(1) $f \in \ell_1^{\mathrm{d}}$.

(2) $f(e_n) = 0$ *holds for all* n.

(3) *There exists a constant* κ *such that*

$$f(x) = \kappa \lim x_n$$

holds for all $x = (x_1, x_2, \ldots) \in c$.

Proof. $(1) \Longrightarrow (2)$ If $x \in \ell_\infty$ satisfies $x \wedge (e_n - x) = 0$, then it should be clear that either $x = e_n$ or $x = 0$. Thus, by Theorem 1.50, we have

$$0 = |f| \wedge e_n(e_n) = \inf\{|f|(x) + e_n(e_n - x) : \ x \wedge (e_n - x) = 0\}$$
$$= \min\{|f|(e_n), 1\}.$$

This implies $|f|(e_n) = 0$, and so $f(e_n) = 0$ for each n.

$(2) \Longrightarrow (3)$ Let $e = (1, 1, \ldots) \in c$. If $x \in c$, then

$$f(x) - f(e) \lim_{n \to \infty} x_n$$
$$= f\left(x_1 - \lim_{n \to \infty} x_n, x_2 - \lim_{n \to \infty} x_n, \ldots\right)$$
$$f\left(0, \ldots, 0, x_k - \lim_{n \to \infty} x_n, x_{k+1} - \lim_{n \to \infty} x_n, \ldots\right) \longrightarrow 0,$$

and so $f(x) = f(e) \lim_{n \to \infty} x_n$ holds.

$(3) \Longrightarrow (1)$ Put $u_n = (1, \ldots, 1, 0, 0, \ldots) = e_1 + \cdots + e_n$, and note that

$$|f|(u_n) = \sup\{|f(v)| : \ |v| \le u_n\} = 0.$$

Thus, if $x = (x_1, x_2, \ldots) \in \ell_1$, then

$$0 \le \big(|f| \wedge |x|\big)(e) \le |f|(u_n) + |x|(e - u_n) = \sum_{i=n+1}^{\infty} |x_i| \longrightarrow 0$$

holds. Since e is an order unit for ℓ_∞, the latter shows that $|f| \wedge |x| = 0$ holds for all $x \in \ell_1$. That is, $f \in \ell_1^{\mathrm{d}}$, as required. ∎

R. S. Phillips [**164**] has shown that w^*-convergence in the dual of ℓ_∞ implies uniform convergence on the closed unit ball of c_0. This result is known as **Phillip's lemma** and is stated next.

Theorem 4.67 (Phillip's Lemma). *If a sequence $\{f_n\} \subseteq \ell'_\infty$ satisfies $f_n \xrightarrow{w^*} 0$, then $\{f_n\}$ converges uniformly to zero on the closed unit ball of c_0.*

Proof. We know that ℓ_1 is a band in $\ell_1'' = \ell_\infty'$. Therefore, we can write $f_n = x_n + g_n$ with $x_n \in \ell_1$ and $g_n \in \ell_1^d$. From Theorem 4.46 we have $x_n \xrightarrow{\;w\;} 0$ in ℓ_1, and so Theorem 4.32 implies $\lim \|x_n\|_1 = 0$. Since $g_n(x) = 0$ holds for all $x \in c_0$ (see Theorem 4.66), it follows that

$$\sup\{|f_n(x)|\colon\ x \in c_0 \text{ and } \|x\|_\infty \le 1\}$$
$$= \sup\{|(x_n + g_n)(x)|\colon\ x \in c_0 \text{ and } \|x\|_\infty \le 1\}$$
$$= \sup\{|x_n(x)|\colon\ x \in c_0 \text{ and } \|x\|_\infty \le 1\}$$
$$= \|x_n\|_1 \longrightarrow 0,$$

and the proof is finished. ∎

We now continue with embeddings of ℓ_1. As previously stated, the Banach space ℓ_1 is embeddable into a Banach space X if and only if X has a sequence that is equivalent to the standard basis $\{e_n\}$ of ℓ_1. It is easy to see that for this to happen it is necessary and sufficient that there exists a sequence $\{x_n\}$ of X and two positive constants K and M satisfying

$$K \sum_{i=1}^{n} |\alpha_i| \le \left\| \sum_{i=1}^{n} \alpha_i x_i \right\| \le M \sum_{i=1}^{n} |\alpha_i|$$

for every n and all choices of scalars $\alpha_1, \ldots, \alpha_n$. Note that if a sequence $\{x_n\}$ in a Banach space satisfies $\|x_n\| \le M$ for all n, then

$$\left\| \sum_{i=1}^{n} \alpha_i x_i \right\| \le \sum_{i=1}^{n} |\alpha_i| \cdot \|x_i\| \le M \sum_{i=1}^{n} |\alpha_i|$$

holds for every choice of scalars $\alpha_1, \ldots, \alpha_n$. Therefore, ℓ_1 is embeddable in a Banach space X if and only if there exist a norm bounded sequence $\{x_n\}$ of X and a positive constant K satisfying

$$K \sum_{i=1}^{n} |\alpha_i| \le \left\| \sum_{i=1}^{n} \alpha_i x_i \right\|$$

for every n and all choices of scalars $\alpha_1, \ldots, \alpha_n$. In this case, an embedding $T\colon \ell_1 \to X$ is given by the formula

$$T(\alpha_1, \alpha_2, \ldots) = \sum_{n=1}^{\infty} \alpha_n x_n .$$

Also, note that by replacing each x_n by $x_n/\|x_n\|$ we can assume that each x_n is a unit vector.

It is important to observe that the standard basis $\{e_n\}$ in ℓ_1 does not have any weak Cauchy subsequences. Indeed, if $\{e_{k_n}\}$ is a subsequence of $\{e_n\}$, then consider $x' = (x_1, x_2, \ldots) \in \ell_\infty = \ell_1'$ defined by $x_{k_{2n}} = 1$ for $n = 1, 2, \ldots$ and $x_i = 0$ otherwise, and note that $\lim x'(e_{k_n})$ does not exist in $\mathbb{R}$. Thus, if a sequence $\{x_n\}$ in a Banach space has a weak Cauchy subsequence, then $\{x_n\}$ cannot be equivalent to the standard basis in ℓ_1.

C. Bessaga and A. Pelczynski [**32**] have shown that ℓ_1 embeds complementably in a Banach space X if and only if c_0 embeds in X'. This result is stated next.

Theorem 4.68 (Bessaga–Pelczynski). *For a Banach space X the following statements are equivalent.*

    (1) ℓ_1 *embeds complementably in* X.

    (2) ℓ_∞ *embeds complementably in* X'.

    (3) ℓ_∞ *embeds in* X'.

    (4) c_0 *embeds in* X'.

Proof. (1) $\Longrightarrow$ (2) If two closed vector subspaces Y and Z of X satisfy $X = Y \oplus Z$, then $X' = Y' \oplus Z'$ holds. So, if ℓ_1 is linearly homeomorphic to Y, then ℓ_∞ is linearly homeomorphic to Y' (which is complemented in X').

(2) $\Longrightarrow$ (3) and (3) $\Longrightarrow$ (4) are obvious.

(4) $\Longrightarrow$ (1) Let $T\colon c_0 \to X'$ be an embedding. Then, an easy application of the Hahn–Banach Theorem 1.25 guarantees that $T'\colon X'' \to c_0' = \ell_1$ is onto. Denote by $S\colon X \to \ell_1$ the restriction of T' on X, and note that for each $x \in X$ we have

$$Sx = \big(Te_1(x), Te_2(x), \ldots\big).$$

The proof will be completed by steps.

STEP 1. There exist a norm bounded sequence $\{x_n\}$ of X and some $M > 1$ such that $Sx_n = \big(Te_1(x_n), Te_2(x_n), \ldots\big)$ satisfies

    (a) $\|Sx_n\|_1 \le M$, and

    (b) $\sum_{i=1}^{n-1} |Te_i(x_n)| < \frac{1}{n}$ and $Te_n(x_n) = 1$.

To see this, let U denote the closed unit ball of X, and let U_1 be the closed unit ball of ℓ_1. Since $T'\colon X'' \to \ell_1$ is onto, it follows from the open mapping theorem that there exists some $r > 0$ so that $U_1 \subseteq rT'(U'')$. If bar denotes the w^*-closure in ℓ_1, then in view of the w^*-denseness of U in U'', we see that $U_1 \subseteq \overline{S(rU)}$. Now for each n let

$$\Phi_n = \big\{(\alpha_1, \ldots, \alpha_n, 0, 0, \ldots)\colon\ \alpha_i = \pm 1 \text{ for each } i = 1, \ldots, n\big\},$$

and note that Φ_n is a finite subset of c_0. Taking into account that $e_n \in U_1$, we see that there exists some $y_n \in rU$ so that

$$\big|\langle \alpha, e_n - S(y_n)\rangle\big| = \left| \sum_{i=1}^{n-1} \alpha_i Te_i(y_n) + \alpha_n[1 - Te_n(y_n)] \right| < \frac{1}{2n}$$

holds for all $\alpha \in \Phi_n$. This implies

$$\sum_{i=1}^{n-1} |Te_i(y_n)| < \tfrac{1}{2n} \quad \text{and} \quad |1 - Te_n(y_n)| < \tfrac{1}{2n} \le \tfrac{1}{2}.$$

The latter inequality implies the existence of some δ_n satisfying $0 < |\delta_n| \le 2$, $x_n = \delta_n y_n$ and $Te_n(x_n) = 1$. Clearly, $\sum_{i=1}^{n-1} |Te_i(x_n)| < \tfrac{1}{n}$ holds. The existence of the constant $M > 0$ now follows from the inequalities

$$\|Sx_n\|_1 \le \|S\| \cdot \|x_n\| \le 2r\|S\|.$$

STEP 2. There exists a sequence $\{k_n\}$ of strictly increasing natural numbers such that if $u_n = \sum_{i=k_n}^{k_{n+1}-1} Te_i(x_{k_n})e_i$, then

$$\|u_n\|_1 \le M \quad \text{and} \quad \|Sx_{k_n} - u_n\|_1 < (2M)^{-1}2^{-n-1}$$

holds for all n.

To see this, let $\delta = (2M)^{-1}2^{-1}$. Pick some $k_1 > 1$ with $\tfrac{1}{k_1} < \delta 2^{-2}$, and choose $k_2 > k_1$ satisfying $\tfrac{1}{k_2} < \delta 2^{-3}$ and $\sum_{i=k_2}^{\infty} |Te_i(x_{k_1})| < \delta 2^{-2}$. Put $u_1 = \sum_{i=k_1}^{k_2-1} Te_i(x_{k_1})e_i$, and note that

$$\begin{aligned}
\|Sx_{k_1} - u_1\|_1 &= \sum_{i=1}^{k_1-1} |Te_i(x_{k_1})| + \sum_{i=k_2}^{\infty} |Te_i(x_{k_1})| \\
&< \tfrac{1}{k_1} + \delta 2^{-2} < (2M)^{-1}2^{-2}.
\end{aligned}$$

Next, choose $k_3 > k_2$ with $\tfrac{1}{k_3} < \delta 2^{-4}$ and $\sum_{i=k_3}^{\infty} |Te_i(x_{k_2})| < \delta 2^{-3}$. Put $u_2 = \sum_{i=k_2}^{k_3-1} Te_i(x_{k_2})e_i$, and note that

$$\begin{aligned}
\|Sx_{k_2} - u_2\|_1 &= \sum_{i=1}^{k_2-1} |Te_i(x_{k_2})| + \sum_{i=k_3}^{\infty} |Te_i(x_{k_2})| \\
&< \tfrac{1}{k_2} + \delta 2^{-3} < (2M)^{-1}2^{-3}.
\end{aligned}$$

Now continue the construction inductively in the obvious manner, and note that

$$\|u_n\|_1 \le \|Sx_{k_n}\|_1 \le M.$$

STEP 3. The sequence $\{u_n\}$ is equivalent to the standard basis of ℓ_1. Moreover, if Y is the closed vector subspace generated by $\{u_n\}$ in ℓ_1, then there exists a continuous projection P on ℓ_1 with range Y and $\|P\| \le M$.

Since the k_nth component of u_n is 1 and $u_n \wedge u_m = 0$ for $n \neq m$, we see that

$$\sum_{i=1}^{k} |\alpha_i| \le \left\| \sum_{i=1}^{k} \alpha_i u_i \right\|_1$$

holds for every choice of scalars $\alpha_1, \ldots, \alpha_k$.

Now consider the operator $P\colon \ell_1 \to \ell_1$ defined by

$$P(\alpha_1, \alpha_2, \ldots) = \sum_{n=1}^{\infty} \alpha_{k_n} u_n \, .$$

Again, since the k_nth component of u_n is 1 and $u_n \wedge u_m = 0$ for $n \neq m$, we see that the k_nth component of $P(\alpha_1, \alpha_n, \ldots)$ equals α_{k_n}. This easily implies that P is a projection whose range is Y. On the other hand, the estimate

$$\|P(\alpha_1, \alpha_2, \ldots)\|_1 \leq M \sum_{n=1}^{\infty} |\alpha_{k_n}| \leq M \|(\alpha_1, \alpha_2, \ldots)\|_1$$

shows that P is continuous and that $\|P\| \leq M$.

STEP 4. The sequence $\{Sx_{k_n}\}$ is equivalent to the standard basis of ℓ_1. Moreover, if Z is the closed vector subspace generated by $\{Sx_{k_n}\}$, then the continuous operator $R\colon Y \to Z$ from Y onto Z defined by

$$R\Big(\sum_{i=1}^{\infty} \alpha_i u_i\Big) = \sum_{i=1}^{\infty} \alpha_i Sx_{k_i}$$

satisfies $\|R\| \leq \frac{3}{2}$.

To see that $\{Sx_{k_n}\}$ is equivalent to the standard basis of ℓ_1 note that

$$\Big\| \sum_{i=1}^{k} \alpha_i Sx_{k_i} \Big\|_1 \geq \Big\| \sum_{i=1}^{k} \alpha_i u_i \Big\|_1 - \sum_{i=1}^{k} |\alpha_i| \cdot \|Sx_{k_i} - u_i\|_1$$

$$\geq \sum_{i=1}^{k} |\alpha_i| - \tfrac{1}{2} \sum_{i=1}^{k} |\alpha_i| = \tfrac{1}{2} \sum_{i=1}^{k} |\alpha_i| \, .$$

For the norm estimate of R, observe that if $u = \sum_{n=1}^{\infty} \alpha_n u_n \in Y$, then

$$\|u - Ru\|_1 = \Big\| \sum_{n=1}^{\infty} \alpha_n (u_n - Sx_{k_n}) \Big\|_1 \leq \sum_{n=1}^{\infty} \Big[\sum_{i=1}^{\infty} |\alpha_i| \Big] \cdot 2^{-1-n}$$

$$= \tfrac{1}{2} \sum_{i=1}^{\infty} |\alpha_i| \leq \tfrac{1}{2} \|u\|_1 \, ,$$

and so $\|Ru\|_1 \leq \frac{3}{2} \|u\|_1$ holds. This proves that $\|R\| \leq \frac{3}{2}$.

STEP 5. Consider the operators $\ell_1 \xrightarrow{P} Y \xrightarrow{R} Z$. Then, the operator $RP\colon Z \to Z$ is an invertible operator.

Indeed, if $z = \sum_{n=1}^{\infty} \alpha_n S x_{k_n} \in Z$, then we have

$$
\begin{aligned}
\|RPz - z\|_1 &= \left\| RP\Big(\sum_{n=1}^{\infty} \alpha_n (S x_{k_n} - u_n)\Big) \right\|_1 \\
&\leq \|R\| \cdot \|P\| \cdot \Big[\sum_{n=1}^{\infty} |\alpha_n|\Big] \cdot \Big[\sum_{n=1}^{\infty} \|S x_{k_n} - u_n\|_1\Big] \\
&\leq \tfrac{3}{2}M \cdot 2\|z\|_1 \cdot (2M)^{-1} \cdot \tfrac{1}{2} = \tfrac{3}{4}\|z\|_1 .
\end{aligned}
$$

Thus, $\|RP - I\| \leq \tfrac{3}{4} < 1$ holds, and this implies that $RP\colon Z \to Z$ is invertible, where, of course, $(RP)^{-1} = \sum_{n=0}^{\infty}(I - RP)^n$.

STEP 6. Let $A\colon Z \to Z$ denote the inverse of $RP\colon Z \to Z$, and consider the operators $\ell_1 \xrightarrow{P} Y \xrightarrow{R} Z \xrightarrow{A} Z$. Then $Q = ARP$ is a continuous projection on ℓ_1 whose range is Z.

Clearly, $ARPz = z$ holds for all $z \in Z$, and so

$$
(ARP)^2 \alpha = ARP(ARP\alpha) = ARP\alpha
$$

holds for all $\alpha \in \ell_1$.

STEP 7. The sequence $\{x_{k_n}\}$ is equivalent to the standard basis of ℓ_1. Moreover, if W is the closed vector subspace generated by $\{x_{k_n}\}$ in X, then W is a complemented copy of ℓ_1 in X.

Note that

$$
\left\| \sum_{i=1}^{k} \alpha_i x_{k_i} \right\| \leq 2r \sum_{i=1}^{k} |\alpha_i| \leq 4r \left\| \sum_{i=1}^{k} \alpha_i S x_{k_i} \right\|_1 \leq 4r\|S\| \cdot \left\| \sum_{i=1}^{k} \alpha_i x_{k_i} \right\|,
$$

and so $\{x_{k_n}\}$ is equivalent to the standard basis of ℓ_1. Moreover, for each $\sum_{n=1}^{\infty} \alpha_n x_{k_n} \in W$ we have $S\big(\sum_{n=1}^{\infty} \alpha_n x_{k_n}\big) = \sum_{n=1}^{\infty} \alpha_n S x_{k_n}$. This shows that S carries W onto Z, and that S as an operator from W to Z has a continuous inverse. If $B\colon Z \to W$ denotes the inverse of $S\colon W \to Z$, then consider the operators $X \xrightarrow{S} \ell_1 \xrightarrow{Q} \xrightarrow{B} W$, and note that BQS is a continuous projection on X whose range is W. The proof of the theorem is now complete. ∎

We are now ready to characterize the lattice embeddability of ℓ_1.

Theorem 4.69. *For a Banach lattice E the following statements are equivalent.*

(1) *ℓ_1 is lattice embeddable in E.*

(2) *There exists a norm bounded disjoint sequence of E^+ which is equivalent to the standard basis of ℓ_1.*

(3) *There exists a norm bounded disjoint sequence of E which does not converge weakly to zero.*

(4) E' *does not have order continuous norm, i.e., E' is not a KB-space.*

(5) E' *is not weakly sequentially complete.*

(6) ℓ_∞ *is lattice embeddable in E'.*

(7) c_0 *is lattice embeddable in E'.*

(8) c_0 *is embeddable in E'.*

(9) ℓ_∞ *embeds complementably in E'.*

(10) ℓ_∞ *embeds in E'.*

(11) ℓ_1 *embeds complementably in E.*

Proof. From Theorems 4.51, 4.61, and 4.68, we see that statements (4)–(11) are mutually equivalent.

(1) $\Longrightarrow$ (2) If $T\colon \ell_1 \to E$ is a lattice embedding, then the sequence $\{T(e_n)\}$ of E^+ is disjoint and equivalent to $\{e_n\}$.

(2) $\Longrightarrow$ (3) If $\{x_n\}$ is a norm bounded disjoint sequence of E^+ that is equivalent to the standard basis of ℓ_1, then $\{x_n\}$ does not converge weakly to zero.

(3) $\Longrightarrow$ (4) Let $\{x_n\}$ be a disjoint sequence of U that does not converge weakly to zero, and assume by way of contradiction that E' has order continuous norm.

Let $0 \le x' \in E'$, and let $\epsilon > 0$ be fixed. By Theorem 4.19, there exists some $y \in E^+$ satisfying

$$x'(|x| - y)^+ < \epsilon$$

for all $x \in U$. Since $\{|x_n| \wedge y\}$ is an order bounded disjoint sequence, we have $|x_n| \wedge y \xrightarrow{w} 0$ in E, and so there exists some k such that $x'(|x_n| \wedge y) < \epsilon$ holds for all $n \ge k$. In particular, for $n \ge k$ we have

$$|x'(x_n)| \le x'(|x_n| - y)^+ + x'(|x_n| \wedge y) < \epsilon + \epsilon = 2\epsilon.$$

This shows that $x_n \xrightarrow{w} 0$ holds in E, which is impossible. Hence, E' does not have order continuous norm.

(4) $\Longrightarrow$ (1) Since E' does not have order continuous norm, it follows from Theorem 4.19 that there exist $0 \le x' \in E'$ and $\epsilon > 0$ such that for each $y \in E^+$ we have $x'(x - y)^+ > 2\epsilon$ for at least one $x \in U^+$. In particular, there exists a sequence $\{x_n\} \subseteq U^+$ satisfying

$$x'\left(x_{n+1} - 4^n \sum_{i=1}^{n} x_i\right)^+ > 2\epsilon$$

for all n.

Let $x = \sum_{n=1}^{\infty} 2^{-n} x_n$. If $u_n = (x_{n+1} - 4^n \sum_{i=1}^{n} x_i - 2^{-n} x)^+$, then it follows from $0 \le u_n \le x_{n+1} \in U^+$ that $\{u_n\} \subseteq U^+$. Also, by Lemma 4.35, the sequence $\{u_n\}$ is disjoint. On the other hand, for each n we have

$$2\epsilon < x'\left(x_{n+1} - 4^n \sum_{i=1}^{n} x_i\right)^+ \le x'(u_n) + 2^{-n} x'(x),$$

and so, by passing to a subsequence of $\{u_n\}$, we can assume that $x'(u_n) > \epsilon$ holds for all n. Now if $\alpha_1, \ldots, \alpha_n$ are arbitrary real numbers, then we have

$$\epsilon \sum_{i=1}^{n} |\alpha_i| \le \sum_{i=1}^{n} |\alpha_i| x'(u_i) = x'\left(\sum_{i=1}^{n} |\alpha_i| u_i\right) = x'\left(\left|\sum_{i=1}^{n} \alpha_i u_i\right|\right)$$

$$\le \|x'\| \cdot \left\|\left|\sum_{i=1}^{n} \alpha_i u_i\right|\right\| = \|x'\| \cdot \left\|\sum_{i=1}^{n} \alpha_i u_i\right\|.$$

From this, it easily follows that the operator $T \colon \ell_1 \to E$ defined by

$$T(\alpha_1, \alpha_2, \ldots) = \sum_{n=1}^{\infty} \alpha_n u_n$$

is a lattice embedding, the proof is finished. ∎

Remark. In the preceding theorem, the equivalence of (1) and (6) was established first by P. Meyer-Nieberg [140] under the assumption that E had order continuous norm. In its general form the equivalence (1) $\iff$ (6) was proven by B. Kühn [107].

The next two theorems characterize the reflexive Banach lattices. The first one is due to T. Ogasawara [157].

Theorem 4.70 (Ogasawara). *A Banach lattice E is reflexive if and only if E and E' are both KB-spaces.*

Proof. If the Banach lattice E is reflexive, then clearly E and E' are both KB-spaces.

For the converse, assume that E and E' are both KB-spaces. Then, by Theorem 4.60 we have $E = (E')_n^{\sim}$. On the other hand, the order continuity of the norm in E' implies $E'' = (E')_n^{\sim}$. Therefore, $E = E''$ holds, and so E is a reflexive Banach lattice. ∎

In terms of embeddings, G. Ya. Lozanovsky [122, 123] characterized the reflexive Banach lattices as follows.

Theorem 4.71 (Lozanovsky). *For a Banach lattice E the following statements are equivalent.*

(1) *E is reflexive.*

(2) *Neither c_0 nor ℓ_1 is lattice embeddable in E.*

(3) *ℓ_1 is not lattice embeddable in either E or E'.*

Proof. The nonembeddability of c_0 in E is equivalent (by Theorem 4.60) to E being a KB-space. The nonlattice embeddability of ℓ_1 in E is equivalent (by Theorem 4.69) to E' being a KB-space. Thus, (2) is equivalent to saying that E and E' are both KB-spaces, which by Theorem 4.70 is equivalent to the reflexivity of E.

To see that (3) is equivalent to the reflexivity of E, note that the nonlattice embeddability of ℓ_1 in E is equivalent to the nonlattice embeddability of c_0 in E'; see Theorem 4.69. Thus, by the previous case, statement (3) is equivalent to the reflexivity of E', which in turn is equivalent to the reflexivity of E. ■

The above theorem was reproved by P. Meyer-Nieberg [**140, 141**] and L. Tzafriri [**187**].

Finally, we shall close the section with an elegant result of H. P. Rosenthal [**170**] regarding embeddings of ℓ_1. Keep in mind that in a Banach space a weak Cauchy sequence cannot be equivalent to the standard basis of ℓ_1.

Theorem 4.72 (Rosenthal). *If $\{x_n\}$ is a norm bounded sequence in a Banach space, then there exists a subsequence $\{y_n\}$ of $\{x_n\}$ satisfying one of the two mutually exclusive alternatives:*

(a) *$\{y_n\}$ is a weak Cauchy sequence.*

(b) *$\{y_n\}$ is equivalent to the standard basis of ℓ_1.*

Proof. Let X be a Banach space, and let Ω denote the closed unit ball of X' equipped with the w^*-topology. Then, X can be considered as a Banach subspace of $C(\Omega)$ where, of course, each $x \in X$ is identified with the function $x(x') = x'(x)$, $x' \in \Omega$. Now note that (by the Riesz representation theorem) a sequence $\{x_n\}$ in $C(\Omega)$ is weakly Cauchy if and only if $\{x_n\}$ is norm bounded and converges pointwise. Thus, in order to establish the theorem it suffices to prove the following result.

- *Let $\ell_\infty(\Omega)$ be the Banach lattice of all bounded real-valued functions on an arbitrary set Ω with the sup norm. If a sequence $\{f_n\}$ of $\ell_\infty(\Omega)$ satisfies $\|f_n\|_\infty \le 1$ and has no pointwise convergent subsequence, then $\{f_n\}$ has a subsequence that is equivalent to the standard basis of ℓ_1.*

The proof of (•) will be done by steps. So, let Ω be an arbitrary nonempty set, and let $\{f_n\}$ be a sequence of $\ell_\infty(\Omega)$ satisfying $\|f_n\|_\infty \le 1$ and with no pointwise convergent subsequence.

STEP 1. A sequence $\{(A_n, B_n)\}$ of pairs of subsets of a set Θ with $A_n \cap B_n = \emptyset$ for all n is said to **converge** on a subset D of Θ whenever each $x \in D$ belongs to at most finitely many A_n or to at most finitely many B_n (i.e., whenever $\lim \chi_{A_n}(x) = 0$ or $\lim \chi_{B_n}(x) = 0$ hold for all $x \in D$).

We claim that there exist $\delta > 0$, $r \in \mathbb{R}$, and a subsequence $\{g_n\}$ of $\{f_n\}$ such that if

$$A_n = \{\omega \in \Omega \colon \ g_n(\omega) > r + 2\delta\} \quad and \quad B_n = \{\omega \in \Omega \colon \ g_n(\omega) < r\},$$

then $\{(A_n, B_n)\}$ has no convergent subsequence on Ω.

To see this, let $\{r_1, r_2, \ldots\}$ be an enumeration of $\mathbb{Q}$, the set of all rational numbers of $\mathbb{R}$. For each n and i let

$$A_n^i = \{\omega \in \Omega \colon \ f_n(\omega) > r_i + \tfrac{1}{i}\} \quad and \quad B_n^i = \{\omega \in \Omega \colon \ f_n(\omega) < r_i\},$$

and note that $A_n^i \cap B_n^i = \emptyset$ holds for all n. Also, for simplicity, if $\{x_n\}$ is a sequence and M is an infinite subset of $\mathbb{N}$, then $\{x_n \colon \ n \in M\}$ will be denote the subsequence of $\{x_n\}$ determined by M. Now assume by way of contradiction that our claim is false.

Then, an easy inductive argument guarantees the existence of a sequence $\mathbb{N}_1, \mathbb{N}_2, \ldots$ of infinite subsets of $\mathbb{N}$ satisfying $\mathbb{N}_{i+1} \subseteq \mathbb{N}_i$ for each i and such that the sequence $\{(A_n^i, B_n^i) \colon \ n \in \mathbb{N}_i\}$ converging in Ω for each i. Fix a sequence $\{k_n\}$ of natural numbers with $k_i \in \mathbb{N}_i$ and $k_{i+1} > k_i$. Next, we claim that the subsequence $\{f_{k_n}\}$ of $\{f_n\}$ converges pointwise. To see this, let $\omega \in \Omega$ be fixed, and then pick an accumulation point r of $\{f_{k_n}(\omega)\}$. Now assume that $r < r_i < r_i + \tfrac{1}{i}$ holds for some i. Since $f_{k_n}(\omega) < r_i$ holds for infinitely many n, we see that ω belongs to $B_{k_n}^i$ for infinitely many n. Taking into account that $\{(A_n^i, B_n^i) \colon \ n \in \mathbb{N}_i\}$ converges on Ω, we obtain that $\omega \in A_n^i$ holds for at most a finite number of $n \in \mathbb{N}_i$. Thus, $f_{k_n}(\omega) \leq r_i + \tfrac{1}{i}$ holds for all n sufficiently large. Similarly, if $r_j < r_j + \tfrac{1}{j} < r$, then $f_{k_n}(\omega) \geq r_j + \tfrac{1}{j}$ holds for all sufficiently large n. Since the set $\{r_n + \tfrac{1}{n} \colon \ n = 1, 2, \ldots\}$ is dense in $\mathbb{R}$, the latter shows that $\lim f_{k_n}(\omega) = r$. Thus, $\{f_{k_n}\}$ converges pointwise, which is a contradiction, and our claim has been established.

STEP 2. *Let $\{(A_n, B_n)\}$ be a sequence of pairs of subsets of a set Θ with $A_n \cap B_n = \emptyset$ for all n. If $\{(A_n, B_n)\}$ has a subsequence with no convergent subsequence on Θ, then there exist some j and some infinite subset M of $\mathbb{N}$ such that $\{(A_n, B_n) \colon \ n \in M\}$ has no convergent subsequence on A_j and also no convergent subsequence on B_j.*

By passing to a subsequence of $\{(A_n, B_n)\}$ we can suppose that $\{(A_n, B_n)\}$ has no convergent subsequence on Θ.

Assume by way of contradiction that the conclusion is false. Put $n_1 = 1$. Then there exists an infinite subset $\mathbb{N}_1$ of $\mathbb{N}$ such that $\{(A_n, B_n) \colon \ n \in \mathbb{N}_1\}$

converges either on A_{n_1} or on B_{n_1}. Next, pick some $n_2 \in \mathbb{N}_1$ with $n_2 > n_1$. Then there exists an infinite subset $\mathbb{N}_2$ of $\mathbb{N}_1$ such that $\{(A_n, B_n): n \in \mathbb{N}_2\}$ converges either on A_{n_2} or on B_{n_2}. Therefore, by induction, there exist $n_1 < n_2 < \cdots$ and infinite subsets $\mathbb{N}_1 \supseteq \mathbb{N}_2 \supseteq \cdots$ of $\mathbb{N}$ such that $n_{i+1} \in \mathbb{N}_i$ and $\{(A_n, B_n): n \in \mathbb{N}_i\}$ converges either on A_{n_i} or on B_{n_i}. By the symmetry of the situation, we can assume that the set

$$ I = \{i \in \mathbb{N} | \{(A_n, B_n): n \in \mathbb{N}_i\} \text{ converges on } A_{n_i}\} $$

is infinite. Consider the infinite subset $M = \{n_i: i \in I\}$ of $\mathbb{N}$, and note that $\{(A_n, B_n): n \in M\}$ converges on $A = \bigcup_{n \in M} A_n$. On the other hand, since (by our hypothesis) $\{(A_n, B_n): n \in M\}$ does not converge on Θ, there exists some $x \in \Theta$ such that $\{n \in M: x \in A_n\}$ and $\{n \in M: x \in B_n\}$ are both infinite. Taking into account that $x \in A_n$ holds for some $n \in M$, we see that $x \in A$, and hence $\{(A_n, B_n): n \in M\}$ does not converge on A, a contradiction.

STEP 3. *Let $\{(A_n, B_n)\}$ be a sequence of pairs of subsets of a set Θ with $A_n \cap B_n = \emptyset$ for all n. Assume that there exist disjoint subsets $D_1, \ldots, D_r$ of the set Θ such that $\{(A_n, B_n)\}$ has no subsequence convergent on each D_i $(1 \le i \le r)$. Then there exist some j and an infinite subset M of $\mathbb{N}$ such that for each $i = 1, \ldots, r$ the sequence $\{(A_n, B_n): n \in M\}$ has no subsequence convergent on $D_i \cap A_j$ and also no subsequence convergent on $D_i \cap B_j$.*

Since (by definition) $\{(A_n, B_n)\}$ converges on the empty set, it follows that each D_i must be nonempty. The proof is by induction on r. For $r = 1$, the validity of the statement follows from Step 2. Now assume that the statement is true for some r, and let $D_1, \ldots, D_r, D_{r+1}$ be disjoint subsets of Θ such that $\{(A_n, B_n)\}$ has no subsequence convergent on each D_i for $i = 1, \ldots, r+1$. Assume by way of contradiction that our claim is false.

By the induction hypothesis there exist n_1 and an infinite subset M_1 of $\mathbb{N}$ such that for each $i = 2, \ldots, r+1$ the sequence $\{(A_n, B_n): n \in M_1\}$ has no subsequence convergent on $D_i \cap A_{n_1}$ and also no subsequence convergent on $D_i \cap B_{n_1}$. It follows that there exists an infinite subset $\mathbb{N}_1$ of M_1 such that $\{(A_n, B_n): n \in \mathbb{N}_1\}$ converges on either $D_1 \cap A_{n_1}$ or on $D_1 \cap B_{n_1}$. We can assume that $\{1, \ldots, n_1\} \cap \mathbb{N}_1 = \emptyset$. Next, pick some $n_2 \in \mathbb{N}_1$, and note that there exists an infinite subset M_2 on $\mathbb{N}_1$ such that for each $i = 2, \ldots, r+1$ the sequence $\{(A_n, B_n): n \in M_2\}$ has no subsequence convergent on $D_i \cap A_{n_2}$ and also no subsequence convergent on $D_i \cap B_{n_2}$. Then there exists an infinite subset $\mathbb{N}_2$ of M_2 such that $\{(A_n, B_n): n \in \mathbb{N}_2\}$ converges either on $D_1 \cap A_{n_2}$ or on $D_1 \cap B_{n_2}$. We can assume that $\{1, \ldots, n_2\} \cap \mathbb{N}_2 = \emptyset$.

Thus, continuing this process, we see that there exist $n_1 < n_2 < \cdots$ and infinite subsets $\mathbb{N}_1 \supseteq \mathbb{N}_2 \supseteq \cdots$ of $\mathbb{N}$ with $n_{i+1} \in \mathbb{N}_i$ and such that $\{(A_n, B_n): n \in \mathbb{N}_i\}$ converges either on $D_1 \cap A_{n_i}$ or on $D_1 \cap B_{n_i}$. Now consider the set $M = \{n_1, n_2, \ldots\}$, and note that the sequence of pairs of subsets $\{(D_1 \cap A_n, D_1 \cap B_n) \colon n \in M\}$ of D_1 has no subsequence convergent on D_1. By Step 2, there exists some $j \in M$ and an infinite subset L of M such that $\{(D_1 \cap A_n, D_1 \cap B_n) \colon n \in L\}$ has no subsequence convergent on $D_1 \cap A_j$ and also no subsequence convergent on $D_1 \cap B_j$, which is impossible. This completes the proof of Step 3.

STEP 4. A sequence $\{(A_n, B_n)\}$ of pairs of subsets of a set Θ with $A_n \cap B_n = \emptyset$ for all n is said to be **independent** whenever for each pair of disjoint finite subsets I and J of $\mathbb{N}$ we have $\left(\bigcap_{i \in I} A_i\right) \cap \left(\bigcap_{j \in J} B_j\right) \neq \emptyset$.

If $(A_n, B_n)\}$ is a sequence of pairs of subsets of a set Θ with $A_n \cap B_n = \emptyset$ for all n and $\{(A_n, B_n)\}$ has no convergent subsequence on Θ, then some subsequence of $\{(A_n, B_n)\}$ is independent.

By Step 2, there exists some n_1 and an infinite subset $\mathbb{N}_1$ of $\mathbb{N}$ such that $\{(A_n, B_n) \colon n \in \mathbb{N}_1\}$ has no subsequence convergent on either A_{n_1} or B_{n_1}. We can assume without loss of generality that $\mathbb{N}_1 \cap \{1, \ldots, n_1\} = \emptyset$. Now suppose that $n_1 < \cdots < n_k$ and infinite subsets $\mathbb{N}_1 \supseteq \cdots \supseteq \mathbb{N}_k$ of $\mathbb{N}$ have been selected with $\mathbb{N}_k \cap \{1, \ldots, n_k\} = \emptyset$ and such that $\{(A_n, B_n) \colon n \in \mathbb{N}_k\}$ has no subsequence convergent on anyone of the 2^k sets $\bigcap_{i=1}^{k} C_i$, where C_i equals A_{n_i} or B_{n_i}. Clearly, these sets are pairwise disjoint, and consequently by Step 3 there is $n_{k+1} \in \mathbb{N}_k$ and an infinite subset $\mathbb{N}_{k+1}$ of $\mathbb{N}_k$ with $\mathbb{N}_{k+1} \cap \{1, \ldots, n_k, n_{k+1}\} = \emptyset$ and such that $\{(A_n, B_n) \colon n \in \mathbb{N}_{k+1}\}$ has no subsequence convergent on any set of the form $\bigcap_{i=1}^{k} C_i \cap A_{n_{k+1}}$ or $\bigcap_{i=1}^{k} C_i \cap B_{n_{k+1}}$, where again C_i equals A_{n_i} or B_{n_i}. Note that $n_{k+1} > n_k$. Thus, continuing this way, we construct an infinite set $M = \{n_1, n_2, \ldots\}$, and an easy argument shows that $\{(A_n, B_n) \colon n \in M\}$ is an independent subsequence of $\{(A_n, B_n)\}$.

STEP 5. *By Steps 2 and 4 there exist a subsequence $\{g_n\}$ of $\{f_n\}$, $r \in \mathbb{R}$ and $\delta > 0$ such that if*

$$A_n = \{\omega \in \Omega \colon g_n(\omega) > r + 2\delta\} \quad and \quad B_n = \{\omega \in \Omega \colon g_n(\omega) < r\},$$

then $\{(A_n, B_n)\}$ is an independent sequence. In this case the sequence $\{g_n\}$ is equivalent (in the sup norm) to the standard basis of ℓ_1.

To see this, let $\alpha_1, \ldots, \alpha_n$ be arbitrary real scalars. Put

$$I = \{i \in \{1, \ldots, n\} \colon \alpha_i \geq 0\} \quad and \quad J = \{j \in \{1, \ldots, n\} \colon \alpha_j < 0\},$$

and then pick

$$s \in \left(\bigcap_{i \in I} A_i\right) \cap \left(\bigcap_{j \in J} B_j\right) \quad and \quad t \in \left(\bigcap_{j \in J} A_j\right) \cap \left(\bigcap_{i \in I} B_i\right).$$

For convenience, we assume that $\bigcap_{i\in\emptyset}C_i = \Omega$. Note that

$$\sum_{i=1}^{n}\alpha_i g_i(s) \geq \sum_{i\in I}|\alpha_i|(r+2\delta) - \sum_{j\in J}|\alpha_i|r$$

and

$$-\sum_{i=1}^{n}\alpha_i g_i(t) \geq -\sum_{i\in I}|\alpha_i|r + \sum_{j\in J}|\alpha_j|(r+2\delta).$$

Therefore,

$$2\Big\|\sum_{i=1}^{n}\alpha_i g_i\Big\|_{\infty} \geq \sum_{i=1}^{n}\alpha_i g_i(s) - \sum_{i=1}^{n}\alpha_i g_i(t) \geq 2\delta\sum_{i=1}^{n}|\alpha_i|,$$

and so

$$\delta\sum_{i=1}^{n}|\alpha_i| \leq \Big\|\sum_{i=1}^{n}\alpha_i g_i\Big\|_{\infty}$$

holds, which shows that $\{g_n\}$ is equivalent to the standard basis of ℓ_1. The proof of the theorem is now complete. ∎

Exercises

1. Show that if a Banach lattice E is lattice embeddable in a Banach lattice with order continuous norm, then E itself has order continuous norm.

2. Show that a Dedekind σ-complete Banach lattice E has order continuous norm if and only if $C[0,1]$ is not lattice embeddable in E.

3. Show that the mapping $T: \ell_1^d \to (\ell_\infty/c_0)'$, defined by $[Tf](\dot{x}) = f(x)$ for $f \in \ell_1^d$ and $x \in \ell_\infty$, is an onto lattice isometry.

4. If a Banach lattice has a separable double dual, then show that it is reflexive.

5. If the double dual of a Banach lattice E has order continuous norm, then show that E is a KB-space.

6. Show that a Banach lattice E is a KB-space if and only if every disjoint sequence of E^+ with norm bounded sequence of partial sums is norm convergent to zero.

7. For a Banach lattice E establish the following properties.
   (a) E has order continuous norm if and only if every disjoint sequence of U' is w^*-convergent to zero.
   (b) E' has order continuous norm if and only if every disjoint sequence of U converges weakly to zero.
   [*Hint*: Consider part (b). Assume that E' has order continuous norm. Let a sequence $\{x_n\} \subseteq U$ be disjoint, let $0 \leq x' \in E'$, and let $\epsilon > 0$. By

Theorem 4.19, there exists some $u \in E^+$ satisfying $x'(|x| - u)^+ < \epsilon$ for all $x \in U$. In particular, we have

$$|x'(x_n)| \leq x'(|x_n|) = x'(|x_n| - u)^+ + x'(|x_n| \wedge u) < \epsilon + x'(|x_n| \wedge u).$$

Since $\{|x_n| \wedge u\}$ is an order bounded disjoint sequence, it follows that $|x_n| \wedge u \xrightarrow{w} 0$, and so $\limsup |x'(x_n)| \leq \epsilon$. This implies $\lim x'(x_n) = 0$.

For the converse assume that every norm bounded disjoint sequence of E is weakly convergent to zero. Since the standard basis $\{e_n\}$ of ℓ_1 is a norm bounded disjoint sequence that does not converge weakly to zero, we see that ℓ_1 is not lattice embeddable in E. By Theorem 4.69 this means that c_0 is not embeddable in E', which by Theorem 4.60 is equivalent to saying that E' is a KB-space.]

8. Let $\{x_n\}$ be a weak Cauchy sequence in a Banach space X. If either
   (a) $\{x_n\}$ converges weakly in X, or
   (b) X is a Banach lattice and $\{x_n\}$ is monotone,
   then show that $\{x_n\}$ has property (u).

9. Let $\{x_n\}$ be a sequence in a Banach space X such that for each $x' \in X'$ we have $\sum_{n=1}^{\infty} |x'(x_n)| < \infty$. Then show that:
   (a) For each $(\alpha_1, \alpha_2, \ldots) \in c_0$ the series $\sum_{n=1}^{\infty} \alpha_n x_n$ is norm convergent.
   (b) The operator $T \colon c_0 \to X$, defined by

   $$T(\alpha_1, \alpha_2, \ldots) = \sum_{n=1}^{\infty} \alpha_n x_n$$

   is continuous.

10. Show that every finite dimensional vector subspace of a Banach space is complemented.

11. If X is a Banach space and $x'_1, \ldots, x'_n \in X'$, then show that the closed vector subspace $M = \{x \in X \colon x'_i(x) = 0 \text{ for } i = 1, \ldots, n\}$ of X has a finite dimensional complement. [*Hint:* We can assume that $x'_1, \ldots, x'_n$ are linearly independent functionals. This implies (by Lemma 3.15) that $\bigcap_{i \neq j} \operatorname{Ker} x'_i \not\subseteq \operatorname{Ker} x'_j$ holds for each j. So, for each $j = 1, \ldots, n$ there exists some $x_j \in X$ with $x'_i(x_j) = 0$ for $i \neq j$ and $x'_j(x_j) = 1$. If N is the finite dimensional vector subspace generated by $\{x_1, \ldots, x_n\}$, then show that $X = M \oplus N$.]

12. If c_0 is embeddable in a Banach lattice E, then show that ℓ_1 is lattice embeddable in E'.

13. For a Banach lattice E with order continuous norm show that the following statements are equivalent:
    (a) E' has order continuous norm, i.e., E' is a KB-space.
    (b) ℓ_1 does not embed in E.
    (c) c_0 does not embed in E'.
    [*Hint:* Use Theorem 4.25.]

14. Let E be an AL- or an AM-space. Then show that E is reflexive if and only if E is finite dimensional. [*Hint:* Let E be a reflexive AM-space with unit. Then E has order continuous norm. Now note that if $\{x_n\}$

is a disjoint sequence of E satisfying $x_n \neq 0$ for all n, then $\{\frac{x_n}{\|x_n\|}$ is an order bounded disjoint sequence of E that does not converge to zero.]

15. Show that ℓ_1 is always lattice embeddable in an infinite dimensional *AL*-space. Also, construct a lattice isometry from ℓ_1 into $L_1[0,1]$. [*Hint*: If E is an infinite dimensional *AL*-space, then by the preceding exercise and Theorem 4.70, we see that E' does not have order continuous norm. Now apply Theorem 4.69.]

16. Show that c_0 is always lattice embeddable in an infinite dimensional *AM*-space. Also construct a lattice isometry from c_0 into $C[0,1]$. [*Hint*: Use Exercise 14 above and Theorem 4.61.]

17. If a norm bounded disjoint sequence $\{x_n\}$ in a Banach lattice does not converge weakly to zero, then show that $\{x_n\}$ has a subsequence with no weakly Cauchy subsequence.

18. (Johnson–Tzafriri [**80**]) If a Banach space X embeds complementably in a Banach lattice, then show that:
    (a) X is weakly sequentially complete if and only if c_0 is not embeddable in X.
    (b) X is reflexive if and only if neither ℓ_1 nor c_0 is embeddable in X.

19. Recall that a Banach space has the **Schur property** whenever $x_n \xrightarrow{w} 0$ implies $\|x_n\| \to 0$. For a Banach space X with the Schur property prove the following:
    (a) X is weakly sequentially complete.
    (b) If X is reflexive, then X is finite dimensional.

20. (Rosenthal [**170**]) Show that for a weakly sequentially complete Banach space X one of the following two mutually exclusive alternatives hold.
    (a) X is reflexive.
    (b) ℓ_1 is embeddable in X.

21. (Rosenthal [**170**]) Show that if a Banach space X has the Schur property, then ℓ_1 is embeddable into every infinite dimensional closed vector subspace of X. [*Hint*: A reflexive Banach space with the Schur property is finite dimensional.]

4.4. Banach Lattices of Operators

The lattice properties of *AM*-spaces were discussed in the previous sections. An *AM*-space with unit, besides being a Banach lattice, has also an f-algebra structure which is very important. In this section, we shall utilize the ring structure of *AM*-spaces to derive some useful "local approximation" properties of operators on Banach lattices.

For an arbitrary pair of Banach spaces X and Y, the symbol $L(X,Y)$ will denote the vector space of all continuous operators from X into Y. For simplicity, instead of $L(X,X)$ we shall write $L(X)$. It is well known that

under the norm
$$\|T\| := \sup\{\|Tx\|: \ \|x\| \le 1\}$$
the vector space $L(X, Y)$ is also a Banach space. Since every positive operator between Banach lattices is continuous (see Theorem 4.3), it follows that $\mathcal{L}_r(E, F) \subseteq L(E, F)$ holds for every pair of Banach lattices E and F. In general, as the next example shows, this inclusion is proper.

Example 4.73. Consider the operator $T: C[0, 1] \to c_0$ defined by
$$Tf = \left(f(1) - f(0), f(\tfrac{1}{2}) - f(0), f(\tfrac{1}{3}) - f(0), \ldots\right).$$
From $|f(\tfrac{1}{n}) - f(0)| \le 2\|f\|_\infty$, we see that $\|Tf\|_\infty \le 2\|f\|_\infty$ holds for each $f \in C[0, 1]$, and so T is a continuous operator. On the other hand, we claim that T is not order bounded.

To see this, assume by way of contradiction that there exists some vector $u = (u_1, u_2, \ldots) \in c_0$ satisfying $|Tf| \le u$ for all $f \in [0, \mathbf{1}]$, where $\mathbf{1}$ denotes the constant function one. For each n pick some $f_n \in [0, \mathbf{1}]$ with $f_n(0) = 0$ and $f_n(\tfrac{1}{n}) = 1$, and note that $1 = |f_n(\tfrac{1}{n}) - f_n(0)| \le u_n$ holds. This shows that $u \notin c_0$, which is a contradiction. Hence, T is not order bounded (and thus is not a regular operator), as claimed. ■

Consider two Banach lattices E and F. If $T: E \to F$ is an operator with modulus, then the **regular norm**, abbreviated as the **r-norm**, is defined by
$$\|T\|_r := \big\||T|\big\| := \sup\{\big\||T|x\big\|: \ \|x\| \le 1\}.$$
Clearly, $\|T\| \le \|T\|_r$ holds.

Now assume that F is also Dedekind complete. Then, it is easy to see that $\|\cdot\|_r$ is a lattice norm on $\mathcal{L}_b(E, F) = \mathcal{L}_r(E, F)$. Remarkably, $\mathcal{L}_b(E, F)$ under the r-norm is also a Banach lattice.

Theorem 4.74. *If E and F are Banach lattices with F Dedekind complete, then $\mathcal{L}_b(E, F)$ under the r-norm is a Dedekind complete Banach lattice.*

Proof. Let $\{T_n\}$ be a $\|\cdot\|_r$-Cauchy sequence of $\mathcal{L}_b(E, F)$. By passing to a subsequence we can assume that
$$\|T_{n+1} - T_n\|_r = \big\||T_{n+1} - T_n|\big\| < \tfrac{1}{2^n}$$
holds for each n. From $\|T_{n+1} - T_n\| \le \|T_{n+1} - T_n\|_r$, we see that $\{T_n\}$ is a Cauchy sequence of $L(E, F)$. Thus, there exists some $T \in L(E, F)$ with $\|T_n - T\| \to 0$.

Next let $x \in E^+$. Then for each $y \in E$ with $|y| \le x$ we have
$$(T - T_n)y = \sum_{i=n}^{\infty} (T_{i+1} - T_i)y \le \sum_{i=n}^{\infty} |T_{i+1} - T_i|x.$$

Therefore, the modulus of $T - T_n$ exists and satisfies

$$|T - T_n|x = \sup\{(T - T_n)y\colon \; |y| \le x\} \le \sum_{i=n}^{\infty} |T_{i+1} - T_i|x \qquad (\star)$$

for all $x \in E^+$. Form $T = (T - T_1) + T_1$, we see that T is a regular operator, i.e., $T \in \mathcal{L}_b(E, F)$. On the other hand, it follows from $(\star)$ that

$$\|T - T_n\|_r \le \sum_{i=n}^{\infty} \|T_{i+1} - T_i\|_r \le 2^{1-n}\,,$$

and so $\lim \|T - T_n\|_r = 0$. That is, $\mathcal{L}_b(E, F)$ under the r-norm is a Banach lattice as claimed. ∎

In view of Example 4.73 we know that in general $\mathcal{L}_r(E, F)$ is a proper vector subspace of $L(E, F)$. However, if F is a Dedekind complete AM-space with unit, then it should be clear that $\mathcal{L}_r(E, F) = L(E, F)$ holds. The next result presents another not so obvious case under which $\mathcal{L}_r(E, F) = L(E, F)$ holds.

Theorem 4.75. *If E is an arbitrary AL-space and F is a KB-space, then $\mathcal{L}_b(E, F) = L(E, F)$ holds. Moreover, in this case we have $\|T\|_r = \|T\|$ for each $T \in L(E, F)$.*

Proof. Let $T \in L(E, F)$, and let $[0, x]$ be an interval of E. Since E' is an AM-space with unit, the adjoint operator $T'\colon F' \to E'$ is an order bounded operator, and hence $T''\colon E'' \to F''$ is likewise order bounded. So, there exists some $x'' \in F''$ satisfying $-x'' \le Ty \le x''$ for all $y \in [0, x]$. Now if P denotes the order projection on F'' onto F (see Theorem 4.60), then $-Px'' \le Ty \le Px''$ holds for all $y \in [0, x]$, and this shows that T is an order bounded operator, i.e., $T \in \mathcal{L}_b(E, F)$. Hence $\mathcal{L}_b(E, F) = L(E, F)$. ∎

Observe, that by the preceding theorem every continuous operator between AL-spaces possesses a modulus.

Recall that an orthomorphism is an order bounded operator on a Riesz space that leaves all bands invariant. When the underlying Riesz space has the structure of a Banach lattice, the orthomorphisms enjoy some extra properties. For instance, Example 2.38 demonstrates that a band preserving operator on a Riesz space need not be order bounded. However, Y. A. Abramovich, A. I. Veksler, and A. V. Koldunov [4] have shown that on a Banach lattice every band preserving operator is necessarily an orthomorphism. This result is stated next.

Theorem 4.76 (Abramovich–Veksler–Koldunov). *Every band preserving operator on a Banach lattice is order bounded—and hence norm bounded.*

Proof. Let $T\colon E \to E$ be a band preserving operator on a Banach lattice. The proof consists of two parts.

Assume first that the operator T is also norm bounded. Then we claim that $|Tx| \le \|T\|x$ holds for all $x \in E^+$, and this will be enough to establish that T is order bounded. To this end, assume by way of contradiction that there exists some $x > 0$ with $(|Tx| - \|T\|x)^+ > 0$. Since E is Archimedean, it is easy to see that there exists some $M > \|T\|$ with $(|Tx| - Mx)^+ > 0$. Let $y = |Tx| - Mx$, and denote by B the band generated by y^-. Clearly, $x \notin B$. (Otherwise, $x \in B$ implies $Tx \in B$, and so $y^+ \in B$, which implies $y^+ = 0$, a contradiction.) Since every band of E is norm closed, E/B is a Banach lattice. Moreover, in view of $T(B) \subseteq B$, it is easy to see that the formula $\dot{T}\dot{x} = (Tx)^{\boldsymbol{\cdot}}$ defines a norm bounded operator on E/B satisfying $\|\dot{T}\| \le \|T\|$. From

$$\big|(Tx)^{\boldsymbol{\cdot}}\big| - M\dot{x} = (|Tx| - Mx)^{\boldsymbol{\cdot}} = (y^+)^{\boldsymbol{\cdot}} - (y^-)^{\boldsymbol{\cdot}} = (y^+)^{\boldsymbol{\cdot}} \ge 0,$$

we see that $|(Tx)^{\boldsymbol{\cdot}}| \ge M\dot{x} \ge 0$, and so $\|\dot{T}\dot{x}\| \ge M\|\dot{x}\|$. Since $\dot{x} \ne 0$, the latter implies $\|\dot{T}\| \ge M > \|T\|$, which is impossible. Thus, $|Tx| \le \|T\|x$ holds for each $x \in E^+$. The preceding proof is due to W. A. J. Luxemburg [126].

Now consider the general case. By Zorn's lemma there exists a maximal collection $\{B_\alpha\colon \alpha \in \mathcal{A}\}$ of pairwise disjoint bands such that $T\colon B_\alpha \to B_\alpha$ is norm bounded (and hence order bounded by the preceding case). Therefore, T is order bounded on the idea $A = \sum_{\alpha \in \mathcal{A}} \oplus B_\alpha$, and we claim that A is order dense. If this is the case, then by Theorem 2.42, the operator T must be order bounded (and hence it must be also norm bounded since it will be the difference of two positive orthomorphisms).

To see that A is order dense in E, assume by way of contradiction that $A^{\mathrm{d}} \ne \{0\}$ holds. Then A^{d} must have infinite dimension. (Otherwise, $T\colon A^{\mathrm{d}} \to A^{\mathrm{d}}$ must be continuous, and so, by incorporating the band A^{d} into $\{B_\alpha\colon \alpha \in \mathcal{A}\}$, we would violate the maximality property of $\{B_\alpha\colon \alpha \in \mathcal{A}\}$.) Therefore, there exists a disjoint sequence $\{u_n\}$ of A^{d} with $u_n > 0$ for each n. Let B_n denote the band generated by u_n. Then, $T\colon B_n \to B_n$ cannot be norm bounded, and so for each n there exists some $x_n \in B_n$ with $\|x_n\| = 1$ and $\|Tx_n\| \ge n^3$. Let $x = \sum_{n=1}^{\infty} \frac{x_n}{n^2} \in E$, and note that if $y_k = x - \frac{x_k}{k^2}$, then $y_k \perp x_k$. Thus, $Tx_k \perp Ty_k$ holds for each k, and so from

$$\frac{|Tx_k|}{k^2} \le |Ty_k| + \frac{|Tx_k|}{k^2} = \left|Ty_k + \frac{Tx_k}{k^2}\right| = |Tx|,$$

we see that $k \le \frac{\|Tx_k\|}{k^2} \le \|Tx\|$ holds for each k, which is impossible. Hence, $A^{\mathrm{d}} = \{0\}$, and so A is order dense in E. The above proof of the second part is due to B. de Pagter. ∎

Now let E be a Banach lattice. By Theorem 2.40 we know that every orthomorphism on E possesses a modulus, and so $\mathrm{Orth}(E)$ under the r-norm is a normed space. The next result, due to A. W. Wickstead [**191**], tells us that $\mathrm{Orth}(E)$ is, in fact, an AM-space with unit the identity operator.

Theorem 4.77 (Wickstead). *If E is a Banach lattice, then*

$$\mathrm{Orth}(E) = \{T \in \mathcal{L}_b(E) \colon \ \exists \lambda > 0 \ \ with \ \ -\lambda I \leq T \leq \lambda I\}$$

and $\|T\|_r = \inf\{\lambda > 0 \colon \ |T| \leq \lambda I\}$ holds for all $T \in \mathrm{Orth}(E)$.

In particular, $\mathrm{Orth}(E)$ under the r-norm is an AM-space with unit the identity operator I.

Proof. Let $0 \leq T \in \mathrm{Orth}(E)$. By Theorem 4.3 we know that $\|T\| < \infty$. Now assume that for some $\lambda > 0$ we have $(T - \lambda I)^+ > 0$. Then there exists some $x \in E^+$ such that $y = (T - \lambda I)^+ x > 0$. Using the fact that $\mathrm{Orth}(E)$ is an f-algebra, it follows from Theorem 2.54 that $(T - \lambda I)y = [(T - \lambda I)^+]y \geq 0$, and so $Ty \geq \lambda y > 0$. This implies $\|Ty\| \geq \lambda\|y\| > 0$, from which it follows that $\|T\| \geq \lambda$ holds. Hence, $(T - \lambda I)^+ = 0$ must hold for all $\lambda > \|T\|$. That is, $T \leq \lambda I$ holds for all $\lambda > \|T\|$, and the conclusions of the first part follow. The above elegant proof is due to W. A. J. Luxemburg [**126**].

Finally, to establish that $\mathrm{Orth}(E)$ under the r-norm is an AM-space with unit I, it remains to be shown that $\mathrm{Orth}(E)$ is $\|\cdot\|_r$-complete. To this end, let $\{T_n\} \subseteq \mathrm{Orth}(E)$ be a $\|\cdot\|_r$-Cauchy sequence. Since every $T \in \mathrm{Orth}(E)$ satisfies $|T|(|x|) = |T(x)|$ for all $x \in E$ (see Theorem 2.40), it follows that $\|T_n - T_m\| = \|T_n - T_m\|_r$, and consequently there exists some $T \in L(E)$ with $\lim \|T_n - T\| = 0$, Now if B is a band of E and $x \in B$, then $\{T_n(x)\} \subseteq B$, $\lim \|T_n(x) - T(x)\| = 0$ and the norm closedness of B imply $T(x) \in B$. Thus, T is a band preserving operator, and so it follows from Theorem 4.76 that $T \in \mathrm{Orth}(E)$. To complete the proof, note that $\|T_n - T\|_r = \|T_n - T\| \to 0$. ∎

Consider a positive operator $T \colon E \to F$ between two Riesz spaces with F Dedekind complete. In Section 2.1, under the assumption that E had the principal projection property, we were able to describe the components of T by employing order projections. Subsequently, the positive operators dominated by T were approximated by linear combinations of operators of the form QTP with Q and P appropriate order projections. However, if E lacks the principal projection property, then these techniques are not applicable. It is, therefore, remarkable that for Banach lattices the positive operators dominated by T can approximated "locally" by linear combinations of operators of the form LTM, where L and M are now orthomorphisms. As we shall see, the reason is that a Banach lattice has an abundance of orthomorphisms to compensate for the loss of order projections. The discussion below will clarify the situation.

Let E be an AM-space with a unit e. By Theorem 2.58 we know that E admits at most one product under which it is an f-algebra having e as its multiplicative unit. On the other hand, by Theorem 4.29, there exists a unique Hausdorff compact topological space Ω and an onto lattice isometry $\pi\colon E \to C(\Omega)$ with $\pi(e) = \mathbf{1}$ (where $\mathbf{1}$ is the constant function one on Ω). Since $C(\Omega)$ is an f-algebra with multiplicative unit $\mathbf{1}$, it follows that E is also an f-algebra having e as its multiplicative unit. Clearly, the product in E is defined by

$$xy = \pi^{-1}\big(\pi(x)\pi(y)\big),$$

form which it follows that

$$|xy| = |x|\,|y|$$

holds for all $x, y \in E$. Thus, E admits a unique product under which it is an f-algebra with e being also its multiplicative unit. In particular, every vector $x \in E$ gives rise to a **multiplication operator** $M\colon E \to E$ defined by the formula

$$M(y) = xy, \quad y \in E.$$

If M is also a positive operator, then it will be referred to as a **positive multiplication operator** (clearly, this is equivalent to saying that the "multiplier" x of M is a positive element). Of course, on an AM-space with unit the multiplication operators are precisely the orthomorphisms; see Theorem 2.62. In sum:

- *Every AM-space E with unit e is an f-algebra having e also as its multiplicative unit, and every $x \in E$ defines a multiplication operator on E by the formula $M(y) = xy$.*

Now let E be a Banach lattice, and let E_x be the ideal generated by some nonzero vector x of E. Then, we know that under the norm

$$\|y\|_\infty = \inf\{\lambda > 0\colon\ |y| \le \lambda|x|\}, \quad y \in E_x,$$

the ideal E_x is an AM-space with unit $|x|$ (see Theorem 4.21), and so E_x has plenty of multiplication operators. In other words, every Banach lattice has the following important "local" behavior:

- *On every principal ideal there is an abundance of multiplication operators.*

Remarkably, quite often these multiplication operators have extensions to the whole space, and this will be very useful to the study of operators on Banach lattices.

Lemma 4.78. *For an element $u > 0$ in a Banach lattice E and a multiplication operator $T\colon E_u \to E_u$ the following statements hold:*

(1) T is continuous with respect to the norm induced on E_u by E.

   (2) *If E is Dedekind σ-complete, then T extends to an orthomorphism on E (which is positive if T is positive).*

Proof. Pick a unique $x \in E_u$ such that $T(y) = xy$ holds for all $y \in E_u$, and then fix some $\lambda > 0$ with $|x| \leq \lambda u$. Then for each $y \in E_u$ we have

$$|T(y)| = |xy| = |x|\,|y| \leq \lambda u|y| = \lambda|y|. \qquad (\star)$$

   (1) From $(\star)$ we have

$$\|Ty\| \leq \lambda\|y\|$$

for each $y \in E_u$, and so T is a continuous operator on E_u for the norm induced by E.

   (2) Let $I\colon E \to E$ be the identity operator. Consider the multiplication operators $T^+, T^-\colon E_u \to E_u$ defined by $T^+(y) = x^+y$ and $T^-(y) = x^-y$. From $(\star)$, we see that

$$0 \leq T^+(y) \leq \lambda I(y) \quad \text{and} \quad 0 \leq T^-(y) \leq \lambda I(y)$$

holds for each $0 \leq y \in E_u$.

   Thus, we can suppose that $0 \leq T(y) \leq \lambda y$ holds for all $0 \leq y \in E_u$. Define $S\colon E^+ \to E^+$ by

$$S(y) = \sup\{T(y \wedge nu)\colon\ n = 1, 2, \ldots\}.$$

Clearly, the supremum exists and $S(y) = T(y)$ holds for all $0 \leq y \in E_u$. Moreover, we have $0 \leq S(y) \leq \lambda y$ for all $y \in E^+$. On the other hand, as in the proof of Theorem 1.22, we see that S is additive on E^+, and so it defines a positive orthomorphism on E. Now note that S is the desired extension of T. ∎

   We know that the natural embedding of a Banach lattice E into its double dual E'' preserves the algebraic, norm, and lattice structures of E. When E is an AM-space with unit, the natural embedding of E into E'' preserves also the ring structure of E. The details follow.

Theorem 4.79. *If E is an AM-space with unit e, then E'' is a Dedekind complete AM-space with unit e containing the f-algebra E as an f-subalgebra.*

Proof. Clearly, E'' is an AM-space with unit. To see that e is also the unit of E'', note that if $x'' \in E''$ satisfies $\|x''\| \leq 1$, then for each $0 \leq x' \in E'$ we have

$$|x''|(x') \leq \|x''\| \cdot \|x'\| \leq \|x'\| = x'(e) \leq e(x'),$$

and so $|x''| \leq e$. Conversely, $|x''| \leq e$ implies $\|x''\| \leq 1$.

Next, we shall show that E is an f-subalgebra of E''. To this end, denote the product of E by $\cdot$ and the product of E'' by $\star$. Fix $x \in E^+$ with $0 \le x \le e$, and define the orthomorphism $T \colon E \to E$ by

$$T(y) = x \cdot y, \quad y \in E.$$

Now consider T as an operator from E into E'', and note that for each $y \in E^+$ we have

$$0 \le T(y) = x \cdot y \le e \cdot y = y.$$

Thus, by Theorem 2.49, T extends to an orthomorphism from E'' into E'', which we denote by T again. Next, according to Theorem 2.62, there exists some $x'' \in E''$ so that $T(y'') = x'' \star y''$ holds for all $y'' \in E''$. In particular, we have

$$x'' = x'' \star e = T(e) = x \cdot e = x.$$

Therefore, $x \cdot y = x \star y$ holds for all $y \in E$, and from this it follows that E is an f-subalgebra of E''. ∎

Let E be an AM-space with unit. Then, it is an immediate consequence of the preceding theorem that every multiplication operator on E defines (in the obvious manner) a multiplication operator on E''. From now on, without any further discussion, every multiplication operator on E also will be considered as a multiplication operator on E''. Although E may not have any nontrivial order projections, the Dedekind completeness of E'' guarantees an abundance of order projections on E''. The next result shows that the order projections of E'' can be approximated "locally" by the multiplication operators of E.

Lemma 4.80. *Let E be an AM-space with unit e, and let P be an order projection on E''. Then for each $0 \le x' \in E'$ and each $\epsilon > 0$ there exists a positive multiplication operator M on E such that:*

(a) *$0 \le M \le I$.*

(b) *The modulus of $P - M$ in E'' satisfies $\langle x', |P - M|e \rangle < \epsilon$.*

Proof. Let P be an order projection on E'', let $0 \le x' \in E'$, and let $\epsilon > 0$. In view of $0 \le Py'' \le y''$ for all $0 \le y'' \in E''$, there exists (by Theorem 2.62) a unique $x'' \in E''$ with $0 \le x'' \le e$ and satisfying $P(y'') = x'' \cdot y''$ for all $y'' \in E''$.

Now note that E' (as an AL-space) has order continuous norm, and so $E'' = (E')^{\sim}_n$ holds. Thus, by Theorem 3.60, E is $|\sigma|(E'', E')$-dense in E'', and therefore there exists some $x \in E^+$ satisfying $\langle x', |x'' - x| \rangle < \epsilon$. Replacing x by $x \wedge e$, we can assume that $0 \le x \le e$ holds.

Next, consider the multiplication operator M on the Banach lattice E defined by $M(y) = xy$ (and, of course, M defines a multiplication operator on E'' by the formula $M(y'') = xy''$). Clearly, $0 \leq M \leq I$ holds on E, and, on the other hand, it follows from Theorem 2.40 that

$$|P - M|e = |(P - M)e| = |(x'' - x)e| = |x'' - x|.$$

Thus, $\langle x', |P - M|e \rangle = \langle x', |x'' - x| \rangle < \epsilon$ holds, as desired. ∎

We now come to an important approximation property of positive operators defined on an AM-space.

Theorem 4.81. *Let E be an AM-space with unit e, let F be a Dedekind complete Banach lattice, and let $S, T \colon E \to F$ be two positive operators such that $0 \leq S \leq T$ holds. Then, for each $0 \leq x' \in F_n^\sim$ and each $\epsilon > 0$ there exist positive multiplication operators $M_1, \ldots, M_k$ on E and order projections $P_1, \ldots, P_k$ on F satisfying*

$$0 \leq \sum_{i=1}^{k} P_i T M_i \leq T \quad and \quad \left\langle x', \left| S - \sum_{i=1}^{k} P_i T M_i \right| e \right\rangle < \epsilon.$$

Proof. Fix $\epsilon > 0$ and $0 \leq x' \in F_n^\sim$. Assume first that S is a component of T, i.e., $S \wedge (T - S) = 0$ holds in $\mathcal{L}_b(E, F)$. Then, by Theorem 1.21, we have

$$\left\{ \sum_{i=1}^{n} Sx_i \wedge (T - S)x_i \colon \ x_1, \ldots, \in E^+ \ and \ \sum_{i=1}^{n} x_i = e \right\} \downarrow 0.$$

Thus, there exist $x_1, \ldots, x_n \in E^+$ with $x_1 + \cdots + x_n = e$ and

$$\left\langle x', \sum_{i=1}^{n} Sx_i \wedge (T - S)x_i \right\rangle < \epsilon. \tag{$\star\star$}$$

Now let M_i be the positive multiplication operators on E determined by x_i (i.e., $M_i(y) = x_i y$ for all $y \in E$), and note that $(M_1 + \cdots + M_n)y = y$ holds for all $y \in E$. Next, let P_i be the order projection of F onto the band generated in F by $(2Sx_i - Tx_i)^+$. Clearly, $P_i(2Sx_i - Tx_i) = (2Sx_i - Tx_i)^+$, and moreover we have $0 \leq \sum_{i=1}^{n} P_i T M_i \leq \sum_{i=1}^{n} T M_i = T$. On the other hand, the modulus of $S - \sum_{i=1}^{n} P_i T M_i$ satisfies

$$\left| S - \sum_{i=1}^{n} P_i T M_i \right| e \leq \left| S - \sum_{i=1}^{n} P_i S M_i \right| e + \left| \sum_{i=1}^{n} P_i (T - S) M_i \right| e$$

$$= \sum_{i=1}^{n} (S - P_i S) M_i e + \sum_{i=1}^{n} P_i (T - S) M_i e$$

$$= \sum_{i=1}^{n} (S - P_i S) x_i + \sum_{i=1}^{n} P_i (T - S) x_i$$

$$= \sum_{i=1}^{n} \left[S x_i - P_i (2 S x_i - T x_i) \right]$$

$$= \sum_{i=1}^{n} \left[S x_i - (2 S x_i - T x_i)^+ \right]$$

$$= \sum_{i=1}^{n} S x_i \wedge (T - S) x_i .$$

Taking into account ($\star\star$), we see that

$$\left\langle x', \left| S - \sum_{i=1}^{n} P_i T M_i \right| e \right\rangle < \epsilon .$$

Now consider the general case. Since $[0, T]$ is an order interval in the Banach lattice $\mathcal{L}_b(E, F)$, it follows from Theorem 3.61 that there exists a convex combination $\sum_{i=1}^{m} \alpha_i S_i$ of components of T such that

$$\left\| \left| S - \sum_{i=1}^{m} \alpha_i S_i \right| \right\| = \left\| S - \sum_{i=1}^{m} \alpha_i S_i \right\|_r < \tfrac{\epsilon}{2\|x'\|} .$$

By the preceding case, for each $i = 1, \ldots, m$ there exist positive multiplication operators $M_1^i, \ldots, M_{m_i}^i$ on E and order projections $P_1^i, \ldots, P_{m_i}^i$ on F satisfying

$$0 \leq \sum_{j=1}^{m_i} P_j^i T M_j^i \leq T \quad \text{and} \quad \left\langle x', \left| S_i - \sum_{j=1}^{m_i} P_j^i T M_j^i \right| e \right\rangle < \tfrac{\epsilon}{2} .$$

Now let $R = \sum_{i=1}^{m} \sum_{j=1}^{m_i} P_j^i T (\alpha_i M_j^i)$. Clearly, $0 \leq R \leq T$, and moreover

$$\left\langle x', |S - R| e \right\rangle \leq \left\langle x', \left| S - \sum_{i=1}^{m} \alpha_i S_i \right| e \right\rangle + \left\langle x', \left| \sum_{i=1}^{m} \alpha_i \left(S_i - \sum_{j=1}^{m_i} P_j^i T M_j^i \right) \right| e \right\rangle$$

$$\leq \tfrac{\epsilon}{2} + \sum_{i=1}^{m} \alpha_i \left\langle x', \left| S_i - \sum_{j=1}^{m_i} P_j^i T M_j^i \right| e \right\rangle < \tfrac{\epsilon}{2} + \tfrac{\epsilon}{2} = \epsilon .$$

The proof of the theorem is now complete. ∎

Now let $S, T: E \to F$ be two positive operators between Banach lattices such that $0 \leq S \leq T$. If F is Dedekind complete, then by Theorem 2.9 the operator S can be approximated uniformly by linear combinations of components of T (the T-step functions). However, if F is not Dedekind complete, then T may not have any nontrivial components, and so such an approximation of S need not be possible. On the other hand, if we consider F as a Banach sublattice of F'' and view T as an operator from E into F'', then T has plenty of components in $\mathcal{L}_b(E, F'')$, and hence every positive operator dominated by T can be approximated uniformly by the $\mathcal{L}_b(E, F'')$-step functions of T. In the sequel, this observation will be used frequently.

In case E is an AM-space with unit, the next result presents an important approximation property of positive operators in $\mathcal{L}_b(E, F'')$ that are dominated by a positive operator of $\mathcal{L}_b(E, F)$. Its proof follows immediately from Theorem 4.81.

Theorem 4.82. *let E be an AM-space with unit e, let F be a Banach lattice, and let $T: E \to F$ be a positive operator. If $S: E \to F''$ is a positive operators satisfying $0 \leq S \leq T$ (in $\mathcal{L}_b(E, F'')$), then given, given $0 \leq x' \in F'$ and $\epsilon > 0$, there exist positive multiplication operators $M_1, \ldots, M_k$ on E and order projections $P_1, \ldots, P_k$ on F'' satisfying*

$$0 \leq \sum_{i=1}^{k} P_i T M_i \leq T \quad \text{and} \quad \left\langle x', \left| S - \sum_{i=1}^{k} P_i T M_i \right| e \right\rangle < \epsilon.$$

When E and F are both AM-spaces with units, then the following version of Theorem 4.81 holds true.

Theorem 4.83. *Let $S, T: E \to F$ be two positive operators between AM-spaces with units such that $0 \leq S \leq T$ holds. If $0 \leq x' \in F'$ and $\epsilon > 0$ are given, then there exist positive multiplication operators $M_1, \ldots, M_k$ on E and positive multiplication operators $L_1, \ldots, L_k$ on F such that*

$$\left\langle x', \left| \left(S - \sum_{i=1}^{k} L_i T M_i \right) x \right| \right\rangle < \epsilon$$

holds for all $x \in E$ with $\|x\| \leq 1$.

Proof. Fix $0 \leq x' \in F'$ and $\epsilon > 0$, and let e denote the unit of E. Consider S and T as operators from E into F'', and note that $0 \leq x' \in (F'')^{\sim}_n$ holds. Then, by Theorem 4.81, there exist positive multiplication operators $M_1, \ldots, M_k$ on E and order projections $P_1, \ldots, P_k$ on F'' satisfying

$$\left\langle x', \left| S - \sum_{i=1}^{k} P_i T M_i \right| e \right\rangle < \frac{\epsilon}{2}.$$

Now for each $i = 1, \ldots, k$ there exists (by Lemma 4.80) a positive multiplication operator L_i on F such that

$$\langle x', |P_i - L_i| T M_i e \rangle < \tfrac{\epsilon}{2k} \, .$$

Therefore, for each $x \in E$ with $\|x\| \le 1$ we have

$$\left\langle x', \left| \left(S - \sum_{i=1}^{k} L_i T M_i \right) x \right| \right\rangle$$

$$\le \left\langle x', \left| \left(S - \sum_{i=1}^{k} P_i T M_i \right) x \right| \right\rangle + \left\langle x', \left| \sum_{i=1}^{k} (P_i - L_i) T M_i x \right| \right\rangle$$

$$\le \left\langle x', \left| S - \sum_{i=1}^{k} P_i T M_i \right| e \right\rangle + \sum_{i=1}^{k} \langle x', |P_i - L_i| T M_i e \rangle$$

$$< \tfrac{\epsilon}{2} + \tfrac{\epsilon}{2} = \epsilon \, ,$$

and the proof is finished. ∎

For Dedekind complete Banach lattices we have the following companion of the preceding result.

Theorem 4.84. *Let E and F be two Dedekind complete Banach lattices, and let $S, T \colon E \to F$ be two positive operators such that $0 \le S \le T$ holds. Then for each $x \in E^+$, $0 \le x' \in F'$, and $\epsilon > 0$ there exist positive orthomorphisms $M_1, \ldots, M_k$ on E and positive orthomorphisms $L_1, \ldots, L_k$ on F such that*

$$\left\langle x', \left| \left(S - \sum_{i=1}^{k} L_i T M_i \right) y \right| \right\rangle < \epsilon$$

holds for all $y \in E$ with $|y| \le x$.

Proof. Let $x \in E^+$, $0 \le x' \in F'$, and $\epsilon > 0$ be fixed. Put $u = Tx$, and consider the ideals E_x and F_u generated by x in E and u in F, respectively. Clearly, S and T carry E_x into F_u. Thus, by Theorem 4.83, there exist positive multiplication operators $M_1 \ldots, M_k$ on E_x and positive multiplication operators $L_1, \ldots, L_k$ on F_u such that $|y| \le x$ implies

$$\left\langle x', \left| \left(S - \sum_{i=1}^{k} L_i T M_i \right) y \right| \right\rangle < \epsilon \, . \tag{†}$$

Now by part (2) of Lemma 4.78 each M_i extends to a positive orthomorphism on E and each L_i extends to a positive orthomorphism on F, and the desired conclusion follows from (†). ∎

To continue our discussion, we need the concept of a quasi-interior point. A positive element u in a normed Riesz space is said to be a **quasi-interior point** whenever the ideal E_u generated by u is norm dense.

The quasi-interior points are characterized as follows.

Theorem 4.85. *For a positive vector u in a normed Riesz space E the following statements are equivalent.*

   (1) *The vector u is a quasi-interior point.*

   (2) *For each $x \in E^+$ we have $\|x - x \wedge nu\| \to 0$.*

   (3) *The vector u is strictly positive on E', i.e., $0 < x' \in E'$ implies $x'(u) > 0$.*[3]

Proof. (1) $\Longrightarrow$ (2) Assume that E_u is norm dense in E. Let $x \in E^+$, and let $\epsilon > 0$. Pick some $y \in E_u$ with $\|x - y\| < \epsilon$. Then the element $z = y^+ \wedge x \in E_u$ satisfies $\|x - z\| \le \|x - y\| < \epsilon$ and $0 \le z \le x$. Pick some k with $z \le ku$, and note that for $n \ge k$ we have $0 \le x - x \wedge nu \le x - x \wedge ku \le x - z$. Therefore, $\|x - x \wedge nu\| < \epsilon$ holds for all $n \ge k$, which shows that $\|x - x \wedge nu\| \to 0$.

   (2) $\Longrightarrow$ (3) Let $0 < x' \in E'$, and assume by way of contradiction that $x'(u) = 0$. Then for each $x \in E^+$ we have $0 \le x \wedge nu \le nu$, and so $x'(x \wedge nu) = 0$ holds for all n and all $x \in E^+$. This implies $x'(x) = 0$ for all $x \in E^+$, which means that $x' = 0$, a contradiction.

   (3) $\Longrightarrow$ (1) If E_u is not norm dense in E, then by the Hahn–Banach Theorem 1.25 there exists some nonzero $f \in E'$ which vanishes on E_u. We can suppose that $f^+ > 0$ holds. Now notice that

$$f^+(u) = \sup\{f(x) : x \in E \text{ and } 0 \le x \le u\}$$
$$= \sup\{f(x) : x \in E_u \text{ and } 0 \le x \le u\} = 0,$$

which is a contradiction. Hence, E_u is norm dense in E. ∎

Here are a few consequences of the preceding result.

   (a) *If u is an order unit in a normed Riesz space, then u is a quasi-interior point.*

   (b) *If u is a quasi-interior point in a normed Riesz space E, then $u + v$ is likewise a quasi-interior point for each $v \in E^+$.*

Indeed, if $0 < x' \in E'$, then $x'(u + v) \ge x'(u) > 0$ holds for each $v \in E^+$.

   (c) *If E is a separable Banach lattice, then E has quasi-interior points.*

---

[3] By virtue of this property a quasi-interior point also is known as a **strictly positive vector**.

To see this, pick a countable dense subset $\{x_1, x_2, \ldots\}$ of nonzero elements, and put $u = \sum_{n=1}^{\infty} |x_n|/2^n \|x_n\|$. Clearly, $0 < x' \in E'$ implies $x'(u) > 0$.

(d) *If u is a quasi-interior point in a normed Riesz space E, then u is a weak order unit.*

Indeed, from $\|x - x \wedge nu\| \to 0$, we see that $x \wedge nu \uparrow x$ holds for all $x \in E^+$.

(e) *A weak order unit need not be a quasi-interior point. However, in a Banach lattice with order continuous norm every weak order unit is a quasi-interior point.*

To see this, consider $u \in C[0, 1]$ defined by $u(t) = t$. Clearly, u is a weak order unit of $C[0, 1]$, and moreover, the equality $\|\mathbf{1} - \mathbf{1} \wedge nu\|_\infty = 1$, shows that u is not a quasi-interior point.

(f) *If u is a positive vector in a normed Riesz space E, then u is a quasi-interior point in $\overline{E}_u$ (the norm closure of the ideal generated by u in E).*

The analogue of Theorem 4.84 for Banach lattices with quasi-interior points is due to W. Haid [75] and is stated next.

Theorem 4.86 (Haid). *Let E and F be Banach lattices with quasi-interior points, and let $S, T \colon E \to F$ be two positive operators such that $0 \le S \le T$ holds. Then for each $x \in E^+$, $0 \le x' \in F'$, and $\epsilon > 0$ there exist positive orthomorphisms $M_1, \ldots, M_k$ on E and positive orthomorphisms $L_1, \ldots, L_k$ on F such that*

$$\left\langle x', \left|\left(S - \sum_{i=1}^{k} L_i T M_i\right)y\right|\right\rangle < \epsilon$$

holds for all $y \in E$ with $|y| \le x$.

Proof. Fix $x \in E^+$, $0 \le x' \in F'$, and $\epsilon > 0$, and let $u \in E^+$ be a quasi-interior point of E. Since $u + x$ is also a quasi-interior point, if we replace u by $u + x$, we can assume that $0 \le x \le u$ holds. Similarly, if v is a quasi-interior point of F, then replacing v by $v + T(u)$, we can suppose that $0 \le T(u) \le v$ holds. Let E_u and F_v denote the ideals generated by u in E and v in F, respectively. Then E_u and F_v with their sup norms are both AM-spaces having units u and v, respectively.

Clearly, S and T carry E_u into F_v, and moreover if we replace S and T restricted to E_u, then $0 \le S \le T$ holds. In addition, x' restricted to F_v is positive and continuous. Thus, by Theorem 4.83, there exist there exist positive multiplication operators $M_1, \ldots, M_k$ on E_u and positive multiplication operators $L_1, \ldots, L_k$ on F_v such that

$$\left\langle x', \left| \left(S - \sum_{i=1}^{k} L_i T M_i \right) y \right| \right\rangle < \epsilon$$

holds for all $y \in E$ with $|y| \leq x$.

Since E_u is dense in E, it follows from part (1) of Lemma 4.78 that each M_i extends to a positive orthomorphism on E (which we denote by M_i again). Similarly, each L_i extends to a positive orthomorphism on F, and the proof of the theorem is finished. ■

Finally, we close this section with one more approximation result.

Theorem 4.87. *Let* $T \colon E \to F$ *be a positive operator from a Banach lattice* E *which is either Dedekind* σ*-complete or has quasi-interior points into a Banach lattice* F *with order continuous norm. If a positive operator* $S \colon E \to F$ *satisfies* $0 \leq S \leq T$, *then given* $x \in E^+$ *and* $\epsilon > 0$ *there exist positive operators* $M_1, \ldots, M_k$ *on* E *and order projections* $P_1, \ldots, P_k$ *on* F *satisfying*

$$0 \leq \sum_{i=1}^{k} P_i T M_i \leq T \quad \text{and} \quad \left\| \left| S - \sum_{i=1}^{k} P_i T M_i \right| x \right\| < \epsilon.$$

Proof. Let $0 < x \in E$ and $\epsilon > 0$ be fixed. By Theorem 4.18 there exists some $0 < y' \in F'$ satisfying

$$(|x'| - y')^+ (Tx) < \epsilon$$

for all $x' \in F'$ with $\|x'\| \leq 1$.

(a) Suppose that E is Dedekind σ-complete. Let $C \colon E \to F$ be a component of T, and let $\delta > 0$. Then, by Theorem 2.3, there exist pairwise disjoint order projections $M_1, \ldots, M_k$ on E and order projections $P_1, \ldots, P_k$ on F such that

$$y' \left(\left| C - \sum_{i=1}^{k} P_i T M_i \right| x \right) < \delta.$$

Since $M_i M_j = 0$ for $i \neq j$, we have $(P_i T M_i) \wedge (P_j T M_j) = (P_i P_j) T (M_i M_j) = 0$ for $i \neq j$, and so

$$0 \leq \sum_{i=1}^{k} P_i T M_i = \bigvee_{i=1}^{k} P_i T M_i \leq T.$$

Now by Theorem 3.61 there exists a convex combination $\sum_{i=1}^{m} \alpha_i C_i$ of components of T such that

$$\left\| \left| S - \sum_{i=1}^{m} \alpha_i C_i \right| \right\| < \frac{\epsilon}{2\|x\| \cdot \|y'\|}.$$

By the preceding case for each $i = 1, \ldots, m$ there exist order projections $M_1^i, \ldots, M_{m_i}^i$ on E and order projections $P_1^i, \ldots, P_{m_i}^i$ on F satisfying

$$0 \le \sum_{j=1}^{m_i} P_j^i T M_j^i \le T \quad \text{and} \quad y'\Big(\Big|C_i - \sum_{j=1}^{m_i} P_j^i T M_j^i\Big|x\Big) < \tfrac{\epsilon}{2}.$$

Let $R = \sum_{i=1}^{m} \sum_{j=1}^{m_i} P_j^i T(\alpha_i M_j^i)$, and note that $0 \le R \le T$ holds. Moreover, we have

$$y'(|S - R|x) \le y'\Big(\Big|S - \sum_{i=1}^{m} \alpha_i C_i\Big|x\Big) + y'\Big(\Big|\sum_{i=1}^{m} \alpha_i \Big(C_i - \sum_{j=1}^{m_i} P_j^i T M_j^i\Big)x\Big|\Big)$$

$$< \tfrac{\epsilon}{2} + \sum_{i=1}^{m} \alpha_i y'\Big(\Big|C_i - \sum_{j=1}^{m_i} P_j^i T M_j^i\Big|x\Big) < \tfrac{\epsilon}{2} + \tfrac{\epsilon}{2} = \epsilon.$$

(b) Suppose that E has quasi-interior points. Pick a quasi-interior point $0 < e \in E$, and note that we can assume that $0 \le x \le e$ holds. Then E_e is an AM-space with unit e, and $\overline{E}_e = E$. By Theorem 4.81 there exist positive multiplication operators $M_1, \ldots, M_k$ on E_e and order projections $P_1, \ldots, P_k$ on F with

$$0 \le \sum_{i=1}^{k} P_i T M_i \le T \quad \text{and} \quad y'\Big(\Big|S - \sum_{i=1}^{k} P_i T M_i\Big|x\Big) < \epsilon.$$

By Lemma 4.78 each M_i is continuous with respect to the norm induced on E_u by E, and so each M_i has a unique continuous extension (which we denote by M_i again) to $\overline{E}_e = E$. This implies that we can suppose that each M_i is defined on all of E.

So, if E is Dedekind σ-complete or has a quasi-interior point, then there exist positive operators $M_1, \ldots, M_k$ on E and order projections $P_1, \ldots, P_k$ on F such that

$$0 \le \sum_{i=1}^{k} P_i T M_i \le T \quad \text{and} \quad y'\Big(\Big|S - \sum_{i=1}^{k} P_i T M_i\Big|x\Big) < \epsilon. \qquad (\star)$$

Also, from $(\star)$, we see that

$$\Big|S - \sum_{i=1}^{k} P_i T M_i\Big| \le S + \sum_{i=1}^{k} P_i T M_i \le 2T.$$

Finally, for $x' \in F'$ with $\|x'\| \le 1$ we have

$$\left| x'\left(\left|S - \sum_{i=1}^{k} P_i T M_i \big| x\right)\right| \right.$$

$$\le (|x'| - y')^+ \left(\left|S - \sum_{i=1}^{k} P_i T M_i \big| x\right) + y'\left(\left|S - \sum_{i=1}^{k} P_i T M_i \big| x\right)\right.$$

$$< (|x'| - y')^+ (2Tx) + \epsilon < 3\epsilon.$$

Thus, $\left\|\left|S - \sum_{i=1}^{k} P_i T M_i \big| x\right\| \le 3\epsilon\right.$ holds, and the proof of the theorem is finished. ∎

Exercises

1. If $T\colon E \to F$ is a positive operator between normed Riesz spaces, then show that
$$\|T\| = \sup\{\|Tx\|\colon\ x \in E^+ \text{ and } \|x\| \le 1\}.$$

2. Let $T\colon E \to F$ be a positive operator between normed Riesz spaces. If E is an AM-space with unit e, then show that $\|T\| = \|T(e)\|$.

3. If E is an AM-space and F is an AL-space, then show that $\mathcal{L}_b(E, F)$ is an AL-space.

4. Consider the operator $u \mapsto T_u$ from ℓ_∞ to $\mathcal{L}_b(\ell_p)$ $(1 \le p \le \infty)$ defined by
$$T_u(x_1, x_2, \ldots) = (u_1 x_1, u_2 x_2, \ldots).$$
Show that $u \mapsto T_u$ is a lattice isometry from ℓ_∞ into $\mathcal{L}_b(\ell_p)$, and from this (and Theorem 4.51) conclude that the Banach lattice $\mathcal{L}_b(\ell_p)$ does not have order continuous norm. Can you show directly that $\mathcal{L}_b(\ell_p)$ does not have order continuous norm?

5. Let E and F be two Banach lattices. If either
   (a) F is a Dedekind complete AM-space with unit, or
   (b) E is an AL-space and F is complemented in F'' (i.e., there exists a positive projection of F'' with range F),
   then show that $\mathcal{L}_b(E, F) = L(E, F)$.

6. If an operator T on a Banach lattice is the pointwise limit of a sequence of orthomorphisms, then show that T is also an orthomorphism.

7. Let E be an AM-space with unit e, and let L be a multiplication operator on E''. Show that for each $0 \le x' \in E'$ and each $\epsilon > 0$ there exists a multiplication operator M on E (which is positive if L is positive) satisfying $\langle x', |L - M|e\rangle < \epsilon$.

8. Let x be a vector in an AM-space E with unit. If M is the multiplication operator on E determined by x (i.e., $M(y) = xy$ holds for all $y \in E$), then show that $M''(y'') = xy''$ also holds for all $y'' \in E''$. [*Hint:* Note that $M''\colon E'' \to E''$ is an orthomorphism.]

9. Let E be a Banach lattice with order continuous norm, and let $0 < u \in E$. If $M: E_u \to E_u$ is a multiplication operator, then show that that there exists a unique orthomorphism T on E such that $T = M$ on E_u and $T = 0$ on E_u^d.

10. Show that a reflexive Banach lattice E has a quasi-interior point if and only if E' has a quasi-interior point. [*Hint:* Use Theorems 4.15 and 4.85.]

11. Let $T: E \to F$ be a continuous operator between normed Riesz spaces. If T has dense range and u is a quasi-interior point of E, then show that $T(u)$ is a quasi-interior point of F. [*Hint:* Note that T' is one-to-one, and so $0 < x' \in F'$ implies $x'(Tu) = T'x'(u) > 0$.]

12. Let F be a Riesz subspace of a normed Riesz space E. If $u \in F^+$ is a quasi-interior point of E, then show that u is also a quasi-interior point of F.

13. Let E be a normed Riesz space, and let $\mathcal{Q}$ denote the collection of all quasi-interior points of E. Show that either $\mathcal{Q} = \emptyset$ or else $\mathcal{Q}$ is norm dense in E^+.

14. Let $S, T: E \to F$ be two positive operators between Banach lattices such that $0 \le S \le T$ holds. Assume that E has quasi-interior points and that F is Dedekind complete.

    Show that given $x \in E^+$, $0 \le x' \in F_n^\sim$, and $\epsilon > 0$, there exist positive operators $M_1, \ldots, M_k$ on E and positive operators $L_1, \ldots, L_k$ on F such that $\langle x', |S - \sum_{i=1}^k L_i T M_i | x \rangle < \epsilon$.

15. Let $S, T: E \to F$ be two positive operators between Banach lattices such that $0 \le S \le T$ holds. Assume that E is Dedekind complete and that F has quasi-interior points.

    Show that, given $x \in E^+$, $0 \le x' \in F'$, and $\epsilon > 0$, there exist positive operators $M_1, \ldots, M_k$ on E and positive operators $L_1, \ldots, L_k$ on F such that $\langle x', |(S - \sum_{i=1}^k L_i T M_i) y| \rangle < \epsilon$ holds for all $y \in E$ with $|y| \le x$.

16. Let E be an AM-space with unit, and let F be a Banach lattice with order continuous norm, and let $S, T: E \to F$ be two positive operators satisfying $0 \le S \le T$. Show that given $\epsilon > 0$ there exist positive operators $M_1, \ldots, M_k$ on E and positive operators $L_1, \ldots, L_k$ on F such that

$$\left\| S - \sum_{i=1}^k L_i T M_i \right\| < \epsilon.$$

17. Let E be a reflexive Banach lattice, let F be an AL-space, and let $S, T: E \to F$ be two positive operators satisfying $0 \le S \le T$. Show that given $\epsilon > 0$ there exist positive operators $M_1, \ldots, M_k$ on E and positive operators $L_1, \ldots, L_k$ on F such that

$$\left\| S - \sum_{i=1}^k L_i T M_i \right\| < \epsilon.$$

18. Let E be an AM-space with unit, and let F be a Banach lattice with order continuous norm. For two positive operators $S, T: E \to F$ show that the following statements are equivalent:

(a) The operator T is in the band generated by S in $\mathcal{L}_b(E, F)$.

(b) For every $\epsilon > 0$ there exists some $\delta > 0$ such that whenever positive orthomorphisms $M_1, \ldots, M_k$ on E and positive orthomorphisms $L_1, \ldots, L_k$ on F satisfy $\left\| \sum_{i=1}^k L_i S M_i \right\| < \delta$, then we have

$$\left\| \sum_{i=1}^k L_i T M_i \right\| < \epsilon.$$

[*Hint*: Mimic the proof of Theorem 2.11.]

19. Let $S, T : E \to F$ be two positive operators between Banach lattices with F Dedekind complete satisfying $0 \leq S \leq T$. Show that for each $\epsilon > 0$ there exists a convex combination $\sum_{i=1}^n \alpha_i C_i$ of components of T such that

$$\left\| S - \sum_{i=1}^n \alpha_i C_i \right\| < \epsilon.$$

[*Hint*: Consider $[0, T]$ in $\mathcal{L}_b(E, F)$ and use Theorem 3.61.]

20. Assume that u is a quasi-interior point in a Banach lattice E. If the order interval $[0, u]$ is weakly compact, then show that E has order continuous norm.

Chapter 5

Compactness Properties of Positive Operators

A compact operator sends an arbitrary norm bounded sequence to a sequence with a norm convergent subsequence. For this reason, when operators are associated with integral equations, the compact operators are the most desirable. Besides being compact, an operator with some type of compactness is more useful than an arbitrary operator.

This chapter studies various compactness properties of operators on Banach spaces. Specifically, the four sections of this chapter deal with compact operators, weakly compact operators, L- and M-weakly compact operators, and Dunford–Pettis operators. Particular emphasis is given to compactness properties of a positive operator dominated by a compact operator. Also, relationships between the ring and order ideals generated by a positive operator are examined. As we shall see, when the ingredient of positivity is added, the results are sharp and amazing.

5.1. Compact Operators

It will be convenient to agree that the phrase *an operator $T\colon X \to Y$ between Banach spaces is positive* will mean that X and Y are both Banach lattices and that T is a positive operator. Also, the expression *an operator S between two Riesz spaces is dominated by another operator T* will simply mean that $S \leq T$. In this terminology, an operator between Riesz spaces is, of course, regular if and only if it is dominated by a positive operator.

Let $T\colon X \to Y$ be an operator between two normed vector spaces. Recall that T is said to be a **compact operator** whenever T maps the closed unit ball U of X onto a norm relatively compact subset of Y (i.e., whenever $\overline{T(U)}$ is a norm compact subset of Y). In other words, T is a compact operator if and only if for every norm bounded sequence $\{x_n\}$ of X the sequence $\{Tx_n\}$ has a norm convergent subsequence in Y. In case Y is a Banach space, T is a compact operator if and only if $T(U)$ is a norm totally bounded subset of Y. Clearly, every compact operator is norm bounded (and hence continuous).

We start our discussion with some basic properties of compact operators. (Recall that $L(X,Y)$ denotes the normed vector space of all continuous operators from X to Y; $L(X)$ stands for $L(X,X)$.)

Theorem 5.1. *For Banach spaces X, Y, and Z we have the following.*

(1) *The set of all compact operators from X to Y is a norm closed vector subspace of $L(X,Y)$.*

(2) *If $X \xrightarrow{\ S\ } Y \xrightarrow{\ T\ } Z$ are continuous operators and either S or T is compact, then TS is likewise a compact operator.*

Proof. (1) Clearly, the collection $\mathcal{K}(X,Y)$ of all compact operators from X to Y is a vector subspace of $L(X,Y)$. To see that $\mathcal{K}(X,Y)$ is also norm closed, let S be in the norm closure of $\mathcal{K}(X,Y)$, and let $\epsilon > 0$. Also, denote by U and V the closed unit balls of X and Y, respectively.

Choose some $T \in \mathcal{K}(X,Y)$ satisfying $\|S - T\| < \epsilon$, and observe that $S(U) \subseteq T(U) + \epsilon V$ holds. Since $T(U)$ is norm totally bounded, it follows from Theorem 3.1 that $S(U)$ is likewise norm totally bounded. That is, $S \in \mathcal{K}(X,Y)$ holds.

(2) Straightforward. ∎

When we consider the continuous operators on a Banach space X, Theorem 5.1 expresses the fact that the compact operators on X form a two sided norm closed ring ideal in $L(X)$.

The identity operator $I\colon X \to X$ on a Banach space is compact if and only if X is finite dimensional. Also, if X and Y are Banach spaces, then each finite rank operator of $X' \otimes Y$ (i.e., each operator $T\colon X \to Y$ of the form $T = \sum_{i=1}^{n} x_i' \otimes y_i$ with $x_1', \ldots, x_n' \in X'$ and $y_1, \ldots, y_n \in Y$) is clearly a compact operator. Thus, the norm closure of $X' \otimes Y$ in $L(X,Y)$ consists of compact operators. For a long time it was an open problem as to whether or not the norm closure of $X' \otimes Y$ was precisely the set of all compact operators. However, in 1972 P. Enflo [61] proved with a famous counterexample that in general this is not the case.

An operator is compact if and only if its adjoint is compact. This useful result is due to J. Schauder [**175**].

Theorem 5.2 (Schauder). *A norm bounded operator $T: X \to Y$ between Banach spaces is compact if and only if its adjoint $T': Y' \to X'$ is likewise a compact operators.*

Proof. Denote by U and B the closed unit balls of X and Y', respectively. To obtain the desired conclusion, put $\mathfrak{S} = \{\epsilon U : \epsilon > 0\}$ and $\mathfrak{I} = \{\delta B : \delta > 0\}$, and then apply Theorem 3.27. ∎

For our discussion, we shall need the following characterization of the norm totally bounded subsets of a Banach space.

Theorem 5.3 (Grothendieck). *A subset of a Banach space is norm totally bounded if and only if it is included in the closed convex hull of a sequence that converges in norm to zero.*

Proof. Let A be a subset of a Banach space X. If there exists a sequence $\{x_n\}$ of X with $\|x_n\| \to 0$ and $A \subseteq \overline{\text{co}}\,\{x_n\}$, then by Theorem 3.4 the set A is norm totally bounded.

For the converse, assume that A is a norm totally bounded set. Let U denote the closed unit ball of X. The desired sequence $\{x_n\}$ will be constructed by an inductive argument as follows. Put $A_0 = A$ and $k_0 = 0$. Pick a finite subset $\Phi_1 = \{x_1, \ldots, x_{k_1}\}$ of $2A_0$ with $2A_0 \subseteq \Phi_1 + 2^{-1}U$. Then, put $A_1 = (2A_0 - \Phi_1) \cap 2^{-1}U$, and note that by Theorem 3.2 the set A_1 is norm totally bounded. Now for the induction argument, assume that $\Phi_n = \{x_{k_{n-1}+1}, \ldots, x_{k_n}\}$ is a finite subset of the totally bounded set $2A_{n-1}$ with $2A_{n-1} \subseteq \Phi_n + 2^{-n}U$. Put $A_n = (2A_{n-1} - \Phi_n) \cap 2^{-n}U$, and note that A_n is a norm totally bounded set. Next, choose a finite subset $\Phi_{n+1} = \{x_{k_n+1}, \ldots, x_{k_{n+1}}\}$ of $2A_n$ with the property $2A_n \subseteq \Phi_{n+1} + 2^{-n-1}U$, and put $A_{n+1} = (2A_n - \Phi_{n+1}) \cap 2^{-n-1}U$.

Now we claim that the sequence $\{x_n\}$ satisfies $A \subseteq \overline{\text{co}}\,\{x_n\}$. To see this, note first that for each $x \in A_n$ we have $\|x\| \leq 2^{-n}$, and so $\|x_n\| \to 0$. On the other hand, if $x \in A$, then an easy argument shows that there exist integers $m_1 < m_2 < \cdots$ with $k_{i-1} < m_i \leq k_i$ such that the element $\frac{x_{m_1}}{2} + \cdots + \frac{x_{m_n}}{2^n} \in \overline{\text{co}}\,\{x_n\}$ satisfies

$$\left\| x - \left(\frac{x_{m_1}}{2} + \cdots + \frac{x_{m_n}}{2^n} \right) \right\| \leq \frac{1}{4^n}.$$

This implies that $x \in \overline{\text{co}}\,\{x_n\}$. That is, $A \subseteq \overline{\text{co}}\,\{x_n\}$ holds, and the proof is finished. ∎

We now pass to a useful characterization of compact operators due to T. Terzioglu [**183**].

Theorem 5.4 (Terzioglu). *An operator $T\colon X \to Y$ between two Banach spaces is compact if and only if there exists a sequence $\{x'_n\}$ of X' with $\|x'_n\| \to 0$ and satisfying*

$$\|Tx\| \leq \sup\{|x'_n(x)|\}$$

for all $x \in X$.

Proof. Assume first that T is a compact operator. Let B denote the closed unit ball of Y'. Then $T'(B)$ is (by Theorem 5.2) a norm totally bounded subset of X', and so by Theorem 5.3 there exists a sequence $\{x'_n\}$ of X' with $\|x'_n\| \to 0$ and $T'(B) \subseteq \overline{\mathrm{co}}\,\{x'_n\}$. Thus, for each $x \in X$ we have

$$\|Tx\| = \sup\{|y'(Tx)|\colon\ y' \in B\} = \sup\{|T'y'(x)|\colon\ y' \in B\} \leq \sup\{|x'_n(x)|\}\,.$$

For the converse, assume that there exists a sequence $\{x'_n\}$ of X' with $\|x'_n\| \to 0$ such that $\|Tx\| \leq \sup\{|x'_n(x)|\}$ holds for all $x \in X$. Next, consider the operator $S\colon X \to c_0$ defined by

$$S(x) = \big(x'_1(x), x'_2(x), \ldots\big)\,.$$

Put $u = \big(\|x'_1\|, \|x'_2\|, \ldots\big) \in c_0$, and note that if U is the closed unit ball of X, then $S(U) \subseteq [-u, u]$ holds. Since the order interval $[-u, u]$ is a compact subset of c_0 (see Exercise 14 of Section 3.2), it follows that $S(U)$ is a norm totally bounded subset of c_0. In particular, $S(U)$ is also a norm totally bounded subset of the normed vector space $Z = S(X)$. Now, define $R\colon Z \to Y$ by

$$R(Sx) = Tx\,.$$

Since $Sx = Sy$ implies $x'_n(x - y) = 0$ for all n, it follows from

$$\|Tx - Ty\| \leq \sup\{|x'_n(x - y)|\} = 0$$

that $Tx = Ty$, and so the operator R is well defined. Moreover, the inequality

$$\big\|R(Sx)\big\| = \|Tx\| \leq \sup\{|x'_n(x)|\} = \|Sx\|_\infty$$

shows that $R\colon Z \to Y$ is also continuous. Thus, $T(U) = R(S(U))$ is a norm totally bounded subset of Y and so T is a compact operator. ∎

In terms of factorizations, the compact operators are characterized as follows.

Theorem 5.5. *An operator $T\colon X \to Y$ between two Banach spaces is a compact operator if and only if T factors with compact factors through a closed vector subspace of c_0.*

Proof. Let $T\colon X \to Y$ be a compact operator between two Banach spaces. By Theorem 5.4, there exists a sequence $\{x'_n\}$ of X' satisfying $\|x'_n\| \to 0$ and

$$\|Tx\| \le \sup\{|x'_n(x)|\} \qquad (\star)$$

for all $x \in X$. We can assume that $x'_n \ne 0$ holds for all n. Put $\alpha_n = \|x'_n\|^{\frac{1}{2}}$ and $y'_n = \|x'_n\|^{-\frac{1}{2}} \cdot x'_n$. Clearly, $\|y'_n\| = \alpha_n \to 0$. Now define the operator $S\colon X \to c_0$ by

$$S(x) = \left(y'_1(x), y'_2(x), \dots\right),$$

and note that $\|Sx\|_\infty = \sup\{|y'_n(x)|\}$ holds. In view of Theorem 5.4, the latter shows that S is a compact operator. Now let Z be the norm closure of $S(X)$ in c_0, and note that $S\colon X \to Z$ is a compact operator.

Next, consider the operator $R\colon S(X) \to Y$ defined by $R(Sx) = Tx$. Note that R is well defined. Indeed, if $Sx = Sy$ holds, then we have $x'_n(x - y) = 0$ for all n, and so by $(\star)$ we see that $Tx = Ty$. On the other hand, if $M = \sup\{|\alpha_n|\}$, then $|x'_n(x)| = |\alpha_n y'_n(x)| \le M|y'_n(x)|$ holds for all n, and so

$$\|R(Sx)\| = \|Tx\| \le \sup\{|x'_n(x)|\} \le M \sup\{|y'_n(x)|\} = M\|Sx\|_\infty,$$

which shows that $R\colon S(X) \to Y$ is also continuous. Hence, R has a continuous extension to all of Z, which we denote by R again. Clearly, $T = RS$ holds, and it remains to be shown that R is a compact operator.

If the linear functional $f_n \in c'_0$ is defined by $f_n(\lambda_1, \lambda_2, \dots) = \alpha_n \lambda_n$, then $\|f_n\| = \alpha_n \to 0$ holds, and moreover $f_n(Sx) = \alpha_n y'_n(x) = x'_n(x)$. Since $\|R(Sx)\| = \|Tx\| \le \sup\{|f_n(Sx)|\}$ holds for all $x \in X$, it follows that $\|R(z)\| \le \sup\{|f_n(z)|\}$ for all $z \in Z$. Hence, by Theorem 5.4, the operator $R\colon Z \to Y$ is compact, as desired. ∎

We now turn our discussion to lattice properties of compact operators between Banach lattices. To start with, we may ask whether or not a compact operator between Banach lattices possesses a modulus. It is somewhat surprising to learn that the answer is negative. The following example (due to U. Krengel [106]) shows that a compact operator need not have a modulus and also that a compact operator may have a modulus which is not compact.

Example 5.6 (Krengel). For each n consider the finite dimensional Banach lattice $E_n = \mathbb{R}^{2^n}$ (pointwise ordering, Euclidean norm). Let A_n be a $2^n \times 2^n$ matrix with ± 1 entries and orthogonal rows. For instance, the matrices A_n can be constructed inductively as follows:

$$A_1 = \begin{bmatrix} 1 & 1 \\ 1 & -1 \end{bmatrix} \quad \text{and} \quad A_{n+1} = \begin{bmatrix} A_n & A_n \\ A_n & -A_n \end{bmatrix}.$$

For each n let $T_n\colon E_n \to E_n$ be the operator whose matrix (with respect to the standard unit vectors of E_n) is $2^{-\frac{n}{2}} A_n$. Since $2^{-\frac{n}{2}} A_n$ is an orthogonal matrix, T_n is an onto isometry, and so $\|T_n\| = 1$ holds for all n. On the other hand, $|A_n|$ is the matrix with all of its entries equal to 1, and from this an easy computation shows that $\| |A_n| \| = 2^n$. Therefore, $\| |T_n| \| = 2^{\frac{n}{2}}$ holds for all n.

Next, consider the Banach lattice $E = (E_1 \oplus E_2 \oplus \cdots)_0$, and note that E is Dedekind complete with order continuous norm. If $\alpha = (\alpha_1, \alpha_2, \ldots) \in \ell_\infty$ is fixed, then consider the operator $T\colon E \to E$ defined by

$$T(x_1, x_2, \ldots) = (\alpha_1 T_1 x_1, \alpha_2 T_2 x_2, \ldots).$$

From

$$\|T(x_1, x_2, \ldots)\| = \sup\{|\alpha_n| \cdot \|T_n x_n\|\} \le \|\alpha\|_\infty \cdot \|(x_1, x_2, \ldots)\|,$$

we see that T is a norm bounded operator. On the other hand, if $\lim \alpha_n = 0$, then T is a compact operator. This can be easily seen by observing that the operators $K_n(x_1, x_2, \ldots) = (\alpha_1 T_1 x_1, \ldots, \alpha_n T_n x_n, 0, 0, \ldots)$ are all compact and that they satisfy $\|T - K_n\| \le \sup\{|\alpha_i|\colon i \ge n\}$.

Thus, if we put $\alpha_n = 2^{-\frac{n}{2}}$, then T is a compact operator, and moreover its modulus exists. In fact, an easy computation shows that

$$|T|(x_1, x_2, \ldots) = (\alpha_1 |T_1| x_1, \alpha_2 |T_2| x_2, \ldots).$$

However, $|T|$ is not a compact operator. To see this, for each n fix some $x_n \in E_n$ with $\|x_n\| = 1$ and $\| |T_n|(x_n)\| = 2^{\frac{n}{2}}$, and let $\widehat{x}_n$ denote the element of E whose n^{th} component is x_n and every other zero. Then $\|\widehat{x}_n\| = 1$ holds in E, and for $n > m$ we have

$$\big\| |T|\widehat{x}_n - |T|\widehat{x}_m \big\| = \big\|(0, \ldots, 0, -\alpha_m |T| x_m, 0, \ldots, 0, \alpha_n |T| x_n, 0, 0, \ldots)\big\| = 1.$$

This shows that $|T|$ is not a compact operator.

Finally, arguing as above, we see that for $\alpha_n = 2^{-\frac{n}{3}}$ the operator T is still compact, but its modulus does not exist. ∎

By virtue of Example 4.73 we know that a continuous operator between two AM-spaces need not be order bounded (and hence need not possess a modulus). However, it is surprising to learn that whenever the range of a compact operator is an AM-space, then the operator possesses a modulus (which is also compact). This interesting result is due to U. Krengel [105].

Theorem 5.7 (Krengel). *Let $T\colon E \to F$ be a compact operator from a Banach lattice to an AM-space. Then the modulus of T exists and is a compact operator. Moreover,*

$$|T|(x) = \sup\{|Ty|\colon |y| \le x\}$$

holds for all $x \in E^+$.

In addition, in this case, the vector space of all compact operators from E to F (with the r-norm) is a Banach lattice.

Proof. If $x \in E^+$, then $T[-x, x]$ is norm totally bounded in F, and so according to Theorem 4.30 the supremum

$$|T|(x) = \sup\{|Ty| : |y| \le x\} = \sup T[-x, x]$$

exists in F. Thus, by Theorem 1.14, the modulus of T exists.

Now let U denote the closed the unit ball of E. To see that $|T|$ is a compact operator, note that if A denotes the set of all finite suprema of $T(U)$, then by Theorem 4.30 the norm closure $\overline{A}$ of A is compact. Therefore, by Theorem 4.30 again, we see that $|T|(x) \in \overline{A}$ holds for all $x \in U^+$. That is, $|T|(U^+) \subseteq \overline{A}$ holds, and from this it follows that $|T|$ is a compact operator.

To see that the vector space of all compact operators from E to F is a Banach lattice, repeat the arguments of the proof of Theorem 4.74. ∎

As an application of the preceding theorem, let us establish the following result dealing with the finite rank operators.

Theorem 5.8. *Every finite rank operator between two Banach lattices has a compact modulus.*

Proof. Let $T \colon E \to F$ be a finite rank operator between two Banach lattices. Pick a basis $\{x_1, \ldots, x_n\}$ for $T(E)$, and put $u = |x_1| + \cdots + |x_n|$. Then F_u (with the sup norm) is an AM-space, and from Theorem 3.28 it is easy to see that the operator $T \colon E \to F_u$ is also compact. By Theorem 5.7 the supremum

$$|T|(x) = \sup\{|Ty| : |y| \le x\}$$

exists in F_u (and hence in F) for each $x \in E^+$, and moreover $|T| \colon E \to F_u$ is a compact operator. Thus, the operator $T \colon E \to F$ possesses a modulus. Since $\|x\| \le \|u\| \cdot \|x\|_\infty$ holds for each $x \in F_u$, it follows that $|T| \colon E \to F$ is also a compact operator. ∎

The next result is a dual of Theorem 5.7, and also is due to U. Krengel [105].

Theorem 5.9 (Krengel). *If E is an AL-space and F is a Banach lattice with order continuous norm, then every order bounded compact operator from E to F has a compact modulus.*

Moreover, in this case, the compact operators of $\mathcal{L}_b(E, F)$ form a Banach lattice.

Proof. Let $T\colon E \to F$ be an order bounded compact operator from an AL-space to a Banach lattice with order continuous norm. Since $T'\colon F' \to E'$ is a compact operator and E' is an AM-space, it follows from Theorem 5.7 that $|T'|$ is a compact operator. On the other hand, the order continuity of the norm of F implies $F' = F_n^\sim$, and so by Theorem 1.76, we have $|T|' = |T'|$. Thus, $|T|'$ is a compact operator, and hence $|T|$ is likewise a compact operator. ∎

An important compactness property of order bounded operators, due to P. G. Dodds and D. H. Fremlin [54], is described in the next theorem.

Theorem 5.10 (Dodds–Fremlin). *Let E and F be two Riesz spaces with F Dedekind complete. If τ is an order continuous locally convex-solid topology on F, then for each $x \in E^+$ the set*

$$\mathcal{B} = \Big\{T \in \mathcal{L}_b(E, F)\colon \ T[0, x] \text{ is } \tau\text{-totally bounded}\Big\}$$

is a band of $\mathcal{L}_b(E, F)$.

Proof. The proof below is a simplified version, due to A. R. Schep, of the original proof of P. G. Dodds and D. H. Fremlin [54]. Let $x \in E^+$ be fixed. Clearly, $\mathcal{B}$ is a vector subspace of $\mathcal{L}_b(E, F)$. The rest of the proof goes by steps.

STEP 1. *If $T \in \mathcal{B}$ and $R, S \in \mathcal{L}_b(E, F)$ satisfy $R + S = T$ and $R \perp S$, then $R, S \in \mathcal{B}$.*

To see this, let V and W be two solid τ-neighborhoods of zero such that $W + W \subseteq V$. Since τ is order continuous, it follows from Theorem 1.21 that there exist $x_1, \ldots, x_n \in E^+$ such that

$$x = \sum_{i=1}^n x_i \quad \text{and} \quad \sum_{i=1}^n |R|x_i \wedge |S|x_i \in W \,.$$

Next, choose another solid τ-neighborhood U of zero with $U + \cdots + U \subseteq W$, where the sum to the left has n summands. In view of $T[0, x_i] \subseteq T[0, x]$, it follows that each $T[0, x_i]$ is a τ-totally bounded set. Therefore, for each $i = 1, \ldots, n$ there exists a finite subset Φ_i of $[0, x_i]$ with $T[0, x_i] \subseteq T(\Phi_i) + U$. Put

$$\Phi = \Big\{\sum_{i=1}^n y_i\colon \ y_i \in \Phi_i \text{ for each } i = 1, \ldots, n\Big\},$$

and note that Φ is a finite subset of $[0, x]$.

Now if $z \in [0, x]$, then use the Riesz decomposition property (see Theorem 1.13) to write $z = \sum_{i=1}^n z_i$ with $0 \le z_i \le x_i$ for each $i = 1, \ldots, n$. By the above, for each i there exists some $y_i \in \Phi_i$ with $T(z_i - y_i) \in U$. Let $y = \sum_{i=1}^n y_i \in \Phi$, and note that $z - y = \sum_{i=1}^n (z_i - y_i) = \sum_{i=1}^n w_i$, where for simplicity we put $w_i = z_i - y_i$. Thus, by taking into account that in a Riesz space $|u| - |u| \wedge |v| \leq |u + v|$ holds, we obtain

$$\left|R(z - y)\right|$$
$$\leq \sum_{i=1}^{n} |R(w_i)|$$
$$= \sum_{i=1}^{n} \left(|R(w_i)| - |R(w_i)| \wedge |S(w_i)|\right) + \sum_{i=1}^{n} |R(w_i)| \wedge |S(w_i)|$$
$$\leq \sum_{i=1}^{n} \left|(R + S)(w_i)\right| + \sum_{i=1}^{n} |R|x_i \wedge |S|x_i$$
$$= \sum_{i=1}^{n} |T(w_i)| + \sum_{i=1}^{n} |R|x_i \wedge |S|x_i \in U + \cdots + U + W \subseteq W + W \subseteq V.$$

Hence, $R(z - y) \in V$. This implies $R[0, x] \subseteq R(\Phi) + V$, which shows that $R[0, x]$ is τ-totally bounded. That is, $R \in \mathcal{B}$. Similarly, $S \in \mathcal{B}$.

STEP 2. *If $T \in \mathcal{B}$, then $T^+ \in \mathcal{B}$. That is, $\mathcal{B}$ is a Riesz subspace of $\mathcal{L}_b(E, F)$.*

The proof follows immediately from $T = T^+ + (-T^-)$ and Step 1 by observing that $|T^+| \wedge | - T^-| = 0$.

STEP 3. *If $\{T_\alpha\} \subseteq \mathcal{B}$ satisfies $T_\alpha \uparrow T$ in $\mathcal{L}_b(E, F)$, then $T \in \mathcal{B}$.*

To see this, let V and W be two solid τ-neighborhoods of zero with $W + W \subseteq V$. Since τ is order continuous and $T_\alpha(x) \uparrow T(x)$ holds in F, there exists some index β with $T(x) - T_\beta(x) \in W$. In view of the τ-total boundedness of $T_\beta[0, x]$, there exists a finite subset Φ of $[0, x]$ such that $T_\beta[0, x] \subseteq T_\beta(\Phi) + W$. Now if $y \in [0, x]$, then choose some $z \in \Phi$ with $T_\beta(y - z) \in W$ and note that the relation

$$\begin{aligned}\left|T(y) - T_\beta(z)\right| &\leq T(y) - T_\beta(y) + |T_\beta(y - z)| \\ &\leq T(x) - T_\beta(x) + \left|T_\beta(y - z)\right| \in W + W \subseteq V\end{aligned}$$

implies $T(y) - T_\beta(y) \in V$. Thus, $T[0, x] \subseteq T_\beta(\Phi) + V$, and hence $T \in \mathcal{B}$.

STEP 4. *If $0 \leq S \leq T$ holds in $\mathcal{L}_b(E, F)$ with $T \in \mathcal{B}$, then $S \in \mathcal{B}$.*

Indeed, by Theorem 2.9, there exists a sequence $\{S_n\}$ of T-step functions with $0 \leq S_n \uparrow S$ in $\mathcal{L}_b(E, F)$. By Step 1 each S_n belongs to $\mathcal{B}$, and by Step 3 it follows that $S \in \mathcal{B}$. The proof of the theorem is now complete. ■

A useful property of operators that map order intervals to totally bounded sets is described in the next result. Recall that the topology

$|\sigma|(F, F')$ on a Banach lattice F is the locally convex-solid topology of uniform convergence on the order intervals of F'.

Theorem 5.11. *Let $S, T\colon E \to F$ be two positive operators between Banach lattices such that $0 \le S \le T$ holds. If $T[0, x]$ is $|\sigma|(F, F')$-totally bounded for each $x \in E^+$, then $S[0, x]$ is likewise $|\sigma|(F, F')$-totally bounded for each $x \in E^+$.*

Proof. Consider S and T as operators from E to F'', and note that the order continuous locally convex-solid topology $|\sigma|(F'', F')$ induces $|\sigma|(F, F')$ on F. Clearly, $T[0, x]$ is $|\sigma|(F'', F')$-totally bounded for each $x \in E^+$, and so, by Theorem 5.10, $S[0, x]$ is $|\sigma|(F'', F')$-totally bounded for each $x \in E^+$. That is, $S[0, x]$ is $|\sigma|(F, F')$-totally bounded for each $x \in E^+$, as desired. ∎

We continue with a useful lattice approximation property of positive operators dominated by compact operators.

Lemma 5.12. *Let $S, T\colon E \to F$ be two positive operators between Banach lattices such that $0 \le S \le T$ holds. If T sends a subset A of E^+ to a norm totally bounded set, then for each $\epsilon > 0$ there exists some $u \in F^+$ such that*

$$\left\| (Sx - u)^+ \right\| < \epsilon$$

holds for all $x \in A$.

Proof. Let $\epsilon > 0$. Since $T(A)$ is a norm totally bounded subset of F, there exist $x_1, \ldots, x_n \in A$ such that for each $x \in A$ we have $\|Tx - Tx_i\| < \epsilon$ for some i. Put $u = T\left(\sum_{i=1}^n x_i\right) \in F^+$. Now if $x \in A$, then pick some x_i such that $\|Tx - Tx_i\| < \epsilon$, and note that

$$0 \le (Sx - u)^+ \le (Tx - u)^+ \le (Tx - Tx_i)^+ \le |Tx - Tx_i|.$$

Thus, $\|(Sx - u)^+\| < \epsilon$ holds for all $x \in A$, as claimed. ∎

We are now in the position to establish some major results concerning positive operators. The next result of the authors [9] states that if a positive operator on a Banach lattice is dominated by a compact operator, then its third power is also a compact operator.

Theorem 5.13 (Aliprantis–Burkinshaw). *If a positive operator S on a Banach lattice is dominated by a compact operator, then S^3 is a compact operator.*

Proof. Let $S, T\colon E \to E$ be two positive operators on a Banach lattice with T compact and $0 \le S \le T$. Since $U \subseteq U^+ - U^+$ holds, it is enough to show that $S^3(U^+)$ is a norm totally bounded set.

To this end, let $\epsilon > 0$. By Lemma 5.12, there exists some $u \in F^+$ such that $\|(Sx - u)^+\| < \epsilon$ holds for all $x \in U^+$. From the identity

$$Sx = Sx \wedge u + (Sx - u)^+,$$

we see that $S(U^+) \subseteq [0, u] + \epsilon U$. therefore,

$$S^2(U^+) \subseteq S[0, u] + \epsilon S(U) \subseteq S[0, u] + \epsilon \|S\| U,$$

and

$$S^3(U^+) \subseteq S^2[0, u] + \epsilon \|S\|^2 U. \qquad (\star)$$

Now observe that if a norm bounded net $\{x_\alpha\}$ satisfies $|x_\alpha| \xrightarrow{w} 0$, then the inequalities $0 \le S|x_\alpha| \le T|x_\alpha|$ and the compactness of T imply that $\lim \|Sx_\alpha\| = 0$. In other words, on every norm bounded subset D of E the function $S \colon D \to E$ is continuous for the $|\sigma|(E, E')$-topology on D and the norm topology on E. Thus, by Theorem 3.3, the operator S carries $|\sigma|(E, E')$-totally bounded sets to norm totally bounded sets. Since $T[0, u]$ is norm totally bounded (and hence $|\sigma|(E, E')$-totally bounded), it follows from Theorem 5.11 that $S[0, u]$ is likewise $|\sigma|(E, E')$-totally bounded. Therefore, by the above discussion, $S^2[0, u] = S(S[0, u])$ must be a norm totally bounded set. Now $(\star)$ (combined with part (4) of Theorem 3.1) shows that $S^3(U^+)$ is a norm totally bounded set, as desired. ∎

A variation of the preceding result is described in the next theorem.

Theorem 5.14 (Aliprantis–Burkinshaw). *If in the scheme of positive operators*

$$E \xrightarrow{S_1} F \xrightarrow{S_2} G \xrightarrow{S_3} H$$

each operator S_i is dominated by a compact operators, then $S_3 S_2 S_1$ is also a compact operator.

Proof. Assume that for each $i = 1, 2, 3$ there is some compact operator T_i satisfying $0 \le S_i \le T_i$. Next, consider the Banach lattice $L = E \oplus F \oplus G \oplus H$, and then define the operators $S, T \colon L \to L$ via the matrices

$$S = \begin{bmatrix} 0 & 0 & 0 & 0 \\ S_1 & 0 & 0 & 0 \\ 0 & S_2 & 0 & 0 \\ 0 & 0 & S_3 & 0 \end{bmatrix} \quad \text{and} \quad \begin{bmatrix} 0 & 0 & 0 & 0 \\ T_1 & 0 & 0 & 0 \\ 0 & T_2 & 0 & 0 \\ 0 & 0 & T_3 & 0 \end{bmatrix}.$$

Clearly, T is a compact operator and $0 \le S \le T$ holds. Now, by Theorem 5.13, the operator

$$S^3 = \begin{bmatrix} 0 & 0 & 0 & 0 \\ 0 & 0 & 0 & 0 \\ 0 & 0 & 0 & 0 \\ S_3 S_2 S_1 & 0 & 0 & 0 \end{bmatrix}$$

must be compact, and therefore $S_3 S_2 S_1$ is also a compact operator. ■

The power of the preceding theorem lies in the fact that no extra conditions were imposed on the Banach lattices. However, when order continuity of some norm is assumed, then one can obtain sharper compactness theorems. These results are stated next. We start with the companion of Theorem 5.14, which is also due to the authors [9].

Theorem 5.15 (Aliprantis–Burkinshaw). *Let E, F, and G be Banach lattices such that either E' or G has order continuous norm. If in the scheme of operators*

$$E \xrightarrow{S_1} F \xrightarrow{S_2} G$$

each S_i is dominated by a compact operator, then $S_2 S_1$ is a compact operator.

Proof. Assume that for each $i = 1, 2$ there exists a compact operator T_i satisfying $0 \le S_i \le T_i$, and let U and V denote the closed unit balls of E and G, respectively. Assume first that G has order continuous, and let $\epsilon > 0$.

By Lemma 5.12, there exists some $u \in F^+$ such that $\|(S_1 x - u)^+\| < \epsilon$ holds for all $x \in U^+$. From the identity $S_1 x = S_1 x \wedge u + (S_1 x - u)^+$, it is easy to see that

$$S_2 S_1 (U^+) \subseteq S_2[0, u] + \epsilon \|S_2\| V. \qquad (\star\star)$$

Since G has order continuous norm, it follows from Theorem 5.10 that $S_2[0, u]$ is norm totally bounded. Therefore, from $(\star\star)$ we see that $S_2 S_1(U^+)$ is a norm totally bounded set, and so $S_2 S_1$ is a compact operator.

To prove the case when E' has order continuous norm, take adjoints and apply the above conclusion to $(S_2 S_1)' = S_1' S_2'$. ■

The following special case of the preceding theorem is also useful.

Corollary 5.16. *Let E be a Banach lattice such that either E or E' has order continuous norm. If a positive operator $S \colon E \to E$ is dominated by a compact operator, then S^2 is also a compact operator.*

Next, we illustrate the limitations of the previous results with some examples taken from [9]. The first one is associated with Theorem 5.15.

Example 5.17. Let $\{r_n\}$ denote the sequence of Rademacher functions on $[0, 1]$, that is, $r_n(x) = \mathrm{Sgn} \sin(2^n \pi x)$ holds for each $x \in [0, 1]$. Define the positive operators

$$\ell_1 \xrightarrow{S_1} L_2[0, 1] \xrightarrow{S_2} \ell_\infty$$

via the formulas

$$S(\alpha_1, \alpha_2, \ldots) = \sum_{n=1}^{\infty} \alpha_n r_n^+$$

and

$$S_2(f) = \left(\int_0^1 f(x) r_1^+(x)\, dx, \ \int_0^1 f(x) r_2^+(x)\, dx, \ldots \right)$$

for each $(\alpha_1, \alpha_2, \ldots) \in \ell_1$ and $f \in L_2[0,1]$. Similarly, define the positive operators

$$\ell_1 \xrightarrow{\ T_1\ } L_2[0,1] \xrightarrow{\ T_2\ } \ell_\infty$$

by

$$T_1(\alpha_1, \alpha_2, \ldots) = \left(\sum_{n=1}^\infty \alpha_n \right) \mathbf{1} \quad \text{and} \quad T_2(f) = \left[\int_0^1 f(x)\, dx \right] (1, 1, \ldots).$$

Clearly, T_1 and T_2 are both compact operators (each is a rank-one operator), and $0 \leq S_1 \leq T_1$ and $0 \leq S_2 \leq T_2$ hold.

Now note that if $\{e_n\}$ denotes the sequence of the basic unit vectors of ℓ_1, then $S_1(e_n) = r_n^+$ holds for each n. It follows that for $n \neq m$ we have

$$\|S_2 S_1 e_n - S_2 S_1 e_m\|_\infty = \|S_2(r_n^+ - r_m^+)\|_\infty$$

$$\geq \left| \int_0^1 [r_n^+(x) - r_m^+(x)] r_n^+(x)\, dx \right| = \tfrac{1}{4}.$$

This shows that $S_2 S_1$ is not a compact operator (and therefore neither are S_1 and S_2). $\blacksquare$

By Theorem 5.13 we know that if a positive operator on a Banach lattice is dominated by a compact operator, then its cube is compact. The next example shows that the power three cannot be reduced to two.

Example 5.18. Consider the operators S_1, S_2, T_1, and T_2 as they were introduced in the preceding example, and let $E = \ell_1 \oplus L_2[0,1] \oplus \ell_\infty$. Note that neither E nor E' has order continuous norm. Now consider the operators $S, T : E \to E$ defined via the matrices

$$S = \begin{bmatrix} 0 & 0 & 0 \\ S_1 & 0 & 0 \\ 0 & S_2 & 0 \end{bmatrix} \quad \text{and} \quad \begin{bmatrix} 0 & 0 & 0 \\ T_1 & 0 & 0 \\ 0 & T_2 & 0 \end{bmatrix}.$$

Clearly, $0 \leq S \leq T$ holds, T is compact, and S is not compact. On the other hand, the matrix

$$S^2 = \begin{bmatrix} 0 & 0 & 0 \\ 0 & 0 & 0 \\ 0 & S_2 S_1 & 0 \end{bmatrix}$$

shows that S^2 is not a compact operator. Note also that $S^3 = 0$. $\blacksquare$

It may happen that E has order continuous norm and two operators S and T on E satisfy $0 \leq S \leq T$ with T compact and S noncompact. The next example is of this type, which also shows that in this respect Corollary 5.16 is the "best" possible.

Example 5.19. Let S_1 and T_1 be as in the Example 5.17, and then consider the Banach lattice $E = \ell_1 \oplus L_2[0,1]$. Clearly, E has order continuous norm. Now define the operators $S, T: E \to E$ via the matrices

$$S = \begin{bmatrix} 0 & 0 \\ S_1 & 0 \end{bmatrix} \quad \text{and} \quad \begin{bmatrix} 0 & 0 \\ T_1 & 0 \end{bmatrix} .$$

Then $0 \leq S \leq T$ holds, and T is compact. However, it is easy to check that S is not a compact operator. ∎

A glance at the preceding example shows that E' does not have order continuous norm. This is not accidental. The following result of P. G. Dodds and D. H. Fremlin [54] asserts that:

- *If a Banach lattice E and its norm dual E' have order continuous norms and two operators $S, T: E \to E$ satisfy $0 \leq S \leq T$ and T is compact, then S also must be compact.*

This result in its general form is as follows.

Theorem 5.20 (Dodds–Fremlin). *Let E and F be two Banach lattices with E' and F having order continuous norms. If a positive operator $S: E \to F$ is dominated by a compact operator, then S is necessarily a compact operator.*

Proof. Assume that the positive compact operator $T: E \to F$ satisfies $0 \leq S \leq T$. Denote by U and V the closed unit balls of E and F, and let $\epsilon > 0$. By Lemma 5.12, there exists some $u \in F^+$ satisfying $\|(Sx - u)^+\| < \epsilon$ for all $x \in U^+$. From $Sx = Sx \wedge u + (Sx - u)^+$, it follows that

$$S(U^+) \subseteq u \wedge S(U^+) + \epsilon V . \qquad (\star)$$

Now observe that (by Theorem 5.2) $T': F' \to E'$ is a compact operator and that $0 \leq S' \leq T'$ holds. Since E' has order continuous norm, it follows from Theorem 5.10 that S' maps order intervals of F' to norm totally bounded subsets of E'. Therefore, Theorem 3.27 (applied with $\mathfrak{S} = \{rU: r > 0\}$ and $\mathfrak{J} = \{[-x', x']: 0 \leq x' \in F'\}$) shows that $S(U^+)$ must be a $|\sigma|(F, F')$-totally bounded set, and hence $u \wedge S(U^+)$ is likewise a $|\sigma|(F, F')$-totally bounded set. Since F has order continuous norm and $u \wedge S(U^+)$ is order bounded, it follows from Theorem 4.17 that $u \wedge S(U^+)$ is also norm totally bounded. By $(\star)$ we see that $S(U^+)$ is a norm totally bounded set, as required. ∎

Remark. Theorem 5.20 has an interesting history. It has certainly influenced most of the current work on positive compact operators. Its

"roots" seem to go back to the paper by W. A. J. Luxemburg and A. C. Zaanen [**131**]. For integral operators it appeared in the Soviet literature
in the book of M. A. Kransoselskii, P. P. Zabreiko, E. I. Pustylnik, and
P. E. Sobolevskii [**100**, Theorem 5.10, p. 279], and the same result (in a
more general context) appeared in the work of R. J. Nagel and U. Schlotterbeck [**145**]; see also [**174**, Proposition 10.2, p. 291]. A few years later,
physical evidence led the mathematical physicists J. Avron, I. Herbst, and
B. Simon [**28**] to conjecture its validity for "good" L_p-spaces, and their conjecture was settled by L. D. Pitt [**165**] . Almost at the same time (and independently) P. G. Dodds and D. H. Fremlin [**54**] established Theorem 5.20
in its present general form. In its own turn, the paper of P. G. Dodds and
D. H. Fremlin inspired the basic works of the authors [**9, 10, 11, 14, 15, 17**]
on compactness of positive operators.

Finally, we close this section by establishing that a positive orthomorphism dominated by a compact operator is itself compact.

Theorem 5.21. *Let $S, T: E \to E$ be two positive operators on a Banach
lattice such that $0 \leq S \leq T$ holds. If T is a compact operator and S is an
orthomorphism, then S is compact.*

Proof. By taking adjoints we can assume without loss of generality that E
also is Dedekind complete.

Note first that if an order projection P satisfies $0 \leq P \leq K$ with $K: E \to E$
compact, then P is also a compact operator. (This follows immediately from
$P = P^3$ and Theorem 5.13.) Since S is an orthomorphism, there exists
(by Theorem 4.77) some $\lambda > 0$ with $0 \leq S \leq \lambda I$. Now by Theorem 2.9
there exists a sequence $\{S_n\}$ of I-step functions satisfying $0 \leq S_n \leq S$ and
$0 \leq S - S_n \leq \frac{1}{n} I$ for each n. By Theorem 1.44 each component of I is
an order projection, and so, by the preceding observations, every S_n is a
compact operator. Since $\|S - S_n\| < \frac{1}{n}$ holds for all n, it is easy to see that
S must be a compact operator. ∎

Exercises

1. Show that a norm bounded subset A of a normed vector space X is norm
   totally bounded if and only if for each $\epsilon > 0$ there exists a finite dimensional vector subspace Y of X satisfying $A \subseteq \epsilon U + Y$. [*Hint*: Assume
   that the condition holds, and let $\epsilon > 0$. Pick a finite dimensional vector
   subspace Y so that $A \subseteq \epsilon U + Y$. Consider the set

   $$B = \{y \in Y: \exists a \in A \text{ and } u \in U \text{ with } y = a - \epsilon u\},$$

and note that B is a norm bounded subset of Y. Since Y is finite dimensional, it follows from Theorem 3.28 that B is norm totally bounded. Now note that $A \subseteq B + \epsilon U$ holds.]

2. Let $X_1, \ldots, X_n$ be arbitrary Banach spaces, and consider the Banach space $X = X_1 \oplus \cdots \oplus X_n$. For each $i, j = 1, \ldots, n$ let $T_{ij} \colon X_j \to X_i$ be an operator.

   (a) Show that the matrix

   $$T = \begin{bmatrix} T_{11} & \cdots & T_{1n} \\ \vdots & & \vdots \\ T_{n1} & \cdots & T_{nn} \end{bmatrix}$$

   defines an operator on X, and that every operator $T \colon X \to X$ can be represented (uniquely) as above by a matrix.

   (b) Show that an operator $T \colon X \to X$ is continuous if and only if each operator T_{ij} is continuous.

   (c) Show that an operator $T \colon X \to X$ is a compact operator if and only if each T_{ij} is a compact operator.

3. Show that every compact operator has a separable range.

4. Let X be a Banach space, and let $T \colon X \to c_0$ be a continuous operator. Then show that T is a compact operator if and only if there exists a sequence $\{x_n'\} \subseteq X'$ with $\|x_n'\| \to 0$ and such that

   $$T(x) = \big(x_1'(x), x_2'(x), \ldots\big)$$

   holds for all $x \in X$. [*Hint*: Use Exercise 14 of Section 3.2.]

5. Let $T \colon X \to Y$ be a compact operator between Banach spaces. Show that

   $$\|T\| = \inf \sup\{\|x_n'\|\},$$

   where the infimum is taken over all null sequences $\{x_n'\} \subseteq X'$ satisfying $\|Tx\| \leq \sup\{|x_n'(x)|\}$ for all $x \in X$.

6. Show that every finite rank operator $T \colon E \to F$ from a Riesz space E to a uniformly complete Riesz space F possesses a modulus. Is the modulus necessarily a finite rank operator?

7. Let $R, S, T \colon E \to F$ be three operators on a Banach lattice such that $R \leq S \leq T$ holds. If R and T are both compact operators, then show that S^3 is a compact operator.

8. Show that on an infinite dimensional Banach lattice the identity operator cannot be dominated by a positive compact operator.

9. Consider the operators $S, T \colon \ell_1 \to \ell_\infty$ defined by

   $$S(x_1, x_2, \ldots) = (x_1, x_2, \ldots) \quad \text{and} \quad T(x_1, x_2, \ldots) = \Big(\sum_{n=1}^{\infty} x_n, \sum_{n=1}^{\infty} x_n, \ldots\Big).$$

   Show that $0 \leq S \leq T$ holds, T is a compact operator, and that S is not a compact operator. Why does this not contradict Theorem 5.20?

10. (Lozanovsky [119]) Consider the classical "Fourier coefficients" operator $T: L_1[0,1] \to c_0$ defined by

$$T(f) = \left(\int_0^1 f(x) \sin x \, dx, \ \int_0^1 f(x) \sin 2x \, dx, \ldots \right).$$

Establish the following remarkable properties of the operator T.
   (a) T is norm bounded.
   (b) T is not order bounded (and hence, by Theorem 5.7, T is not compact and its modulus does not exist).
   (c) The adjoint operator $T': \ell_1 \to L_\infty[0,1]$ is order bounded (and hence T is an example of a non-order bounded operator whose adjoint is order bounded.)
   (d) T is σ-order continuous (and hence T is an example of a σ-order continuous operator which is not order bounded; compare this with Exercise 1 of Section 1.5).

[*Hint:* To see (d), let $u_k \xrightarrow{o} 0$ in $L_1[0,1]$. Since $L_1[0,1]$ has order continuous norm, we have $\lim \|u_k\|_1 = 0$, and so the set $\Omega = \{0, u_1, u_2, \ldots\}$ is a norm compact subset of $L_1[0,1]$. Now the sequence of functions $\{\sin(nx)\}$ is norm bounded in $L_\infty[0,1]$ ($=$ the norm dual of $L_1[0,1]$), and hence it can be considered as an equicontinuous sequence on $L_1[0,1]$. In particular, $\{\sin(nx)\}$ can be considered as a norm bounded equicontinuous sequence of $C(\Omega)$, which satisfies $\lim \int_0^1 u(x) \sin(nx) \, dx = 0$ for each $u \in \Omega$. By the Ascoli–Arzelá theorem, the sequence $\{\sin(nx)\}$ converges uniformly to zero on Ω. Thus, if v_k is the sequence whose nth component is $\sup\{|\int_0^1 u_i(x) \sin(nx) \, dx| : i \geq k\}$, then $v_k \in c_0$ holds for each k, and from $u_k \xrightarrow{o} 0$ we see that $v_k \downarrow 0$ holds in c_0. Now note that $|Tu_k| \leq v_k$ holds for all k.]

11. Consider two positive operators $S, T: E \to E$ on a Banach lattice such that $0 \leq S \leq T$ holds. Assume that T maps the order intervals of E to norm totally bounded sets and that for each $\epsilon > 0$ there exists some $u \in E^+$ such that $\|(|Tx| - u)^+\| < \epsilon$ holds for all $\|x\| \leq 1$. Then, show that T^2 and S^3 are both compact operators.

12. Give an alternate proof of Theorem 5.10 using Theorem 4.81.

13. Consider an operator $T: E \to E$ on a Banach lattice having a compact modulus. Then show that:
   (a) T^3 is a compact operator.
   (b) If either E or E' has order continuous norm, then T^2 is a compact operator.
   (c) If both E and E' have order continuous norms, then T is a compact operator.

14. Let E be an atomless Banach lattice, and let $T: E \to E$ be a positive compact operator. Show that if an orthomorphism $S: E \to E$ satisfies $0 \leq S \leq T$, then $S = 0$.

15. Let E and F be two Banach lattices with E' and F having order continuous norms. Show that the set

$$C(E,F) = \{T \in \mathcal{L}_b(E,F) : \exists K_1, K_2 \geq 0 \text{ compact with } T = K_1 - K_2\}$$

is an ideal of $\mathcal{L}_b(E,F)$.

   Give an example of two Banach lattices E and F with F Dedekind complete for which the vector subspace $C(E,F)$ is not an ideal of $\mathcal{L}_b(E,F)$.

16. Assume that E is a Banach lattice such that E' has order continuous norm, and let F be an AL-space. Show that the collection of all order bounded compact operators from E to F is a band of $\mathcal{L}_b(E,F)$.

17. Let (E,τ) be a locally convex-solid Riesz space, and let $S, T: E \to E$ be two positive continuous operators satisfying $0 \leq S \leq T$. If T maps bounded sets to totally bounded sets, then show that S^3 likewise maps bounded sets to totally bounded sets.

18. Show that for a Dedekind complete Banach lattice E the following statements are equivalent.
   (a) For each $0 < T \in \mathcal{L}_b(E)$ there exists a positive compact operator $K: E \to E$ with $0 < K \leq T$.
   (b) There exists a net $\{K_\alpha\}$ of positive compact operators on E with $0 \leq K_\alpha \uparrow I$ in $\mathcal{L}_b(E)$ (where, of course, $I: E \to E$ is the identity operator).
   (c) E is a discrete Banach lattice (i.e., for each $0 < u \in E$ there exists a discrete vector $v \in E$ with $0 < v \leq u$).

19. If a positive operator $T: C[0,1] \to C[0,1]$ satisfies $\|Tf\|_\infty \leq \|f\|_1$ for all $f \in C[0,1]$, then show that T^2 is a compact operator.

5.2. Weakly Compact Operators

Recall that if X is a normed vector space, then the weak topology $\sigma(X, X')$ on X is denoted by w and the weak* topology $\sigma(X', X)$ on X' by w^*. We start the section with a useful characterization of norm bounded operators.

Theorem 5.22. *For an operator $T: X \to Y$ between two normed vector spaces the following statements are equivalent:*

   (1) *T is norm continuous, i.e., $\|x_n\| \to 0$ implies $\|Tx_n\| \to 0$.*
   (2) *T is weakly continuous, i.e., $x_\alpha \xrightarrow{w} 0$ implies $Tx_\alpha \xrightarrow{w} 0$.*

Proof. (1) $\Longrightarrow$ (2) Let $x_\alpha \xrightarrow{w} 0$ in X, and let $y' \in Y'$. Since T is norm continuous, $T'y' \in X'$, and so the relation

$$\langle Tx_\alpha, y' \rangle = \langle x_\alpha, T'y' \rangle \longrightarrow 0$$

shows that $Tx_\alpha \xrightarrow{w} 0$ holds in Y.

(2) $\Longrightarrow$ (1) Suppose that T is not norm continuous. Then, there exists a sequence $\{x_n\} \subseteq X$ of unit vectors such that $\|Tx_n\| \geq n^2$ holds for all n. From $\left\|\frac{1}{n}x_n\right\| = \frac{1}{n} \to 0$ and our hypothesis, it follows that $\frac{1}{n}Tx_n \xrightarrow{w} 0$ in Y. In particular, the sequence $\left\{\frac{1}{n}Tx_n\right\}$ is weakly bounded in Y, and hence a norm bounded sequence. However, the latter contradicts the inequality $\left\|\frac{1}{n}Tx_n\right\| \geq n$, and so T is norm continuous. ∎

Let $T \colon X \to Y$ be an operator between two Banach spaces. Then, T is said to be **weakly compact** whenever T carries the closed unit ball of X to a weakly relatively compact subset of Y. Thus, according to the Eberlein–Šmulian Theorem 3.40, T is weakly compact if and only if for every norm bounded sequence $\{x_n\}$ of X the sequence $\{Tx_n\}$ has a weakly convergent subsequence in Y.

Clearly, every compact operator is weakly compact. Also, it should be clear that every weakly compact operator is continuous. In [68] V. Gantmacher characterized the weakly compact operators as follows.

Theorem 5.23 (Gantmacher). *If $T \colon X \to Y$ is a continuous operator between Banach spaces, then the following statements are equivalent:*

(1) *The operator T is weakly compact.*

(2) *The range of the double adjoint operator $T'' \colon X'' \to Y''$ is included in Y, i.e., $T''(X'') \subseteq Y$.*

(3) *The operator $T' \colon (Y', w^*) \to (X', w)$ is continuous.*

(4) *The operator T' is weakly compact.*

Proof. (1) $\Longrightarrow$ (2) Let U and U'' denote the closed unit balls of X and X'', respectively. If $\overline{U}$ denotes the w^*-closure of U in X'', then by Theorem 3.32 we have $\overline{U} = U''$. Taking into account that $T'' \colon (X'', w^*) \to (Y'', w^*)$ is continuous and that (by hypothesis) the w^*-closure of $T(U)$ in Y'' lies in Y, we see that

$$T''(U'') = T''(\overline{U}) \subseteq \overline{T''(U)} = \overline{T(U)} \subseteq Y.$$

Hence, $T''(X'') \subseteq Y$ holds.

(2) $\Longrightarrow$ (3) Assume that $y'_\alpha \xrightarrow{w^*} 0$ in Y', and let $x'' \in X''$. By our hypothesis we have $T''x'' \in Y$, and so

$$\langle T'y'_\alpha, x'' \rangle = \langle y'_\alpha, T''x'' \rangle \longrightarrow 0.$$

Therefore, $T'y'_\alpha \xrightarrow{w} 0$ holds in X'.

(3) $\Longrightarrow$ (4) Let V denote the closed unit ball of Y'. By Alaoglu's Theorem 3.20 we know that V is w^*-compact. Therefore, $T(V)$ is a weakly compact subset of X', and so T' is a weakly compact operator.

(4) $\implies$ (1) From the above established implications, it follows immediately that $T'': (X'', w^*) \to (Y'', w)$ is continuous. Thus, $T''(U'')$ is a weakly compact subset of Y''. On the other hand, since Y is norm closed in Y'', we see that Y is also weakly closed in Y'' (see Theorem 3.13), and hence $T''(U'') \cap Y$ is a weakly compact subset of Y''. In view of $\sigma(Y'', Y') \subseteq \sigma(Y'', Y''')$, it follows that $T''(U'') \cap Y$ is also $\sigma(Y'', Y')$-compact. Now from $T(U) = T''(U) \subseteq T''(U'') \cap Y$ and the fact that $\sigma(Y'', Y')$ and $\sigma(Y, Y')$ agree on Y, it follows that $T(U)$ is a weakly relatively compact subset of Y. That is, T is a weakly compact operator, and the proof of the theorem is finished. ∎

An immediate application of the preceding result is the following.

Theorem 5.24. *If one of the Banach spaces X and Y is reflexive, then every continuous operator from X to Y is weakly compact.*

The weakly compact operators on a Banach space X exhibit ring properties similar to those of compact operators. They form a norm closed two-sided ring ideal of $L(X)$.

Theorem 5.25. *If X, Y, and Z are Banach spaces, then:*

(1) *The collection of all weakly compact operators from X to Y is a norm closed vector subspace of $L(X, Y)$.*

(2) *Whenever $X \xrightarrow{S} Y \xrightarrow{T} Z$ are continuous operators and either S or T is weakly compact, TS is likewise weakly compact.*

Proof. (1) Let $\{T_n\} \subseteq L(X, Y)$ be a sequence of weakly compact operators such that $\|T_n - T\| \to 0$ holds in $L(X, Y)$. Clearly,

$$\left\| T_n'' - T'' \right\| = \|T_n - T\| \to 0.$$

Now let $x'' \in X''$. Since (by Theorem 5.23) $T_n''(X'') \subseteq Y$ holds for each n, we have $\{T_n'' x''\} \subseteq Y$, and so from $\left\| T_n'' x'' - T'' x'' \right\| \to 0$, we see that $T'' x'' \in Y$. Thus, $T''(X'') \subseteq Y$ holds, and so by Theorem 5.23 the operator T is weakly compact.

(2) Routine. ∎

The next theorem demonstrates how to construct weakly compact operators from weakly convergent sequences. This result will play an important role later and guarantees an abundance of weakly compact operators.

Theorem 5.26. *Let X be a Banach space, and let the sequences $\{x_n\} \subseteq X$ and $\{x_n'\} \subseteq X'$ satisfy $x_n \xrightarrow{w} 0$ in X and $x_n' \xrightarrow{w} 0$ in X'. Then the operators*

$$S: \ell_1 \to X \quad and \quad T: X \to c_0$$

defined by

$$S(\alpha_1, \alpha_2, \ldots) = \sum_{n=1}^{\infty} \alpha_n x_n \quad \text{and} \quad T(x) = \big(x_1'(x), x_2'(x), \ldots\big)$$

are both weakly compact.

Proof. The weak compactness of the operator S follows immediately from Corollary 3.43.

For T note that $T' : \ell_1 \to X'$ satisfies $T'(\alpha_1, \alpha_2, \ldots) = \sum_{n=1}^{\infty} \alpha_n x_n'$, and so by the preceding case T' is weakly compact. By Theorem 5.23 the operator T is itself weakly compact. ∎

A. Grothendieck proved in [72] that every continuous operator from an AM-space to a weakly sequentially complete Banach space is weakly compact. This result was extended by N. Ghoussoub and W. B. Johnson [69] as follows.

Theorem 5.27 (Grothendieck–Ghoussoub–Johnson). *Let $T : E \to X$ be a continuous operator from a Banach lattice E to a Banach space X. If E' has order continuous norm and c_0 does not embed in X (in particular, if X is weakly sequentially complete), then T is weakly compact.*

Proof. By Theorem 4.63, the operator T admits a factorization

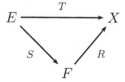

with F a KB-space and the factor S positive. To establish the weak compactness of T, it suffices to show that the positive operator $S : E \to F$ is weakly compact.

To this end, note first that (by Theorem 1.73) the adjoint operator $S' : F' \to E'$ is order continuous. This easily implies that $S''(E')_n^\sim \subseteq (F')_n^\sim$ holds. On the other hand, the order continuity of the norm of E' implies $(E')_n^\sim = E''$, and in addition we have $(F')_n^\sim = F$; see Theorem 4.60. Thus, $S''(E'') \subseteq F$ holds, which in view of Theorem 5.23 shows that S is a weakly compact operator, as desired. ∎

By virtue of Theorem 5.20 we know that if E and F are Banach lattices such that E' and F have order continuous norms, then every positive operator from E to F dominated by a compact operator is itself compact. The assumption that E' and F have order continuous norms is quite strong. As we shall see next, positive operators between this type of Banach lattices are almost weakly compact.

Theorem 5.28. *Let $T\colon E \to F$ be a positive operator between two Banach lattices such that E' and F have order continuous norms. If $\{x_n\}$ is a norm bounded sequence, then $\{Tx_n\}$ has a weak Cauchy subsequence.*

Proof. Let $\{x_n\}$ be a norm bounded sequence of the Banach lattice E and put $x = \sum_{n=1}^{\infty} 2^{-n}|Tx_n|$. Then, by Theorem 4.15, there exists some $0 \le \phi \in F'$ that is strictly positive on the ideal F_x generated by x in F.

Since E' has order continuous norm, it follows from Theorem 4.24 that there exists a subsequence $\{y_n\}$ of $\{x_n\}$ and some element $x'' \in E''$ such that $\lim_{n\to\infty} x'(y_n) = x''(x')$ holds for all $x' \in [0, T'\phi]$. But then for $0 \le f \le \phi$ we have

$$\lim_{n\to\infty} f(Ty_n) = \lim_{n\to\infty} [T'f](y_n) = x''(T'f), \qquad (\star)$$

and thus $\{Ty_n\}$ converges pointwise to $T''x''$ on $[0, \phi]$.

Now let A be the ideal generated by ϕ in F', and let B be the band generated by ϕ. Since $T'\colon F' \to E'$ is an order continuous operator (see Theorem 1.73), A is order dense in B, and E' has order continuous norm, it follows that $T'(B) \subseteq \overline{T'(A)}$ holds. The latter inclusion combined with $(\star)$ implies that for each $f \in B$ we have

$$\lim_{n\to\infty} f(Ty_n) = [T''x''](f).$$

On the other hand, we have $F' = B \oplus B^{\mathrm{d}}$, and we claim that $g(Tx_n) = 0$ holds for all $g \in B^{\mathrm{d}}$ and all n. To see this, let $0 \le g \in B^{\mathrm{d}}$. Since F has order continuous norm, it follows from Theorem 1.67 that $C_\phi \subseteq N_g$ holds. Taking into account that ϕ is strictly positive on F_x, we see that $F_x \cap N_\phi = \{0\}$, and so $F_x \subseteq C_\phi \subseteq N_g$ must hold. This implies $g(Tx_n) = 0$ for all n. Hence, $\lim g(Tx_n) = 0$ holds for all $g \in B^{\mathrm{d}}$. In view of $F' = B \oplus B^{\mathrm{d}}$, we see that $\lim f(Ty_n)$ exists in $\mathbb{R}$ for all $f \in F'$, and so $\{Ty_n\}$ is a weak Cauchy subsequence of $\{Tx_n\}$. This completes the proof of the theorem. ∎

In terms of weakly compact operators the reflexive Banach lattices were characterized by B. Kühn [**107**] as follows.

Theorem 5.29 (Kühn). *For a Banach lattice E the following statements hold:*

(1) *E is reflexive if and only if every positive operator from ℓ_1 to E is weakly compact.*

(2) *E' has order continuous norm (i.e., E' is a KB-space) if and only if every positive operator from E to ℓ_1 is weakly compact—and hence compact.*

Proof. (1) The "only if" part follows from Theorem 5.24. For the "if" part assume that every positive operator from ℓ_1 to E is weakly compact. We first show that E is a KB-space.

To this end, let $0 \le x_n \uparrow$ satisfy $\sup\{\|x_n\|\} < \infty$. The operator $T: \ell_1 \to E$ defined by $T(\alpha_1, \alpha_2, \ldots) = \sum_{n=1}^{\infty} \alpha_n x_n$ is positive, and so it must be weakly compact. From $T(e_n) = x_n$, it follows that $\{x_n\}$ has a weakly convergent subsequence. If x is a weak limit of a subsequence of $\{x_n\}$, then it is easy to see that $x_n \uparrow x$ and $x_n \xrightarrow{w} x$ must hold. By Theorem 3.52 we have $\lim \|x_n - x\| = 0$, and so E is a KB-space.

On the other hand, since ℓ_1 cannot be lattice embeddable in E (otherwise ℓ_1 ought to be reflexive), it follows from Theorem 4.69 that E' is also a KB-space. The reflexivity of E now follows from Theorem 4.70.

(2) Assume that E' has order continuous norm. If $T: E \to \ell_1$ is a positive operator, then by Theorem 5.27 the operator T is weakly compact.

Now assume that every positive operator from E to ℓ_1 is weakly compact (and hence compact by Theorem 4.32). To establish that E' has order continuous norm, it suffices to show that every order bounded disjoint sequence of E' is norm convergent to zero; see Theorem 4.14. To this end, let $\{x_n'\}$ be a disjoint sequence of E' satisfying $0 \le x_n' \le x'$ for all n. Next, define the positive operator $T: E \to \ell_1$ by

$$T(x) = (x_1'(x), x_2'(x), \ldots).$$

From $\sum_{n=1}^{k} |x_n'(x)| \le \sum_{n=1}^{k} x_n'(|x|) \le x'(|x|) < \infty$, we see that indeed $T(x) \in \ell_1$. From our hypothesis, it follows that T is a compact operator, and so if U is the closed unit ball of E, then $T(U)$ is a norm totally bounded subset of ℓ_1. Thus, given $\epsilon > 0$, there exists (by Theorem 4.33) some k such that $\sum_{n=k}^{\infty} |x_n'(x)| < \epsilon$ for all $x \in U$. So, $\|x_n'\| = \sup\{|x_n'(x)|: x \in U\} \le \epsilon$ holds for all $n \ge k$, and thus $\lim \|x_n'\| = 0$. Therefore, E' has also order continuous norm. ∎

We now turn our attention to the following problem:

- Let $S, T: E \to F$ be two positive operators between Banach lattices such that $0 \le S \le T$. If T is weakly compact, then what effect does the weak compactness of T have on S?

Before giving some answers to this question, we shall present an example to show that in general two positive operators can satisfy $0 \le S \le T$ with T weakly compact and S not a weakly compact operator.

Example 5.30. Let $\{r_n\}$ denote the sequence of Rademacher functions on $[0, 1]$, i.e., $r_n(t) = \operatorname{Sgn} \sin(2^n \pi t)$. Consider the operators $S, T: L_1[0, 1] \to \ell_\infty$

defined by

$$S(f) = \left(\int_0^1 f(x) r_1^+(x)\, dx, \int_0^1 f(x) r_2^+(x)\, dx, \ldots \right)$$

and

$$T(f) = \left(\int_0^1 f(x)\, dx, \int_0^1 f(x)\, dx, \ldots \right).$$

Then T is compact (it has rank one) and $0 \le S \le T$ holds. We claim that S is not weakly compact.

To establish this, observe first that the sequence $\{u_n\}$, where we let $u_n = (0, \ldots, 0, 1, 1, 1, \ldots)$ with the 0 occupying the first n positions, has no subsequence that converges weakly in ℓ_∞. Indeed, if some subsequence $\{w_n\}$ of $\{u_n\}$ converges weakly in ℓ_∞, then its weak limit must be zero, and so $w_n \downarrow 0$ and $\|w_n\|_\infty = 1$ contradict Theorem 3.52. Now consider the sequence $\{f_n\}$ of $L_1[0,1]$ defined by $f_n = 2^n \chi_{[0, 2^{-n}]}$. Clearly, $\|f_n\|_1 = 1$ holds for each n, and an easy computation shows that

$$S(f_n) = \left(1, \ldots, 1, \tfrac{1}{2}, \tfrac{1}{2}, \tfrac{1}{2}, \ldots \right),$$

with the 1 occupying the first n positions. On the other hand, it is easy to see that the only possible weak limit of any subsequence of $\{S(f_n)\}$ is $e = (1, 1, 1, \ldots)$. However, the relation $S(f_n) - e = -\tfrac{1}{2} u_n$ shows that no subsequence of $\{S(f_n)\}$ can converge weakly, and thus S is not a weakly compact operator.   ∎

The next result, due to A. W. Wickstead [193], tells us when every positive operator dominated by a weakly compact operator is always weakly compact.

Theorem 5.31 (Wickstead). *For a pair of Banach lattices E and F the following statements are equivalent.*

(1) *Either E' or F has order continuous norm.*

(2) *Every positive operator from E to F dominated by a weakly compact operator is weakly compact.*

Proof. (1) $\Longrightarrow$ (2) Let $S, T: E \to F$ be two positive operators satisfying $0 \le S \le T$ with T weakly compact.

Assume first that F has order continuous norm. From Theorem 4.9 we know that F is an ideal of F''. Therefore, if $0 \le x'' \in E''$, then it follows from $0 \le S''x'' \le T''x'' \in F$ that $S''x'' \in F$. That is, $S''(E'') \subseteq F$ holds, and so, by Theorem 5.23, the operator S is weakly compact.

Now assume that E' has order continuous norm. Then the operators $S', T' \colon F' \to E'$ satisfy $0 \le S' \le T'$ and T' is weakly compact (see Theorem 5.23). By the previous case, S' is weakly compact, and, by applying Theorem 5.23 once more, we see that S is likewise weakly compact.

$(2) \implies (1)$ Assume by way of contradiction that neither E' nor F has order continuous norm. To finish the proof, we have to construct two positive operators $S, T \colon E \to F$ with T weakly compact, S not weakly compact, and $0 \le S \le T$.

Since the norm of E' is not order continuous, there exists a positive order bounded disjoint sequence $\{f_n\}$ of E' satisfying $\|f_n\| = 1$ for all n (see Theorem 4.14). Let $f = \bigvee_{n=1}^{\infty} f_n$ in E'. Also, since F does not have order continuous norm there exists some $0 \le u \in F$ so that the order interval $[0, u]$ is not weakly compact; see Theorem 4.9. Thus, by Theorem 3.40, there exists a sequence $\{u_n\} \subseteq [0, u]$ without any weakly convergent subsequences.

Now consider the operators $S, T \colon E \to F$, defined by

$$T(x) = f(x)u = [f \otimes u](x) \quad \text{and} \quad S(x) = \sum_{n=1}^{\infty} f_n(x) u_n$$

for each $x \in E$. Note that in view of

$$\sum_{n=1}^{\infty} \|f_n(x) u_n\| \le \sum_{n=1}^{\infty} f_n(|x|) \|u\| = f(|x|) \|u\|,$$

the series defining S converges in norm for each x. Clearly, $0 \le S \le T$ holds and T is a compact operator (it has rank one). On the other hand, we claim that S is not weakly compact.

To see this, let $g_n = \bigvee_{i \ne n} f_i$, and note that $g_n \wedge f_n = 0$ and $f_n + g_n = f$. Since $\|f_n\| = 1$ holds, there exists some $x_n \in E^+$ with $\|x_n\| = 1$ and $1 - \frac{1}{2^n} < f_n(x_n) \le 1$. From $f_n \wedge g_n = 0$, it follows (from Theorem 1.18) that there exist $v_n, w_n \in [0, x_n]$ with $v_n + w_n = x_n$ and $f_n(v_n) + g_n(w_n) < \frac{1}{2^n}$. Hence,

$$0 \le 1 - f_n(w_n) = 1 - f_n(x_n) + f_n(v_n) < 2 \cdot \frac{1}{2^n},$$

and clearly $\|w_n\| \le 1$. Now from the inequalities

$$
\begin{aligned}
\|S w_n - u_n\| &= \left\| [f_n(w_n) - 1] u_n + \sum_{i \ne n} f_i(w_n) u_i \right\| \\
&\le 2 \cdot \frac{1}{2^n} \|u\| + \left[\sum_{i \ne n} f_i(w_n) \right] \cdot \|u\| \\
&= 2 \cdot \frac{1}{2^n} \|u\| + g_n(w_n) \|u\| < 3 \cdot \frac{1}{2^n} \|u\|
\end{aligned}
$$

and the fact that $\{u_n\}$ has no weakly convergent subsequences, we see that $\{Sw_n\}$ has no weakly convergent subsequences. That is, S is not a weakly compact operator, and the proof of the theorem is finished. ∎

The authors have shown in [**10**] that if a positive operator S on a Banach lattice is dominated by a weakly compact operators, then S^2 is weakly compact. The details follow.

Theorem 5.32 (Aliprantis–Burkinshaw). *If a positive operator on a Banach lattice is dominated by a weakly compact operator, then its square is weakly compact.*

Proof. Assume that $S, T\colon E \to E$ are two positive operators on a Banach lattice such that $0 \le S \le T$ holds and with T weakly compact. Let A denote the ideal generated by E in E''. The weak compactness of T implies (by Theorem 5.23) that $T''(E'') \subseteq E$. Thus, if $0 \le x'' \in E''$, then (in view of $0 \le S'' \le T''$) we have $0 \le S'' x'' \le T'' x'' \in E$, and so

$$S''(E'') \subseteq A \,.$$

Next, we claim that S'' also satisfies

$$S''(A) \subseteq E \,. \qquad\qquad (\star)$$

If $(\star)$ is established, then we see that

$$(S^2)''(E'') = S''\big[S''(E'')\big] \subseteq S''(A) \subseteq E \,,$$

which means that S^2 is a weakly compact operator.

The rest of the proof is devoted to establishing $(\star)$. To this end, let $0 \le x'' \in A$. Pick some $y \in E$ with $0 \le x'' \le y$. Since T' is a weakly compact operator, it follows that $T'(U')$ is a weakly compact subset of E'. Now let $\epsilon > 0$. Then, by Theorem 4.37, there exists some $0 \le g \in E'$ such that for each $f \in U'$ we have

$$\big(|T'f| - g\big)^{+}(y) < \epsilon \,.$$

Also, according to Theorem 3.60, there exists some element $x \in E$ satisfying $0 \le x \le y$ and $g(|x'' - x|) < \epsilon$. Clearly, $|x'' - x| \le y$. Moreover, for each $f \in U'$ we have

$$
\begin{aligned}
|\langle f, S''(x - x'')\rangle| &\le \langle |S'f|, |x - x''|\rangle \le \langle T'|f|, |x - x''|\rangle \\
&\le (T'|f| - g)^{+}(|x - x''|) + g(|x - x''|) \\
&< \epsilon + \epsilon = 2\epsilon \,,
\end{aligned}
$$

from which it follows that $\|S(x) - S''(x'')\| \le 2\epsilon$. This shows that $S''(x'')$ is in the norm closure of E in E''. Since E is a Banach space, E is closed in E'', and so $S''(x'') \in E$. Thus, $S''(A) \subseteq E$ holds, as desired. ∎

The previous theorem can be stated in a more general form as follows.

Theorem 5.33. *If in the scheme of positive operators*

$$F \xrightarrow{\ S_1\ } G \xrightarrow{\ S_2\ } H$$

each S_i is dominated by a weakly compact positive operator, then $S_2 S_1$ is weakly compact.

Proof. Assume that $0 \le S_i \le T_i$ holds for $i = 1, 2$ with each T_i being weakly compact. Consider the Banach lattice $E = F \oplus G \oplus H$, and define the operators $S, T \colon E \to E$ via the matrices

$$S = \begin{bmatrix} 0 & 0 & 0 \\ S_1 & 0 & 0 \\ 0 & S_2 & 0 \end{bmatrix} \quad \text{and} \quad T = \begin{bmatrix} 0 & 0 & 0 \\ T_1 & 0 & 0 \\ 0 & T_2 & 0 \end{bmatrix}.$$

Then $0 \le S \le T$ holds, T is weakly compact and

$$S^2 = \begin{bmatrix} 0 & 0 & 0 \\ 0 & 0 & 0 \\ S_2 S_1 & 0 & 0 \end{bmatrix}.$$

By Theorem 5.32 the operator S^2 is weakly compact, and so $S_2 S_1$ is likewise a weakly compact operator. ∎

For weakly compact operators the analogue of Theorem 5.21 is as follows.

Theorem 5.34. *A positive orthomorphism on a Banach lattice dominated by a weakly compact operator is itself weakly compact.*

Proof. Let $S, T \colon E \to E$ be two positive operators on a Banach lattice such that $0 \le S \le T$ holds. Assume that S is an orthomorphism and that T is weakly compact. By taking adjoints, we can assume that E is also Dedekind complete. In addition, we can assume that $0 \le S \le I$ holds.

Now let $\epsilon > 0$. Then, by Theorem 2.8, there exists an I-step function $\sum_{i=1}^{n} \alpha_i P_i$ with $0 \le \sum_{i=1}^{n} \alpha_i P_i \le S$ and $\left\| S - \sum_{i=1}^{n} \alpha_i P_i \right\| < \epsilon$. We can suppose that $\alpha_i > 0$ holds for each i. From $0 \le P_i \le \frac{1}{\alpha_i} T$ and Theorem 5.32, it follows that each $P_i = P_i^2$ is weakly compact, and so $\sum_{i=1}^{n} \alpha_i P_i$ is also weakly compact. By Theorem 5.25 the operator S is weakly compact. ∎

When does a weakly compact operator have a weakly compact modulus? The next result provides an answer.

Theorem 5.35. *Every weakly compact operator from an AL-space to a KB-space has a weakly compact modulus.*

Proof. Let $T\colon E \to F$ be a weakly compact operator from an AL-space E to a KB-space F. By Theorem 4.75 we know that $|T|$ exists. Now for $x \in E^+$ and $0 \le x' \in F' = F_n^{\sim}$, it follows from Theorem 1.21 that

$$x'(|T|x) = \sup\Big\{x'\Big(\sum_{i=1}^{n}|Tx_i|\Big)\colon\ x_i > 0 \text{ and } \sum_{i=1}^{n}x_i = x\Big\}$$

$$= \sup\Big\{\sum_{i=1}^{n}\|x_i\|x'\Big(\frac{|Tx_i|}{\|x_i\|}\Big)\colon\ x_i > 0 \text{ and } \sum_{i=1}^{n}x_i = x\Big\}$$

$$\le \Big[\sup\Big\{\sum_{i=1}^{n}\|x_i\|\colon\ x_i > 0;\ \sum_{i=1}^{n}x_i = x\Big\}\Big]\cdot\big[\sup\{x'(|Ty|)\colon\ \|y\| \le 1\}\big].$$

Therefore,

$$x'(|T|x) \le \|x\|\sup\{x'(|Ty|)\colon\ \|y\| \le 1\} \tag{$\star\star$}$$

holds for all $x \in E^+$ and all $0 \le x' \in F'$.

To prove that $|T|$ is weakly compact, it suffices to show that $|T|''(E'') \subseteq F$ holds (see Theorem 5.23). So, let $0 \le x'' \in E''$. Since F is a KB-space, we have $F = (F')_n^{\sim}$, and so, in order to prove that $|T|''(x'') \in F$, it suffices to show that $|T|''(x'')$ is order continuous on F'.

To this end, let $x'_\alpha \downarrow 0$ in F', and let U denote the closed unit ball of E. Since $T(U)$ is a weakly relatively compact subset of F, it follows from Theorem 4.38 that $\{x'_\alpha\}$ converges uniformly to zero on the solid hull of $T(U)$. From $(\star\star)$, we see that $\{x'_\alpha\}$ converges uniformly to zero on $|T|(U^+)$ or, equivalently, that $\{|T|'(x'_\alpha)\}$ converges uniformly to zero on U^+, i.e., $\big\||T|'(x'_\alpha)\big\| \downarrow 0$ holds. In particular, we have

$$\langle x'_\alpha, |T|''(x'')\rangle = \langle |T|'(x'_\alpha), x''\rangle \downarrow 0,$$

which shows that $|T|''(x'') \in (F')_n^{\sim}$, as required. ∎

As an immediate consequence of the preceding result we have the following.

Corollary 5.36. *The weakly compact operators from an AL-space E to a KB-space F form a norm closed order ideal of $\mathcal{L}_b(E, F)$ (and hence a Banach lattice in their own right).*

It should be noted that, in general, the weakly compact operators from an AL-space E to a KB-space F do not form a band in $\mathcal{L}_b(E, F)$. For instance, if $E = F = \ell_1$ and $T_n(\alpha_1, \alpha_2, \ldots) = (\alpha_1, \ldots, \alpha_n, 0, 0, \ldots)$, then $\{T_n\}$ is a sequence of positive compact operators on ℓ_1 satisfying $0 \le T_n \uparrow I$ in $\mathcal{L}_b(\ell_1)$. However, the identity operator $I\colon \ell_1 \to \ell_1$ is not weakly compact.

We now turn our attention to factoring weakly compact operators. Recall that a continuous operator $T\colon X \to Y$ between two Banach spaces is said to **factor through a Banach space** Z whenever there exist two continuous operators $X \xrightarrow{R} Z \xrightarrow{S} Y$ such that $T = SR$ holds. The continuous operators R and S are called **factors** of T.

Clearly, every continuous operator that factors through an arbitrary reflexive Banach space is weakly compact. Remarkably, W. J. Davis, T. Figiel, W. B. Johnson and A. Pelczynski [51] have shown that this factorization property characterizes the weakly compact operators. The proof will be based upon the following basic result.

Theorem 5.37 (Davis–Figiel–Johnson–Pelczynski). *Let X be a Banach space with closed unit ball U, and let W be a convex, circled, norm bounded subset of X. For each n let $U_n = 2^n W + 2^{-n} U$, and denote by $\| \cdot \|_n$ the Minkowski functional of U_n, i.e.,*

$$\|x\|_n := \inf\{\lambda > 0 \colon \ x \in \lambda U_n\}.$$

Put $\Psi = \{x \in X \colon \ |||x||| = [\sum_{n=1}^{\infty} \|x\|_n^2]^{\frac{1}{2}} < \infty\}$, and let $J \colon \Psi \to X$ denote the natural inclusion. Then:

(1) *$(\Psi, ||| \cdot |||)$ is a Banach space and J is continuous.*

(2) *W is a subset of the closed unit ball of $(\Psi, ||| \cdot |||)$.*

(3) *$J'' \colon \Psi'' \to X''$ is one-to-one and satisfies $(J'')^{-1}(X) = \Psi$.*

(4) *$(\Psi, |||\cdot|||)$ is reflexive if and only if W is a weakly relatively compact subset of the Banach space X.*

Proof. (1) Fix some $M > 1$ satisfying $\|w\| \le M$ for all $w \in W$. If $x \in \lambda U_n$, then pick $w \in W$ and $u \in U$ with $x = \lambda(2^n w + 2^{-n} u)$ and note that the inequality $\|x\| \le \lambda(2^n M + 2^{-n}) \le 2^{2n} M \lambda$ implies $\|x\| \le 2^{2n} M \|x\|_n$. Also, $x \in \|x\| U \subseteq 2^n \|x\| U_n$ implies $\|x\|_n \le 2^n \|x\|$, and so

$$2^{-n} \|x\|_n \le \|x\| \le 2^{2n} M \|x\|_n$$

holds for each $x \in X$. Therefore, $\| \cdot \|_n$ is a norm on X which is equivalent to the original norm of X. Let X_n denote the Banach space X equipped with $\| \cdot \|_n$. Now consider the vector subspace Y of $(X_1 \oplus X_2 \oplus \cdots)_2$ consisting of the "diagonal" elements, i.e.,

$$Y = \{(x_1, x_2, \ldots) \in (X_1 \oplus X_2 \oplus \cdots)_2 \colon \ x_n = x_1 \text{ for all } n\}.$$

It is easy to see that Y is a closed vector subspace of $(X_1 \oplus X_2 \oplus \cdots)_2$, and hence Y is a Banach space. On the other hand, the mapping $x \mapsto (x, x, \ldots)$ from Ψ onto Y is a linear isomorphism, and hence the norm $||| \cdot |||$ makes Ψ a Banach space.

The continuity of the natural inclusion $J: \Psi \to X$ follows from the inequality

$$15M^2|||x|||^2 = 15M^2 \sum_{n=1}^{\infty} \|x\|_n^2 \geq 15M^2 \sum_{n=1}^{\infty} M^{-2}2^{-4n}\|x\|^2 = \|x\|^2.$$

(2) Let $w \in W$. Then $w \in 2^{-n}(2^n W) \subseteq 2^{-n}U_n$, and so $\|w\|_n \leq 2^{-n}$ holds for all n. Therefore,

$$|||w||| = \Big[\sum_{n=1}^{\infty} \|w\|_n^2\Big]^{\frac{1}{2}} \leq \Big[\sum_{n=1}^{\infty} 2^{-2n}\Big]^{\frac{1}{2}} \leq 1,$$

which shows that W is a subset of the closed unit ball of Ψ.

(3) Consider the operator $T: \Psi \to Z = (X_1 \oplus X_2 \oplus \cdots)_2$ defined by

$$T(x) = (x, x, \ldots) = (Jx, Jx, \ldots).$$

We claim that $T'': \Psi'' \to Z''$ is given by

$$T''(x'') = (J''x'', J''x'', \ldots). \tag{$\star$}$$

To see this, let $x'' \in \Psi''$. According to Theorem 4.6 we can write

$$T''x'' = (x_1'', x_2'', \ldots) \in Z'' = (X_1'' \oplus X_2'' \oplus \cdots)_2.$$

Clearly, $x_n'' \in X_n'' = X''$ holds for each n. Now fix n, let $x' \in X_n' = X'$ be arbitrary, and let $z' = (0, \ldots, 0, x', 0, 0, \ldots)$ (where x' occupies the nth position). Then for each $x \in \Psi$ we have

$$T'z'(x) = z'(Tx) = x'(Jx) = J'x'(x),$$

and so $T'z' = J'x'$. On the other hand, by Theorem 4.6 we have

$$x_n''(x') = \langle z', T''x'' \rangle = \langle T'z', x'' \rangle = x''(J'x') = J''x''(x').$$

Since $x' \in X'$ is arbitrary, we see that $x_n'' = J''x''$ holds for each n, and thus the validity of $(\star)$ has been established.

Now note that since T is an isometry, T'' is likewise an isometry (see Exercise 17 of Section 3.2). In particular, since T'' is one-to-one, it follows from $(\star)$ that $J'': \Psi'' \to X''$ is one-to-one (which is, of course, equivalent to saying that $J'(X')$ is dense in Ψ').

Finally, to establish that $(J'')^{-1}(X) = \Psi$ holds, note first that the inclusion $\Psi \subseteq (J'')^{-1}(X)$ is trivially true. On the other hand, if $x'' \in (J'')^{-1}(X)$, then $J''x'' \in X$, and so it follows from $(\star)$ that $J''x'' \in \Psi$ and that $T(J''x'') = T''(x'')$ holds. Since T'' is one-to-one, we see that $x'' = J''x'' \in \Psi$. Therefore, $(J'')^{-1}(X) \subseteq \Psi$ also holds, and so $(J'')^{-1}(X) = \Psi$.

(4) By (2) we know that W is a norm bounded subset of Ψ. So, if Ψ is reflexive, then the continuity of $J: \Psi \to X$ implies that W is a weakly relatively compact subset of X.

For the converse, assume that W is a weakly relatively compact subset of X. Denote by C the closed unit ball of Ψ, i.e., $C = \{x \in \Psi \colon \||x\|| \le 1\}$, and by C'' the closed unit ball of Ψ''. By Alaoglu's Theorem 3.20, C'' is w^*-compact. Since C is w^*-dense in C'' and J'' is w^*-continuous, it follows that $J(C)$ is w^*-dense in the w^*-compact set $J''(C'')$. If $\overline{W}$ denotes the weak closure of W in X, then $\overline{W}$ is weakly compact. Thus, if U'' is the closed unit ball of X'' and

$$V_n = 2^n\overline{W} + 2^{-n}U'',$$

then each V_n is a w^*-compact subset of X''. Also, note that $J(C) \subseteq V_n$ holds for each n. To see this, assume that $x \in C$. Then we have $\||x\|| \le 1$, and, in particular, we have $\|x\|_n \le 1$ for each n. Thus, for each $\delta > 1$ there exists some $1 < \lambda_n < \delta$ with $x \in \lambda_n U_n \subseteq \lambda_n V_n$, and from this it follows that $x \in V_n$. Now the w^*-denseness of $J(C)$ in $J''(C'')$ implies $J''(C'') \subseteq V_n$ for all n, and so

$$J''(C'') \subseteq \bigcap_{n=1}^{\infty} V_n \subseteq \bigcap_{n=1}^{\infty}(X + 2^{-n}U'') = X.$$

The latter implies $J''(\Psi'') \subseteq X$, and hence $\Psi'' \subseteq (J'')^{-1}(X) = \Psi$ holds. This shows that $\Psi'' = \Psi$, proving that Ψ is reflexive. The proof of the theorem is now complete. ∎

An important application of the preceding result is that every weakly compact operator factors through a reflexive Banach space.

Theorem 5.38 (Davis–Figiel–Johnson–Pelczynski). *An operator between two Banach spaces is weakly compact if and only if it factors through a reflexive Banach space.*

Proof. Let U be the closed unit ball of Y, and let $T \colon Y \to X$ be a weakly compact operator. Put $W = T(U)$, and note that W is a convex, circled, and weakly relatively compact subset of X. Thus, the Banach space Ψ of Theorem 5.37 is a reflexive Banach space. Now if $S \colon Y \to \Psi$ is defined by $S(y) = T(y)$, then S is a continuous operator, and the diagram

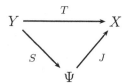

completes the proof of the theorem. ∎

Every compact operator factors with compact factors through a reflexive Banach space. This is due to T. Figiel [62] and W. B. Johnson [79].

Theorem 5.39 (Figiel–Johnson). *An operator between two Banach spaces is compact if and only if it factors with compact factors through a separable reflexive Banach space.*

Proof. Let $T\colon Y \to X$ be a compact operator between two Banach spaces. The proof is based upon the following diagram which will be explained below.

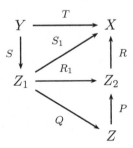

By Theorem 5.5 the operator T factors through a Banach space Z_1 with S and S_1 compact. By the same theorem S_1 factors through a Banach space Z_2 with R_1 and R compact. Now by Theorem 5.38 the operator R_1 factors through a reflexive Banach space Z. Then $T = (RP)(QS)$ provides a factorization of T with compact factors through the reflexive Banach space Z. If we replace Z with the closure of $QS(Y)$, then Z can also be taken to be separable. ∎

The next result presents a useful connection between the totally bounded subsets of X and Ψ.

Theorem 5.40. *Let W be a convex, circled, norm bounded subset of a Banach space X, and let $(\Psi, ||| \cdot |||)$ be the Banach space of Theorem 5.37 determined by W. Then a subset of W is totally bounded in X if and only if it is totally bounded in Ψ.*

In particular, if a compact operator $T\colon Y \to X$ maps the closed unit ball of Y into W, then T considered as an operator from Y to Ψ is also compact.

Proof. Let A be a subset of W. Assume that A is totally bounded in X, and let $\epsilon > 0$. Since $x, y \in W$ implies
$$x - y \in W + W = 2W = 2^{1-n}(2^n W) \subseteq 2^{1-n}U_n,$$
we see that $\|x - y\|_n := \inf\{\lambda > 0\colon\ x - y \in \lambda U_n\} \leq 2^{1-n}$ holds for all n, and so there exists some k satisfying $\sum_{n=k}^{\infty} \|x - y\|_n^2 < \epsilon^2$ for all $x, y \in W$. Pick $x_1, \ldots, x_m \in A$ such that $A \subseteq \{x_1, \ldots, x_m\} + \epsilon 2^{-2k}U$ (where U is the closed unit ball of X).

Now let $x \in A$ be fixed. Choose some x_i with $\|x - x_i\| < \epsilon 2^{-2k}$. Then for each $n = 1, \ldots, k$ we have
$$x - x_i \in \epsilon 2^{-2k}U \subseteq \epsilon 2^{-k}2^{-n}U \subseteq \epsilon 2^{-k}U_n,$$

and so $\|x - x_i\|_n \leq \epsilon 2^{-k}$ holds for each $n = 1, \ldots, k$. Therefore,

$$|||x - x_i|||^2 \leq \sum_{n=1}^{k} \|x - x_i\|_n^2 + \sum_{n=k}^{\infty} \|x - x_i\|_n^2 < k\epsilon^2 2^{-2k} + \epsilon^2 < \epsilon^2 + \epsilon^2 < 4\epsilon^2$$

holds, and so $|||x - x_i||| < 2\epsilon$, from which it follows that A is also a totally bounded subset of Ψ. ∎

When X is a Banach lattice and W is also a solid set, then the Banach space Ψ constructed in Theorem 5.37 is itself a Banach lattice.

Theorem 5.41. *Let W be a convex, solid, and norm bounded subset of a Banach lattice E, and let $(\Psi, ||| \cdot |||)$ be the Banach space of Theorem 5.37 determined by W. Then we have the following:*

(1) *$(\Psi, ||| \cdot |||)$ is a Banach lattice and Ψ is an ideal of E.*

(2) *The natural inclusion $J \colon \Psi \to E$ is an interval preserving lattice homomorphism.*

(3) *$J' \colon E' \to \Psi'$ is also an interval preserving lattice homomorphism.*

(4) *If E has order continuous norm, then Ψ also has order continuous norm.*

Proof. (1) Let $|x| \leq |y|$ hold in E with $y \in \Psi$. If $y \in \lambda U_n$, then (in view of the solidness of $U_n = 2^n W + 2^{-n} U$) we see that $x \in \lambda U_n$, and so $\|x\|_n \leq \|y\|_n$ holds for all n. Therefore, $|||x||| \leq |||y|||$. This shows that Ψ is an ideal of E and $(\Psi, ||| \cdot |||)$ is a Banach lattice.

(2) Obvious.

(3) This follows immediately from Theorems 2.19 and 2.20.

(4) Assume that E has order continuous norm. Then, E is Dedekind complete, and so Ψ (as an ideal of E) is likewise Dedekind complete. Let $x_n \downarrow 0$ in Ψ. According to Theorem 4.9, we have to show that $|||x_n||| \downarrow 0$.

To this end, let $\epsilon > 0$. Choose some m with $\sum_{k=m}^{\infty} \|x_1\|_k^2 < \epsilon^2$, and note that

$$|||x_n|||^2 \leq \sum_{k=1}^{m} \|x_n\|_k^2 + \sum_{k=m}^{\infty} \|x_1\|_k^2 \leq \sum_{k=1}^{\infty} \|x_n\|_k^2 + \epsilon^2 . \qquad (\star)$$

Since $x_n \downarrow 0$ also holds in E, and each $\| \cdot \|_k$ is equivalent to the norm of E, it follows that $\|x_n\|_k \downarrow 0$ holds for each k. The latter combined with $(\star)$ shows that $\limsup |||x_n||| \leq \epsilon$ holds. Since $\epsilon > 0$ is arbitrary, we see that $|||x_n||| \downarrow 0$, and the proof is finished. ∎

We have seen so far that a weakly compact operator between Banach spaces factors through a reflexive Banach space. Can we conclude something stronger whenever we have a weakly compact operator between Banach lattices? For instance, can we factor it through a reflexive Banach lattice? The following question comes naturally.

When does a weakly compact operator factor through a reflexive Banach lattice?

The rest of the section provides some answers to this question. As a first application of the preceding result let us prove that a weakly compact operator whose range is in a KB-space factors through a reflexive Banach lattice.

Theorem 5.42. *Every weakly compact operator $T\colon X \to E$ from a Banach space X to a KB-space E factors through a reflexive Banach lattice.*

If, in addition, T is a positive operator, then the factors can be taken to be positive operators.

Proof. Let U be the closed unit ball of X, and let W be the convex solid hull of $T(U)$. By Theorems 4.60 and 4.39, we know that W is weakly relatively compact, and so the Banach lattice $(\Psi, ||| \cdot |||)$ is reflexive. If $S\colon X \to \Psi$ is defined by $S(x) = T(x)$, then a glance at the diagram

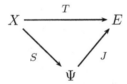

completes the proof of the theorem. ∎

The following result of the authors [**15**] will be very basic for the factorization of compact and weakly compact operators through reflexive Banach lattices. For its proof we shall invoke the following simple property:

- *A continuous operator $T\colon X \to Y$ between Banach spaces has a dense range if and only if $T'\colon Y' \to X'$ is one-to-one.*

Its proof: $T(X)$ is weakly dense (and hence norm dense) in Y if and only if $y' \in Y'$ and $y'(Tx) = T'y'(x) = 0$ for all $x \in X$ imply $y' = 0$ (i.e., if and only if $y' \in Y'$ and $T'y' = 0$ imply $y' = 0$).

Theorem 5.43 (Aliprantis–Burkinshaw). *Assume that W is the convex solid hull of a weakly relatively compact subset of a Banach lattice E. If $(\Psi, ||| \cdot |||)$ is the Banach lattice determined by W, then Ψ' has order continuous norm (and hence it is a KB-space).*

Proof. By Theorem 5.41 the operator $J'\colon E' \to \Psi'$ is interval preserving, and hence $J'(E')$ is an ideal of Ψ'. Also, since J'' is one-to-one, we see that

$J'(E')$ is norm dense in Ψ'. Therefore, by Theorem 4.11, in order to establish that Ψ' has order continuous norm, it is enough to show that $J'x'_n \downarrow 0$ in $J'(E')$ implies $|||J'x'_n||| \downarrow 0$.

To see this, let $J'x'_n \downarrow 0$ in $J'(E')$. Put $x' = \inf\{|x'_n|\}$ in E'. Using the fact that J' is a lattice homomorphism, we see that

$$0 \leq J'x' \leq J'|x'_n| = |J'x'_n| = J'x'_n$$

holds for all n, and so $J'x' = 0$. On the other hand, if $y'_n = \bigwedge_{i=1}^{n}(|x'_i| - x')$, then $y'_n \downarrow 0$ in E' and $J'y'_n = T'x'_n$ holds for each n. Thus, replacing $\{x'_n\}$ by $\{y'_n\}$, we can assume that $x'_n \downarrow 0$ holds in E'.

Now let $x \in \Psi$ satisfy $|||x||| \leq 1$. Then, $\|x\|_k \leq 1$ also holds for each k, and so $x \in 2(2^kW + 2^{-k}U)$. Therefore,

$$|J'x'_n(x)| = |x'_n(x)| \leq 2^{k+1}\sup\{|x'_n(w)|: \ w \in W\} + 2^{1-k}\|x'_1\|$$

holds for each k, and thus

$$|||J'x'_n||| \leq 2^{k+1}\sup\{|x'_n(w)|: \ w \in W\} + 2^{1-k}\|x'_1\| \qquad (\star)$$

holds for all k. Since W is the convex solid hull of a weakly relatively compact subset of E, it follows from Theorem 4.38 that $\{x'_n\}$ converges uniformly to zero on W. The latter combined with $(\star)$ shows that

$$\limsup |||J'x'_n||| \leq 2^{1-k}\|x'_1\|$$

holds for all k. Thus, $\lim |||J'x'_n||| = 0$, as desired. $\blacksquare$

We continue with a special factorization theorem for weakly compact operators due to the authors [15].

Theorem 5.44 (Aliprantis–Burkinshaw). *Assume that E is a Banach lattice such that E' has order continuous norm. If $T: E \to X$ is a compact or a weakly compact operator from E to a Banach space X, then there exist a reflexive Banach lattice F and a factorization of T*

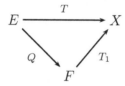

such that:

(1) *Q is a lattice homomorphism (and hence, a positive operator).*

(2) *T_1 is positive if T is positive.*

(3) *T_1 is compact if T is compact.*

Moreover, if X is a Banach lattice and another operator $S\colon E \to X$ satisfies $0 \le S \le T$, then S admits a factorization through the reflexive Banach lattice F

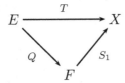

with $0 \le S_1 \le T_1$.

Proof. Clearly, $T'\colon X' \to E'$ is weakly compact, and by Theorem 4.59 the Banach lattice E' is a KB-space. Thus, if B is the closed unit ball of X', then the convex solid hull W of $T'(B)$ is a weakly relatively compact subset of E'. Let $(\Psi, ||| \cdot |||)$ be the reflexive Banach lattice of Theorems 5.37 and 5.41 determined by W (where Ψ is also an ideal of E'). Next, define the operator $R\colon X' \to (\Psi, ||| \cdot |||)$ by $R(x') = T'(x')$, and note that R is continuous, that R is positive if T is positive, and that (by Theorem 5.40) R is compact if T is compact. Thus, we have the diagrams:

Since $J'(E'')$ is dense in Ψ' and $T''(E'') \subseteq X$ holds true, we see that $R'(\Psi') \subseteq X$. Now consider the reflexive Banach lattice $F = \Psi'$ and the continuous operators $E \xrightarrow{Q} F \xrightarrow{T_1} X$, where $T_1(u) = R'(u)$ for $u \in F$ and $Q(x) = J'(x)$ for $x \in E$. By Theorem 5.41, Q is a lattice homomorphism. Also, T_1 is positive if T is positive, and T_1 is compact if T is compact. Now note that $T = T_1Q$ holds.

Next, assume that $0 \le S \le T$. Then $S'(X') \subseteq \Psi$ holds. On the other hand, by Theorem 5.31 the operator S is also weakly compact, and hence $S''(E'') \subseteq X$ also holds. As above, this implies that the operator $P\colon X' \to \Psi$ defined by $P(x') = S'(x')$ satisfies $P'(\Psi') \subseteq X$. Define $S_1\colon \Psi' \to X$ by $S_1(y') = P'(y')$, and note that $0 \le S_1 \le T_1$ and $S = S_1Q$ hold. The proof of the theorem is now complete. ∎

We now come to a very important factorization theorem. It is due to the authors [**15**].

Theorem 5.45 (Aliprantis–Burkinshaw). *Consider the scheme of operators*

$$X \xrightarrow{T_1} E \xrightarrow{T_2} Y$$

between Banach spaces with each T_i $(i = 1, 2)$ compact or weakly compact. If E is a Banach lattice, then there exist a reflexive Banach lattice F and a factorization of $T_2 T_1$

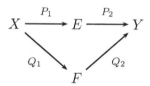

such that for each S_i has the same compactness property as T_i and is positive if T_i is positive.

Moreover, if X and Y are Banach lattices and another scheme of operators

$$X \xrightarrow{P_1} E \xrightarrow{P_2} Y$$

satisfies $0 \le P_i \le T_i$ $(i = 1, 2)$, then there exists a factorization of $P_2 P_1$ through the reflexive Banach lattice F

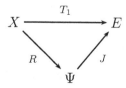

such that $0 \le Q_i \le S_i$ holds for each i.

Proof. Denote by U the closed unit ball of X. Assume that T_1 is compact or weakly compact, and let W denote the convex solid hull of the weakly relatively compact set $T_1(U)$. Then (by Theorem 5.43), Ψ' has order continuous norm, and T_1 admits the factorization

$$X \xrightarrow{T_1} E$$

where $R(x) = T_1(x)$ for all $x \in X$. Note that J is a positive operator and that R is positive if T_1 is positive.

The proof will be based upon the following diagram:

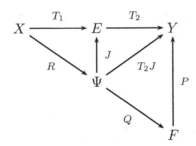

Since J is a positive operator, $T_2 J$ is positive if and only if T_2 is positive. Also, $T_2 J$ is compact or weakly compact, and so by Theorem 5.44 it admits a factorization through a reflexive Banach lattice F such that Q is positive, P is positive if T_2 is positive, and P is compact if T_2 is compact. Now note that the desired factorization is through the reflexive Banach lattice F with factors $S_1 = QR$ and $S_2 = P$.

Next, assume that X and Y are Banach lattices and that another scheme of operators $X \xrightarrow{P_1} E \xrightarrow{P_2} Y$ satisfies $0 \le P_i \le T_i$ $(i = 1, 2)$. From $0 \le P_1 \le T_1$, we see that $P_1(X) \subseteq \Psi$ holds. Thus, if $A \colon X \to \Psi$ is defined by $A(x) = P_1(x)$, then $0 \le A \le R$ holds, and so $Q_1 = QA$ satisfies $0 \le Q_1 \le S_1$. Now note that $0 \le P_2 J \le T_2 J$ holds, and therefore by Theorem 5.44 the operator $P_2 J$ admits a factorization $P_2 J = Q_2 Q$ through F with $0 \le Q_2 \le P = S_2$. to complete the proof, note that $P_2 P_1 = Q_2 Q_1$ holds. ∎

An immediate consequence of the preceding result is that the square of a weakly compact operator on a Banach lattice factors through a reflexive Banach lattice.

Corollary 5.46. *If $T \colon E \to E$ is a weakly compact operator on a Banach lattice, then T^2 factors (with positive factors if T is positive) through a reflexive Banach lattice.*

The corresponding result for compact operators is the following.

Corollary 5.47. *If $T \colon E \to E$ is a compact operator on a Banach lattice, then T^2 factors with compact factors through a reflexive Banach lattice.*

If T is also positive, then the factors can be taken to be positive compact operators.

The next consequence of Theorem 5.45 generalizes Theorem 5.33.

Theorem 5.48. *Consider the scheme of positive operators*

$$E \xrightarrow{S_1} F \xrightarrow{S_2} H$$

between Banach lattices. If each S_i is dominated by a weakly compact operator, then $S_2 S_1$ factors through a reflexive Banach lattice (and hence $S_2 S_1$ is weakly compact).

Using the main factorization theorem, we can also generalize Theorem 5.15 as follows.

Theorem 5.49. *Consider the scheme of positive operators*

$$E \xrightarrow{S_1} G \xrightarrow{S_2} H$$

between Banach lattices. If H has order continuous norm, S_1 is dominated by a weakly compact operator, and S_2 is dominated by a compact operator, then $S_2 S_1$ is a compact operator.

Dually, if E' has order continuous norm, S_1 is dominated by a compact operator, and S_2 is dominated by a weakly compact operator, then $S_2 S_1$ is a compact operator.

Proof. By Theorem 5.45 there exist a reflexive Banach lattice F and a factorization

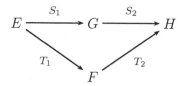

with T_2 positive and dominated by a compact operator. By Theorem 5.20, T_2 is compact, and hence $S_2 S_1 = T_2 T_1$ is also compact. ∎

We continue with a variation (due to W. Haid [**75**]) of Theorem 5.14.

Theorem 5.50 (Haid). *Consider the scheme of positive operators*

$$E \xrightarrow{S_1} F \xrightarrow{S_2} G \xrightarrow{S_3} H$$

between Banach lattices. If S_1 and S_3 are dominated by weakly compact operators and S_2 is dominated by a compact operator, then $S_3 S_2 S_1$ is a compact operator.

Proof. The proof is based upon the following diagram:

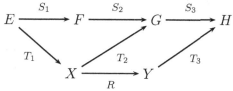

According to Theorem 5.45 the scheme of operators $E \xrightarrow{S_1} F \xrightarrow{S_2} G$ factors through a reflexive Banach lattice X with T_2 positive and dominated by a compact operator. Similarly, the scheme of operators $X \xrightarrow{T_2} G \xrightarrow{S_3} H$ factors through a reflexive Banach lattice Y with R positive and dominated by a compact operator. By Theorem 5.20 the operator R must be a compact operator, and so $S_3 S_2 S_1 = T_3 R T_1$ must be likewise compact. ∎

Finally, we close the section with an example (due to M. Talagrand [**182**]) of a positive weakly compact operator that does not factor through any reflexive Banach lattice.

Example 5.51 (Talagrand). The construction of the operator is based upon the following remarkable result:

- *There exists a weakly compact subset W of $C[0,1]^+$ such that whenever E is a reflexive Banach lattice and $S\colon E \to C[0,1]$ is a continuous operator, then $S(U)$ does not include W (i.e., $W \nsubseteq S(U)$).*

For the proof of the existence of the set W see [**182**].

The construction of the operator now goes as follows. Fix a weakly compact subset W of $C[0,1]^+$ with the above property, and let $\{x_n\}$ be a norm dense sequence in W. Now define the positive operator $T\colon \ell_1 \to C[0,1]$ by

$$T(\alpha_1, \alpha_2, \ldots) = \sum_{n=1}^{\infty} \alpha_n x_n \,.$$

Clearly, $Te_n = x_n$ holds for each n. An easy application of Theorem 3.42 shows that T is also weakly compact, and we claim that T cannot be factored through any reflexive Banach lattice.

To see this, assume by way of contradiction that T admits a factorization

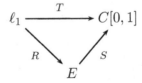

through a reflexive Banach lattice E. Replacing R by $R/\|R\|$ and S by $\|R\|S$, we can assume that $\|R\| = 1$. Thus, $\|Re_n\| \le 1$ holds for all n, and from $SRe_n = Te_n = x_n$, we see that $\{x_n\} \subseteq S(U)$ (where U is the closed unit ball of E). The latter implies $W \subseteq S(U)$, contrary to the property of W. Therefore, the positive weakly compact operator T does not factor through any reflexive Banach lattice. ∎

A consequence of Example 5.51 is that the preceding results on factorizations of positive weakly compact operators are the best possible.

Exercises

1. Show that a Banach space X is a Grothendieck space (i.e., that weak* and weak convergence of sequences in X' coincide) if and only if every continuous operator from X to c_0 is weakly compact.

2. For a Banach space X prove the following statements:

(a) An operator $T\colon X \to \ell_1$ is weakly compact if and only if T is compact.

(b) An operator $T\colon c_0 \to X$ is weakly compact if and only if T is compact.

3. Show that every continuous operator from $C[0,1]$ to ℓ_1 is compact.

4. (Lozanovsky [**122, 123**]) Show that a Banach lattice E is a KB-space if and only if every positive operator from c_0 to E is weakly compact.

5. Show that a Dedekind σ-complete Banach lattice E has order continuous norm if and only if every positive operator from ℓ_∞ to E is weakly compact.

6. Show that the operator T of Example 4.73 is not weakly compact. [*Hint:* If $f_n \in C[0,1]$ satisfies $0 \le f_n \le 1$, $f_n(x) = 0$ for $0 \le x \le \frac{1}{n-1}$, and $f_n(x) = 1$ for $\frac{1}{n} \le x \le 1$, then note that $Tf_n = (1, \ldots, 1, 0, 0, \ldots)$, where the 1 occupy the first n positions.]

7. Let $T\colon E \to F$ be a positive operator from an AM-space E into a Banach lattice F with order continuous norm. If either E has a unit or F is a KB-space, then show that T is a weakly compact operator.

8. Let E be a Banach lattice such that E' has order continuous norm, and let F be an AM-space with order continuous norm. If $T\colon E \to F$ is an order bounded weakly compact operator, then show that its modulus exists and is weakly compact. [*Hint:* Use Theorems 5.35 and 1.76.]

9. If the modulus of an operator $T\colon E \to E$ on a Banach lattice exists and is weakly compact, then show that T^2 is also weakly compact.

10. (Buhvalov [**45**]) Let W be a convex, circled, norm bounded subset of a Banach space X, and let $(\Psi, |||\cdot|||)$ be the Banach space of Theorem 5.37 determined by W. If $\|\cdot\|$ denotes the norm of X, then show that $\|\cdot\|$ and $|||\cdot|||$ induce the same topology on W.

11. Let W and $(\Psi, |||\cdot|||)$ be as in the preceding exercise. Then, show that $\sigma(\Psi, \Psi')$ and $\sigma(X, X')$ agree on every norm bounded subset of Ψ. [*Hint:* Use the fact that $J\colon \Psi \to X$ is continuous and that $J'(X')$ is dense in Ψ'.]

12. Generalize Theorem 5.43 as follows: *Let E be a Banach lattice, let A be the closed convex solid hull of a weakly relatively compact subset of E, and let W be a convex solid subset of A. If $(\Psi, |||\cdot|||)$ is the Banach lattice determined by W, then show that Ψ' is a KB-space.*

13. Let W be a convex, solid, and norm bounded subset of a Banach lattice E, and let Ψ be the Banach lattice determined by W as in Theorem 5.41. If E is a KB-space, then show that Ψ is likewise a KB-space. [*Hint:* Let $\{x_n\}$ be a norm bounded sequence of Ψ^+ satisfying $0 \le x_n \uparrow$. It follows that $\{x_n\}$ is a norm bounded sequence of E, and so $0 \le x_n \uparrow x$ holds in E. It suffices to show that $x \in \Psi$. Since each norm $\|\cdot\|_k$ is order continuous, we have $\|x_n\|_k \uparrow \|x\|_k$ for each k. Thus, for each m we have

$$\sum_{k=1}^{m} \|x\|_k^2 = \lim_{n\to\infty} \left[\sum_{k=1}^{m} \|x_n\|_k^2 \right] \le \lim_{n\to\infty} |||x_n|||^2 < \infty,$$

and from this it follows that $x \in \Psi.$]

14. Let W be a convex, solid, and norm bounded subset of a Banach lattice E, and let Ψ be the Banach lattice determined by W as in Theorem 5.41. If E' is a KB-space, then show that Ψ' is likewise a KB-space. [*Hint*: If $x'_n \downarrow 0$ holds in E', then $\{x'_n\}$ converges uniformly to zero on W. Now repeat the proof of Theorem 5.43.]

15. Consider a positive operator $T \colon E \to F$ between Banach lattices. If E' and F have order continuous norms, then show that T admits a factorization through a Banach lattice G

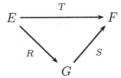

such that G and G' both have order continuous norms with R and S being positive operators. [*Hint*: Denote by U the closed unit ball of E. Let W be the convex solid hull of $T(U)$, and let Ψ be the Banach lattice generated by W as in Theorem 5.41. According to this theorem, Ψ has order continuous norm. Now by Theorem 1.73 the operator $T' \colon F' \to E'$ is order continuous, and so $x'_n \downarrow 0$ in F' implies $\|T'x'_n\| \downarrow 0$. In particular, if $x'_n \downarrow 0$ holds in F', then $\{x'_n\}$ converges uniformly to zero on W, and, by repeating the proof of Theorem 5.43, we see that $\||J'x'_n\|| \downarrow 0$. This implies that Ψ' has order continuous norm.]

16. Let $T \colon E \to F$ be a positive weakly compact operator between two Banach lattices. If F has order continuous norm, then show that T factors with positive factors through a reflexive Banach lattice. [*Hint*: Use Theorems 4.39(1) and 5.41.]

17. Let $T \colon E \to X$ be a weakly compact operator from a Banach lattice to a Banach space. Then, show that T admits a factorization through a KB-space F

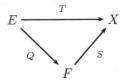

with Q a lattice homomorphism and with the factor S weakly compact (and with S positive if T is also positive). [*Hint*: Consider the weakly compact operator $T' \colon X' \to E'$, and let W be the convex solid hull of $T'(U')$. By Theorem 5.43 the norm dual of $(\Psi, \||\cdot\||)$ is a KB-space, and moreover we have the factorization

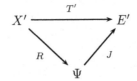

where $R(x') = T'(x')$. From Exercise 11 of this section, it follows that R is weakly compact. Since $T''(E'') \subseteq X$ holds and $J'(E'')$ is dense in Ψ', we see that $R'(\Psi') \subseteq X$. Now, to complete the proof consider the diagram

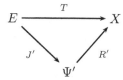

and note that J' is a lattice homomorphism.]

18. For a Banach lattice E with order continuous norm establish the following statements:

    (a) The convex solid hull of a norm totally bounded subset of E is weakly relatively compact.

    (b) Every compact operator $T\colon X \to E$ admits a factorization through a separable reflexive Banach lattice F

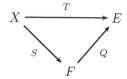

    such that:

    (i) Q is a lattice homomorphism (and hence, a positive operator).
    (ii) S is a compact operator.
    (iii) S is positive if T is also positive.

    Prove a similar result if T is a positive operator dominated by a compact operator.

19. (Ghoussoub–Johnson [69]) Generalize Theorem 5.27 as follows: *Assume that $T\colon E \to X$ is a continuous operator from a Banach lattice E to a Banach space X. If E' has order continuous norm and c_0 does not embed in X, then show that T factors through a reflexive Banach lattice.*

20. A Banach space X is said to be **weakly compactly generated** whenever X has a weakly relatively compact subset whose linear span is dense in X.

    Show that a Banach space X is weakly compactly generated if and only if there exists a continuous operator $T\colon Y \to X$ from a reflexive Banach space Y to X whose range is dense in X. [*Hint:* Let A be a weakly relatively compact subset of X whose linear span is dense in X. By Theorem 3.42 the convex circled hull W of A is weakly relatively compact. Now, consider the operator $J\colon \Psi \to X$.]

21. Let $T\colon E \to E$ be a weakly compact operator on a Banach lattice. If the ideal generated by the range of T coincides with E, then show that T factors (with positive factors if T is also positive) though a reflexive Banach lattice.

5.3. L- and M-weakly Compact Operators

Thus far, we have dealt with weakly compact operators in a general setting. However, many additional useful results hold true for special classes of weakly compact operators. The main objective of this section is to study the following three classes of operators: operators that carry order intervals to weakly relatively compact sets, L- and M-weakly compact operators, and semicompact operators.

As noted previously, among the important subsets of a Banach lattice are its order intervals. Therefore, operators that carry order intervals to weakly relatively compact sets arise naturally. These operators will be discussed next.

Theorem 5.52. *If $T: E \to X$ is a continuous operator from a Banach lattice to a Banach space and $x \in E^+$, then $T[0, x]$ is a weakly relatively compact subset of X if and only if for every disjoint sequence $\{x_n\}$ of $[0, x]$ we have $\lim \|Tx_n\| = 0$.*

Proof. Consider the ideal E_x generated by x in E. Then, E_x with the norm $\|y\|_\infty = \inf\{\lambda > 0\colon |y| \le \lambda x\}$ is an AM-space having x as unit and $[-x, x]$ as its closed unit ball.

The relative weak compactness of $T[0, x]$ is equivalent to saying that the restriction operator $T: E_x \to X$ is a weakly compact operator, and this in turn is equivalent to $T': X' \to E'_x$ being weakly compact. On the other hand, if U' denotes the closed unit ball of X', then $T'(U')$ is (by Theorem 4.41) weakly relatively compact if and only if every disjoint sequence of $[0, x]$ converges uniformly to zero on $T'(U')$, i.e., if and only if $\lim \|Tx_n\| = 0$ holds for each disjoint sequence $\{x_n\}$ of $[0, x]$. ∎

If a positive operator T carries an order interval $[0, x]$ to a weakly relatively compact set, then every positive operator dominated by T also carries $[0, x]$ to a weakly relatively compact set.

Corollary 5.53. *Let $S, T: E \to F$ be two positive operators between Banach lattices such that $0 \le S \le T$ holds. If for some $x \in E^+$ the set $T[0, x]$ is weakly relatively compact, then $S[0, x]$ is likewise a weakly relatively compact subset of F.*

Proof. Let $\{x_n\}$ be a disjoint sequence of $[0, x]$. By Theorem 5.52 we have $\lim \|Tx_n\| = 0$ and so, from $\|Sx_n\| \le \|Tx_n\|$, it follows that $\lim \|Sx_n\| = 0$. By Theorem 5.52, the set $S[0, x]$ is weakly relatively compact. ∎

In terms of disjoint sequences, the weakly compact order intervals are characterized as follows.

Corollary 5.54. *For a positive vector x in a Banach lattice E the order interval $[0, x]$ is weakly compact if and only if every disjoint sequence of $[0, x]$ is norm convergent to zero.*

Proof. Note first that $[0, x]$ is a weakly closed set, and then apply Theorem 5.52 to the identity operator $I: E \to E$. ∎

The next result (due to P. Meyer-Nieberg [**140**]) presents a useful sufficient condition for a set to have a weakly relatively compact solid hull.

Theorem 5.55 (Meyer-Nieberg). *Let A be a norm bounded subset of a Banach lattice. If every disjoint sequence in the solid hull $\mathrm{Sol}\,(A)$ of A is norm convergent to zero, then $\mathrm{Sol}\,(A)$ is weakly relatively compact.*

Proof. Let A be a norm bounded subset of a Banach lattice E such that every disjoint sequence in the solid hull of A is norm convergent to zero, and let $\epsilon > 0$.

Applying Theorem 4.36 to the identity operator $I: E \to E$, $\rho(x) = \|x\|$, and the solid hull of A, we see that there exists some $u \in E^+$ lying in the ideal generated by A such that $\|(|x| - u)^+\| < \epsilon$ holds for all $x \in A$. From the identity $|x| = |x| \wedge u + (|x| - u)^+$ and Theorem 1.13, we see that

$$\mathrm{Sol}\,(A) \subseteq [-u, u] + \epsilon U\,. \qquad (\star)$$

Now pick $\alpha > 0$ and vectors $u_1, \ldots, u_n \in A$ with $0 \le u \le \alpha \sum_{i=1}^n |u_i|$, and note that by the Riesz decomposition property (Theorem 1.13) we have

$$[0, u] \subseteq \alpha[0, |u_1|] + \cdots + \alpha[0, |u_n|]\,. \qquad (\star\star)$$

Since (by our hypothesis) every disjoint sequence in any $[0, |u_i|]$ is norm convergent to zero, it follows from Corollary 5.54 that each order interval $[0, |u_i|]$ is weakly compact. Thus, $\alpha[0, |u_1|] + \cdots + \alpha[0, |u_n|]$ is weakly compact, and so from $(\star\star)$ we see that $[0, u]$ is weakly compact. This implies that $[-u, u] = -u + 2[0, u]$ is weakly compact, and so $(\star)$ combined with Theorem 3.44 shows that $\mathrm{Sol}\,(A)$ is a weakly relatively compact set. ∎

The converse of the preceding theorem is true for *AL*-spaces.

Theorem 5.56. *A norm bounded subset A of an AL-space is weakly relatively compact if and only if every disjoint sequence in the solid hull of A is norm convergent to zero.*

Proof. The "if" part is just Theorem 5.55. For the "only if" part assume that $\{x_n\}$ is a disjoint sequence in the solid hull of a weakly relatively compact subset of an *AL*-space E. Then, by Theorem 4.34, we have $|x_n| \xrightarrow{w} 0$. On the other hand, since E' is an *AM*-space with unit e', it follows that

$$\|x_n\| = e'(|x_n|) \to 0\,,$$

and the proof is finished. ∎

Following P. G. Dodds [**53**], we shall say that a continuous operator $T: E \to X$ from a Banach lattice to a Banach space is **order weakly compact** (abbreviated as o-**weakly compact**) whenever $T[0, x]$ is a relatively weakly compact subset of X for each $x \in E^+$.

The o-weakly compact operators have been characterized in [**53**] as follows.

Theorem 5.57 (Dodds). *For a continuous operator $T: E \to X$ from a Banach lattice to a Banach space the following statements are equivalent:*

(1) *The operator T is o-weakly compact.*

(2) *If $\{x_n\}$ is any order bounded disjoint sequence of E, then we have $\lim \|Tx_n\| = 0$.*

(3) *For each $x \in E^+$ and each $\epsilon > 0$ there exist $0 \le x' \in E'$ and $\delta > 0$ such that $|y| \le x$ and $x'(|y|) < \delta$ imply $\|Ty\| < \epsilon$.*

(4) *If A denotes the ideal generated by E in E'', then $T''(A) \subseteq X$ holds.*

Proof. (1) $\Longrightarrow$ (2) This follows immediately from Theorem 5.52.

(2) $\Longrightarrow$ (3) Fix $\epsilon > 0$ and $x \in E^+$. Consider $T': X' \to E'$, and let U' denote the closed unit ball of X'. By (2), we see that every disjoint sequence of $[0, x]$ converges uniformly to zero on $T'(U')$. Then, by Theorem 4.40, there exists some $0 \le x' \in E'$ such that $(|T'f| - x')^+(x) < \frac{\epsilon}{3}$ holds for all $f \in U'$. Now put $\delta = \frac{\epsilon}{3}$, and let $|y| \le x$ satisfy $x'(|y|) < \delta$. Then for each $f \in U'$ we have

$$|f(Ty)| \le |T'f|(|y|) = (|T'f| - x')^+(|y|) + [x' \wedge |T'f|](|y|)$$
$$\le (|T'f| - x')^+(x) + x'(|y|) < \tfrac{\epsilon}{3} + \tfrac{\epsilon}{3} = \tfrac{2}{3}\epsilon.$$

The latter implies that $\|Ty\| < \epsilon$ holds, as desired.

(3) $\Longrightarrow$ (4) Let $0 \le x'' \in A$. Then, there exists some $x \in E^+$ with $0 \le x'' \le x$. Since $A \subseteq (E')_n^\sim$ holds, it follows from Theorem 3.60 that E is $|\sigma|(E'', E')$-dense in A. Thus, there exists a net $\{x_\alpha\} \subseteq [0, x]$ with $|x_\alpha - x''| \xrightarrow{w^*} 0$. Taking into account that $T'': E'' \to X''$ is w^*-continuous, we see that $Tx_\alpha \xrightarrow{w^*} T''x''$ also holds.

Now let $\epsilon > 0$. Choose $\delta > 0$ and $0 \le x' \in E'$ such that $|y| \le x$ and $x'(|y|) < 2\delta$ imply $\|Ty\| < \epsilon$. Next, pick some α_0 so that $\langle x', |x'' - x_\alpha| \rangle < \delta$ holds for all $\alpha \succeq \alpha_0$. Clearly, $|x_\alpha - x_\beta| \le x$ and $x'(|x_\alpha - x_\beta|) < 2\delta$ holds for all $\alpha, \beta \succeq \alpha_0$. Fix some $\beta \succeq \alpha_0$, and note that if $g \in X'$ satisfies $\|g\| \le 1$, then

$$|g(T''x'' - Tx_\beta)| = \lim_{\alpha \succeq \alpha_0} |g(T(x_\alpha - x_\beta))| \le \limsup_\alpha \|T(x_\alpha - x_\beta)\| \le \epsilon$$

holds, from which it follows that $\|T''x'' - Tx_\beta\| \le \epsilon$. This shows that $T''x''$ lies in the norm closure of X in X''. Since X is a Banach space, we see that $T''x'' \in X$. Therefore, $T''(A) \subseteq X$ holds.

(4) $\Longrightarrow$ (1) Let $x \in E^+$. Put $[0, x] = \{x'' \in E'': 0 \le x'' \le x\}$, and note that $[\![0, x]\!]$ is a w^*-compact subset of E''. Since $T'': E'' \to X''$ is w^*-continuous, it follows that $T''[\![0, x]\!]$ is a w^*-compact subset of X''. On the other hand, the weak* topology on X'' induces the weak topology on X, and so, from $T''[\![0, x]\!] \subseteq X$, we see that $T''[\![0, x]\!]$ is a weakly compact subset of X. Finally, the inclusion $T[0, x] \subseteq T''[\![0, x]\!]$ implies that $T[0, x]$ is a weakly relatively compact subset of X. ∎

Clearly, every weakly compact operator from a Banach lattice to a Banach space is o-weakly compact. Also, every continuous operator from a Banach lattice with order continuous norm into a Banach space must be o-weakly compact. This is true because if a Banach lattice has order continuous norm, then its order intervals are weakly compact (see Theorem 4.9). Therefore, the identity operator $I: L_1[0, 1] \to L_1[0, 1]$ is an example of an o-weakly compact operator which is not weakly compact.

On the other hand, consider a continuous operator $T: E \to X$ from a Banach lattice to a Banach space that admits a factorization through a Banach lattice F

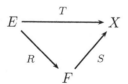

with the factor R positive. If F has order continuous norm, then it should be clear that T is o-weakly compact. Remarkably, the converse of the latter is also true. This is due to the authors [**17**].

Theorem 5.58 (Aliprantis–Burkinshaw). *If* $T: E \to X$ *is an o-weakly compact operator, then* T *admits a factorization through a Banach lattice* F *with order continuous norm*

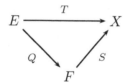

such that:

(1) *Q is a lattice homomorphism.*

(2) *S is positive if T is also positive.*

Proof. The first part of the proof is precisely the first part of the proof of Theorem 4.63. Define a lattice seminorm ρ on E for each $x \in E$ by

$$\rho(x) = \sup\{\|Ty\|\colon \ |y| \le |x|\}\,.$$

Let $N = \{x \in E\colon \ \rho(x) = 0\}$ be the null ideal of ρ, and let $Q\colon E \to E/N$ denote the canonical projection, i.e., $Q(x) = \dot{x}$. Then, E/N under the quotient norm

$$\|\dot{x}\| = \rho(x)\,,$$

is a normed Riesz space, and so its norm completion F of $(E/N, \|\cdot\|)$ is a Banach lattice. Note that

$$\|Tx\| \le \|\dot{x}\|$$

holds for all $x \in E$. In particular, this implies that the formula $S(\dot{x}) = Tx$ gives rise to a well defined continuous operator from E/N to X. Denote by S again the unique continuous linear extension of S to all of F. Thus, we have the factorization:

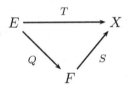

Clearly, Q is a lattice homomorphism, and S is positive if T is also positive.

It remains to be shown that F has order continuous norm. By Theorem 4.11, it suffices to show that if $0 \le \dot{x}_n \uparrow \le \dot{x}$ holds in E/N, then $\{\dot{x}_n\}$ is a Cauchy sequence. So, let $0 \le \dot{x}_n \uparrow \le \dot{x}$ hold in E/N. Since Q is a lattice homomorphism, replacing x by x^+ and each x_n by $\left(\bigvee_{i=1}^n x_i^+\right) \wedge x^+$, we can suppose that $0 \le x_n \uparrow \le x$ also holds in E. Assume by way of contradiction that $\{\dot{x}_n\}$ is not a Cauchy sequence of E/N. Then, by passing to a subsequence, we can suppose that there exists some $\epsilon > 0$ satisfying $\|\dot{x}_{n+1} - \dot{x}_n\| > \epsilon$ for all n. Thus, for each n there exists some $y_n \in E$ with $|y_n| \le x_{n+1} - x_n$ and

$$\|Ty_n\| > \epsilon\,. \qquad\qquad (\star)$$

From $0 \le \sum_{i=1}^n |y_i| \le \sum_{i=1}^n (x_{i+1} - x_i) \le x$, it easily follows that $|y_n| \xrightarrow{w} 0$. Next, by Theorem 5.57 (Condition 3) there exist $0 \le x' \in E'$ and $\delta > 0$ such that $|y| \le x$ and $x'(|y|) < \delta$ imply $\|Ty\| < \epsilon$. Since $\lim x'(|y_n|) = 0$ and $|y_n| \le x$, we see that $\|Ty_n\| < \epsilon$ must hold for all sufficiently large n. However, the latter contradicts $(\star)$, and the proof is finished. ∎

We now turn our attention to two special classes of weakly compact operators. They are the classes of L- and M-weakly compact operators and were introduced by P. Meyer-Nieberg in [**142**]. As we shall see, these operators are in a duality with each other.

Definition 5.59 (Meyer-Nieberg). *A continuous operator $T: E \to X$ from a Banach lattice to a Banach space is said to be **M-weakly compact** if $\lim \|Tx_n\| = 0$ holds for every norm bounded disjoint sequence $\{x_n\}$ of E.*

*Similarly, a continuous operator $T: X \to E$ from a Banach space to a Banach lattice is said to be **L-weakly compact** whenever $\lim \|y_n\| = 0$ holds for every disjoint sequence $\{y_n\}$ in the solid hull of $T(U)$, where U is the closed unit ball of the Banach space X.*

In the sequel, we shall see that the L- and M-weakly compact operators enjoy many interesting features whose proofs require a deep understanding of the properties of disjoint sequences. We begin with the following lattice approximation properties.

Theorem 5.60 (Meyer-Nieberg). *For a Banach lattice E and a Banach space X the following statements hold:*

(1) *If $T: E \to X$ is an M-weakly compact operator, then for each $\epsilon > 0$ there exists some $u \in E^+$ such that*

$$\left\| T((|x| - u)^+) \right\| < \epsilon$$

holds for all $x \in E$ with $\|x\| \leq 1$.

(2) *If $T: X \to E$ is an L-weakly compact operator, then for each $\epsilon > 0$ there exists some $u \in E^+$ lying in the ideal generated by $T(X)$ satisfying*

$$\left\| (|Tx| - u)^+ \right\| < \epsilon$$

for all $x \in X$ with $\|x\| \leq 1$.

Proof. (1) Let A be the closed unit ball of E, and let $\rho(x) = \|x\|$. Then, $\lim \rho(Tx_n) = 0$ holds for each disjoint sequence $\{x_n\}$ in A. So, by Theorem 4.36, there exists some $u \in E^+$ satisfying $\|T(|x| - u)^+\| < \epsilon$ for all $x \in A$.

(2) Let U denote the closed unit ball of X, and let A be the solid hull of $T(U)$. Since T is L-weakly compact, every disjoint sequence in A is norm convergent to zero.

Now if $I: E \to E$ is the identity operator and $\rho(x) = \|x\|$, then it follows from Theorem 4.36 that there exists some $u \in E^+$ lying in the ideal generated by A such that $\|I(|y| - u)^+\| < \epsilon$ holds for all $y \in A$. In particular, we have $\|(|Tx| - u)^+\| < \epsilon$ for all $x \in U$. ∎

P. Meyer-Nieberg [**142**] has shown that L-weakly compact and M-weakly compact operators are indeed weakly compact operators.

Theorem 5.61 (Meyer-Nieberg). *L- and M-weakly compact operators are weakly compact.*

Proof. Assume first that $T\colon E \to X$ is an M-weakly compact operator. Denote by U and V the closed unit balls of E and X, respectively, and let $\epsilon > 0$. By part (1) of Theorem 5.60 there exists some $u \in E^+$ such that $\|T(|x| - u)^+\| < \epsilon$ holds for all $x \in U$, and consequently from the lattice identity $|x| = |x| \wedge u + (|x| - u)^+$ we see that

$$T(U^+) \subseteq T[0, u] + \epsilon V. \tag{$\star$}$$

On the other hand, if $\{u_n\}$ is a disjoint sequence of $[0, u]$, then it follows from our hypothesis that $\lim \|Tu_n\| = 0$, and thus by Theorem 5.52 the set $T[0, u]$ is weakly relatively compact. Now $(\star)$ combined with Theorem 3.44 shows that $T(U^+)$ (and hence $T(U)$) is weakly relatively compact, and so T is a weakly compact operator.

Next, assume that $T\colon X \to E$ is an L-weakly compact operator. If U is the closed unit ball of X, then every disjoint sequence in the solid hull of $T(U)$ is norm convergent to zero. By Theorem 5.55 the solid hull of $T(U)$ (and hence $T(U)$ itself) is weakly relatively compact. Therefore, T is a weakly compact operator. ∎

A compact (and hence a weakly compact) operator between Banach lattices need not be L- or M-weakly compact. For instance, consider the operator $T\colon \ell_1 \to \ell_\infty$ defined by

$$T(\alpha_1, \alpha_2, \ldots) = \left(\sum_{n=1}^{\infty} \alpha_n, \sum_{n=1}^{\infty} \alpha_n, \ldots \right) = \left[\sum_{n=1}^{\infty} \alpha_n \right](1, 1, 1, \ldots).$$

Clearly, T is a compact operator (it has rank one). The sequence $\{e_n\}$ of the standard unit vectors is a norm bounded disjoint sequence of ℓ_1 satisfying $Te_n = (1, 1, 1, \ldots)$ for each n. This shows that T is not M-weakly compact. On the other hand, if U is the closed unit ball of ℓ_1, then it is easy to see that $\{e_n\}$ is also a disjoint sequence in the solid hull of $T(U)$. From $\|e_n\|_\infty \not\to 0$, we see that T fails to be L-weakly compact.

Theorem 5.62. *A continuous operator $T\colon E \to X$ from an AM-space to a Banach space is weakly compact if and only if it is M-weakly compact.*

Similarly, a continuous operator $T\colon X \to E$ from a Banach space to an AL-space is weakly compact if and only if it is L-weakly compact.

Proof. By Theorem 5.61 we know that L- and M-weakly compact operators are weakly compact. Assume first that $T\colon E \to X$ is a weakly compact operator from an AM-space to a Banach space, and let $\{x_n\}$ be a norm bounded disjoint sequence of E. Then $T''\colon E'' \to X''$ is weakly compact, E'' is an AM-space with unit, and $\{x_n\}$ is an order bounded disjoint sequence of E''. Since T'' is o-weakly compact, it follows from Theorem 5.52 that $\|Tx_n\| = \|T''x_n\| \to 0$, and so T is M-weakly compact.

Next, suppose that $T: X \to E$ is a weakly compact operators from a Banach space to an AL-space, and let U denote the closed unit ball of X. Then $T(U)$ is weakly relatively compact, and since E is an AL-space, it follows that every disjoint sequence in the solid hull of $T(U)$ is norm convergent to zero (see Theorem 5.56). So, T is L-weakly compact. $\blacksquare$

We continue with a duality theorem involving disjoint sequences that is due to O. Burkinshaw and P. G. Dodds [49].

Theorem 5.63 (Burkinshaw–Dodds). *If E is an arbitrary Banach lattice and $A \subseteq E$ and $B \subseteq E'$ are two nonempty norm bounded sets, then the following statements are equivalent:*

(1) *Each disjoint sequence in the solid hull of A converges uniformly to zero on B.*

(2) *Each disjoint sequence in the solid hull of B converges uniformly to zero on A.*

Proof. For simplicity write

$$\rho_B(x) = \sup\{|x'(x)|: \ x' \in B\} \quad \text{and} \quad \rho_A(x') = \sup\{|x'(x)|: \ x \in A\}.$$

$(1) \implies (2)$ Let $\{x'_n\}$ be a disjoint sequence in the solid hull of B. Then, we have to show that $\lim \rho_A(x'_n) = 0$.

First, we claim that $\lim |x'_m|(u) = 0$ holds for every u in the ideal generated by A. To see this, let $u \in E^+$ be in the ideal generated by A, and let $\epsilon > 0$. Since (by our hypothesis) every disjoint sequence in $[0, u]$ converges uniformly to zero on B, there exists by Theorem 4.40 some $0 \leq y' \in E'$ such that for each $x' \in B$ we have

$$(|x'| - y')^+(u) < \epsilon.$$

Note that since $\{|x'_n| \wedge y'\}$ is an order bounded disjoint sequence, we have $\lim(|x'_n| \wedge y')(u) = 0$. On the other hand, from

$$|x'_n|(u) = (|x'_n| - y')^+(u) + (|x'_n| \wedge |y'|)(u) < \epsilon + (|x'_n| \wedge y')(u),$$

we see that $\limsup |x'_n|(u) \leq \epsilon$. Since $\epsilon > 0$ is arbitrary, the latter implies that $\lim |x'_n|(u) = 0$, as claimed.

Next, consider the lattice seminorm ρ on E defined by

$$\rho(x) = \sup\{|x'|(|x|): \ x' \in B\}.$$

Then, we claim that $\lim \rho(x_n) = 0$ holds for each disjoint sequence in the solid hull of A. To see this, let $\{x_n\}$ be a disjoint sequence in the solid hull of A. For each n choose some $x'_n \in B$ with

$$\rho(x_n) \leq 2|x'_n|(|x_n|) = 2\sup\{|x'_n(y)|: \ |y| \leq |x_n|\}.$$

It follows that for each n there exists some $|y_n| \leq |x_n|$ with

$$|x'_n|(|x_n|) \leq 2|x'_n(y_n)| \leq 2\rho_B(y_n),$$

and so $\rho(x_n) \leq 4\rho_B(y_n)$ holds for each n. Since $\{y_n\}$ is a disjoint sequence in the solid hull of A, it follows from our hypothesis that $\lim \rho_B(y_n) = 0$, and hence $\lim \rho(x_n) = 0$ holds, as desired.

Now let $\epsilon > 0$. Theorem 4.36 applied to the identity operator $I \colon E \to E$, $\mathrm{Sol}\,(A)$ and ρ, shows that there exists some u in the ideal generated by A satisfying $\rho((|x| - u)^+) < \epsilon$ for all $x \in A$. In particular, for each $x \in A$ we have

$$
\begin{aligned}
|x'_n(x)| &\leq |x'_n|(|x|) \leq |x'_n|(|x| - u)^+ + |x'_n|(u) \\
&\leq \rho((|x| - u)^+) + |x'_n|(u) < \epsilon + |x'_n|(u),
\end{aligned}
$$

and so $\rho_A(x'_n) \leq \epsilon + |x'_n|(u)$ holds. Taking into account that $\lim |x'_n|(u) = 0$, we see that $\limsup \rho_A(x'_n) \leq \epsilon$. Since $\epsilon > 0$ is arbitrary, the latter implies that $\lim \rho_A(x'_n) = 0$, as required.

(2) $\implies$ (1) Let $\{x_n\}$ be a disjoint sequence in the solid hull of A. We have to show that $\lim \rho_B(x_n) = 0$ holds.

Pick a sequence $\{y_n\} \subseteq A$ with $|x_n| \leq |y_n|$ for all n. Also, for each n choose some $y'_n \in B$ with

$$\rho_B(x_n) \leq 2|y'_n(x_n)|.$$

Since $\{x_n\}$ is a disjoint sequence of E, there exists a disjoint sequence $\{z'_n\}$ of E' satisfying $|z'_n| \leq |y'_n|$ and $z'_n(x_n) = y'_n(x_n)$ for each n (see Exercise 22 of Section 1.5). From statement 3 of Theorem 1.23 there exists a sequence $\{x'_n\}$ of E' such that $|x'_n| \leq |z'_n|$ and $x'_n(y_n) = |z'_n|(|y_n|)$ holds for all n. Clearly, $\{x'_n\}$ is a disjoint sequence lying in the solid hull of B, and so by our hypothesis we have $\lim \rho_A(x'_n) = 0$. Now from

$$
\begin{aligned}
\rho_B(x_n) &\leq 2|y'_n(x_n)| = 2|z'_n(x_n)| \leq 2|z'_n|(|x_n|) \\
&\leq 2|z'_n|(|y_n|) = 2x'_n(y_n) \leq 2\rho_A(x'_n),
\end{aligned}
$$

it easily follows that $\lim \rho_B(x_n) = 0$, and the proof is finished. ∎

We are now in the position to prove that the notions of L- and M-weakly compact operators are in duality to each other. This result is due to P. Meyer-Nieberg [142].

Theorem 5.64 (Meyer-Nieberg). *For a Banach lattice E and a Banach space X the following statements hold:*

(1) *An operator $T \colon E \to X$ is M-weakly compact if and only if its adjoint $T' \colon X' \to E'$ is L-weakly compact.*

(2) *An operator $T: X \to E$ is L-weakly compact if and only if its adjoint $T': E' \to X'$ is M-weakly compact.*

Proof. (1) Consider a continuous operator $T: E \to X$, and denote by U' the closed unit ball of X'. Now let A be the closed unit ball of E (i.e., $A = \{x \in E: \|x\| \leq 1\}$), and let $B = T'(U')$. Then, T is M-weakly compact if and only if $\lim \|Tx_n\| = 0$ holds for each disjoint sequence $\{x_n\}$ of A. Thus, T is M-weakly compact if and only if every disjoint sequence of A converges uniformly to zero on B. By Theorem 5.63, this is equivalent to saying that every disjoint sequence in the solid hull of $B = T'(U')$ converges uniformly to zero on A (i.e., is norm convergent to zero). In other words, T is M-weakly compact if and only if $T': X' \to E'$ is L-weakly compact.

(2) Let U be the closed unit ball of X, and let $T: X \to E$ be a continuous operator. Put $A = T(U)$, and $B = \{x' \in E': \|x'\| \leq 1\}$. Thus, T is L-weakly compact if and only if every disjoint sequence in the solid hull of A converges uniformly to zero on B. The latter is (by Theorem 5.63) equivalent to saying that every disjoint sequence $\{x'_n\}$ of B converges uniformly to zero (i.e., $\lim \|T'x'_n\| = 0$ holds). In other words, T is L-weakly compact if and only if T' is M-weakly compact. ∎

The norm limit of a sequence of L-weakly compact (resp. M-weakly compact) operators is again L-weakly compact (resp. M-weakly compact). The details are included in the next result.

Theorem 5.65. *For a Banach lattice E and a Banach space X the following statements hold:*

(1) *The set of all M-weakly compact operators from E to X is a closed vector subspace of $L(E, X)$.*

(2) *The set of all L-weakly compact operators from X to E is a closed vector subspace of $L(X, E)$.*

Proof. (1) Clearly, the set of all M-weakly compact operators from E to X is a vector subspace of $L(E, X)$. To see that it is also a closed vector subspace of $L(E, X)$, let T be in the closure of the set of all M-weakly compact operators of $L(E, X)$.

Assume that $\{x_n\}$ is a disjoint sequence of E satisfying $\|x_n\| \leq 1$ for all n. We have to show that $\lim \|Tx_n\| = 0$. To this end, let $\epsilon > 0$. Pick an M-weakly compact operator $S: E \to X$ with $\|T - S\| < \epsilon$, and note that it follows from the inequalities

$$\|Tx_n\| \leq \|(T - S)x_n\| + \|Sx_n\| < \epsilon + \|Sx_n\|$$

that $\limsup \|Tx_n\| \leq \epsilon$. Since $\epsilon > 0$ is arbitrary, we see that $\lim \|Tx_n\| = 0$ holds, as desired.

(2) To see that the sum of two L-weakly compact operators is L-weakly compact, let $S, T \in L(X, E)$ be two L-weakly compact operators. Then part (1) and $(S + T)' = S' + T'$ coupled with Theorem 5.64 show that $S + T$ is L-weakly compact. Thus, the set of all L-weakly compact operators is a vector subspace of $L(X, E)$.

Finally, to see that the vector subspace of all L-weakly compact operators of $L(X, E)$ is norm closed, let $\{T_n\} \subseteq L(X, E)$ be a sequence of L-weakly compact operators satisfying $\lim \|T_n - T\| = 0$ in $L(X, E)$. Then, $\lim \|T_n' - T'\| = 0$ holds in $L(E', X')$. Since (by Theorem 5.64) each T_n' is M-weakly compact, it follows from part (1) that T' is also M-weakly compact. By Theorem 5.64 again, we see that T is L-weakly compact. ∎

It is useful to know that the range of an L-weakly compact operator is included in a Banach lattice with order continuous norm.

Theorem 5.66. *Let $T: X \to E$ be an L-weakly compact operator. If A is the ideal generated by $T(X)$, then $\overline{A}$ (the norm closure of A) is a Banach lattice with order continuous norm.*

Proof. By Theorems 4.13 and 4.11, it suffices to show that every order bounded disjoint sequence in A is norm convergent to zero. So, let $\{x_n\}$ be a disjoint sequence satisfying $0 \leq x_n \leq x$ for all n and some $x \in A$.

Pick $y_1, \ldots, y_k \in X$ with $x \leq \sum_{i=1}^{k} |Ty_i|$, and then use Theorem 1.13 to write $x_n = x_n^1 + \cdots + x_n^k$ with $0 \leq x_n^i \leq |Ty_i|$ for each n and each $i = 1, \ldots, k$. Clearly, for each i the sequence $\{x_n^i\}$ is disjoint, and so from the L-weak compactness of T it follows that $\lim \|x_n^i\| = 0$ holds for each i. Thus, $\lim \|x_n\| = 0$, and the proof is finished. ∎

The concepts of L- and M-weakly compact operators may coincide. The next result of P. G. Dodds and D. H. Fremlin [**54**] presents some conditions for this to happen.

Theorem 5.67 (Dodds–Fremlin). *Let E and F be a pair of Banach lattices such that E' and F have order continuous norms. Then, for an order bounded operator $T: E \to F$ the following statements are equivalent:*

(1) *T is L-weakly compact.*

(2) *T is M-weakly compact.*

(3) *T' is L-weakly compact.*

(4) *T' is M-weakly compact.*

(5) *For each pair $\{x_n\}$ and $\{x_n'\}$ of norm bounded disjoint sequences of E^+ and F_+', respectively, we have $\lim x_n'(Tx_n) = 0$.*

Proof. By Theorem 5.64 we know that $(1) \Longleftrightarrow (4)$ and $(2) \Longleftrightarrow (3)$.

$(4) \Longrightarrow (5)$ Let $\{x_n\} \subseteq E^+$ and $\{x_n'\} \subseteq F_+'$ be a pair of norm bounded disjoint sequences. If $\|x_n\| \leq M$ holds for all n, then from

$$\left|x_n'(Tx_n)\right| = \left|T'x_n'(x_n)\right| \leq \left\|T'x_n'\right\| \cdot \|x_n\| \leq M\left\|T'x_n'\right\| \to 0$$

we see that $\lim x_n'(Tx_n) = 0$.

$(5) \Longrightarrow (2)$ Let $\{x_n\}$ be a norm bounded disjoint sequence of E^+. We have to show that $\lim \|Tx_n\| = 0$. To this end, assume by way of contradiction that $\lim \|Tx_n\| \neq 0$. Then, there exists some $\epsilon > 0$ and a subsequence $\{y_n\}$ of $\{x_n\}$ satisfying $\|Ty_n\| > 2\epsilon$ for all n.

Since E' has order continuous norm, it follows that $y_n \xrightarrow{w} 0$ (see Exercise 7 of Section 4.3), and so $|T|y_n \xrightarrow{w} 0$. The inequality $|Ty_n| \leq |T|y_n$ implies $|Ty_n| \xrightarrow{w} 0$. Now an easy inductive argument shows that there exist a subsequence $\{z_n\}$ of $\{y_n\}$ and a sequence $\{x_n'\} \subseteq F_+'$ with $\|x_n'\| = 1$ for each n such that

$$x_n'\left(|Tz_n|\right) > 2\epsilon \quad \text{and} \quad \left(4^n \sum_{i=1}^{n} x_i'\right)\left(|Tz_{n+1}|\right) < \tfrac{1}{n}.$$

Put $\phi = \sum_{n=1}^{\infty} 2^{-n} x_n'$ and $\phi_n = \left(x_{n+1}' - 4^n \sum_{i=1}^{n} x_i' - 2^{-n}\phi\right)^+$. By Lemma 4.35 the sequence $\{\phi_n\}$ is disjoint. From

$$\phi_n\left(|Tz_{n+1}|\right) \geq \left(x_{n+1}' - 4^n \sum_{i=1}^{n} x_i' - 2^{-n}\phi\right)\left(|Tz_{n+1}|\right)$$
$$> 2\epsilon - \tfrac{1}{n} - 2^{-n}\phi\left(|Tz_{n+1}|\right),$$

we see that

$$\phi_n\left(|Tz_{n+1}|\right) > \epsilon \tag{$\star$}$$

must hold for all n sufficiently large. For each n pick $|\psi_n| \leq \phi_n$ with $\phi_n(|Tz_{n+1}|) = \psi_n(Tz_{n+1})$ (see Theorem 1.23), and note that $\{\psi_n\}$ is a disjoint sequence. On the other hand, from our hypothesis it follows that

$$\phi_n(|Tz_{n+1}|) = \psi_n(Tz_{n+1}) \longrightarrow 0,$$

which contradicts $(\star)$. Therefore, $\lim \|Tx_n\| = 0$ must hold true.

$(2) \Longrightarrow (1)$ Let U and V be the closed unit balls of E and F, respectively, and consider a disjoint sequence $\{y_n\} \subseteq E^+$ lying in the solid hull of $T(U)$. We need to show that $\lim \|y_n\| = 0$.

To this end, let $\epsilon > 0$. For each n choose some vector $x_n \in U$ with $0 \leq y_n \leq |Tx_n|$. Now by Theorem 5.60 there exists some $u \in E^+$ satisfying

$\|T(|x| - u)^+\| < \epsilon$ for all $x \in U$. From

$$
\begin{aligned}
0 \;\le\; & y_n \le |Tx_n| \le |Tx_n^+| + |Tx_n^-| \\
\le\; & |T(x_n^+ - u)^+| + |T|u + |T(x_n^- - u)^+| + |T|u \in 2\epsilon V + 2\big[0, |T|u\big],
\end{aligned}
$$

it follows that for each n there exists $0 \le u_n \in 2\epsilon V$ and $v_n \in 2\big[0, |T|u\big]$ with $y_n = u_n + v_n$. Since F has order continuous norm and $\{v_n\}$ is an order bounded disjoint sequence, we have $\lim \|v_n\| = 0$. The inequality

$$\|y_n\| \le \|u_n\| + \|v_n\| \le 2\epsilon + \|v_n\|,$$

implies $\limsup \|y_n\| \le 2\epsilon$. Since $\epsilon > 0$ is arbitrary, the latter yields $\lim \|y_n\| = 0$, and the proof is finished. ∎

We continue now with the introduction of operators having order continuous norms. Consider two Banach lattices E and F with F Dedekind complete, and let $T \colon E \to F$ be an order bounded operator. Then T is said to have **order continuous norm** whenever every sequence of positive operators with $|T| \ge T_n \downarrow 0$ in $\mathcal{L}_b(E, F)$ satisfies $\|T_n\| \downarrow 0$. A glance at Theorems 4.9, 4.11, and 4.13 shows that the following statements are equivalent:

(1) T has order continuous norm.

(2) $|T| \ge T_\alpha \downarrow 0$ in $\mathcal{L}_b(E, F)$ implies $\|T_\alpha\| \downarrow 0$.

(3) $0 \le T_\alpha \uparrow \le |T|$ in $\mathcal{L}_b(E, F)$ implies that $\{T_\alpha\}$ is an r-norm Cauchy net.

(4) $0 \le T_n \uparrow \le |T|$ in $\mathcal{L}_b(E, F)$ implies that $\{T_n\}$ is an r-norm Cauchy sequence.

(5) Every disjoint sequence of $\big[0, |T|\big]$ is norm convergent to zero.

(6) If $\mathcal{A}_T$ is the ideal generated by T in $\mathcal{L}_b(E, F)$ equipped with the r-norm, then the norm completion of $\mathcal{A}_T$ is a Banach lattice with order continuous norm.

The positive operators that have order continuous norms are precisely the positive operators which are simultaneously L- and M-weakly compact. This is due to P. G. Dodds and D. H. Fremlin [54].

Theorem 5.68 (Dodds–Fremlin). *A positive operator $T \colon E \to F$ between Banach lattices with F Dedekind complete has order continuous norm if and only if T is both L- and M-weakly compact.*

Proof. Assume first that T has order continuous norm. We shall show next that T is M-weakly compact.

Let $\{x_n\}$ be a disjoint sequence of E^+ satisfying $\|x_n\| \leq 1$ for all n. For each n, the formula

$$T_n(x) = \sup\{T(x \wedge kx_n): \ k = 1, 2, \ldots\}, \quad x \in E^+, \qquad (\star\star)$$

defines a positive operator $T_n: E \to F$ satisfying the properties: $0 \leq T_n \leq T$, $T_n(x_n) = T(x_n)$, and $T_n(y) = 0$ for $y \perp x_n$ (see Theorem 1.22). Moreover, it follows from $(\star\star)$ that $T_n(x) = T(x)$ whenever $0 \leq x \leq kx_n$ holds for some k. On the other hand, we claim that $T_n \perp T_m$ for $n \neq m$. To see this, let $n \neq m$, and let $x \in E^+$. From $x \wedge kx_n \perp x_m$, we see that $T_m(x \wedge kx_n) = 0$ and so from

$$\begin{aligned}
0 \ \leq \ & [T_n \wedge T_m](x) \leq T_n(x - x \wedge kx_n) + T_m(x \wedge kx_n) \\
= \ & T_n(x - x \wedge kx_n) = T_n(x) - T(x \wedge kx_n) \downarrow_k 0
\end{aligned}$$

it follows that $[T_n \wedge T_m](x) = 0$. Therefore, $T_n \wedge T_m = 0$ holds. Since T has order continuous norm, we infer that $\lim \|T_n\| = 0$, and so from

$$\|T(x_n)\| = \|T_n(x_n)\| \leq \|T_n\|$$

it follows that $\lim \|T(x_n)\| = 0$. That is, T is M-weakly compact.

Now, to show that T is L-weakly compact, let $\{y_n\} \subseteq F^+$ be a disjoint sequence in the solid hull of the image under T of the closed unit ball of E. Pick a sequence $\{x_n\} \subseteq E^+$ with $\|x_n\| \leq 1$ and $0 \leq y_n \leq Tx_n$ for all n. Denote by P_n the order projection of F onto the band generated by y_n, and note that $P_n \perp P_m$ implies $P_n(x) \perp P_m(x)$ for all $x \in F$ ($n \neq m$). Next, define the operators $T_n: E \to F$ by $T_n(x) = P_n T(x)$, and note that $0 \leq T_n \leq T$ holds for each n. On the other hand, for $n \neq m$ and $x \in E^+$ we have

$$0 \leq [T_n \wedge T_m](x) \leq P_n(Tx) \wedge P_m(Tx) = 0,$$

and so $\{T_n\}$ is a disjoint sequence. Since T has order continuous norm, the latter implies $\lim \|T_n\| = 0$, and from

$$\|y_n\| = \|P_n(y_n)\| \leq \|P_n T(x_n)\| = \|T_n(x_n)\| \leq \|T_n\|$$

we see that $\lim \|y_n\| = 0$. Therefore, T is also L-weakly compact.

For the converse, suppose that T is both L- and M-weakly compact. Let $\{T_n\} \subseteq \mathcal{L}_b(E, F)$ satisfy $T \geq T_n \downarrow 0$. We need to show that $\lim \|T_n\| = 0$. To this end, let $\epsilon > 0$. By Theorem 5.60 there exists some $u \in E^+$ satisfying $\|T(|x| - u)^+\| < \epsilon$ for all $x \in E$ with $\|x\| \leq 1$. From

$$|T_n(x)| \leq T_n(|x|) \leq T_n(|x| - u)^+ + T_n(u) \leq T(|x| - u)^+ + T_n(u)$$

we see that

$$\|T_n(x)\| \leq \|T(|x| - u)^+\| + \|T_n(u)\| < \epsilon + \|T_n(u)\|$$

holds for all $x \in E$ with $\|x\| \leq 1$. Therefore,

$$\|T_n\| \leq \epsilon + \|T_n(u)\|$$

holds for all n. On the other hand, $T_n(u) \downarrow 0$ holds in the ideal generated by $T(E)$, and by Theorem 5.66 the ideal generated by $T(E)$ has order continuous norm. Thus, $\|T_n(u)\| \downarrow 0$, and so we see that $\limsup \|T_n\| \le \epsilon$. Since $\epsilon > 0$ is arbitrary, the latter shows that $\lim \|T_n\| = 0$, and the proof is finished. ∎

Now consider a continuous operator $T \colon X \to Y$ between two Banach spaces, and let $\mathrm{Ring}(T)$ be the norm closure in $L(X, Y)$ of the vector subspace consisting of all operators of the form $\sum_{i=1}^{n} R_i T S_i$, where $S_1, \ldots, S_n$ belong to $L(X)$ and $R_1, \ldots, R_n \in L(Y)$. That is,

$$\mathrm{Ring}(T) := \Big\{ S \colon \ S = \sum_{i=1}^{n} R_i T S_i \text{ with } S_i \in L(X) \text{ and } R_i \in L(Y) \Big\}^{-}.$$

Definition 5.69. *The closed vector subspace* $\mathrm{Ring}(T)$ *of* $L(X, Y)$ *is called the* **ring ideal** *generated by the operator* T.

Next, let E and F be two Banach lattices with F Dedekind complete. Then, we know that $\mathcal{L}_b(E, F) \subseteq L(E, F)$ holds. If $T \colon E \to F$ is an order bounded operator, then we can consider the (order) ideal $\mathcal{A}_T$ generated by T in $\mathcal{L}_b(E, F)$, i.e.,

$$\mathcal{A}_T = \big\{ S \in \mathcal{L}_b(E, F) \colon \exists \, n \text{ with } |S| \le n|T| \big\},$$

and ask the following question: *Is there any relationship between the order and ring ideals generated by an operator?*

H. Leinfelder proved in [**111**] that for a positive compact operator T between L_p-spaces we have $\mathcal{A}_T \subseteq \mathrm{Ring}(T)$. This was generalized by B. de Pagter in [**159**] for an operator with order continuous norm between certain Banach lattices. These results are special cases of the following theorem.

Theorem 5.70. *Let* E *be a Banach lattice which is either Dedekind σ-complete or has a quasi-interior point, and let* F *be a Dedekind complete Banach lattice. If a positive operator* $T \colon E \to F$ *has order continuous norm, then* $\mathcal{A}_T \subseteq \mathrm{Ring}(T)$.

Proof. By Theorem 5.68, T is both L- and M-weakly compact. Let A be the ideal generated by $T(E)$. Then, by Theorem 5.66, $\overline{A}$ is a Banach lattice with order continuous norm. Now consider an operator $S \colon E \to F$ satisfying $0 \le S \le T$. Clearly, $S(E) \subseteq \overline{A}$ holds.

Let $\epsilon > 0$. Since T is M-weakly compact, there exists some $u \in E^+$ satisfying $\|T(|x| - u)^+\| < \epsilon$ for all $x \in E$ with $\|x\| \le 1$. Also, since $\overline{A}$ has order continuous norm, there exist by Theorem 4.87 positive operators

$M_1, \ldots, M_k$ on E and order projections $P_1, \ldots, P_k$ on $\overline{A}$ such that

$$0 \leq \sum_{i=1}^{k} P_i T M_i \leq T \quad \text{and} \quad \left\| \left| S - \sum_{i=1}^{k} P_i T M_i \right| u \right\| < \epsilon.$$

Next, note that each $P_i \colon \overline{A} \to F$ is bounded by the identity operator $I \colon F \to F$, and so by Theorem 1.26 each P_i extends to a positive operator (which we shall denote by P_i again) to all of F satisfying $0 \leq P_i \leq I$. Clearly, $\sum_{i=1}^{k} P_i T M_i \in \mathrm{Ring}(T)$ and

$$\left| S - \sum_{i=1}^{k} P_i T M_i \right| \leq S + \sum_{i=1}^{k} P_i T M_i \leq 2T$$

holds in $\mathcal{L}_b(E, F)$. On the other hand, for each $x \in E$ we have

$$\left| \left[S - \sum_{i=1}^{k} P_i T M_i \right] x \right| \leq \left| S - \sum_{i=1}^{k} P_i T M_i \right| (|x| - u)^+ + \left| S - \sum_{i=1}^{k} P_i T M_i \right| u$$

$$\leq 2T (|x| - u)^+ + \left| S - \sum_{i=1}^{k} P_i T M_i \right| u,$$

and so for each $x \in E$ with $\|x\| \leq 1$ we have

$$\left\| \left[S - \sum_{i=1}^{k} P_i T M_i \right] x \right\| \leq 2 \| T (|x| - u)^+ \| + \left\| \left| S - \sum_{i=1}^{k} P_i T M_i \right| u \right\| \leq 3\epsilon.$$

This easily implies $\left\| S - \sum_{i=1}^{k} P_i T M_i \right\| \leq 3\epsilon$ for each $\epsilon > 0$, from which it follows that $S \in \mathrm{Ring}(T)$. ∎

If two operators $S, T \colon E \to E$ on a Banach lattice satisfy $0 \leq S \leq T$, then under what conditions is some power of S in the ring ideal generated by T?

Before giving some answers to this question we need to introduce the class of semicompact operators. Following A. C. Zaanen [197] we say that a continuous operator $T \colon X \to E$ from a Banach space to a Banach lattice is **semicompact** whenever for each $\epsilon > 0$ there exists some $u \in E^+$ satisfying

$$\| (|Tx| - u)^+ \| < \epsilon$$

for all $x \in X$ with $\|x\| \leq 1$.

Consider a continuous operator $T \colon X \to E$ from a Banach space to a Banach lattice, and let U and V denote the closed unit balls of X and E, respectively. From the identity $|Tx| = |Tx| \wedge u + (|Tx| - u)^+$ and Theorem 1.13, it is easy to see that T is a semicompact operator if and only if for each $\epsilon > 0$ there exists some $u \in E^+$ such that

$$T(U) \subseteq [-u, u] + \epsilon V .$$

The discussion below will establish that many of the operators encountered previously were in fact semicompact operators.

Theorem 5.71. *If an operator $T : X \to E$ from a Banach space to a Banach lattice is either*

(a) *compact, or*

(b) *L-weakly compact,*

then T is a semicompact operator.

Proof. (a) Denote by U the closed unit ball of X. Assume that T is compact, and let $\epsilon > 0$. Pick a finite subset $\{u_1, \ldots, u_n\}$ of E such that for each $x \in U$ we have $\|Tx - u_i\| < \epsilon$ for at least one i. Put $u = |u_1| + \cdots + |u_n| \in E^+$.

Now let $x \in U$. Choose some u_i with $\|Tx - u_i\| < \epsilon$, and note that the inequalities

$$\left(|Tx| - u\right)^+ \leq \left(|Tx| - |u_i|\right)^+ \leq \left||Tx| - |u_i|\right| \leq |Tx - u_i|$$

imply $\|(|Tx| - u)^+\| \leq \|Tx - u_i\| < \epsilon$. This shows that T is a semicompact operator.

(b) This is part (2) of Theorem 5.60. ∎

An operator dominated by a semicompact operator is also semicompact.

Theorem 5.72. *If a positive operator $T : E \to F$ between two Banach lattices is either*

(a) *M-weakly compact, or*

(b) *dominated by a semicompact operator,*

then T is semicompact.

Proof. (a) Denote by U the closed unit ball of E. Assume that T is M-weakly compact, and let $\epsilon > 0$. By Theorem 5.60 there exists some $w \in E^+$ satisfying $\|T(|x| - w)^+\| < \epsilon$ for all $x \in U$. Put $u = Tw \in F^+$, and note that

$$\begin{aligned}
\left(|Tx| - u\right)^+ &= \left(|Tx| - Tw\right)^+ \leq \left(T|x| - Tw\right)^+ \\
&= \left(T(|x| - w)\right)^+ \leq T\left((|x| - w)^+\right) .
\end{aligned}$$

Thus, $\|(|Tx| - u)^+\| \leq \|T(|x| - w)^+\| < \epsilon$ holds for all $x \in U$.

(b) Assume that a semicompact operator $R\colon E \to F$ satisfies $0 \leq T \leq R$. Given $\epsilon > 0$ pick some $u \in F^+$ with $\||(|Rx| - u)^+\|| < \epsilon$ for all $x \in U$, and note that

$$\||(|Tx| - u)^+\|| \leq \||(T|x| - u)^+\|| \leq \||(R|x| - u)^+\|| < \epsilon$$

holds for all $x \in U$. ∎

A semicompact operator need not be compact, weakly compact, L-weakly compact, or M-weakly compact. For instance, the identity operator $I\colon \ell_\infty \to \ell_\infty$ is semicompact, but it does not have any one of the above mentioned compactness properties.

We now pass to two basic properties of semicompact operators. The first one tells us that the range of a semicompact operator is always included in a Banach lattice with a quasi-interior point.

Theorem 5.73. *If $T\colon X \to E$ is a semicompact operator, then there exists some $y \in E^+$ such that the ideal E_y generated by y satisfies $T(X) \subseteq \overline{E}_y$.*

Proof. For each n choose some $0 < u_n \in E$ such that $\||(|Tx| - u_n)^+\|| < \frac{1}{n}$ holds for all $x \in X$ with $\|x\| \leq 1$. Put $y = \sum_{n=1}^\infty 2^{-n} u_n / \|u_n\|$, and let E_y be the ideal generated by y in E. Clearly, $\{u_n\} \subseteq E_y$.

Now let $x \in X$ satisfy $\|x\| \leq 1$. From

$$\||Tx| - |Tx| \wedge u_n\|| = \||(|Tx| - u_n)^+\|| < \tfrac{1}{n}$$

and $|Tx| \wedge u_n \in E_y$, we see that $|Tx| \in \overline{E}_y$. Since the norm closure of E_y is an ideal, it follows that $Tx \in \overline{E}_y$. Consequently, $T(X) \subseteq \overline{E}_y$ holds. ∎

A continuous operator with a semicompact adjoint is necessarily an o-weakly compact operator.

Theorem 5.74. *If a continuous operator $T\colon E \to F$ between Banach lattices has a semicompact adjoint, then T is o-weakly compact.*

Proof. Let $0 < x \in E$, and let $\epsilon > 0$. Pick $0 \leq \phi \in E'$ such that

$$\||(|T'x'| - \phi)^+\|| < \tfrac{\epsilon}{\|x\|}$$

holds for all $x' \in F'$ with $\|x'\| \leq 1$.

If $y \in E$ satisfies $|y| \leq x$ and $\phi(|y|) < \epsilon$, then for each $x' \in F'$ with $\|x'\| \leq 1$ we have

$$\begin{aligned} |x'(Ty)| &\leq |T'x'|(|y|) \leq (|T'x'| - \phi)^+(|y|) + \phi(|y|) \\ &\leq \||(|T'x'| - \phi)^+\|| \cdot \|x\| + \phi(|y|) < \epsilon + \epsilon = 2\epsilon, \end{aligned}$$

and so $\|Ty\| \leq 2\epsilon$ holds. Hence, T satisfies condition (3) of Theorem 5.57, and consequently T is o-weakly compact. ∎

For a semicompact operator there are some subtle relationships between the ring ideal and the order ideal generated by the operator. The next result of the authors [17] presents such a relationship.

Theorem 5.75 (Aliprantis–Burkinshaw). *Let $S, T\colon E \to E$ be two positive operators on a Banach lattice such that $0 \le S \le T$ holds. If S and its adjoint S' are both semicompact, then S^3 belongs to* $\mathrm{Ring}(T)$.

Proof. Without loss of generality we can assume that $\|T\| \le 1$. This implies $\|S\| \le 1$. Let $\epsilon > 0$ be fixed.

Choose $0 \le u \in E$ and $0 \le \phi \in E'$ such that

$$\big\|(|Sx| - u)^+\big\| < \epsilon \quad \text{and} \quad \big\|(|S'x'| - \phi)^+\big\| < \epsilon$$

hold for all $x \in U$ and $x' \in U'$. By Theorem 5.73 there exists some $y \in E^+$ with $S(E) \subseteq \overline{E}_y$ (where $\overline{E}_y$ is the norm closure of E_y in E). Replacing y by $y + u$ we can assume without loss of generality that $u \in \overline{E}_y$.

Next, restrict S and T to E_y and consider $S, T\colon E_y \to E''$. Since E_y is an AM-space with unit, it follows from Theorem 4.82 that there exist positive multiplication operators $M_1, \ldots, M_k$ on E_y and order projections $P_1, \ldots, P_k$ on E'' with

$$\phi\Big(\Big|S - \sum_{i=1}^{k} P_i T M_i\Big|u\Big) < \epsilon \quad \text{and} \quad 0 \le \sum_{i=1}^{k} P_i T M_i \le T \text{ on } E_y. \qquad (\star)$$

By Lemma 4.78 each $M_i\colon E_y \to E_y$ is continuous for the norm induced by E, and hence each M_i extends uniquely to a continuous operator (which we denote by M_i again) from $\overline{E}_y$ to $\overline{E}_y$. From $(\star)$ we see that

$$0 \le \sum_{i=1}^{k} P_i T M_i \le T$$

also holds on $\overline{E}_y$.

Since P_i carries E into the ideal generated by E in E'' and S is o-weakly compact (see Theorem 5.74), it follows that $S'' P_i$ carries E into E. Let $R_i\colon E \to E$ denote the restriction of $S'' P_i$ to E. Also, from $E \xrightarrow{S} \overline{E}_y \xrightarrow{M_i} E$, we see that $S_i = M_i S$ is a continuous operator from E to E.

Now consider the operator $R = S - \sum_{i=1}^{k} P_i T M_i\colon \overline{E}_y \to E''$, and note that since E_y and $\overline{E}_y$ are both ideals of E, the moduli of the operators $R\colon E_y \to E''$ and $R\colon \overline{E}_y \to E''$ agree on E_y. In particular, note that

$$|R| \le S + \sum_{i=1}^{k} P_i T M_i \le 2T$$

holds of $\overline{E}_y$, and so in view of $S(E) \subseteq \overline{E}_y$ we have

$$\|RSx\| \leq \||R|(|Sx|)\| \leq 2\|T\|\cdot\|Sx\| \leq 2\|T\|\cdot\|S\| \leq 2$$

for all $x \in U$. Therefore, if $x \in U$ and $x' \in U'$, then

$$\left|\left\langle x', \left(S^3 - \sum_{i=1}^{k} R_i T S_i\right)x\right\rangle\right| = |\langle x', S''RSx\rangle| \leq \langle |S'x'|, |RSx|\rangle$$

$$\begin{aligned}
&\leq (|S'x'| - \phi)^+(|RSx|) + [|S'x'| \wedge \phi](|RSx|) \\
&\leq \|(|S'x'| - \phi)^+\|\cdot\|RSx\| + [|S'x'| \wedge \phi](|RSx|) \\
&\leq 2\epsilon + (|S'x'| \wedge \phi)\left[|R|((|Sx| - u)^+ + |Sx| \wedge u)\right] \\
&\leq 2\epsilon + |S'x'|\left(2T([|Sx| - u]^+)\right) + \phi(|R|u) \\
&\leq 2\epsilon + \|S'\|\cdot\|x'\|\cdot2\|T\|\cdot\|(|Sx| - u)^+\| + \epsilon \\
&\leq 2\epsilon + 2\epsilon + \epsilon = 5\epsilon,
\end{aligned}$$

from which it follows that

$$\left\|S^3 - \sum_{i=1}^{k} R_i T S_i\right\| \leq 5\epsilon.$$

Since $\epsilon > 0$ is arbitrary, the latter shows that $S^3 \in \mathrm{Ring}(T)$, and the proof of the theorem is finished. ∎

An immediate consequence of the preceding result is the following.

Corollary 5.76. *Assume that a pair of positive operators $S, T: E \to E$ on a Banach lattice satisfies $0 \leq S \leq T$. If anyone of the operators S or T is L- or M-weakly compact, then S^3 belongs to $\mathrm{Ring}(T)$.*

Proof. By Theorems 5.71, 5.72, and 5.64, we see that S and S' are both semicompact. The conclusion now follows from Theorem 5.75. ∎

The next immediate consequence of Theorem 5.75 generalizes a result of B. de Pagter [159] and presents an alternative proof of Theorem 5.13.

Corollary 5.77. *If a pair of positive operators $S, T: E \to E$ on a Banach lattice satisfies $0 \leq S \leq T$ and T is compact, then S^3 belongs to $\mathrm{Ring}(T)$.*

Proof. The conclusion follows from Theorem 5.75 by observing that S and S' are both semicompact operators. ∎

The last result of the section presents a condition for the square of an operator to lie in the ring ideal generated by another operators.

Theorem 5.78. *Let $S, T: E \to E$ be two positive operators on a Banach lattice such that $0 \leq S \leq T$ holds. If E has order continuous norm and S is semicompact, then S^2 belongs to $\mathrm{Ring}(T)$.*

Proof. Without loss of generality we can assume that $\|T\| \leq 1$ holds. Let $\epsilon > 0$. Pick some $u \in E^+$ such that $\|(|Sx| - u)^+\| < \epsilon$ holds for all $x \in U$. By Theorem 4.87 there exist positive operators $M_1, \ldots, M_k$ on E and order projections $P_1, \ldots, P_k$ on E satisfying

$$0 \leq \sum_{i=1}^{k} P_i T M_i \leq T \quad \text{and} \quad \left\| \left| S - \sum_{i=1}^{k} P_i T M_i \right| u \right\| < \epsilon.$$

Put $R = S - \sum_{i=1}^{k} P_i T M_i$. Clearly, $\||R|u\| < \epsilon$ and $|R| \leq 2T$ hold.

Now for $x \in U$ we have

$$\left\| \left(S^2 - \sum_{i=1}^{k} P_i T M_i S \right) x \right\| = \|RSx\| \leq \||R|(|Sx| - u]^+)\| + \||R|u\|$$

$$\leq 2\|T\| \cdot \|(|Sx| - u)^+\| + \epsilon < 2\epsilon + \epsilon = 3\epsilon,$$

and consequently

$$\left\| S^2 - \sum_{i=1}^{k} P_i T M_i S \right\| \leq 3\epsilon.$$

Since $\epsilon > 0$ is arbitrary and $\sum_{i=1}^{k} P_i T M_i S$ belongs to $\text{Ring}(T)$, we see that indeed $S^2 \in \text{Ring}(T)$. ∎

It is interesting to note that the preceding result can be used to provide an alternative proof of Corollary 5.16.

Exercises

1. Give an example of an o-weakly compact operator whose adjoint is not o-weakly compact. [*Hint*: Consider the identity operator $I: \ell_1 \to \ell_1$.]

2. Let $S, T: E \to F$ be two positive operators between Banach lattices such that $0 \leq S \leq T$ holds. Establish the following statements.
   (a) If T is o-weakly compact, then S is o-weakly compact.
   (b) If T is M-weakly compact, then S is M-weakly compact.
   (c) If T is L-weakly compact, then S is L-weakly compact.

3. If $T: E \to X$ is a continuous operator from a Banach lattice to a Banach space, then show that the following statements are equivalent.
   (a) T is o-weakly compact.
   (b) If $0 \leq x_n \uparrow \leq x$ holds in E, then $\{Tx_n\}$ is norm convergent in X.
   (c) For each order bounded sequence $\{x_n\} \subseteq E^+$ with $x_n \xrightarrow{w} 0$ we have $\|Tx_n\| \to 0$.
   (d) The set $T'(U')$ is $\sigma(E', A)$-compact, where U' is the closed unit ball of X' and A is the ideal generated by E in E''.

4. For a positive operator $T: E \to F$ between Banach lattices establish the following statements.

    (a) If T is an L-weakly compact lattice homomorphism, then T is also M-weakly compact.

    (b) If T is M-weakly compact and interval preserving, then T is also L-weakly compact.

5. Let $T: E \to F$ be a positive operator from an AM-space with unit to a Banach lattice with order continuous norm. Show that T has order continuous norm and that $\mathcal{A}_T \subseteq \mathrm{Ring}(T)$. [*Hint:* If e is the unit of E and $S: E \to F$ is any positive operator, then note that $\|S\| = \|S(e)\|$.]

6. Show that a continuous operator from a Dedekind σ-complete Banach lattice to a separable Banach space is o-weakly compact. [*Hint:* Use Theorem 4.43.]

7. Show that a positive operator between Banach lattices has order continuous norm if and only if its adjoint likewise has order continuous norm.

8. (Dodds [**53**]) If $T: E \to X$ is a continuous operator from a Banach lattice to a Banach space, then show that each one of the following two statements implies that T is o-weakly compact.

    (a) c_0 is not embeddable in X.

    (b) E is Dedekind σ-complete and ℓ_∞ is not embeddable in X.

[*Hint:* For (a) use Theorem 4.63.]

9. Show that an L-weakly compact operator $T: X \to E$ admits a factorization through a reflexive Banach lattice F

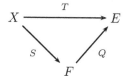

such that:

    (a) Q is an L-weakly compact interval preserving lattice homomorphism.

    (b) S is L-weakly compact.

    (c) S is positive if T is positive.

[*Hint:* Let U be the closed unit ball of X, and let W be the convex solid hull of $T(U)$. By Theorem 5.55 the set W is a weakly relatively compact subset of E. Now let Ψ be the reflexive Banach lattice of Theorems 5.37 and 5.41. Then we have the factorization

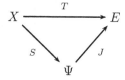

where $S(x) = T(x)$ and $J: \Psi \to E$ is the natural inclusion (which is an interval preserving lattice homomorphism). Since the norm topologies of Ψ and E agree on W (see Exercise 10 of Section 5.2), it follows that S is L-weakly compact. To see that J is also L-weakly compact, let $\{x_n\}$ be a disjoint sequence of Ψ^+ satisfying $\||x_n\|| \le 1$ for each n. Fix k and note that $\|x_n\|_k \le 1$ holds for each n. Thus, for each n there exist

$0 \leq w_n \in W$ and $0 \leq y_n \in E$ with $\|y_n\| \leq 1$ and $x_n = 2(2^k w_n + 2^{-k} y_n)$. Clearly, $\{w_n\}$ is a disjoint sequence, and therefore $\lim \|w_n\| = 0$. From $\|x_n\| \leq 2^{k+1} \|w_n\| + 2^{1-k}$, we see that $\limsup \|x_n\| \leq 2^{1-k}$ holds. Since k is arbitrary, we have $\lim \|Jx_n\| = 0$.]

10. Show that an M-weakly compact operator $T \colon E \to X$ admits a factorization through a reflexive Banach lattice F

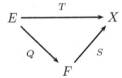

such that:
   (a) Q is an M-weakly compact lattice homomorphism.
   (b) S is M-weakly compact.
   (c) S is positive if T is also positive.
   [*Hint*: Use the preceding exercise by observing that $T' \colon X' \to E'$ is L-weakly compact.]

11. (Aliprantis–Burkinshaw [**17**]) This exercise is the dual to Theorem 5.58. Let $T \colon X \to E$ be a continuous operator from a Banach space to a Banach lattice. If T' is o-weakly compact, then show that T admits a factorization through a Banach lattice F

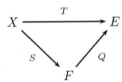

such that:
   (a) F' is a KB-space.
   (b) Q is an interval preserving lattice homomorphism.
   (c) S is positive if T is also positive.
   [*Hint*: Let W be the convex solid hull of $T(U)$, and consider the Banach lattice Ψ of Theorem 5.41. Now using the fact that $x'_n \downarrow 0$ implies $\lim \|T'x'_n\| = 0$, parallel the proof of Theorem 5.43.]

12. Let $T \colon E \to E$ be a positive L-weakly compact operator on a Banach lattice such that $T[0, x]$ is norm totally bounded for each $x \in E^+$. If another positive operator $S \colon E \to E$ satisfies $0 \leq S \leq T$, then show that:
   (a) $S[0, x]$ is norm totally bounded for each $x \in E^+$.
   (b) S^2 is a compact operator.

13. Let $T \colon E \to E$ be a positive M-weakly compact operator on a Banach lattice such that $T[0, x]$ is norm totally bounded for each $x \in E^+$. If another operator $S \colon E \to E$ satisfies $0 \leq S \leq T$, then show that S^2 is a compact operator.

14. Let $T \colon E \to F$ be a positive operator between two Banach lattices. Then show that T is both L- and M-weakly compact if and only if for each

$\epsilon > 0$ there exist $u \in E^+$ and $0 \le y' \in F'$ such that

$$\left\| T(|x| - u)^+ \right\| < \epsilon \quad \text{and} \quad \left\| T'(|x'| - y')^+ \right\| < \epsilon$$

holds for all $x \in E$ and $x' \in F'$ with $\|x\| \le 1$ and $\|x'\| \le 1$.

15. Establish the following properties:
   (a) Every continuous operator from a Banach space to an AM-space with unit is semicompact.
   (b) Every regular operator from an AM-space with unit to a Banach lattice is semicompact.

16. Show that the identity operator $I: \ell_2 \to \ell_2$ is not a semicompact operator (and conclude from this that the adjoint of an o-weakly compact operator need not be semicompact).

17. Let $E = c_0$ or $E = \ell_p$ for some $1 \le p < \infty$. Then show that an operator $T: E \to E$ is compact if and only if T is semicompact.

18. Show that an operator from a Banach space to a Banach lattice with order continuous norm is semicompact if and only if it is L-weakly compact.

19. Generalize Theorem 5.75 as follows: *Let $S, T: E \to E$ be two positive operators on a Banach lattice such that $0 \le S \le T$ holds. If for some $n \in \mathbb{N}$ the operators S^n and $(S^n)'$ are both semicompact, then S^{2n+1} belongs to* $\mathrm{Ring}(T)$.

20. Let $T: E \to E$ be a positive L- or M-weakly compact operator on a Dedekind complete Banach lattice. If $T_1, T_2, T_3 \in \mathcal{A}_T$, then show that $T_1 T_2 T_3 \in \mathrm{Ring}(T)$.

21. Generalize Theorem 5.49 as follows: *Consider the scheme of positive operators*

$$E \xrightarrow{S_1} G \xrightarrow{S_2} H$$

between Banach lattices such that H has order continuous norm. If S_2 is dominated by a compact operator and $(S_1)'$ is o-weakly compact, then show that $S_2 S_1$ is a compact operator. [*Hint*: Use Exercise 11 of this section.]

22. Let $T: E \to X$ be a continuous operator from a locally convex-solid Riesz space to a topologically complete locally convex space. Then show that the following statements are equivalent:
   (a) T maps order intervals to weakly relatively compact subsets of the space X.
   (b) For every order bounded disjoint sequence $\{x_n\}$ of E the sequence $\{Tx_n\}$ converges to zero in X.
   (c) $T''(A) \subseteq X$, where A is the ideal generated by E in E''.

23. If $0 \le x = (x_1, x_2, \dots) \in \ell_\infty$, then show that the order interval $[0, x]$ is weakly compact if and only if $\lim x_n = 0$ (in which case $[0, x]$ is also norm compact).

5.4. Dunford–Pettis Operators

In a remarkable paper [**56**], N. Dunford and P. J. Pettis proved (among other things) that a weakly compact operator from $L_1(\mu)$ to itself carries weakly convergent sequences to norm convergent sequences. This prompted A. Grothendieck [**72**] to call every operator with this property a Dunford–Pettis operator. Accordingly, following A. Grothendieck we shall say that an operator $T\colon X \to Y$ between two Banach spaces is a **Dunford–Pettis operator** (or that T has the **Dunford–Pettis property**) whenever $x_n \xrightarrow{w} 0$ in X implies $\|Tx_n\| \to 0$ or, equivalently, whenever $x_n \xrightarrow{w} x$ in X implies $\lim \|Tx_n - Tx\| = 0$.

Every Dunford–Pettis operator is continuous. (Indeed, if T is a Dunford–Pettis operator and $\|x_n\| \to 0$, then $x_n \xrightarrow{w} 0$, and so $\|Tx_n\| \to 0$.) A compact operator is necessarily a Dunford–Pettis operator. From Theorem 3.40, it is easy to see that whenever X is a reflexive Banach space, then an operator with domain X is Dunford–Pettis if and only if it is compact. On the other hand, a Dunford–Pettis operator need not be a compact operator, and its adjoint may fail as well to have the Dunford–Pettis property. For instance, the identity operator $I\colon \ell_1 \to \ell_1$ is a Dunford–Pettis operator (see Theorem 4.32) which is not weakly compact. Moreover, its adjoint is the identity operator $I\colon \ell_\infty \to \ell_\infty$, which is not a Dunford–Pettis operator.

In terms of weak Cauchy sequences the Dunford–Pettis operators are characterized as follow.

Theorem 5.79. *An operator $T\colon X \to Y$ between two Banach spaces is a Dunford–Pettis operator if and only if T carries weakly Cauchy sequences of X to norm convergent sequences of Y.*

Proof. Assume that T is a Dunford–Pettis operator, and let $\{x_n\}$ be a weak Cauchy sequence of X. If $\{Tx_n\}$ is not a norm Cauchy sequence of Y, then there exist some $\epsilon > 0$ and a subsequence $\{y_n\}$ of $\{x_n\}$ satisfying $\|T(y_{2n} - y_{2n-1})\| > \epsilon$ for all n. Since $y_{2n} - y_{2n-1} \xrightarrow{w} 0$ holds in X, it follows that $\lim \|T(y_{2n} - y_{2n-1})\| = 0$, which is impossible. Thus, $\{Tx_n\}$ is a norm Cauchy sequence, and hence is norm convergent in Y.

For the converse assume that T carries weakly Cauchy sequences to norm convergent sequences, and let $x_n \xrightarrow{w} 0$ in X. Then $\{x_n\}$ is clearly a weak Cauchy sequence of X, and so $\{Tx_n\}$ is norm convergent in Y. Since T is continuous (why?), $Tx_n \xrightarrow{w} 0$ also holds in Y, from which it follows that $\|Tx_n\| \to 0$. ∎

From the preceding result and Theorem 4.72, the following theorem should be immediate.

Theorem 5.80. *If ℓ_1 does not embed in a Banach space X, then every Dunford–Pettis operator from X to an arbitrary Banach space is compact.*

In terms of compact and weakly compact operators the Dunford–Pettis operators are characterized as follows.

Theorem 5.81. *For a continuous operator $T\colon X \to Y$ between two Banach spaces the following statements are equivalent.*

(1) *T is a Dunford–Pettis operator.*

(2) *T carries weakly relatively compact subsets of X to norm totally bounded subsets of Y.*

(3) *For an arbitrary Banach space Z and every weakly compact operator $S\colon Z \to X$, the operator TS is a compact operator.*

(4) *For every weakly compact operator $S\colon \ell_1 \to X$, the operator TS is compact.*

Proof. (1) $\Longrightarrow$ (2) Let W be a weakly relatively compact subset of X. If $\{x_n\}$ is a sequence of W, then there exists a subsequence $\{y_n\}$ of $\{x_n\}$ satisfying $y_n \xrightarrow{w} y$ in X, and so $\lim \|Ty_n - Ty\| = 0$. This shows that $T(W)$ is a norm totally bounded set.

(2) $\Longrightarrow$ (3) and (3) $\Longrightarrow$ (4) are obvious.

(4) $\Longrightarrow$ (1) Let $x_n \xrightarrow{w} 0$ in X and define the operator $S\colon \ell_1 \to X$ by

$$S(\alpha_1, \alpha_2, \ldots) = \sum_{n=1}^{\infty} \alpha_n x_n \,.$$

By Theorem 5.26, the operator S is weakly compact, and so by our hypothesis TS is a compact operator. Now let $\{e_n\}$ denote the sequence of basic vectors of ℓ_1. Clearly, $TS(e_n) = Tx_n \xrightarrow{w} 0$ holds. On the other hand, since TS is compact, we see that every subsequence of $\{TSe_n\}$ has a subsequence converging in norm to zero. Therefore, $\|Tx_n\| \to 0$ holds, and this proves that T is a Dunford–Pettis operator. ∎

The Dunford–Pettis operators are related with the Dunford–Pettis property of Banach spaces. Following A. Grothendieck [72], we say that a Banach space X has the **Dunford–Pettis property** whenever $x_n \xrightarrow{w} 0$ in X and $x_n' \xrightarrow{w} 0$ in X' imply $\lim x_n'(x_n) = 0$. Again, Grothendieck's definition has been inspired by the paper of N. Dunford and P. J. Pettis [56]. From the identity

$$x_n'(x_n) = x_n'(x) + x'(x_n - x) + (x_n' - x')(x_n - x) \,,$$

it should be obvious that a Banach space X has the Dunford–Pettis property if and only if $x_n \xrightarrow{w} x$ in X and $x_n' \xrightarrow{w} x'$ in X' imply $x_n'(x_n) \to x'(x)$.

J. W. Brace [43] and A. Grothendieck [72] have characterized the Dunford–Pettis property as follows.

Theorem 5.82 (Brace–Grothendieck). *For a Banach space X the following statements are equivalent.*

(1) X *has the Dunford–Pettis property.*

(2) *Every weakly compact operator from X to an arbitrary Banach space maps weakly compact sets to norm compact sets.*

(3) *Every weakly compact operator from X to an arbitrary Banach space is a Dunford–Pettis operator.*

(4) *Every weakly compact operator from X to c_0 is a Dunford–Pettis operator.*

Proof. $(1) \Longrightarrow (2)$ Let $T: X \to Y$ be a weakly compact operator (where Y is a Banach space), and let W be a weakly compact subset of X. Clearly, $T(W)$ is norm closed. So, in order to establish that $T(W)$ is norm compact it suffices to show that every sequence in $T(W)$ has a norm convergent subsequence.

To this end, let $\{x_n\} \subseteq W$. By passing to a subsequence, we can assume that $x_n \xrightarrow{w} x$ holds in X. We claim that $\{Tx_n\}$ has a subsequence that converges in norm to Tx. To see this, for each n pick some $f_n \in Y'$ with $\|f_n\| = 1$ and

$$\left\|T(x_n - x)\right\| \le 2\left|f_n(T(x_n - x))\right|. \qquad (\star)$$

Since $T': Y' \to X'$ is a weakly compact operator, there exists a subsequence $\{f_{k_n}\}$ of $\{f_n\}$ such that $T'f_{k_n} \xrightarrow{w} f$ holds in X'. On the other hand, the Dunford–Pettis property of X implies

$$\lim_{n\to\infty} \left|T'f_{k_n}(x_{k_n} - x)\right| = |f(0)| = 0.$$

Therefore, it follows from $(\star)$ that $\lim \|T(x_{k_n} - x)\| = 0$. That is, the subsequence $\{Tx_{k_n}\}$ of $\{Tx_n\}$ converges in norm to Tx.

$(2) \Longrightarrow (3)$ This follows immediately from Theorem 5.81.

$(3) \Longrightarrow (4)$ Obvious.

$(4) \Longrightarrow (1)$ Assume $x_n \xrightarrow{w} 0$ in X and $x'_n \xrightarrow{w} 0$ in X'. Consider the operator $T: X \to c_0$ defined by

$$T(x) = \left(x'_1(x), x'_2(x), \dots\right).$$

By Theorem 5.26 the operator T is weakly compact, and so by our hypothesis T is also a Dunford–Pettis operator. In particular, $x_n \xrightarrow{w} 0$ in X implies $\|Tx_n\|_\infty \to 0$. Now a glance at the inequality $|x'_n(x_n)| \le \|Tx_n\|_\infty$ shows that $\lim x'_n(x_n) = 0$ holds. That is, X has the Dunford–Pettis property, and the proof of the theorem is finished. ∎

Theorem 5.83. *A reflexive Banach space with the Dunford–Pettis property is finite dimensional.*

Proof. If X is a reflexive Banach space with the Dunford–Pettis property, then the identity operator $I \colon X \to X$ is weakly compact, and so I must be a Dunford–Pettis operator. Since X is reflexive, the latter implies that I is a compact operator, and this guarantees that X is finite dimensional. ∎

If the dual of a Banach space has the Dunford–Pettis property, then the Banach space has the Dunford–Pettis. This is due to A. Grothendieck [**72**].

Theorem 5.84 (Grothendieck). *If the norm dual X' of a Banach space X has the Dunford–Pettis property, then X itself has the Dunford–Pettis property.*

Proof. Let X' have the Dunford–Pettis property, and consider two weakly compact operators $Z \overset{S}{\to} X \overset{T}{\to} Y$ (where Y and Z are Banach spaces). Taking adjoints we have $Y' \overset{T'}{\to} X' \overset{S'}{\to} Z'$ with S' and T' weakly compact. Since X' has the Dunford–Pettis property, Theorem 5.82 shows that S' is a Dunford–Pettis operator, and so $S'T' = (TS)'$ is a compact operator. Thus, TS is a compact operator, and by Theorem 5.81 the operator T must be a Dunford–Pettis operator. Now by Theorem 5.82 the Banach space X must have the Dunford–Pettis property. ∎

A. Grothendieck [**72**] also has shown that AL- and AM-spaces have the Dunford–Pettis property.

Theorem 5.85 (Grothendieck). *Every AL-space and every AM-space has the Dunford–Pettis property.*

Proof. Since the norm dual of an AL-space and the double norm dual of an AM-space are AM-spaces with units, according to Theorem 5.84, it is enough to establish the result when E is an AM-space with unit.

To this end, let E be an AM-space with unit e, let $x_n \overset{w}{\to} 0$ in E, and let $x'_n \overset{w}{\to} 0$ in E'. Pick some $M > 0$ such that $\|x_n\| \le M$ holds for all n. Let $\epsilon > 0$. Now the set $\{x'_1, x'_2, \ldots\}$ is a weakly relatively compact subset of E', and so by Theorem 4.41 there exists some $0 \le y' \in E'$ satisfying

$$\left\| (|x'_n| - y')^+ \right\| < \tfrac{\epsilon}{M}$$

for all n. Since the lattice operations of E are weakly sequentially continuous (see Theorem 4.31), we have $|x_n| \overset{w}{\to} 0$, and thus there exists some k

satisfying $y'(|x_n|) < \epsilon$ for all $n \geq k$. In particular, for $n \geq k$ we have

$$|x'_n(x_n)| \leq |x'_n|(|x_n|) = (|x'_n| - y')^+(|x_n|) + (|x'_n| \wedge y')(|x_n|)$$
$$\leq M \cdot \|(|x'_n| - y')^+\| + y'(|x_n|) < M \cdot \tfrac{\epsilon}{M} + \epsilon = 2\epsilon,$$

which shows that $x'_n(x_n) \to 0$ holds, as required. ∎

As an application of the preceding theorem, let us establish that the topological structures of AL- and AM-spaces differ in an essential manner.

Theorem 5.86. *An AL-space is linearly homeomorphic to an AM-space if and only if it is finite dimensional.*

Proof. Let E be an AM-space which is linearly homeomorphic to an AL-space. This implies that E is weakly sequentially complete, and so E is a KB-space. Since E' is an AL-space, it follows that E is a reflexive Banach lattice (Theorem 4.70). Now note that a reflexive Banach space with the Dunford–Pettis property is finite dimensional. ∎

The square of a weakly compact operator on a Banach space with the Dunford–Pettis property is a compact operator. The details are included in the next theorem.

Theorem 5.87. *Let $X \xrightarrow{S} Y \xrightarrow{T} Z$ be two weakly compact operators between Banach spaces. If Y has the Dunford–Pettis property, then TS is a compact operator.*

Proof. By Theorem 5.82 the operator T is Dunford–Pettis, and so TS is a compact operator. ∎

As a consequence of the preceding theorem we have the following classical result of N. Dunford and P. J. Pettis [56].

Corollary 5.88 (Dunford–Pettis). *If T is a weakly compact operator on an AL- or an AM-space, then T^2 is a compact operator.*

We now come to the following question first studied by the authors in [11]: *If a positive operator S between Banach lattices is dominated by a Dunford–Pettis operator, is then S necessarily Dunford–Pettis?*

In general, the answer is negative. For an example, consider the positive operators $S, T \colon L_1[0, 1] \to \ell_\infty$ defined by

$$S(f) = \left(\int_0^1 f(x) r_1^+(x)\, dx, \int_0^1 f(x) r_2^+(x)\, dx, \ldots \right)$$

and

$$T(f) = \left(\int_0^1 f(x)\, dx, \int_0^1 f(x)\, dx, \ldots \right),$$

where $\{r_n\}$ denotes the sequence of Rademacher functions on $[0,1]$. Clearly, $0 \leq S \leq T$ holds and T is a compact (and hence a Dunford–Pettis) operator. On the other hand, we have $r_n \xrightarrow{w} 0$ in $L_1[0,1]$ (why?). Therefore, in view of $\|Sr_n\|_\infty \geq \int_0^1 r_n(x)r_n^+(x)\,dx = \frac{1}{2}$, we see that S is not a Dunford–Pettis operator.

In the sequel, we shall investigate what effect a Dunford–Pettis operator has on the operators it dominates. Recall that a Banach lattice is said to have *weakly sequentially continuous lattice operations* whenever $x_n \xrightarrow{w} 0$ implies $|x_n| \xrightarrow{w} 0$. Every AM-space has this property (Theorem 4.31). Also, any Banach lattice with the *Schur property* (i.e., $x_n \xrightarrow{w} 0$ implies $\|x_n\| \to 0$) has weakly sequentially continuous lattice operations. Thus, for example, the Banach lattices $C[0,1]$, ℓ_1, and $\ell_1 \oplus C[0,1]$ all have weakly sequentially continuous lattice operations.

Theorem 5.89. *Let $S,T\colon E \to F$ be two positive operators between Banach lattices such that $0 \leq S \leq T$. If E has weakly sequentially continuous lattice operations and T is Dunford–Pettis, then S is likewise Dunford–Pettis.*

Proof. If $x_n \xrightarrow{w} 0$ holds in E, then $|x_n| \xrightarrow{w} 0$ also holds in E, and so $\lim \|T|x_n|\| = 0$. Using the inequalities $|Sx_n| \leq S|x_n| \leq T|x_n|$, we see that $\|Sx_n\| \leq \|T|x_n|\|$ for all n, from which we get $\lim \|Sx_n\| = 0$. $\blacksquare$

If a positive Dunford–Pettis operator has its range in a Banach lattice with order continuous norm, then every positive operator that it dominates is also Dunford–Pettis. This result is stated next, and is due to N. J. Kalton and P. Saab [83].

Theorem 5.90 (Kalton–Saab). *Let $S\colon E \to F$ be a positive operator between two Banach lattices such that F has order continuous norm. If S is dominated by a Dunford–Pettis operator, then S itself is Dunford–Pettis.*[1]

Proof. Assume that F has order continuous norm and that $T\colon E \to F$ is a Dunford–Pettis operator satisfying $0 \leq S \leq T$. Let $x_n \xrightarrow{w} 0$ in E, and let $\epsilon > 0$. Put $x = \sum_{n=1}^\infty 2^{-n}|x_n|$, and let E_x be the ideal generated by x in E. Also, let W denote the solid hull of the weakly relatively compact subset $\{x_1, x_2, \ldots\}$ of E. Clearly, $W \subseteq E_x$ holds. Next, note that if $\{y_n\}$ is a disjoint sequence of W, then (by Theorem 4.34) we have $y_n \xrightarrow{w} 0$ in E, and so $\|Ty_n\| \to 0$. Thus, by Theorem 4.36, there exists some $0 \leq u \in E_x$ such that

$$\|T(|x_n| - u)^+\| < \epsilon$$

holds for all n.

---

[1] This theorem also was proved by W. Haid [75] and B. de Pagter [159] under some extra assumptions.

Next, consider the operators $S, T \colon \overline{E}_x \to F$. Then, by Theorem 4.87, there exist operators $M_1, \ldots, M_k$ on $\overline{E}_x$ and positive operators $L_1, \ldots, L_k$ on F satisfying

$$\left\| \left(S - \sum_{i=1}^{k} L_i T M_i \right) u \right\| \leq \epsilon \quad \text{and} \quad 0 \leq \sum_{i=1}^{k} L_i T M_i \leq T \text{ on } \overline{E}_x .$$

Since each $M_i \colon \overline{E}_x \to \overline{E}_x$ is continuous for the norm induced by E, it is easy to see that $M_i(x_n) \xrightarrow{w} 0$ holds in E. Thus, using the fact that T is a Dunford–Pettis operator, we see that $\lim \| \sum_{i=1}^{k} L_i T M_i(x_n) \| = 0$. Pick some m such that $\| \sum_{i=1}^{k} L_i T M_i(x_n) \| < \epsilon$ holds for all $n \geq m$. Now note that for $n \geq m$ we have

$$
\begin{aligned}
\| S x_n \| &\leq \left\| \left(S - \sum_{i=1}^{k} L_i T M_i \right) x_n \right\| + \left\| \sum_{i=1}^{k} L_i T M_i(x_n) \right\| \\
&\leq \left\| \left| S - \sum_{i=1}^{k} L_i T M_i \right| (|x_n| - u)^+ \right\| + \left\| \left(S - \sum_{i=1}^{k} L_i T M_i \right) u \right\| + \epsilon \\
&\leq 2 \| T(|x_n| - u)^+ \| + \epsilon + \epsilon < 4\epsilon ,
\end{aligned}
$$

and so $\| S x_n \| \to 0$ holds, as desired. ∎

It is interesting to know that the preceding theorem can be used to prove Theorem 5.20. To see this, assume that E and F are two Banach lattices such that E' and F have order continuous norms, and let $S \colon E \to F$ be a positive operator dominated by a compact operator. Then, according to Theorem 5.44, the operator S factors through a reflexive Banach lattice G

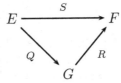

with R positive and dominated by a compact operator. By Theorem 5.90 the operator R is Dunford–Pettis, and hence since G is a reflexive Banach lattice, R is a compact operator. Therefore, $S = RQ$ is a compact operator.

Every Dunford–Pettis operator maps order intervals to weakly relatively compact sets.

Theorem 5.91. *Every Dunford–Pettis operator $T \colon E \to X$ from a Banach lattice to a Banach space is o-weakly compact (i.e., it carries order intervals to weakly relatively compact sets).*

Proof. If $\{x_n\}$ is an order bounded disjoint sequence of E, then $x_n \xrightarrow{w} 0$ holds in E, and so $\lim \|Tx_n\| = 0$. The conclusion now follows immediately from Theorem 5.57. ∎

An immediate consequence of the preceding theorem is that the square of a Dunford–Pettis operator on a Banach lattice carries order intervals to norm totally bounded sets. In a more general context we have the following result.

Corollary 5.92. *Let* $E \xrightarrow{T} X \xrightarrow{S} Y$ *be two Dunford–Pettis operators between Banach spaces. If* E *is a Banach lattice, then* $ST[0, x]$ *is a norm totally bounded subset of* Y *for each* $x \in E^+$.

The next result is due to N. J. Kalton and P. Saab [**83**].

Theorem 5.93 (Kalton–Saab). *Consider the scheme of operators*

$$E \xrightarrow{S_1} F \xrightarrow{S_2} X ,$$

where E *and* F *are Banach lattices and* S_1 *is a positive operator. If* S_1 *is dominated by a Dunford–Pettis operator and* S_2 *is o-weakly compact, then* $S_2 S_1$ *is a Dunford–Pettis operator.*

Proof. By Theorem 5.58 the operator S_2 admits a factorization through a Banach lattice G with order continuous norm

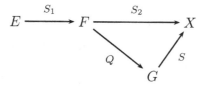

such that Q is a lattice homomorphism. Clearly, the positive operator $QS_1 \colon E \to G$ is dominated by a Dunford–Pettis operator. Thus, by Theorem 5.90, the operator QS_1 is also Dunford–Pettis, and consequently $S_2 S_1 = S(QS_1)$ is likewise a Dunford–Pettis operator. ∎

The authors proved in [**11**] that if a positive operator on a Banach lattice is dominated by a Dunford–Pettis operator, then its third power is a Dunford–Pettis operator. The next consequence of Theorem 5.93 is an improvement of this result and is due to N. J. Kalton and P. Saab [**83**].

Corollary 5.94. *If a positive operator* S *on a Banach lattice is dominated by a Dunford–Pettis operator, then* S^2 *also is a Dunford–Pettis operator.*

The next result is taken from [**15**] and presents another variation of Theorem 5.14 (see also Theorem 5.50).

Theorem 5.95. *Consider the scheme of positive operators*

$$E \xrightarrow{S_1} F \xrightarrow{S_2} G \xrightarrow{S_3} H$$

between Banach lattices. If S_1 is dominated by a weakly compact operator, S_2 by a compact operator, and S_3 by a Dunford–Pettis operator, then $S_3 S_2 S_1$ is a compact operator.

Proof. The proof is based upon the following diagram:

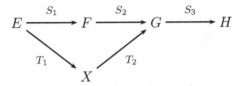

According to Theorem 5.45, the scheme of operators $E \xrightarrow{S_1} F \xrightarrow{S_2} G$ factors through a reflexive Banach lattice X with T_2 positive and dominated by a positive compact operator. By Theorem 5.93 the operator $S_3 T_2$ is Dunford–Pettis, and hence $S_3 S_2 S_1 = (S_3 T_2) T_1$ must be a compact operator. ∎

In the preceding theorem, the domination properties of the first and last operators cannot be reversed. For instance, if S_1 and S_2 are the operators of Example 5.17, then in the scheme of operators

$$\ell_1 \xrightarrow{I} \ell_1 \xrightarrow{S_1} L_2[0,1] \xrightarrow{S_2} \ell_\infty,$$

the identity operator I is a positive Dunford–Pettis operator and S_1 and S_2 are both positive and dominated by compact operators. However, as in Example 5.17, it is easy to see that $S_2 S_1 I$ is not a compact operator.

We continue with a useful weak continuity property of regular operators. It is due to the authors [11].

Theorem 5.96 (Aliprantis–Burkinshaw). *Let $T \colon E \to F$ be a regular operator between two Banach lattices such that $T[0,x]$ is $|\sigma|(F,F')$-totally bounded for each $x \in E^+$. If $x_n \xrightarrow{w} 0$ holds in E, then $|Tx_n| \xrightarrow{w} 0$ in F.*

Proof. Assume $x_n \xrightarrow{w} 0$ in E. Let $0 \le f \in F'$, and fix $\epsilon > 0$. By Theorem 4.37 there exists some $x \in E^+$ such that

$$|T'|f(|x_n| - x)^+ < \epsilon$$

holds for all n. Also, from Theorem 3.27, it follows that the set $T'[-f,f]$ is $|\sigma|(E',E)$-totally bounded (see Exercise 8 of Section 3.3). So, there exists a finite set $\{g_1, \ldots, g_k\} \subseteq [-f,f]$ such that for each $g \in [-f,f]$ we have $|T'(g - g_i)|(x) < \epsilon$ for at least one $1 \le i \le k$. Since $x_n \xrightarrow{w} 0$ holds in E, there exists some m with $|g_i(Tx_n)| < \epsilon$ for each $i = 1, \ldots, k$ and all $n \ge m$.

Now let $g \in [-f, f]$. Choose $1 \le i \le k$ with $|T'(g - g_i)|(x) < \epsilon$, and note that $|g - g_i| \le f$ holds. In particular, for $n \ge m$ we have

$$
\begin{aligned}
|g(Tx_n)| &\le |T'(g - g_i)(x_n)| + |g_i(Tx_n)| \\
&\le |T'(g - g_i)|(|x_n| - x)^+ + |T'(g - g_i)|(x) + \epsilon \\
&\le |T'|f(|x_n| - x)^+ + 2\epsilon \le 3\epsilon.
\end{aligned}
$$

In view of Theorem 1.23, the latter shows that

$$
f(|Tx_n|) = \max\{|g(Tx_n)|: \; -f \le g \le f\} \le 3\epsilon
$$

holds for all $n \ge m$. That is, $|Tx_n| \xrightarrow{\;w\;} 0$ holds in F, as desired. ∎

P. G. Dodds and D. H. Fremlin [54] and J. Bourgain [42] proved that the Dunford–Pettis operators on an AL-space form a band. This was generalized by the authors in [11] as follows.

Theorem 5.97 (Aliprantis–Burkinshaw). *Let E be a Banach lattice with order continuous norm, and let F be an AL-space. Then, a regular operator $T: E \to F$ is a Dunford–Pettis operator if and only if T maps order intervals to norm compact sets.*

In particular, in this case, the vector space of all Dunford–Pettis operators of $\mathcal{L}_b(E, F)$ is a band.

Proof. Assume first that T is Dunford–Pettis operator. Since E has order continuous norm, we know that the order intervals of E are weakly compact, and so T must map order intervals to norm compact sets.

For the converse, let T map order intervals of E to norm compact subsets of F, and let $x_n \xrightarrow{\;w\;} 0$ in E. By Theorem 5.96 we have $|Tx_n| \xrightarrow{\;w\;} 0$ in F. Now if e' is the unit of F', then $\|Tx_n\| = e'(|Tx_n|) \to 0$, and so T is a Dunford–Pettis operator. The last part follows immediately from Theorem 5.10. ∎

Another class of operators related to the Dunford–Pettis operators is the class of weak Dunford–Pettis operators which was introduced by the authors in [11]. An operator $T: X \to Y$ between two Banach spaces is said to be a **weak Dunford–Pettis operator** whenever $x_n \xrightarrow{\;w\;} 0$ in X and $y'_n \xrightarrow{\;w\;} 0$ in Y' imply $\lim\langle Tx_n, y'_n \rangle = 0$ (or, equivalently, whenever $x_n \xrightarrow{\;w\;} x$ in X and $y'_n \xrightarrow{\;w\;} y'$ in Y' imply $\lim\langle Tx_n, y'_n \rangle = \langle Tx, y' \rangle$).

Clearly, a Dunford–Pettis operator is a weak Dunford–Pettis operator. If Y is reflexive, then an easy application of Theorem 3.40 shows that the notions of *weak Dunford–Pettis* and *Dunford–Pettis* operator coincide. Also, if X has the Dunford–Pettis property, then every continuous operator from X to Y is a weak Dunford–Pettis operator. The weak Dunford–Pettis operators enjoy properties similar to those of compact and Dunford–Pettis operators. For instance, the weak Dunford–Pettis operators on a Banach space X form a closed ring ideal of $L(X)$.

A weak Dunford–Pettis operator need not be a Dunford–Pettis operator. For example, let $\{r_n\}$ denote the sequence of Rademacher functions of $[0, 1]$, and let $S \colon L_1[0,1] \to \ell_\infty$ be the positive operator defined by

$$S(f) = \left(\int_0^1 f(x) r_1^+(x)\, dx, \int_0^1 f(x) r_2^+(x)\, dx, \ldots \right).$$

From $r_n \xrightarrow{w} 0$ and $\|Sr_n\| \ge \int_0^1 r_n(x) r_n^+(x) = \frac{1}{2}$, we see that S is not a Dunford–Pettis operator. However, since $L_1[0,1]$ has the Dunford–Pettis property, it is automatically true that S is a weak Dunford–Pettis operator.

Our next objective is to connect the weak Dunford–Pettis operators with the class of Dunford–Pettis sets. Following K. T. Andrews [25] we say that a norm bounded subset A of a Banach space X is a **Dunford–Pettis set** whenever every weakly compact operator from X to an arbitrary Banach space carries A to a norm totally bounded set. The Dunford–Pettis sets were characterized by K. T. Andrews in [25] as follows.

Theorem 5.98 (Andrews). *For a norm bounded subset A of a Banach space X the following statements are equivalent.*

(1) *A is a Dunford–Pettis set.*

(2) *Every weakly compact operator from X to c_0 carries A to a norm totally bounded set.*

(3) *Every sequence $\{x_n'\} \subseteq X'$ satisfying $x_n' \xrightarrow{w} 0$ in X' converges uniformly to zero on the set A.*

Proof. (1) $\Longrightarrow$ (2) Obvious.

(2) $\Longrightarrow$ (3) Let $x_n' \xrightarrow{w} 0$ in X'. Consider the operator $T \colon X \to c_0$ defined by

$$T(x) = \left(x_1'(x), x_2'(x), \ldots \right).$$

By Theorem 5.26 the operator T is weakly compact. But then, according to our hypothesis, $T(A)$ is a norm totally bounded subset of c_0, and from this it follows that $\sup\{|x_n'(x)| \colon x \in A\} \to 0$ (see Exercise 14 of Section 3.2).

(3) $\Longrightarrow$ (1) Let Y be a Banach space, and let $T \colon X \to Y$ be a weakly compact operator. By Theorem 5.38, we can assume without loss of generality that Y is reflexive.

Now assume by way of contradiction that $T(A)$ is not a norm totally bounded subset of Y. Then, since Y is reflexive, there exist a sequence $\{x_n\}$ in A, some $y \in Y$, and some $\epsilon > 0$ satisfying $Tx_n \xrightarrow{w} y$ and $\|Tx_n - y\| > \epsilon$ for all n. For each n choose some $y_n' \in Y'$ with $\|y_n'\| = 1$ and

$y_n'(Tx_n - y) > \epsilon$. Since Y' is reflexive, by passing to a subsequence, we can assume that $y_n' \overset{w}{\longrightarrow} y'$ holds in Y'. Thus, $T'y_n' \overset{w}{\longrightarrow} T'y'$ holds in X', and from our hypothesis we see that

$$(y_n' - y')(Tx_n) = (T'y_n' - T'y')(x_n) \to 0.$$

Consequently, we have

$$0 < \epsilon < y_n'(Tx_n - y) = (y_n' - y')(Tx_n) + y'(Tx_n - y) + (y' - y_n')(y) \to 0,$$

which is impossible. Therefore, $T(A)$ is a norm totally bounded subset of Y, and the proof of the theorem is finished. ∎

The next result characterizes the weak Dunford–Pettis operators and is the analogue of Theorem 5.81.

Theorem 5.99. *For a continuous operator $T: X \to Y$ between two Banach spaces the following statements are equivalent:*

(1) *T is a weak Dunford–Pettis operator.*

(2) *T carries weakly compact subsets of X to Dunford–Pettis subsets of Y.*

(3) *If S is a weakly compact operator from Y to an arbitrary Banach space, then ST is a Dunford–Pettis operator.*

Proof. (1) $\implies$ (2) Let W be a weakly compact subset of X, and let $y_n' \overset{w}{\longrightarrow} 0$ in Y'. If $\{y_n'\}$ does not converge uniformly to zero on $T(W)$, then there exist a sequence $\{x_n\}$ of W, a subsequence of $\{y_n'\}$ (which we shall denote by $\{y_n'\}$ again), and some $\epsilon > 0$ satisfying $|y_n'(Tx_n)| > \epsilon$ for all n. Since W is weakly compact, we can assume that $x_n \overset{w}{\longrightarrow} x$ holds in X. Then $Tx_n \overset{w}{\longrightarrow} Tx$ holds in Y and so, since T is a weak Dunford–Pettis operator, we have

$$0 < \epsilon < |y_n'(Tx_n)| \to 0,$$

which is impossible. Thus, $\{y_n'\}$ converges uniformly to zero on $T(W)$, and so by Theorem 5.98 it is a Dunford–Pettis set.

(2) $\implies$ (3) Let Z be a Banach space, and let $S: Y \to Z$ be a weakly compact operator. Also, assume $x_n \overset{w}{\longrightarrow} 0$ in X. Since the set $A = \{0, x_1, x_2, \ldots\}$ is weakly compact, our hypothesis implies that $T(A)$ is a Dunford–Pettis set, and so $ST(A)$ is a norm totally bounded subset of Z. In view of $STx_n \overset{w}{\longrightarrow} 0$ in Z, it follows that $\|STx_n\| \to 0$, and so ST is a Dunford–Pettis operator.

(3) $\implies$ (1) Let $x_n \overset{w}{\longrightarrow} 0$ in X, and let $y_n' \overset{w}{\longrightarrow} 0$ in Y'. Consider the operator $S: Y \to c_0$ defined by

$$S(y) = (y_1'(y), y_2'(y), \ldots).$$

Then S is weakly compact (see Theorem 5.26), and so by our hypothesis ST is a Dunford–Pettis operator. Thus, $\lim \|STx_n\|_\infty = 0$ and the desired conclusion follows from the inequality $|y_n'(Tx_n)| \le \|STx_n\|_\infty$. ∎

The positive weak Dunford–Pettis operators enjoy some interesting lattice properties.

Theorem 5.100. *Let $T\colon E \to F$ be a positive weak Dunford–Pettis operator between Banach lattices. If $W \subseteq E$ and $V \subseteq F'$ are two weakly relatively compact sets, then the following hold:*

(1) *For every disjoint sequence $\{x_n\}$ in the solid hull of W, the sequence $\{Tx_n\}$ converges uniformly to zero on the solid hull of V.*

(2) *For each $\epsilon > 0$ there exists some $u \in E^+$ satisfying*

$$|f|\big(T(|x| - u)^+\big) < \epsilon$$

for all $x \in W$ and all $f \in V$.

Proof. (1) Let $\{x_n\} \subseteq E^+$ be a disjoint sequence in the solid hull of W, and let $\epsilon > 0$. We claim that there exist $0 \le g \in F'$ and a natural number k such that

$$(|f| - g)^+ (Tx_n) < \epsilon \qquad\qquad (\star)$$

holds for all $f \in V$ and all $n > k$.

To see this, assume by way of contradiction that $(\star)$ is false. That is, assume that for each $0 \le g \in F'$ and each k there exist $f \in V$ and $m > k$ with $(|f| - g)^+ (Tx_m) \ge \epsilon$. An easy inductive argument shows that there exist a sequence $\{f_n\} \subseteq V$ and a subsequence $\{y_n\}$ of $\{x_n\}$ such that

$$\left(|f_{n+1}| - 4^n \sum_{i=1}^{n} |f_i|\right)^+ (Ty_n) \ge \epsilon$$

holds for all n. Let $f = \sum_{n=1}^{\infty} 2^{-n}|f_n|$ and $h_n = \left(|f_{n+1}| - 4^n \sum_{i=1}^{n} |f_i|\right)^+$. Clearly, $h_n(Ty_n) \ge \epsilon$ holds for all n. Next, put

$$g_n = \left(|f_{n+1}| - 4^n \sum_{i=1}^{n} |f_i| - 2^{-n}f\right)^+,$$

and note that (by Lemma 4.35) the sequence $\{g_n\}$ is disjoint and lies in the solid hull of V. Thus, by Theorem 4.34, we see that $g_n \xrightarrow{\;w\;} 0$ holds in F'. Also, by Theorem 4.34 we have $x_n \xrightarrow{\;w\;} 0$ in E. This implies $y_n \xrightarrow{\;w\;} 0$ in E and $Ty_n \xrightarrow{\;w\;} 0$ in F. Since T is a weak Dunford–Pettis operator, it follows that $\lim g_n(Ty_n) = 0$. On the other hand, the inequality $0 \le h_n \le g_n + 2^{-n}f$ implies

$$0 < \epsilon \le h_n(Ty_n) \le g_n(Ty_n) + 2^{-n}f(Ty_n) \to 0,$$

which is impossible. Therefore, $(\star)$ is true.

Next, pick $0 \leq g \in F'$ and k so that $(\star)$ is valid, and then choose $m > k$ such that $g(Tx_n) < \epsilon$ holds for all $n \geq m$. Now if $h \in \text{Sol}(V)$, then pick some $f \in V$ with $|h| \leq |f|$, and note that

$$|h(Tx_n)| \leq |h|(Tx_n) \leq |f|(Tx_n) \leq (|f| - g)^+(Tx_n) + g(Tx_n) \leq \epsilon + \epsilon = 2\epsilon$$

holds for all $n \geq m$. This shows that $\{Tx_n\}$ converges uniformly to zero on the solid hull of V.

(2) Consider the seminorm ρ on E defined by

$$\rho(x) = \sup\{|f|(|x|) : f \in V\}, \quad x \in E.$$

Clearly, ρ is a continuous seminorm on E. Also, by part (1) we have $\lim \rho(Tx_n) = 0$ for each disjoint sequence $\{x_n\}$ in the solid hull of W. Therefore, by Theorem 4.36 there exists some $u \in E^+$ satisfying

$$|f|(T(|x| - u)^+) < \epsilon$$

for all $f \in V$ and all $x \in W$. ∎

The weak Dunford–Pettis property of a positive operator is inherited by the positive operators it dominates. This is due to N. J. Kalton and P. Saab [83].

Theorem 5.101 (Kalton–Saab). *If a positive operator S is dominated by a weak Dunford-Pettis operator, then S is a weak Dunford-Pettis operator.*

Proof. Let $S, T : E \to F$ be two positive operators between Banach lattices such that $0 \leq S \leq T$ holds and T is weak Dunford–Pettis. Let $x_n \xrightarrow{w} 0$ in E and $f_n \xrightarrow{w} 0$ in F'.

Put $x = \sum_{n=1}^{\infty} 2^{-n}|x_n|$, and consider the ideal E_x generated by x. Then $x_n \xrightarrow{w} 0$ holds in $\overline{E}_x$ (the norm closure of E_x). Thus, by restricting S and T on $\overline{E}_x$, we can assume without loss of generality that E has a quasi-interior point.

Now let $\epsilon > 0$. By Theorem 5.100 there exists some $u \in E^+$ such that

$$|f_n|(T(|x_n| - u)^+) < \epsilon$$

for all n. Also, since $f_n \xrightarrow{w} 0$ holds in F', there exists (by Theorem 4.37) some $0 \leq \phi \in F'$ satisfying

$$(|f_n| - \phi)^+(Tu) < \epsilon$$

for all n. Next, consider the operators $S, T : E \to F''$ and note that $0 \leq S \leq T$ holds in $\mathcal{L}_b(E, F'')$. Thus, by Theorem 4.82 there exist positive operators $M_1, \ldots, M_k$ on E and order projections $P_1, \ldots, P_k$ on F'' with

$$0 \leq \sum_{i=1}^{k} P_i T M_i \leq T \quad \text{and} \quad \phi\left(\left|S - \sum_{i=1}^{k} P_i T M_i\right| u\right) < \epsilon.$$

Put $R = \left| S - \sum_{i=1}^{k} P_i T M_i \right|$, and note that $0 \leq R \leq 2T$. Clearly,

$$
\begin{aligned}
|f_n|(R|x_n|) &\leq |f_n|(R(|x_n| - u)^+) + |f_n|(Ru) \\
&\leq 2|f_n|(T(|x_n| - u)^+) + (|f_n| - \phi)^+(2Tu) + \phi(Ru) \\
&< 2\epsilon + 2\epsilon + \epsilon = 5\epsilon
\end{aligned}
$$

holds for all n. The latter inequality easily implies

$$
\left| f_n(Sx_n) \right| < 5\epsilon + \sum_{i=1}^{k} |f_n(P_i T M_i x_n)| \qquad (\star\star)
$$

for all n.

Now for each $i = 1, \ldots, k$ define an operator $R_i \colon F' \to F'$ by the formula

$$
[R_i f](y) = f(P_i y), \quad f \in F' \text{ and } y \in F.
$$

From the inequalities

$$
|R_i f(y)| \leq \|f\| \cdot \|P_i y\| \leq \|f\| \cdot \|P_i\| \cdot \|y\| \leq \|f\| \cdot \|y\|,
$$

we see that $\|R_i f\| \leq \|f\|$ holds for all $f \in F'$, and so each R_i is a continuous operator. In particular, $f_n \xrightarrow{w} 0$ in F' implies $R_i f_n \xrightarrow{w} 0$ in F' for each $i = 1, \ldots, k$. Also, for each i we have $M_i x_n \xrightarrow{w} 0$ in E, and hence, taking into account that T is a weak Dunford–Pettis operator, we infer that

$$
f_n(P_i T M_i x_n) = R_i f_n(T M_i x_n) \to 0.
$$

Consequently, it follows from $(\star\star)$ that $\limsup |f_n(Sx_n)| \leq 5\epsilon$. Since $\epsilon > 0$ is arbitrary, the latter implies $\lim f_n(Sx_n) = 0$, as desired. ∎

Finally, we close our discussion with the reciprocal Dunford–Pettis property. by Theorem 5.82, we know that a Banach space X has the Dunford–Pettis property if and only if every weakly compact operator from X to any Banach space is a Dunford–Pettis operator. Following A. Grothendieck [72], we say that a Banach space X has the **reciprocal Dunford–Pettis property** whenever every Dunford–Pettis operator from X to any Banach space is weakly compact.

The Banach lattices with the reciprocal Dunford–Pettis property are precisely the Banach lattices whose duals have order continuous norms. This is implicitly contained in the paper of C. P. Niculescu [154].

Theorem 5.102. *A Banach lattice E has the reciprocal Dunford–Pettis property if and only if E' has order continuous norm.*

Proof. Assume first that E has the reciprocal Dunford–Pettis property. If $T \colon E \to \ell_1$ is a continuous operator, then it follows from Theorem 4.32 that T is a Dunford–Pettis operator, and so by our hypothesis T is weakly compact. By part (2) of Theorem 5.29, the Banach lattice E' has order continuous norm.

For the converse, assume that E' has order continuous norm. Let X be a Banach space, and let $T\colon E \to X$ be a Dunford–Pettis operator. We claim that T is an M-weakly compact operator. To see this, let $\{x_n\}$ be a norm bounded disjoint sequence of E. Then $x_n \xrightarrow{w} 0$ holds (see Exercise 7 of Section 4.3), and so using that T is Dunford–Pettis we infer that $\lim \|Tx_n\| = 0$. Hence, T is M-weakly compact, and so by Theorem 5.61 the operator T is weakly compact. So, E has the reciprocal Dunford–Pettis property. ∎

It is interesting to observe that AM-spaces have the reciprocal Dunford–Pettis property, while only the finite dimensional AL-spaces have the reciprocal Dunford–Pettis property. For some applications to functional analysis of Dunford–Pettis operators we refer the reader to the book by J. J. Diestel and J. J. Uhl, Jr. [**52**].

Exercises

1. If $X_1, \ldots, X_n$ are Banach spaces with the Dunford–Pettis property, then show that $X_1 \oplus \cdots \oplus X_n$ also has the Dunford–Pettis property. Does the L_p-sum of a sequence of Banach spaces with the Dunford–Pettis property have the Dunford–Pettis property?

2. Show that a Banach space X has the Dunford–Pettis property if and only if every weakly compact subset of X' is relatively $\tau(X', X)$-compact.

3. For a Banach space X with the Dunford–Pettis property prove the following statements.
   (a) If $x_n \xrightarrow{w} 0$ holds in X, then $\{x_n\}$ converges uniformly to zero on every weakly compact subset of X'.
   (b) If $x'_n \xrightarrow{w} 0$ holds in X', then $\{x'_n\}$ converges uniformly to zero on every weakly compact subset of X.

4. Give an example of a continuous operator $T\colon X \to Y$ between Banach spaces that is not Dunford–Pettis while its adjoint T' is a Dunford–Pettis operator. [*Hint:* Consider the identity operator $I\colon c_0 \to c_0$.]

5. Let E be a Banach lattice such that E' has order continuous norm. If E has the Dunford–Pettis property, then show that:
   (a) $x_n \xrightarrow{w} 0$ in E implies $|x_n| \xrightarrow{w} 0$.
   (b) $|x'_n| \xrightarrow{w} 0$ in E' implies $\|x'_n\| \to 0$.

6. Let E be a Banach lattice such that E and E' both have order continuous norms. Then show that every Dunford–Pettis operator from E into an arbitrary Banach space is compact.

7. If X is Banach space, then for an operator $T\colon c_0 \to X$ show that the following statements are equivalent.

    (a) T is weakly compact.
    (b) T is Dunford–Pettis.
    (c) T is compact.

8. Let X be a Banach space. Then, show that every continuous operator from ℓ_1 to X and every continuous operator from X to ℓ_1 is Dunford–Pettis.

9. Let $T\colon E \to F$ be a positive Dunford–Pettis operator between two Banach lattices such that E' and F both have order continuous norms. Show that:
    (a) T has order continuous norm.
    (b) If E is either Dedekind σ-complete or has a quasi-interior point, then $\mathcal{A}_T \subseteq \mathrm{Ring}(T)$.

10. Consider the scheme of positive operators

$$E \xrightarrow{\ S_1\ } F \xrightarrow{\ S_2\ } G$$

between Banach lattices. If S_1 is dominated by a Dunford–Pettis operator and S_2 by a weakly compact operator, then show that $S_2 S_1$ is Dunford–Pettis.

11. Consider the scheme of operators

$$E \xrightarrow{\ S_1\ } F \xrightarrow{\ S_2\ } X$$

between Banach spaces such that E and F are Banach lattices. If E' and F have order continuous norms, S_1 is positive, and S_2 is Dunford–Pettis, then show that $S_2 S_1$ is a compact operator. [*Hint*: Combine Theorems 5.79 and 5.28.]

12. For a continuous operator $T\colon X \to Y$ between Banach spaces establish the following statements.
    (a) If X has the Dunford–Pettis property, then the operator T is weak Dunford–Pettis.
    (b) If Y is reflexive, then T is Dunford–Pettis if and only if T is weak Dunford–Pettis.

13. (Aliprantis–Burkinshaw [**9, 11**]) If $S, T\colon E \to E$ are two positive operators on a Banach lattice such that $0 \le S \le T$ holds and T is compact, then show that:
    (a) S^3 is a compact operator (although S^2 need not be compact).
    (b) S^2 is Dunford–Pettis and weakly compact (although S need not be).
    (c) S is a weak Dunford–Pettis operator.

14. Establish the following properties about weak Dunford–Pettis operators.
    (a) If $X \xrightarrow{\ T\ } Y \xrightarrow{\ S\ } Z$ are continuous operators between Banach spaces and either T or S is weak Dunford–Pettis, then ST is likewise a weak Dunford–Pettis operator.
    (b) If X and Y are two arbitrary Banach spaces, then the set of all weak Dunford–Pettis operators from X to Y is a norm closed vector subspace of $L(X, Y)$.

15. Following N. J. Kalton and A. Wilansky [**84**], we say that an operator
    $T : X \to Y$ between Banach spaces is a **Tauberian operator** whenever
    $(T'')^{-1}(Y) = X$ holds (i.e., if $x'' \in X''$ and $T''(x'') \in Y$ imply $x'' \in X$).
      Establish the following properties about Tauberian operators.
    (a) A continuous operator T on a Banach space is Tauberian if and only
        if T^n is Tauberian for each n.
    (b) A Banach space X is reflexive if and only if there exists a Tauberian
        weakly compact operator on X.
    (c) Let E be a Banach lattice. If $T : E \to X$ is a Tauberian Dunford–
        Pettis operator from E into a Banach space X, then E is a KB-
        space. (Note that E need not be reflexive; for instance, the identity
        operator on ℓ_1 is both Tauberian and Dunford–Pettis.)
    (d) Let $S, T : E \to F$ be two positive operators between Banach lattices
        such that $0 \leq S \leq T$ holds. If S is Tauberian and T is Dunford–
        Pettis, then E is a KB-space.

16. (Aliprantis–Burkinshaw [**17**]) This exercise generalizes Theorems 5.14,
    5.50, and 5.95. Consider the scheme of positive operators
    $$E \xrightarrow{S_1} F \xrightarrow{S_2} G \xrightarrow{S_3} H$$
    between Banach lattices. If S_2 is dominated by a compact operator and
    $(S_1)'$ and S_3 are both o-weakly compact, then show that $S_3 S_2 S_1$ is a com-
    pact operator. [*Hint*: Use Theorem 5.58 and Exercise 11 of Section 5.4.]

Bibliography

1. Y. A. Abramovich, Injective envelopes of normed lattices, *Soviet Math. Dokl.* **12** (1971), 511–514. MR **44** # 7257

2. Y. A. Abramovich, Weakly compact sets in topological K-spaces, *Teor. Funkciĭ Funkcional. Anal. i Priložen* **15** (1972), 27–35 (Russian). MR **46** # 5982

3. Y. A. Abramovich, Multiplicative representation of disjointness preserving operators, *Indag. Math.* **45** (1983), 265–279. MR 85f: 47040

4. Y. A. Abramovich, A. I. Veksler, and V. Koldunov, On operators preserving disjointness, *Soviet Math. Dokl.* **20** (1979), 1089–1093. MR 81e: 47034

5. L. Alaoglu, Weak compactness of normed linear spaces, *Ann. of Math.* **41** (1940), 252–267. MR **1**, 241

6. C. D. Aliprantis, On order properties of order bounded transformations, *Canadian J. Math.* **27** (1975), 666–678. MR **51** # 11183

7. C. D. Aliprantis and O. Burkinshaw, *Locally Solid Riesz Spaces with Applications to Economics*, Math Surveys and Monographs, Volume # 105, American Math. Society, 2003. MR 2005b:46010.

8. C. D. Aliprantis and O. Burkinshaw, *Principles of Real Analysis*, 3$^{\rm rd}$ Edition, Academic Press, New York and London, 1998. MR 00d:28001

9. C. D. Aliprantis and O. Burkinshaw, Positive compact operators on Banach lattices, *Math. Z.* **174** (1980), 289–298. MR 81m: 47053

10. C. D. Aliprantis and O. Burkinshaw, On weakly compact operators on Banach lattices, *Proc. Amer. Math. Soc.* **83** (1981), 573–578. MR 82j: 47057

11. C. D. Aliprantis and O. Burkinshaw, Dunford–Pettis operators on Banach lattices, *Trans. Amer. Math. Soc.* **274** (1982), 227–238. MR 824b: 47045

12. C. D. Aliprantis and O. Burkinshaw, On positive order continuous operators, *Indag. Math.* **45** (1983), 1–6. MR 84e: 47053

13. C. D. Aliprantis and O. Burkinshaw, Some remarks on orthomorphisms, *Colloq. Math.* **47** (1982), 255–265. MR 85b: 47039

14. C. D. Aliprantis and O. Burkinshaw, The components of a positive operators, *Math. Z.* **184** (1983), 245–257. MR 85b: 47040

15. C. D. Aliprantis and O. Burkinshaw, Factoring compact and weakly compact operators through reflexive Banach lattices, *Trans. Amer. Math. Soc.* **283** (1984), 369–381. MR 85e: 47025

16. C. D. Aliprantis and O. Burkinshaw, Projecting onto the band of kernel operators, *Houston J. Math.* **11** (1985), 7–13. MR 86d: 47045

17. C. D. Aliprantis and O. Burkinshaw, On the ring ideal generated by a positive operator, *J. Funct. Anal.* **67** (1986), 60–72. MR 87h: 47085

18. C. D. Aliprantis, O. Burkinshaw, and P. Kranz, On lattice properties of the composition operator, *Manuscripta Math.* **36** (1981), 19–31. MR 83b: 47048

19. I. Amemiya, A general spectral theory in semi-ordered linear spaces, *J. Fac. Sci. Hokkaido Univ., Ser. I,* **12** (1953), 111–156. MR **15**, 137

20. T. Andô, Positive linear operators in semi-ordered linear spaces, *J. Fac. Sci. Hokkaido Univ., Ser. I,* **13** (1957), 214–228. MR **19**, 1067

21. T. Andô, On compactness of integral operators, *Indag. Math.* **24** (1962), 235–239. MR **25** # 2456

22. T. Andô, Extensions of linear functionals on Riesz spaces, *Indag. Math.* **27** (1965), 388–395. MR **31** # 3848

23. T. Andô, Contractive projections in L_p-spaces, *Pacific J. Math.* **17** (1966), 391–405. MR **33** # 566

24. T. Andô, Banachverbänden und positive projection, *Math. Z.* **109** (1969), 121–130. Zbl **174**, 168

25. K. T. Andrews, Dunford–Pettis sets in the space of Bochner integrable functions, *Math. Ann.* **241** (1979), 35–41. MR 80f: 46041

26. W. Arendt, *Über das spektrum regularer Operatoren*, Ph.D. Dissertation, University of Tübingen, 1979.

27. R. Arens, Duality in linear spaces, *Duke Math. J.* **14** (1947), 787–794. MR **9**, 241

28. J. Avron, I. Herbst, and B. Simon, Schrödinger operators with magnetic fields, I. General interactions, *Duke Math. J.* **45** (1978), 847–883. MR 80k: 35054

29. I. A. Bahtin, M. A. Krasnoselkii, and V. Y. Stecenko, On the continuity of positive linear operators, *Sibirsk. Mat. Z.* **3** (1962), 156–160 (Russian). MR **25** # 2451

30. S. Banach, *Théorie des Opérations Linéaires*, Monografje Matematyczne, Warsaw, 1932. (Reprinted by Chelsea Publishing Co., New York, 1955.) Zbl **5**, 209

31. S. J. Bernau, Orthomorphisms of Archimedean vector lattices, *Math. Proc. Cambridge Philos. Soc.* **89** (1981), 119–128. MR 82b: 47043

32. C. Bessaga and A. Pelczynski, On bases and unconditional convergence of series in Banach spaces, *Studia Math.* **17** (1958), 151–164. MR **22** # 5872

33. A. Bigard, Les othomorphismes d'un espace réticulé Archimédien, *Indag. Math.* **34** (1972), 236–246. MR **46** # 7115

34. A. Bigard and K. Keimel, Sur les endomorphismes conversant les polaires d'un group réticulé Archimédien, *Bull. Soc. Math. France* **97** (1969), 381–398. MR **41** # 6747

35. A. Bigard, K. Keimel, and S. Wolfenstein, *Groupes et Anneaux Réticulés*, Lecture Notes in Mathematics, # **608**, Springer–Verlag, Berlin and New York, 1977. MR **58** # 27688

36. G. Birkhoff, Dependent probabilities and the space (L), *Proc. Nat. Acad. Sci. U.S.A.* **24** (1938), 154–159. Zbl **18**, 264

37. G. Birkhoff, *Lattice Theory*, 3rd Edition, Amer. Math. Soc. Colloq. Publ., # **25**, Providence, RI, 1967. MR **37** # 2638.

38. G. Birkhoff and R. S. Pierce, Lattice-ordered rings, *An. Acad. Brasil Ciênc.* **28** (1956), 41–69. MR **18**, 191

39. S. Bochner and R. S. Phillips, Additive set functions and vector lattices, *Ann. of Math.* **42** (1941), 316–324. MR **2**, 315

40. H. F. Bohnenblust, An axiomatic characterization of L_p-spaces, *Duke Math. J.* **6** (1940), 627–640. MR **2**, 102

41. H. F. Bohnenblust and S. Kakutani, Concrete representations of (M)-spaces, *Ann. of Math.* **42** (1941), 1025–1028. MR **3**, 206

42. J. Bourgain, Dunford–Pettis operators on L^1 and the Radon–Nikodym property, *Israel J. Math.* **37** (1980), 34–47. MR 82k: 47047a

43. J. W. Brace, *Transformations on Banach Spaces*, Ph.D. Dissertation, Cornell University, Ithaca, New York, 1953.

44. R. C. Buck, Multiplication operators, *Pacific J. Math.* **11** (1961), 95–103. MR **23** # A513

45. A. V. Buhvalov, Factorization of compact operators and an example of a reflexive Banach lattice without the approximation property, *Soviet Math. Dokl.* **17** (1976), 423–426. MR **54** # 3369

46. A. V. Buhvalov, A. I. Veksler, and V. A. Geiler, Normed lattices, *J. Soviet Math.* **18** (1982), 516–551. MR 82b: 46019

47. A. V. Buhvalov, A. I. Veksler, and G. Ya. Lozanovsky, Banach lattices-Some Banach aspects of their theory, *Russian Math. Surveys* **34** (1979), 159–212. MR 80f: 46019

48. O. Burkinshaw, Weak compactness in the order dual of a vector lattice, *Trans. Amer. Math. Soc.* **187** (1974), 183–201. MR **52** # 14904

49. O. Burkinshaw and P. G. Dodds, Disjoint sequences, compactness and semireflexivity in locally convex Riesz spaces, *Illinois J. Math.* **21** (1977), 759–775. MR **57** # 3814

50. P. F. Conrad and J. E. Diem, The ring of polar preserving endomorphisms of an Abelian lattice-ordered group, *Illinois J. Math.* **15** (1971), 222–240. MR **44** # 2680

51. W. J. Davis, T. Figiel, W. B. Johnson, and A. Pelczynski, Factoring weakly compact operators, *J. Funct. Anal.* **17** (1974), 311–327. MR **50** # 8010

52. J. J. Diestel and J. J. Uhl, Jr., *Vector Measures*, Math. Surveys, # 15, Amer. Math. Soc., Providence, RI, 1977. MR **56** # 12216

53. P. G. Dodds, *o*-weakly compact mappings of Riesz spaces, *Trans. Amer. Math. Soc.* **214** (1975), 389–402. MR **52** # 6489

54. P. G. Dodds and D. H. Fremlin, Compact operators in Banach lattices, *Israel J. Math.* **34** (1979), 287–320. MR 81g: 47037

55. M. Duhoux and M. Meyer, A new proof of the lattice structure of orthomorphisms, *J. London Math. Soc.* (2) **25** (1982), 375–378. MR 83d: 46006

56. N. Dunford and P. J. Pettis, Linear operations on summable functions, *Trans. Amer. Math. Soc.* **47** (1940), 323–392. MR **1**, 338

57. N. Dunford and J. T. Schwartz, *Linear Operators I*, Wiley (Interscience), New York, 1958. MR **22**, 8302

58. W. F. Eberlein, Weak compactness in Banach spaces, I, *Proc. Nat. Acad. Sci. U.S.A.* **33** (1947), 51–53. MR **9**, 42

59. P. van Eldik, The integral component of an order bounded transformation, *Quaestiones Math.* **1** (1976), 135–144. MR **56** # 12974

60. A. J. Ellis, Extreme positive operators, *Quart. J. Math. Oxford Ser.* (2) **15** (1964), 342–344. MR **30** # 14157

61. P. Enflo, A counterexample to the approximation property in Banach spaces, *Acta Math.* **130** (1973), 309–317. MR **53** # 6288

62. T. Figiel, Factorization of compact operators and applications to the approximation property, *Studia Math.* **45** (1973), 191–210. MR **49** # 1070

63. H. O. Flösser, G. Gierz, and K. Keimel, Structure spaces and the center of vector lattices, *Quart. J. Math. Oxford Ser.* (2) **29** (1978), 415–426. MR 80c: 46012

64. D. H. Fremlin, *Topological Riesz Spaces and Measure Theory*, Cambridge University Press, London and New York, 1974. MR **56** # 12824

65. D. H. Fremlin, Riesz spaces with the order continuity property I, *Math. Proc. Cambridge Philos. Soc.* **81** (1977), 31–42. MR **54** # 13526

66. D. H. Fremlin, Riesz spaces with the order continuity property II, *Math. Proc. Cambridge Philos. Soc.* **83** (1978), 211–223. MR **56** # 16318

67. H. Freudenthal, Teilweise geordnete Moduln, *Nederl. Akad. Wetensch. Proc. Ser. A* **39** (1936), 641–651. Zbl **14**, 313

68. V. Gantmacher, Über schwache totalstetige Operatoren, *Mat. Sb.* (*N.S.*) **7** (49) (1940), 301–308. MR **2**, 224

69. N. Ghoussoub and W. B. Johnson, Factoring operators through Banach lattices not containing $C(0,1)$, *Math. Z.* **194** (1987), 153–171. MR 88b: 47048

70. C. Goffman, Compatible seminorms in a vector lattice, *Proc. Nat. Acad. Sci. U.S.A.* **42** (1956), 536–538. MR **18**, 52

71. H. H. Goldstine, Weakly complete Banach spaces, *Duke Math. J.* **4** (1938), 125–131. Zbl **18**, 313

72. A. Grothendieck, Sur les applications linéaires faiblement compactes d'espaces du type $C(K)$, *Canad. J. Math.* **5** (1953), 129–173. MR **15**, 438

73. A. Grothendieck, *Topological Vector Spaces*, Gordon and Breach, New York, 1973. MR **17**, 1110

74. H. Hahn, Über lineare gleichungssysteme in linearen räumen, *J. Reine Angew. Math.* **157** (1927), 214–229.

75. W. Haid, *Sätze vom Radon–Nikodym-typ für Operatoren auf Banachverbänden*, Ph.D. Dissertation, University of Tübingen, 1982.

76. J. Horváth, *Topological Vector Spaces and Distributions*, I, Addison-Wesley, Reading, Massachusetts, 1966. MR **34** # 4863

77. C. B. Huijsmans and B. de Pagter, Ideal theory in f-algebras, *Trans. Amer. Math. Soc.* **269** (1982), 225–245. MR 83k: 06020

78. G. Jameson, *Ordered Linear Spaces*, Lecture Notes in Mathematics, # **141**, Springer–Verlag, Berlin and New York, 1970. MR **55** # 10996

79. W. B. Johnson, Factoring compact operators, *Israel J. Math.* **9** (1971), 337–345. MR **44** # 7318

80. W. B. Johnson and L. Tzafriri, Some more Banach spaces which do not have local unconditional structure, *Houston J. Math.* **3** (1977), 55–60. MR **55** # 3758

81. S. Kakutani, Concrete representations of abstract (L)-spaces and the mean ergodic theorem, *Ann. of Math.* **42** (1941), 523–537. MR **2**, 318

82. S. Kakutani, Concrete representations of abstract (M)-spaces, *Ann. of Math.* **42** (1941), 994–1024. MR **3**, 205

83. N. J. Kalton and P. Saab, Ideal properties of regular operators between Banach lattices, *Illinois J. Math.* **29** (1985), 382–400. MR 87a: 47064

84. N. J. Kalton and A. Wilansky, Tauberian operators on Banach spaces, *Proc. Amer. Math. Soc.* **57** (1976), 251–255. MR **57** # 13555

85. L. V. Kantorovich, On partially ordered linear spaces and their applications in the theory of linear operators, *Dokl. Akad. Nauk SSSR* **4** (1935), 13–16 (Russian). Zbl **13**, 168

86. L. V. Kantorovich, Sur les propriétés des espaces semi-ordonnés linéaires, *C. R. Acad. Sci. Paris Ser. A–B* **202** (1936), 813–816. Zbl **13**, 268

87. L. V. Kantorovich, Les formes génerales des opérations linéaires qui transforment quelques espaces classiques das un espace semi-ordonné linéaire arbitraire, *C. R. Acad. Sci. Paris Ser. A–B* **202** (1936), 1251–1253. Zbl **13**, 309

88. L. V. Kantorovich, Concerning the general theory of operations in partially ordered spaces, *Dokl. Akad. Nauk SSSR* **1** (1936), 283–286 (Russian). Zbl **14**, 67

89. L. V. Kantorovich, Lineare halbgeordnete räume, *Math. Sbornik* **44** (1937), 121–168. Zbl **16**, 405

90. L. V. Kantorovich, On the moment problem for a finite interval, *Dokl. Akad. Nauk SSSR* **14** (1937), 531–537 (Russian). Zbl **16**, 353

91. L. V. Kantorovich, Linear operators in semi-ordered spaces, *Math. Sbornik* **49** (1940), 209–284. MR **2**, 317; Zbl **23**, 328

92. L. V. Kantorovich, B. Z. Vulikh, and A. G. Pinsker, *Functional Analysis in Partially Ordered Spaces*, Gosudarstv. Izdat. Tecn.–Teor. Lit., Moscow and Leningrad, 1950. MR **12**, 340; Zbl **37**, 72

93. S. Kaplan, On the second dual of the space of continuous functions, *Trans. Amer. Math. Soc.*, Note I **86** (1957), 70–90, MR **19**, 868; Note II **93** (1959), 329–350, MR **22** # 2888; Note III **101** (1961), 34–51, MR **24** # A1598; Note IV **113** (1964), 512–546, MR **30** # 444.

94. S. Kaplan, An example in the space of bounded operators from $C(X)$ to $C(Y)$, *Proc. Amer. Math. Soc.* **38** (1973), 595–597. MR **47** # 7505

95. S. Karlin, Positive operators, *J. Math. Mech.* **8** (1959), 907–937. MR **22** # 4965

96. J. L. Kelley, Note on a theorem of Krein and Milman, *J. Osaka Inst. Sci. Tech., Part I* **3** (1951), 1–2. MR **13**, 249

97. J. Kim, The characterization of a lattice homomorphism, *Canadian J. Math.* **27** (1975), 172–175. MR **50** # 10750

98. P. P. Korovkin, On convergence of linear positive operators in the space of continuous functions, *Doklady Akad. Nauk SSSR (N. S.)* **90** (1953), 961–964 (Russian). MR **15**, 236

99. G. Köthe, *Topological Vector Spaces*, Springer–Verlag, Heidelberg and New York, Vol. I, 1969, MR **24** # A411; Vol. II, 1979, MR 81g: 46001.

100. M. A. Krasnoselskii, P. P. Zabreiko, E. I. Pustylnik, and P. E. Sobolevskii, *Integral Operators in Spaces of Summable Functions*, Noordhoff International Publishing, Leiden, Netherlands, 1976. (Translated from the Russian by T. Andô.) MR **34** # 6568

101. M. G. Krein and S. G. Krein, On an inner characteristic of the set of all continuous functions defined on a bicompact Hausdorff space, *Dokl. USSR* **27** (1940), 427–431. MR **2**, 222

102. M. G. Krein and D. Milman, On extreme points of regular convex sets, *Studia Math.* **9** (1940), 133–138. MR **3**, 90

103. M. G. Krein and M. A. Rutman, Linear operators leaving invariant a cone in a Banach space, *Uspekhi Mat. Nauk (N. S.)* **3** (1948), 3–95 (Russian). Also, *Amer. Math. Soc. Transl. Ser.*, # **26**, Providence, RI, 1950. MR **10**, 256

104. M. G. Krein and V. Šmulian, On regularly convex sets in the space conjugate to a Banach space, *Ann. of Math.* **41** (1940), 556–583. MR **1**, 335

105. U. Krengel, Über den Absolutbertrag stetiger linearer Operatoren und seine Anwendung auf ergodische Zerlegungen, *Math. Scand.* **13** (1963), 151–187. MR **31** # 310

106. U. Krengel, Remark on the modulus of compact operators, *Bull. Amer. Math. Soc.* **72** (1966), 132–133. MR **32** # 8162

107. B. Kühn, Banachverbänden mit ordungsstetiger dualnorm, *Math. Z.* **167** (1979), 271–277. MR 80e: 46016

108. S. S. Kutateladze, Support set for sublinear operators, *Soviet Math. Dokl.* **17** (1976), 1428–1431. MR **54** # 8225

109. K. K. Kutty and J. Quinn, Some characterizations of the projection property in Archimedean Riesz spaces, *Canadian J. Math.* **24** (1972), 306–311. MR **49** # 7728

110. H. E. Lacey, *The Isometric Theory of Classical Banach Spaces*, Springer–Verlag, Berlin and New York, 1974. MR **58** # 12308.

111. H. Leinfelder, A remark on a paper of Loren D. Pitt, *Bayreuth. Math. Schr.* No. 11 (1982), 57–66. MR 85c: 47019

112. J. Lindenstrauss and L. Tzafriri, *Classical Banach Spaces I*, Springer–Verlag, Berlin and New York, 1977. MR **58** # 17766

113. J. Lindenstrauss and L. Tzafriri, *Classical Banach Spaces II*, Springer–Verlag, Berlin and New York, 1979. MR 81c: 46001

114. Z. Lipecki, Extensions of positive operators and extreme points II, *Colloq. Math.* **42** (1979), 285–289. MR 82k: 47056

115. Z. Lipecki, Extension of vector-lattice homomorphisms, *Proc. Amer. Math. Soc.* **79** (1980), 247–248. MR 81b: 46009

116. Z. Lipecki, D. Plachky, and W. Thomsen, Extension of positive operators and extreme points I, *Colloq. Math.* **42** (1979), 279–284. MR 82k: 47055

117. H. P. Lotz, Minimal and reflexive Banach lattices, *Math. Ann.* **209** (1974), 117–126. MR **50** # 10751

118. H. P. Lotz, Extensions and liftings of positive linear mappings on Banach lattices, *Trans. Amer. Math. Soc.* **211** (1975), 85–100. MR **52** # 4022

119. G. Ya. Lozanovsky, Two remarks concerning operators in partially ordered spaces, *Vestnik Leningrad Univ. Mat. Mekh. Astronom.* (1965), no. 19, 159–160 (Russian). MR **32** # 6227; Zbl **154**, 158

120. G. Ya. Lozanovsky, On almost integral operators in KB-spaces, *Vestnik Leningrad Univ. Mat. Mekh. Astronom.* (1966), no. 7, 35–44 (Russian). MR **34** # 8185

121. G. Ya. Lozanovsky, On the limit of a sequence of functionals in partially ordered spaces, *Vestnik Leningrad Univ. Mat. Mekh. Astronom.* (1967), no. 1, 148–149 (Russian). MR **35** # 830

122. G. Ya. Lozanovsky, On Banach lattices and bases, *Functional Anal. Appl.* **1** (1967), 249. MR **36** # 3110

123. G. Ya. Lozanovsky, Some topological properties of Banach lattices and reflexivity conditions on them, *Soviet Math. Dokl.* **9** (1968), 1415–1418. MR **38** # 3710

124. G. Ya. Lozanovsky and A. A. Mekler, Completely linear functionals and reflexivity in normed linear lattices, *Izv. Vysš. Učebn. Zaved. Matematika* **66** (1967), 47–53 (Russian). MR **36** # 3111; Zbl **153**, 439

125. W. A. J. Luxemburg, Notes on Banach function spaces, *Nederl. Akad. Wetensch. Proc. Ser. A*, Note XIV **68** (1965), 229–248; Note XV **68** (1965), 415–446; Note XVI **68** (1965), 646–667. MR **32** # 6202

126. W. A. J. Luxemburg, *Some Aspects of the Theory of Riesz Spaces*, University of Arkansas Lecture Notes in Mathematics, # **4**, Fayetteville, Arkansas, 1979. MR 83f: 46010

127. W. A. J. Luxemburg and L. C. Moore, Jr., Archimedean quotient Riesz spaces, *Duke Math. J.* **34** (1967), 725–739. MR **36** # 651

128. W. A. J. Luxemburg and A. R. Schep, A Radon–Nikodym theorem for positive operators and a dual, *Indag. Math.* **40** (1978), 357–375. MR 80a: 47058

129. W. A. J. Luxemburg and A. R. Schep, An extension theorem for Riesz homomorphisms, *Indag. Math.* **41** (1979), 145–154. MR 80i: 47051

130. W. A. J. Luxemburg and A. C. Zaanen, Notes on Banach function spaces, *Nederl. Akad. Wetensch. Proc. Ser. A*, Note I **66** (1963), 135–147; Note II **66** (1963), 148–153; Note III **66** (1963), 239–250; Note IV **66** (1963), 251–263; Note V **66** (1963), 496–504; Note VI **66** (1963), 655–668; Note VII **66** (1963), 669–681; Note VIII **67** (1964), 104–119; Note IX **67** (1964), 360–376; Note X **67** (1964), 493–506; Note XI **67** (1964), 507–518; Note XII **67** (1964), 519–529; Note XIII **67** (1964), 530–543. MR **26** # 6723*ab*; MR **27** # 5119*ab*; MR **28** # 5324*ab*; MR **30** # 3381*ab*

131. W. A. J. Luxemburg and A. C. Zaanen, Compactness of integral operators in Banach function spaces, *Math. Ann.* **149** (1963), 150–180. MR **26** # 2905

132. W. A. J. Luxemburg and A. C. Zaanen, *Riesz Spaces*, I, North-Holland, Amsterdam, 1971. MR **58** # 23483

133. G. W. Mackey, On convex topological linear spaces, *Trans. Amer. Math. Soc.* **60** (1946), 519–537. MR **8**, 519

134. D. Maharam, The representation of abstract integrals, *Trans. Amer. Math. Soc.* **75** (1953), 154–184. MR **14**, 1071

135. D. Maharam, On kernel representation of linear operators, *Trans. Amer. Math. Soc.* **79** (1955), 229–255. MR **16**, 1031

136. S. Mazur, Über die kleinste konvexe Menge, die eine gegebene kompakte Menge enthalt, *Studia Math.* **2** (1930), 7–9.

137. M. Meyer, Le stabilisateur d'un espace vectoriel réticulé, *C. R. Acad. Sci. Paris Sér. A* **283** (1976), 249–250. MR **55** # 6170

138. M. Meyer, Richesses du centre d'un espace vectoriel réticulé, *Math. Ann.* **236** (1978), 147–169. MR 80b: 46012

139. M. Meyer, Quelques propriétés des homomorphisms d'espaces vectoriels réticulés, Equipe d'Analyse-Université Paris VI, Preprint no. 131, 1979.

140. P. Meyer-Nieberg, Zur schwachen kompaktheit in Banachverbänden, *Math. Z.* **134** (1973), 303–315. MR **48** # 9341

141. P. Meyer-Nieberg, Charakterisierung einiger topologischer und ordnungstheretischer Eigenschaften von Banachverbänden mit Hilfe disjunkter Folgen, *Arch. Math. (Basel)* **24** (1973), 640–647. MR **49** # 5771

142. P. Meyer-Nieberg, Über klassen schwach kompakter Operatoren in Banachverbänden, *Math. Z.* **138** (1974), 145–159. MR **50** # 5539

143. E. H. Moore, On the foundations of the theory of linear equations, *Bull. Amer. Math. Soc.* **18** (1912), 334–362.

144. L. C. Moore, Jr., Strictly increasing Riesz norms, *Pacific J. Math.* **37** (1971), 171–180. MR **46** # 5988

145. R. J. Nagel and U. Schlotterbeck, Kompaktheit von Integraloperatoren auf Banachverbänden, *Math. Ann.* **202** (1973), 301–306. MR **48** # 12135

146. M. Nakamura, Notes on Banach space (X): Vitali–Hahn–Sak's Theorem and K-spaces, *Tôhoku Math. J.* (2) **1** (1949), 100–108. MR **11**, 186

147. M. Nakamura, Notes on Banach space (XI): Banach lattices with positive basis, *Tôhoku Math. J.* (2) **2** (1950), 135–141. MR **13**, 361

148. H. Nakano, Teilweise geordnete algebra, *Japan. J. Math.* **17** (1941), 425–511. MR **3**, 210

149. H. Nakano, *Modern Spectral Theory*, Maruzen Co., Tokyo, 1950. MR **12**, 419

150. H. Nakano, *Modulared Semi-ordered Linear Spaces*, Maruzen Co., Tokyo, 1950. MR **12**, 420

151. H. Nakano, Product spaces of semi-ordered linear spaces, *J. Fac. Sci. Hokkaido Univ. Se. I* **12** (1953), 163–210. MR **16**, 49

152. H. Nakano, *Semi-ordered Linear Spaces*, Japan Society for the promotion of science, Tokyo, 1955. MR **17**, 387

153. I. Namioka, Partially ordered linear topological spaces, *Mem. Amer. Math. Soc.*, # **24**, Providence, RI, 1957. MR **20**, 1193

154. C. P. Niculescu, Weak compactness in Banach lattices, *J. Operator Theory* **6** (1981), 217–231. MR 83d: 47044

155. T. Ogasawara, Compact metric Boolean algebras and vector lattices, *J. Sci. Hiroshima Univ. Ser. A* **11** (1942), 125–128. MR **10**, 46

156. T. Ogasawara, Vector lattices I and II, *J. Sci. Hiroshima Univ. Ser. A* **12** (1942), 37–100 and **13** (1944), 41–161 (Japanese). MR **10**, 545

157. T. Ogasawara, Some general theorems and convergence theorems in vector lattices, *J. Sci. Hiroshima Univ. Ser. A* **11** (1949), 14–25. MR **13**, 361

158. B. de Pagter, *f-Algebras and Orthomorphisms*, Ph.D. Dissertation, University of Leiden, 1981.

159. B. de Pagter, The components of a positive operator, *Indag. Math.* **45** (1983), 229–241. MR 85c: 47036

160. B. de Pagter, A note on disjointness preserving operators, *Proc. Amer. Math. Soc.* **90** (1984), 543–549. MR 85e: 47057

161. A. Pelczynski, A connection between weakly unconditional convergence and weak completeness of Banach spaces, *Bull. Acad. Polon. Sci. Ser. Sci. Math. Astronom. Phys.* **6** (1958), 251–253. MR **22** # 5875

162. A. L. Peressini, *Ordered Topological Vector Spaces*, Harper and Row, New York and London, 1967. MR **37** # 3315

163. R. R. Phelps, Extreme positive operators and homomorphisms, *Trans. Amer. Math. Soc.* **108** (1963), 265–274. MR **27** # 6153

164. R. S. Phillips, On linear transformations, *Trans. Amer. Math. Soc.* **48** (1940), 516–541. MR **2**, 318

165. L. D. Pitt, A compactness condition for linear operators on function spaces, *J. Operator Theory* **1** (1979), 49–54. MR 80b: 47044

166. F. Riesz, Sur la décomposition des opérations linéaires, *Atti. Congr. Internaz. Mat. Bologna* **3** (1930), 143–148.

167. F. Riesz, Sur quelques notions fondamentals dans la theorie générale des opérations linéaires, *Ann. of Math.* **41** (1940), 174–206. (This work was published first in 1937 in Hungarian.) MR **1**,147

168. G. T. Roberts, Topologies in vector lattices, *Math. Proc. Cambridge Phil. Soc.* **48** (1952), 533–546. MR **14**, 395

169. A. P. Robertson and W. Robertson, *Topological Vector Spaces*, 2nd Edition, Cambridge Univ. Press, London, 1973. MR **28** # 5318

170. H. P. Rosenthal, A characterization of Banach spaces containing ℓ_1, *Proc. Nat. Acad. Sci. U.S.A.* **71** (1974), 2411–2413. MR **50** # 10773

171. H. H. Schaefer, Halbgeordnete lokalkonvexe Vektorräume, *Math. Ann.*, Note I **135** (1958), 115–141, MR **21**, 5134; Note II **138** (1959), 259–286, MR **21**, 5135; Note III **141** (1960), 113–142, MR **22**, 11265.

172. H. H. Schaefer, Weak convergence of measures, *Math. Ann.* **193** (1971), 57–64. MR **44** # 5759

173. H. H. Schaefer, *Topological Vector Spaces*, Springer–Verlag, Berlin and New York, 1974. MR **33** # 1689

174. H. H. Schaefer, *Banach Lattices and Positive Operators*, Springer–Verlag, Berlin and New York, 1974. MR **54** # 11023

175. J. Schauder, Über lineare vollstetige Funktionaloperatoren, *Studia Math.* **2** (1930), 183–196.

176. A. R. Schep, Order continuous components of operators and measures, *Indag. Math.* **40** (1978), 110–117. MR **57** # 17378

177. A. R. Schep, Positive diagonal and triangular operators, *J. Operator Theory* **3** (1980), 165–178. MR 81g: 47040

178. W. Sierpiński, Sur les fonctions développables en séries absolument convergentes de fonctions continues, *Fund. Math.* **2** (1921), 15–27.

179. V. L. Šmulian, Über lineare topologische räume, *Math. Sbornik* **7** (49) (1940), 425–448. MR **2**, 102

180. M. H. Stone, The theory of representations for Boolean algebras, *Trans. Amer. Math. Soc.* **40** (1936), 37–111. Zbl **14**, 340

181. J. Synnatzschke, On the adjoint of a regular operator and some of its applications to the question of complete continuity and weak continuity of regular operators, *Vestnik Leningrad Univ. Math. Mekh. Astronom.* **5** (1978), 71–81. MR **47** # 876

182. M. Talagrand, Some weakly compact operators between Banach lattices do not factor through reflexive Banach lattices, *Proc. Amer. Math. Soc.* **96** (1986), 95–102.

183. T. Terzioglu, A characterization of compact linear mappings, *Arch. Math. (Basel)* **22** (1971), 76–78. MR **45** # 954

184. C. T. Tucker, Homomorphisms of Riesz spaces, *Pacific J. Math.* **55** (1974), 289–300. MR **51** # 5443

185. C. T. Tucker, Concerning σ-homomorphisms of Riesz spaces, *Pacific J. Math.* **57** (1975), 585–589. MR **52** # 6376

186. C. T. Tucker, Riesz homomorphisms and positive linear maps, *Pacific J. Math.* **69** (1977), 551–556. MR **55** # 11099

187. L. Tzafriri, Reflexivity in Banach lattices and their subspaces, *J. Funct. Anal.* **10** (1972), 1–18. MR **50** # 10769

188. A. I. Veksler, On the homomorphisms between the classes of regular operators in K-lineals and their completions, *Izv. Vysš. Učebn. Zaved. Matematika*, no. 1 (14) (1960), 48–57 (Russian). MR **26** # 599

189. B. Z. Vulikh, *Introduction to the Theory of Partially Ordered Spaces*, Wolters–Noordhoff, Groningen, Netherlands, 1967. (English translation from the Russian.) MR **24** # A3494

190. R. J. Whitley, An elementary proof of the Eberlein–Šmulian theorem, *Math. Ann.* **172** (1967), 116–118. MR **35** # 3419

191. A. W. Wickstead, Representation and duality of multiplication operators on Archimedean Riesz spaces, *Compositio Math.* **35** (1977), 225–238. MR **56** # 12976

192. A. W. Wickstead, Extensions of orthomorphisms, *J. Austral. Math. Soc. Ser. A* **29** (1980), 87–98. MR 81b: 47049

193. A. W. Wickstead, Extremal structure of cones of operators, *Quart. J. Math. Oxford Ser.* (2) **32** (1981), 239–253. MR 82i: 47069

194. A. Wilansky, *Modern Methods in Topological Vector Spaces*, McGraw-Hill, New York, 1978. MR 81d: 46001

195. W. Wils, The ideal center of partially ordered vectors spaces, *Acta Math.* **127** (1971), 41–77. MR **57** # 3819

196. A. C. Zaanen, Examples of orthomorphisms, *J. Approx. Theory* **13** (1975), 192–204. MR **50** # 8001

197. A. C. Zaanen, *Riesz Spaces II*, North-Holland, Amsterdam, 1983. MR 86b: 46001

Monographs

This is a list of monographs that were published after the original publication of this book and contain material on positive operators.

(1) Y. A. Abramovich and C. D. Aliprantis, *An Invitation to Operator Theory*, Graduate Texts in Mathematics, Volume # 50, American Mathematical Society, Providence, RI, 2002. MR 2003h:47072.

(2) Y. A. Abramovich and C. D. Aliprantis, *Problems in Operator Theory*, Graduate Texts in Mathematics, Volume # 51, American Mathematical Society, Providence, RI, 2002. MR 2003h:47073.

(3) C. D. Aliprantis and K. C. Border, *Infinite Dimensional Analysis: A Hitchhikers Guide*, 3rd Edition, Springer–Verlag, Heidelberg and New York, 2006. MR 00k:46001

(4) P. Meyer-Nieberg, *Banach Lattices*, Springer–Verlag, Berlin and New York, 1991. MR 93f: 46025

(5) A. G. Kusraev, *Dominated Operators*, Mathematics and its Applications, Volume # 519, Kluwer Academic Publishers, Dordrecht and London, 2000. MR 2002b:47077

(6) W. Wnuk, *Banach Lattices with Order Continuous Norms*, Advanced Topics in Mathematics, Polish Scientific Publishers PWN, Warszawa, 1999.

(7) A. C. Zaanen, *Introduction to Operator Theory in Riesz Spaces*, Springer–Verlag, Berlin and New York, 1997, MR 2000c:47074.

Index